Calcium and Contractility

Contemporary Biomedicine

Calcium and Contractility: Smooth Muscle
Edited by ***A. K. Grover and E. E. Daniel,*** 1984

Monoclonal Antibodies in Cancer
Edited by ***Stewart Sell and Ralph A. Reisfield,*** 1984

Carcinogenesis and Mutagenesis Testing
Edited by ***J. F. Douglas,*** 1984

The Human Teratomas: *Experimental and Clinical Biology*
Edited by ***Ivan Damjanov, Barbara B. Knowles,*** **and** ***Davor Solter,*** 1983

Human Cancer Markers
Edited by ***Stewart Sell*** **and** ***Britta Wahren,*** 1982

Cancer Markers: *Diagnostic and Developmental Significance*
Edited by ***Stewart Sell,*** 1980.

Calcium and Contractility

Smooth Muscle

Edited by

A. K. Grover and E. E. Daniel

McMaster University,
Hamilton, Ontario, Canada

Humana Press · Clifton, New Jersey

Library of Congress Cataloging in Publication Data

Main entry under title:
Calcium and contractility: smooth muscle.

(Contemporary biomedicine)
Includes bibliographies and index.
1. Muscle contraction. 2. Calcium—Physiological effect. 3. Smooth muscle. I. Grover, A. K.
II. Daniel, E. E. III. Series.
QP321.C29 1985 596′.01852 84-28841
ISBN 0-89603-066-0

Crescent Manor
PO Box 2148
Clifton, NJ 07015

Printed in the United States of America

CONTENTS

Chapter 3

CALCIUM COMPARTMENTS AND MOBILIZATION DURING CONTRACTION OF SMOOTH MUSCLE

Roger Loutzenhiser, Paul Leyten, Kooichi Saida, and Cornelis van Breemen

Chapter 4

MECHANISMS OF SMOOTH MUSCLE RELAXATION

T. Godfraind and R. C. Miller

Chapter 5

SMOOTH MUSCLE RELAXANTS

Charles V. Jackson and John H. McNeill

Chapter 8

STUDIES ON SKINNED FIBER PREPARATIONS
Kaushik D. Meisheri, J. Casper Rüegg, and Richard J. Paul

Chapter 9

SMOOTH MUSCLE SUBCELLULAR FRACTIONATION
A. M. Kidwai

Chapter 10

CALCIUM-HANDLING STUDIES USING ISOLATED SMOOTH MUSCLE MEMBRANES

A. K. Grover

Chapter 11

RECEPTOR BINDING STUDIES ON SMOOTH MUSCLE SUBCELLULAR FRACTIONS

D. J. Crankshaw

Chapter 12

CALCIUM-HANDLING DEFECTS AND SMOOTH MUSCLE PATHOPHYSIOLOGY

C. Y. Kwan

Chapter 13

ADRENERGIC INTERACTIONS IN SMOOTH MUSCLE CONTACTILITY

C. R. Triggle

Chapter 14

CHOLINERGIC INTERACTIONS AND VASCULAR SMOOTH MUSCLE TONE

Tony J-F. Lee

Chapter 15

NONADRENERGIC, NONCHOLINERGIC (NANC) NEURONAL INHIBITORY INTERACTIONS WITH SMOOTH MUSCLE

E. E. Daniel

Chapter 16

NONADRENERGIC, NONCHOLINERGIC (NANC) NEURONAL EXCITATORY INTERACTIONS WITH SMOOTH MUSCLE

E. E. Daniel

Chapter 17

PEPTIDE CALCIUM INTERACTIONS IN SMOOTH MUSCLE

J. E. T. Fox

Preface

In organizing the present volume, we had two intentions. The first was to present the best current understanding of the mechanisms of calcium mobilization during excitation–contraction coupling of smooth muscle at a level suited to the needs of professionals interested in smooth muscle pharmacology and pathophysiology, while remaining appreciable by graduate and medical students. The second intention was to provide insight into present-day controversies, as well as the latest advances achieved by researchers in this field. Thus, we have thoroughly discussed both the techniques and the concepts derived from their application.

An attempt has also been made here to answer a number of profoundly important questions: What are the mechanisms and agents responsible for the control of contractility? What are the accompanying changes in the state of intracellular calcium ions and the mechanisms responsible for them? How does the regulation of contractility occur directly at the level of the actomyosin activity? What role do gap junctions play in cell-to-cell coupling? What are the roles of cholinergic, adrenergic, peptidergic, and nonadrenergic noncholinergic interactions in calcium mobilization in smooth muscle? What changes occur in hypertension? The impact of these recent techniques on future research is also reflected upon. For instance, the application of immunological techniques to the identification of control peptides and their cellular sites of interaction, and the structural analysis of these peptides by peptide mapping and recombinant DNA techniques have just begun and will allow rapid advances in the understanding of the intricate control mechanisms in the future.

The contraction and relaxation of vascular and nonvascular smooth muscle requires mobilization of calcium ions and their subsequent withdrawal from the cytoplasm. These ions are thus of key importance in the normal physiological functioning, as well as in the pathophysiology, of various smooth muscles. The normal signals responsible for changes in the contractile state may originate from cholinergic, adrenergic, noncholinergic, and nonadrenergic nerves, from circulating substances, and from local processes producing active substances. Several neural or circulating peptides have also been shown to be stimuli that regulate smooth muscle contractility. The excitation of smooth muscles to contract usually involves calcium mobilization arising from the interaction of signaling substances with the various receptors located in the cellular membrane of the smooth muscle; these may activate channels for calcium entry or release sequestered calcium. Changes in the electrical potential of the muscle cells directly, as well as those from receptor occupation, can cause calcium mobilization or the opening of calcium channels. Relaxation is not simply a reversal of these processes, but involves several mechanisms to turn off electrical events, as well as a withdrawal of the mobilized calcium.

In recent years, we have seen a vital and growing interest in the area of control of internal calcium; insight into these processes is important to the proper understanding, and thus control, of normal smooth muscle motor function and its alteration in various abnormal states. Changes in the motility of the gastrointestinal tract in disease and in uterine smooth muscle during normal and abnormal gestation, as well as the alteration of contractility of blood vessels in hypertension and asthma—all involve altered control of internal calcium. Because of this interest, various smooth muscles have been studied in vivo in humans, as well as in animal models, and in vitro at various levels of oganization, i.e., in tissue strips, cultured or isolated single-cell preparations, skinned fibers, and subcellular preparations. Each of these approaches has advantages and limitations.

In parallel with our advancing knowledge of the effects of calcium on contractility, a number of techniques have been developed that allow the dissection and characterization of these systems. These techniques include the localization of calcium deposits by electron microprobe and X-ray analyses, immunohistochemical techniques for the localization of neural, hormonal, and other control substances and their receptors, the discovery of a number of calcium antagonists, the development of skinning techniques, and advances in subcellular fractionation methods that allow the use of isolated membrane fractions for studies on membrane transport and receptor binding. The use of long-established techniques in conjunction with these newer ones has led to a greater depth of knowledge of the smooth muscle. The wide interest in the contractility of smooth muscle, development of the new techniques, and the consequent changes in the concepts set the need for a book on this subject.

It is evident that a wealth of information has accumulated on smooth muscle in recent years and that the process is continuing. We very much hope we have distilled this vast knowledge into a comprehensible form. The various chapters were each written by scientists with special experience and expertise in the fields considered. Despite this, and because of rapid developments in smooth muscle research, it is understandable that controversies still exist. As editors, we have encouraged the presentation of such controversies to make the book thought-provoking for the researcher.

The editors thank Christel Bandomir for the constant secretarial help she has given during the various stages of this book's preparation.

A.K. Grover
E.E. Daniel

CONTRIBUTORS

SAMUEL CHACKO • *Department of Pathobiology, University of Pennsylvania, Philadelphia, Pennsylvania*

D. J. CRANKSHAW • *Department of Obstertrics and Gynecology, McMaster University, Hamilton, Ontario, Canada*

E. E. DANIEL • *Department of Neurosciences, McMaster University, Hamilton, Ontario, Canada*

J. E. T. FOX • *Department of Neurosciences, McMaster University, Hamilton, Ontario, Canada*

R. E. GARFIELD • *Department of Neurosciences, McMaster University, Hamilton, Ontario, Canada*

T. GODFRAIND • *Laboratoire de Pharmacodynamie Generale et de Pharmacologie Universite Catholique de Louvain, Bruxelles, Belgium*

A. K. GROVER • *Department of Neurosciences, McMaster University, Hamilton, Ontario, Canada*

CHARLES V. JACKSON • *University of British Columbia, Vancouver, British Columbia, Canada*

R. A. JANIS • *Miles Institute for Preclinical Pharmacology, New Haven, Connecticut*

A. M. KIDWAI • *Industrial Toxicology Research Centre, Lucknow, India*

C. Y. KWAN • *Department of Neurosciences, McMaster University, Hamilton, Ontario, Canada*

TONY J.-F. LEE • *Department of Pharmacology, Southern Illinois University School of Medicine, Springfield, Illinois*

PAUL LEYTEN • *Department of Pharmacology, University of Miami, Miami, Florida*

RODGER LOUTZENHISER • *Nephrology Section, Veterans Administration Medical Center, Miami, Florida*

JOHN H. MCNEILL • *University of British Columbia, Vancouver, British Columbia, Canada*

KAUSHIK D. MEISHERI • *Physiologisches Institut, Universitat Heidelberg, Heidelberg, Germany*

R. C. MILLER • *Laboratoire de Pharmacodynamie Generale et de Pharmacologie Universite Catholique de Lauvain, Bruxelles, Belgium*

RICHARD J. PAUL • *Department of Physiology, University of Cincinnati, Cincinnati, Ohio*

ARLINE ROSENFIELD • *Department of Pathobiology, University of Pennsylvania, Philadelphia, Pennsylvania*

KOOICHI SAIDA • *Department of Pharmacology, University of Miami, Miami, Florida*

A. P. SOMLYO • *Departments of Physiology and Pathology, University of Pennsylvania, Philadelphia, Pennsylvania*

GEORGE THOMAS • *Department of Pathobiology, University of Pennsylvania, Philadelphia, Pennsylvania*

C. R. TRIGGLE • *Memorial University of Newfoundland, St. John's Newfoundland*

D. J. TRIGGLE • *Department of Biochemical Pharmacology, State University of New York, Buffalo, New York*

J. CASPER RÜEGG • *Physiologisches Institut, Universitat Heidelberg, Heidelberg, Germany*

CORNELIS VAN BREEMAN • *Department of Pharmacology, University of Miani, Miami, Florida*

Chapter 1

Structure of Smooth Muscle

R. E. Garfield and A. P. Somlyo

1. Introduction

An understanding of the structural bases of function is an essential prerequisite for understanding biological, and perhaps all physical, processes. Just as organ physiology followed the development of anatomy, and the recognition of the cell as the fundamental unit of organs was the result of light microscopy, modern cell physiology owes much of its existence to electron microscopy and allied disciplines. Electron microscopy of smooth muscle, over the last few decades, revealed neuromuscular relationships, junctional structures responsible for cell-to-cell communication, filaments mediating contraction and the intracellular organelles involved in cellular metabolism, regulation and replication (for review, *see* refs. *14, 15, 33, 42, 78, 79, 84, 95, 98*). In addition to these problems solved, at least partially, by conventional transmission electron microscopy, several questions related to the distribution of ions in smooth muscle lend themselves to analytical electron microscopy: electron probe X-ray microanalysis (EDS) and electron energy loss spectroscopy (EELS). The application of these methods to skeletal muscle has been valuable in determining the movement of ions (e.g., Ca^{2+}, Mg^{2+}) between subcellular compartments (*87, 96*). Because smooth muscle cells are relatively small and embedded in a connective tissue matrix having variable ion-binding properties, analytical electron microscopy has also been useful, and in some instances essential, for distinguishing cellular from extracellular ion content. More specifically, the contractile apparatus of smooth, as well as that of striated, muscle is activated by a rise in cytoplasmic-free Ca^{2+} that may originate, variably, from extracellular or intracellular sources (for review, *see* ref. *51*). Therefore it was apparent that the unambiguous demonstration of the intracellular sources of activator calcium can be most readily accomplished by electron optical methods that permit the detection of Ca^{2+} and other elements at electron microscopic resolution.

The functional (pharmacological and physiological) as well as structural heterogeneity of smooth muscle does not lend itself to facile generalizations between different smooth muscles in the same species or between different species. In the following, although we will attempt to emphasize some general aspects of smooth muscle structure, it is not our objective to reproduce extensive recent reviews dealing largely with intestinal (*33*) or vascular smooth muscle (*84, 98*). Instead, we shall place some emphasis on studies dealing with uterine smooth muscles, a system of considerable experimental interest, as it responds dramatically to variations in steroid hormones by changing both structure and function, and has been a major research interest of one of the authors (REG).

2. Organization and Arrangement of Smooth Muscle

Organs and tissues containing predominantly smooth muscle are usually not homogenous, but include various other types of cells in addition to smooth muscle. A knowledge of the various cells and their organization, arrangement, interrelationships and interactions, is required to understand the structure and function of smooth muscle. Smooth muscle is usually found around ducts or tubes such as the wall of the intestine, blood vessels, airways, uterus, vas deferens, oviduct, ureter, and so on. There are many exceptions, but generally the overall structure is similar. The walls of the conduit organs are usually composed of several layers including an inner epithelial/endothelial layer of cells that line the lumen and one or more surrounding muscle layers.

The walls of blood vessels have been most extensively studied (*64, 65*). The thickness of the walls of arteries varies greatly as a function of the diameter of the blood vessel. The thickness of the walls of veins is smaller than the arteries and their cross-sectional diameters larger. The vessel diameter and thickness of the walls change depending upon the state of contraction. There is also great variation in arrangement of smooth muscle in cells and layers in different parts of the vascular tree. It is evident from the arrangement of the muscle cells into circular, longitudinal and spiral layers that contraction produces both a constriction and shortening.

2.1. Organization of the Uterus and Arrangement of the Myometrium

The uterus consists of one or two hollow tubes or horns, depending upon the species, which open on one end from the oviducts and terminate on the other end into the cervix and vagina. The uterine wall is usually composed of at least three distinct layers. The innermost layer, the endometrium, lines the lumen of the organ and consists of columnar epithelium and underlying connective tissue. The myometrium makes up the other two layers: the outer longitudinal muscle layer and the inner circular layer. Figure 1 shows a light micrograph of a cross-section of the rat uterus showing the endometrium and myometrium.

The outer longitudinal muscle layer of the myometrium consists of a network of bundles of smooth muscle cells that are generally oriented in the long axis of the uterus (22). The bundles interconnect and form a network over the surface of the uterus. An isolated portion of the longitudinal muscle layer from pregnant rat myometrium showing the bundle arrangement is illustrated in Fig. 2. Each bundle is composed of smooth muscle cells arranged in the long axis of the bundle. Figure 3 shows a scanning electron micrograph of the longitudinal muscle from the myometrium showing the cell arrangement. A low magnification electron micrograph of a cross-section through portions of several bundles of longitudinal muscle layer is shown in Fig. 4. Contraction of the longitudinal muscle layer would tend to shorten the uterus and constrict the lumen.

A low magnification electron micrograph showing the entire wall of the uterus from a pregnant rat is shown in Fig. 5. Muscle cells of both muscle layers occupy a major part of the uterine wall. Muscle cells of the circular muscle layer are arranged concentrically around the longitudinal axis of the uterus. The muscle cells are arranged more diffusely and the bundle arrangement, if present, is not as apparent as that of the longitudinal layer. There are functional and structural studies indicating that the longitudinal layer is continuous with the circular layer (57). The extracellular

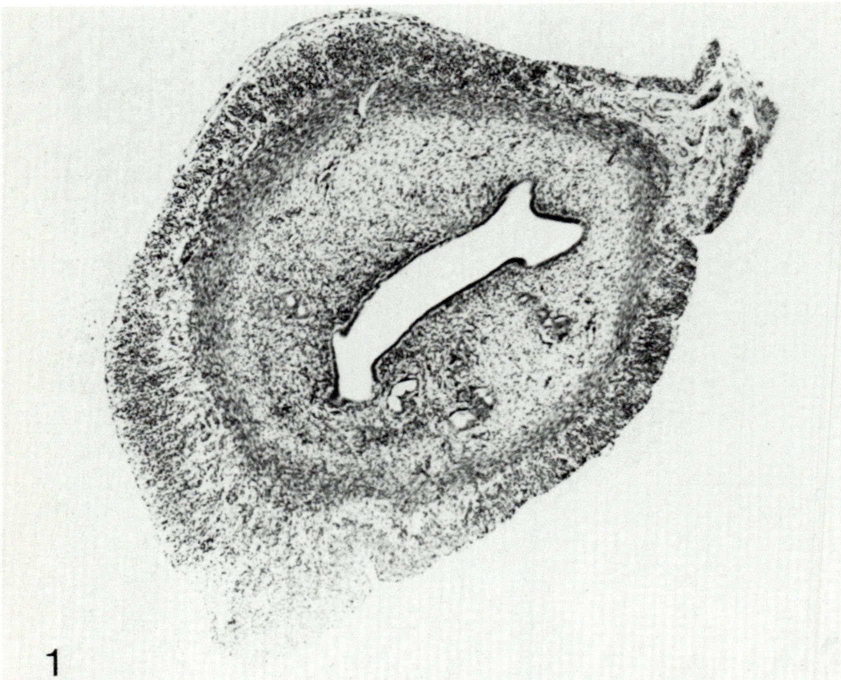

Fig. 1. Light micrograph of cross section through the uterus of rat. ×70.

space between the muscle cells is occupied by collagen and other cells including fibroblasts, blood and lymphatic vessels and nerves. The volume of the extracellular space in smooth muscles estimated by stereological methods and extracellular markers varies from 10 to 60% (*78, 98*).

3. Morphology of the Smooth Muscle Cell

3.1. Size and Shape of Muscle Cells

Smooth muscle cells are generally thought to be long, spindle shaped cells (see Fig. 3), but may, in fact, be irregularly shaped, for example, in large elastic arteries and even in some small blood vessels (*26, 105*). The cells of the myometrium are largest during the later stages of gestation and their size and number is thought to be regulated by steroid hormones and distention (*4, 8, 63*). The cells range in size from 5 to 10 μm in diameter in the center of the cell and from 300 to 600 μm in length (*21, 30*). Contracted cells are

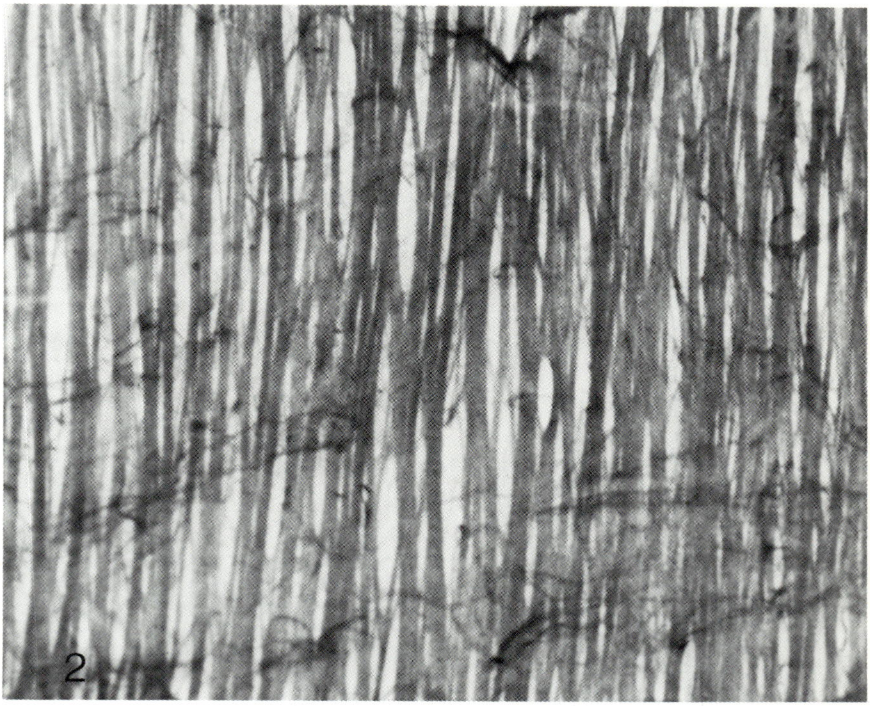

Fig. 2. Photograph of a *whole mount* of isolated longitudinal muscle from rat myometrium (endometrium and circular muscle removed), taken through a dissection microscope after fixation and staining with trypan blue. Note the network arrangement of the bundles of the muscle layer oriented in the long axis (top to bottom of photo) of the uterus. ×50.

shorter, but larger in the middle and relaxed and distended cells are long and narrow. Generally the cells are round or oval in the relaxed state when viewed in the transverse orientation and appear serrated when contracted isotonically (*33*). The average volume of a myometrial smooth muscle cell has been estimated at 21,000 μm^3 (*21*) which would translate into a surface area estimate of about 23,000 μm^2 based upon a volume to surface area ratio of 0.9. It should be mentioned that the size of smooth muscle cells probably varies considerably between different tissues and species. Gabella (*33*) has estimated the volume of intestinal muscle cells at about 3500 μm^3, which includes consideration of the caveolae and a volume to surface ratio of about 0.4. Increases in cell size have been observed in other smooth muscles in response to distention-induced hypertrophy and are associated with polyploidy (*5, 32, 58*).

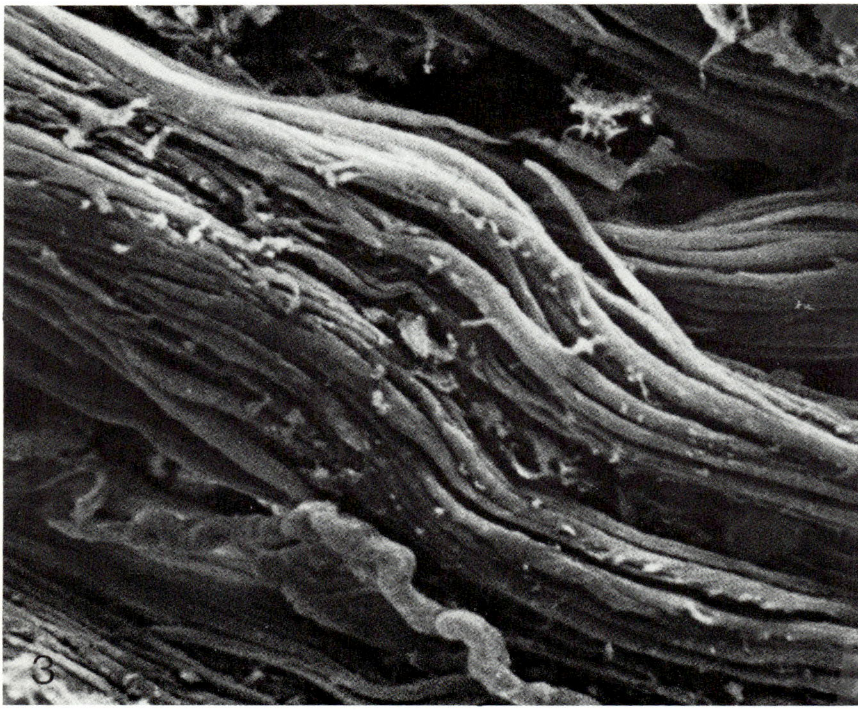

Fig. 3. Electron micrograph taken through a scanning electron microscope of longitudinal muscle of rat uterus showing size and arrangement of muscle cells. ×2600.

3.2. Plasma Membrane

The plasma membrane of smooth muscle cells is a trilaminar structure of approximately 8 nm in thickness as in other cells (*74*). Intramembranous particles about 9 nm in diameter are seen in freeze-fracture replicates of the plasma membrane (Fig. 6). These particles are more numerous on the P-face (protoplasmic), than on the E-face (external) of the membrane of myometrial cells as in other types of smooth muscle (*28, 33, 43*). The particles are thought to be structural and functional proteins that may be sites for binding and transport.

Surface vesicles or caveolae (50–80 nm in diameter) populate the surface of the smooth muscle cells (Figs. 6 and 7). These flask-shaped invaginations of the plasma membrane contribute substantially to the cell surface area. The vesicles are thought to occupy about 30% of the surface area (*33, 35, 80*). The vesicles are not randomly scattered over the surface of the cells, but are

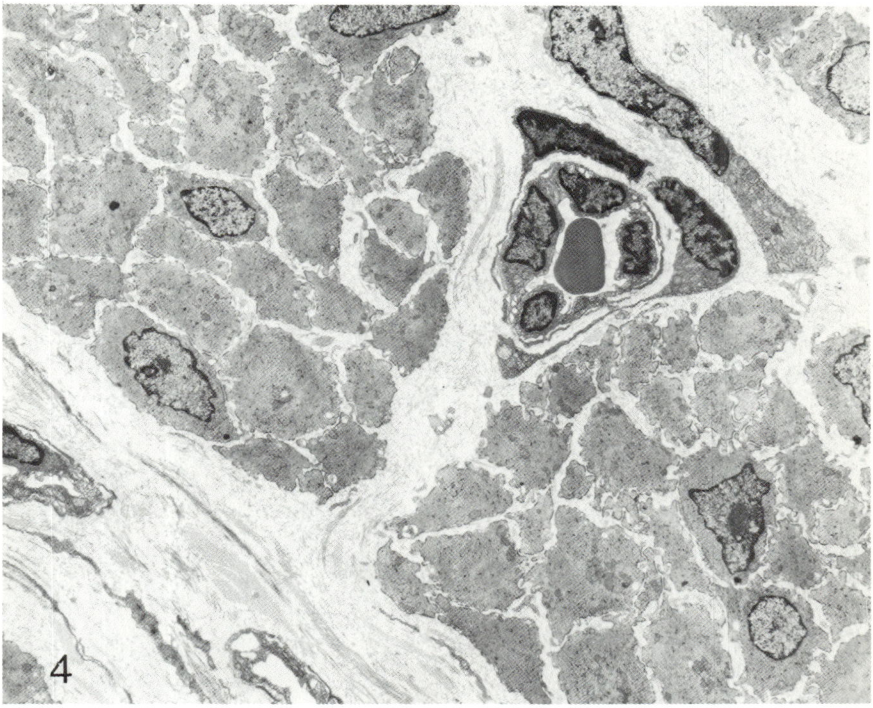

Fig. 4. Electron micrograph of portions of two bundles of the longitudinal muscle layer from pregnant rabbit myometrium in transverse section. The muscle cells are cut in cross section indicating they are oriented in the long axis of the muscle bundle. ×5100.

arranged in longitudinal rows down the cell surface as seen by freeze-fracture microscopy (*28, 33*).

The function of the vesicles is not known, although they have been suggested to be sites for ion transport and binding (*34, 35, 46, 60, 93*).The vesicles are often seen in close association to mitochondria and the sarcoplasmic reticulum (*see* Fig. 8 and refs. *27, 32, 93, 109*). The vesicles are not thought to participate in any pinocytotic or endocytotic function, as they have never been shown to be interiorized. Electron dense tracers readily penetrate into the vesicles when added either before or after fixation (*98*), suggesting that they are all open to the exterior. The caveolae are not to be confused with the coated pits that serve as receptors for lipoproteins, and perhaps other substances and can be interiorized as coated vesicles (*12, 45*).

The plasma membranes of smooth muscle cells form various types of cell-to-cell contacts including gap junctions. The concept

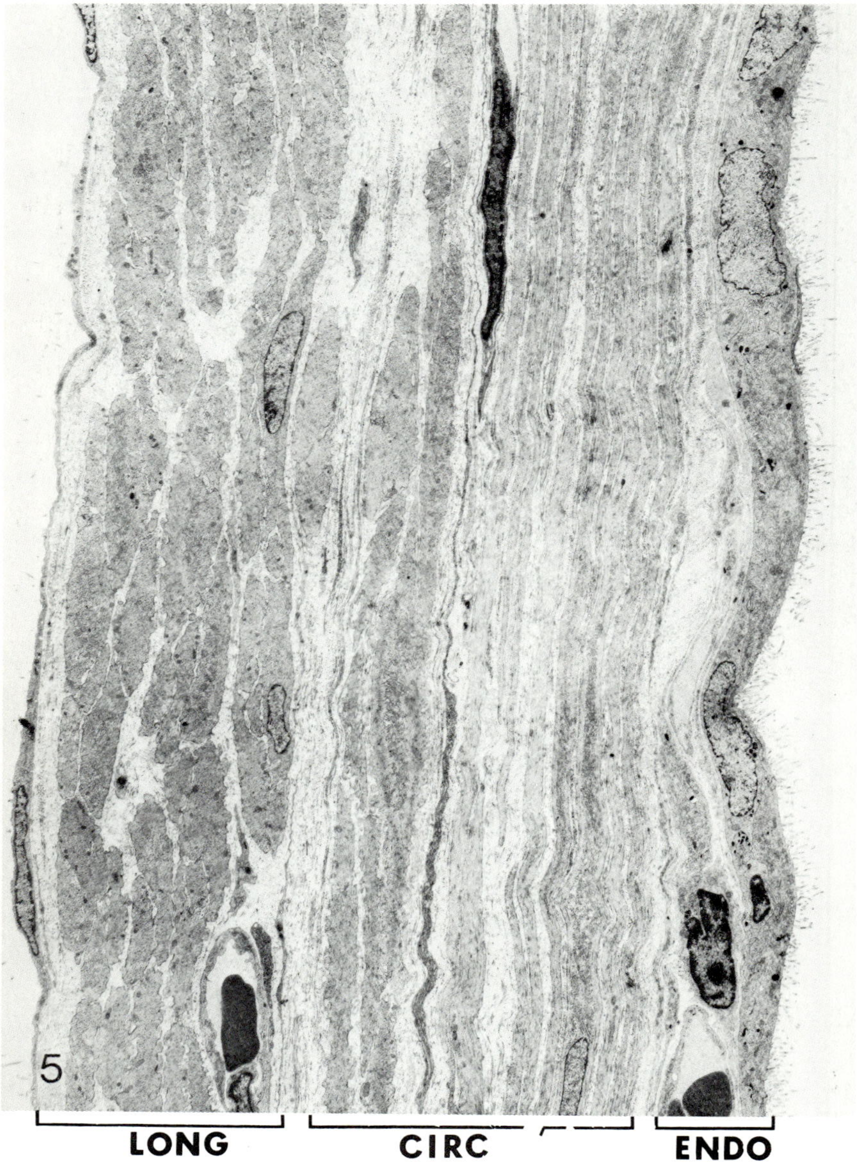

Fig. 5. Low magnification electron micrograph through the uterine wall from a 20-d pregnant rat. The uterine lumen is to the right of the field, and is bordered by the endometrium (ENDO) composed of epithelium, connective tissue and blood vessels. The circular muscle layer (CIRC) is shown in longitudinal section and the longitudinal muscle (LONG) in transverse orientation. The total wall thickness in this distended, term uterus, is approximately, 65 μm. ×4800.

Fig. 6. Electron micrograph of freeze fracture replicate of smooth muscle cell from rat uterus showing surface vesicles (large arrows) and membrane particles or proteins (small arrows). ×145,000.

of cell-to-cell communication is discussed in more detail in Chapter 6.

3.3. Sarcoplasmic Reticulum

All smooth muscle cells have a system of sarcoplasmic reticulum (*26, 80, 98*) consisting of a network of tubules and sacs within the cytoplasm (Figs. 7 and 8). The volume of the sarcoplasmic reticulum is variable in different smooth muscles, and ranges from 2 to 7.5% of the cell volume (*98*). The granular reticulum and agranular reticulum are continuous and the agranular reticulum makes close contact with surface vesicles, plasma membrane and gap junctions (Figs. 7 and 8).

The major known functions of the sarcoplasmic reticulum are those of a storage site of activator calcium (see below; smooth endoplasmic reticulum) and a site for protein synthesis (rough endoplasmic reticulum). The increased prominence of the rough

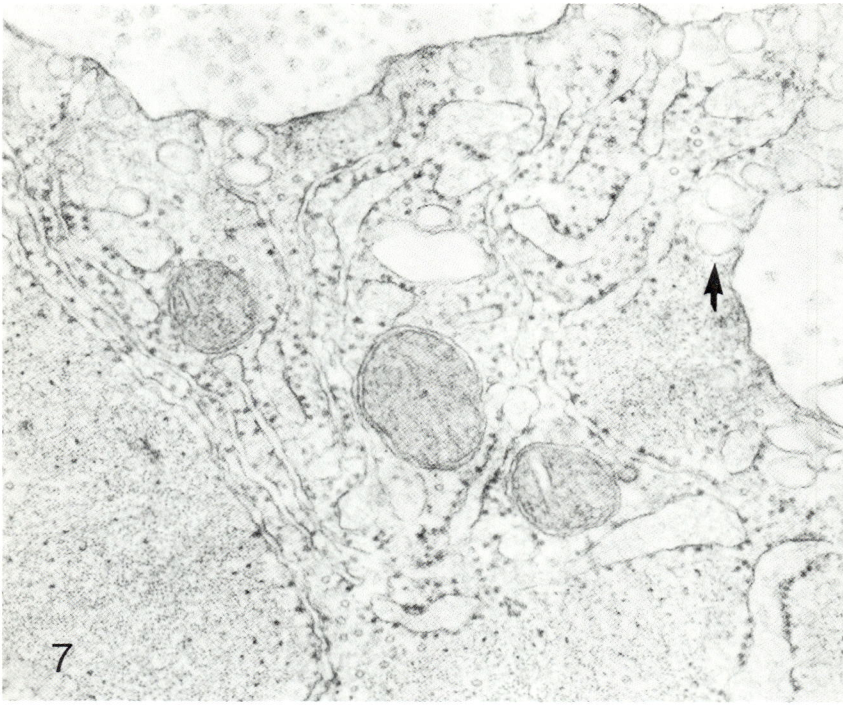

Fig. 7. Transverse section through a myometrial cell from pregnant guinea pig uterus. Note the continuity between vesicles (arrow), reticulum, plasma membrane and mitochondria. ×50,000.

endoplasmic reticulum observed in uterine smooth muscle following estrogen treatment or pregnancy presumably reflects increased protein synthesis (*4, 68*). It is likely that the greater development of rough endoplasmic reticulum in large elastic arteries that contain an extensive system of extracellular connective tissue (elastin, collagen, glycosaminoglycans) also reflects the morphogenetic function of smooth muscle in synthesizing these extracellular proteins (*26, 66, 67*).

Isolated enriched fractions of microsomes, plasma membranes, and sarcoplasmic reticulum membranes from uterine smooth muscle and other smooth muscle accumulate calcium (*76a*). The fact that smooth muscles contract in vitro when placed in calcium-free solutions and accumulate Ca^{2+} and Sr^{2+} *in situ* (*27, 94*) supports the concept of an internal source of calcium for contraction.

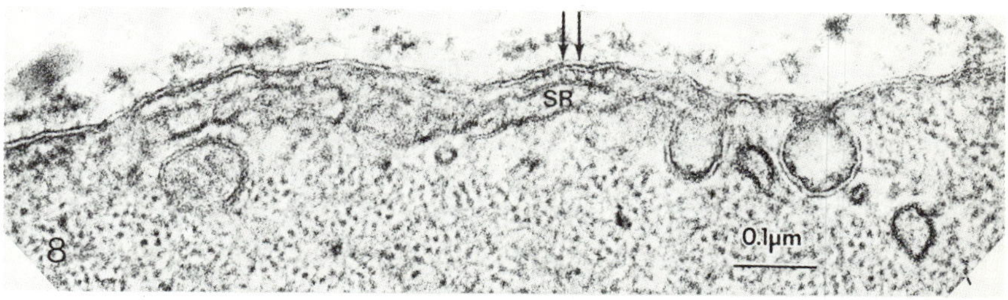

Fig. 8. Transverse section from the main pulmonary artery of the rabbit showing surface coupling between junctional SR and surface membrane and membrane bridges (arrows) crossing the junctional gap. Bar, 0.1 µm. From ref. *97*.

Certain regions of the sarcoplasmic reticulum show structural specialization: the junctional SR is connected to the surface membrane (Fig. 8) through quasi-periodic electron opaque "bridging structures," resembling those found at the triadic junctions between T-tubules and terminal cisternae of sarcoplasmic reticulum in striated muscle (*80, 97, 98*). These "surface couplings" between the junctional sarcoplasmic reticulum and the plasma membrane are structurally suitable sites for electrical activity (depolarization, action potential) of the membrane or the action of drugs on receptors to trigger calcium release. The methods used for the demonstration of calcium in the sarcoplasmic reticulum in smooth muscle and Ca^{2+} release from these sites during contraction, will be discussed in greater detail below.

3.4. Mitochondria and Other Organelles

The mitochondria are the site of oxidative metabolism in smooth muscle, as they are in other cells (*19, 52*). The mitochondria occupy a volume of approximately 3–9% of the smooth muscle cell volume (*33, 98*). Although mitochondria isolated from uterine and other smooth muscles can accumulate Ca^{2+} through an active transport mechanism (*107, 108*), several considerations suggest that mitochondria *do not* play a physiological role in the regulation of cytoplasmic free Ca^{2+} but only accumulate it under pathological conditions. Thus, in the presence of physiological concentrations of Mg^{2+}, the apparent affinity of mitochondria isolated from either uterine (*108*) or vascular smooth muscle (*107*) is too low (10–17 μM) to permit Ca^{2+} accumulation at the levels of cytoplasmic-free Ca^{2+} normally reached during even a maximal contraction. In agreement with this conclusion, electron probe analysis showed no evidence of any calcium uptake by mitochondria in vascular smooth muscles that had maintained maximal tension for 30 min. (*88*). Massive mitochondrial calcium accumulation has been found in damaged cells (*88*) or in smooth muscles that were chemically skinned with saponin and exposed to high ($10^{-5}M$) calcium for 30 min. (*91*), and the calcium content of mitochondria isolated from atherosclerotic (bovine) blood vessels is increased (*87*), supporting the idea that mitochondrial calcification could play a role in cell pathology. An interesting feature of mitochondria isolated from (vascular) smooth muscle is the presence of a very active Mg^{2+} transport system (*75*), albeit its physiological significance is not known.

Uterine smooth muscle cells have an extensive Golgi system consisting of stacked lamellar elements and vacuoles usually located perinuclear (Fig. 9). The function of the Golgi in smooth muscle has not been extensively studied. In vascular smooth

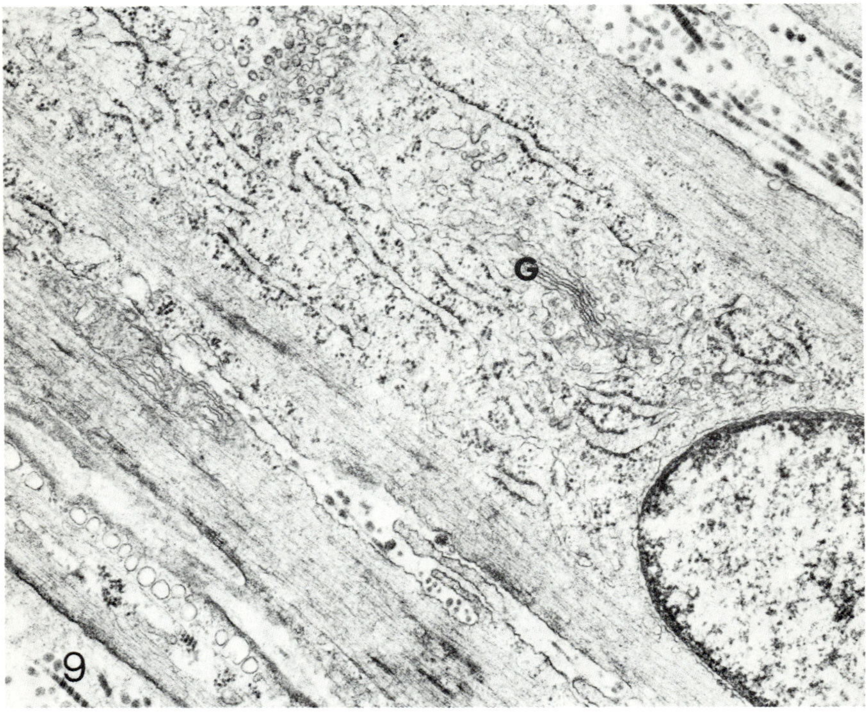

Fig. 9. Longitudinal section from pregnant rat myometrium showing extensive granular sarcoplasmic reticulum of muscle cells and Golgi (G) located perinuclear. ×22,000.

muscle the Golgi is thought to be involved in glycosylation, and synthetic and degradative processes as it is in other cells (*3*). We have demonstrated that the ionophore X537A dilates the Golgi, but this effect may not be due to an action of Ca^{2+}, but rather an effect on glycoprotein synthesis (*37, 83*).

There have also been few studies of lysosomes (*24, 25*) in uterine smooth muscle. Lysosomes appear as electron-dense bodies within the cytoplasm of the muscle cells and their presence seems to be dependent upon estrogen; estrogen promotes lysosomal formation (*76*). Lysosomes appear to be particularly prominent in the postpartum uterus during involution of uterine muscle (*13*). Lysosomes are also abundant in cultured smooth muscle cells (*18*). The electron density, of at least some, lysosomes (dark bodies) presumably reflects largely protein content; dark bodies in cultured vascular smooth muscle contain relatively high concentrations of sulfur, probably associated with sulfated proteins (*49, 50*).

3.5. Myofilaments

Contraction of smooth muscle cells is thought to occur, as in skeletal muscle, through the interaction of myosin and actin filaments (*33, 81, 84, 85, 98*). At least three distinct types of filaments have been identified in smooth muscle (*see* refs. *33* and *98*). These include the thick (15 nm diameter, myosin), thin (6–8 nm, actin) and intermediate (10 nm, desmin or vimentin filaments, Fig. 10). Microtubules are also prominent within the cytoplasm of smooth muscle cells (Fig. 10).

The inability to demonstrate myosin filaments in smooth muscle in early electron microscopy studies (*69, 70*) has been related to difficulties in preparation and solubility. The myosin filaments in vascular smooth muscle are about 2.2 μm in length (*2*), which is longer than those present in skeletal muscle (about 1.6 μm). Myosin filaments are present in both relaxed and contracted smooth muscle, regardless of whether the myosin light chains are phosphorylated or not (*99*). Interest in the effect of light chain phosphorylation on myosin filament assembly was generated through experiments indicating that the conformation of single myosin molecules in vitro and the solubility of *synthetic* myosin filaments was influenced by their state of phosphorylation (*77, 101, 106*). The finding of unphosphorylated myosin filaments in intact muscle suggests that the physiological significance of the in vitro observations, if any, is at a level of structural organization not recognizable by current electron microscopic methods.

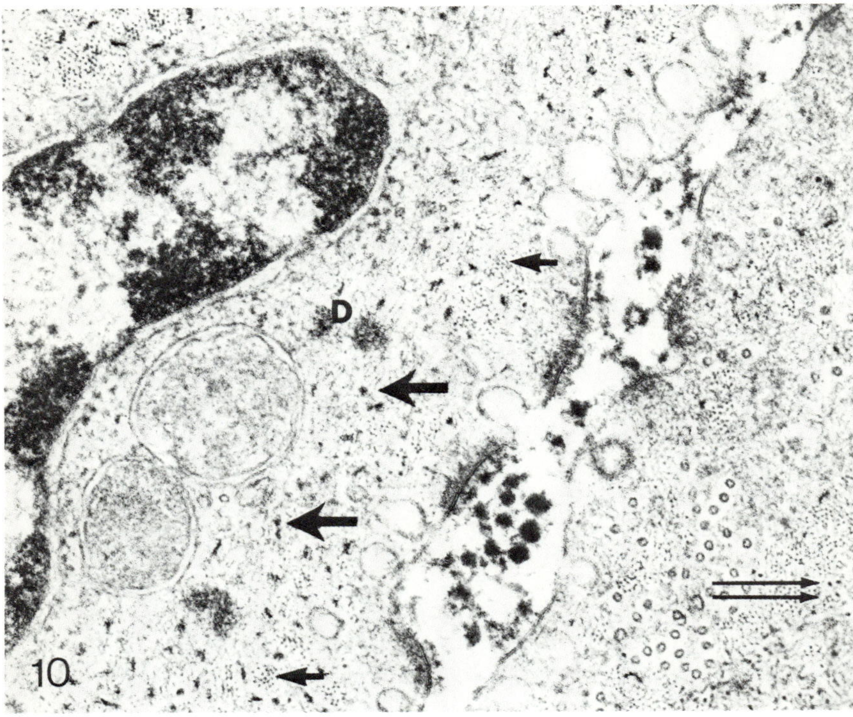

Fig. 10. Transverse section of a portion of a myometrial cell from nonpregnant human uterus showing contractile filaments. Actin (small arrows), myosin (large arrows) and intermediate or 100 Å (double arrows), filaments and dense bodies (D) present in the cytoplasm. ×61,000.

Dense bodies are attachment sites for actin filaments (Fig. 11), analogous to the Z-lines in striated muscle, and may appear "free-floating" in the cytoplasm or attached to the plasma membrane (*98*). The characteristic arrowhead configuration found on actin filaments decorated with the S-1 subfragment of myosin with arrowhead pointing away from either side of the dense body, is consistent with the actin filaments forming functional attachments at these sites (*11*). There are approximately 15 actin filaments to every myosin filament in some types of smooth muscle (*2*).

In early studies by Csapo (*20*) the content of uterine contractile myofilaments (actomyosin) was shown to be dependent upon estrogen and distention. Actomyosin concentrations were higher during pregnancy and lowest after ovariectomy. The intermediate filaments were first recognized in nonmuscle cells as structures 10

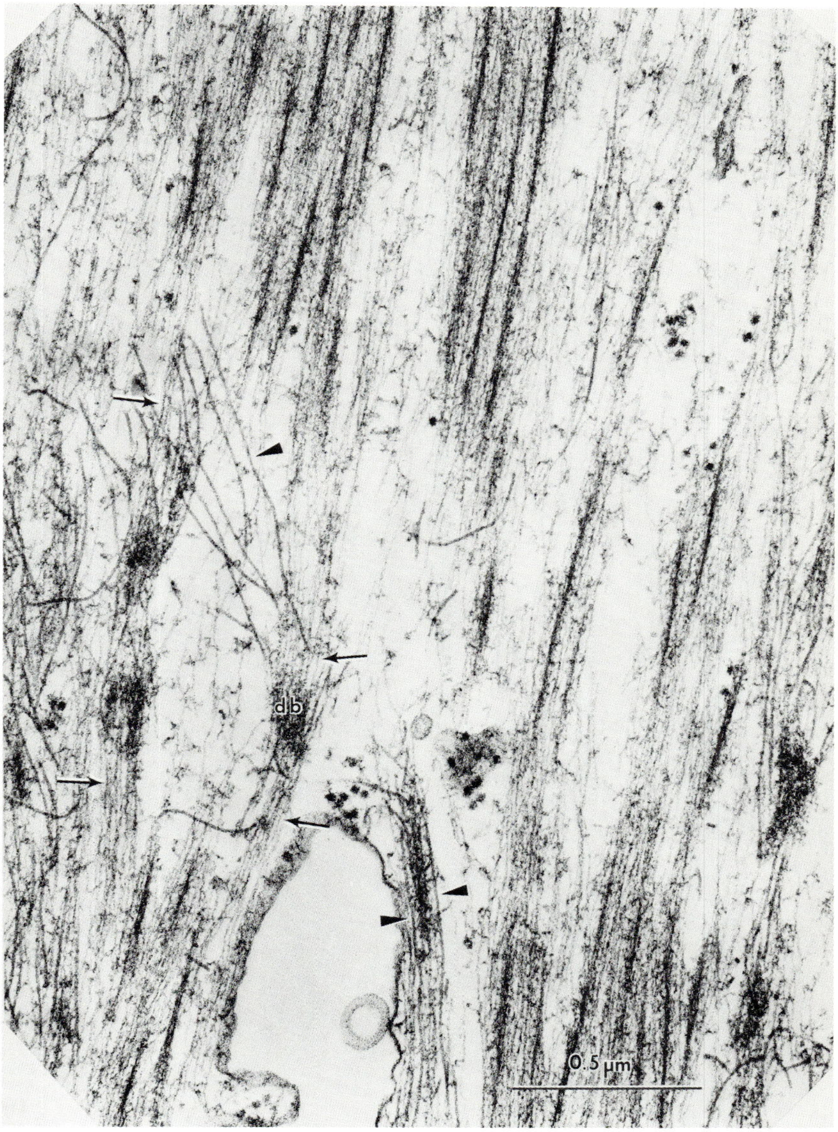

Fig. 11. Longitudinal section of a saponin-skinned PAMV smooth muscle cell. The filaments are splayed out in some of the cells, as shown, which reveals the relationship of the dense bodies with associated actin to the neighboring myosin filaments. Thin filaments (indicated by arrows) which emerge from cytoplasmic dense bodies (db) can be traced to the myosin filaments on either side forming an I-band. The 10 nm filaments (arrowheads) do not run parallel to the sarcomeres. Connections between 10 nm filaments and a cytoplasmic dense body occur at the lower center part of the figure. Bar, 0.5 μm. From ref. *11*.

nm in diameter (otherwise known as 10 nm filaments) that, unlike actin filaments, were not decorated by myosin subfragments (*48*). A large body of work on these filaments has shown that such filaments are ubiquitous in most muscle and nonmuscle cell systems. The major constituent proteins of the intermediate filaments in smooth muscle are desmin and vimentin. Although one of these proteins may be predominantly, or solely, present in some smooth muscles, like vimentin in aortic (Fig. 12) or desmin in intestinal smooth muscle, in some smooth muscles both desmin and vimentin coexist (Fig. 13; *62*). An extensive increase in the number of intermediate filaments occurs in hypertrophied intestinal (*32*) and vascular smooth muscle (Fig. 14).

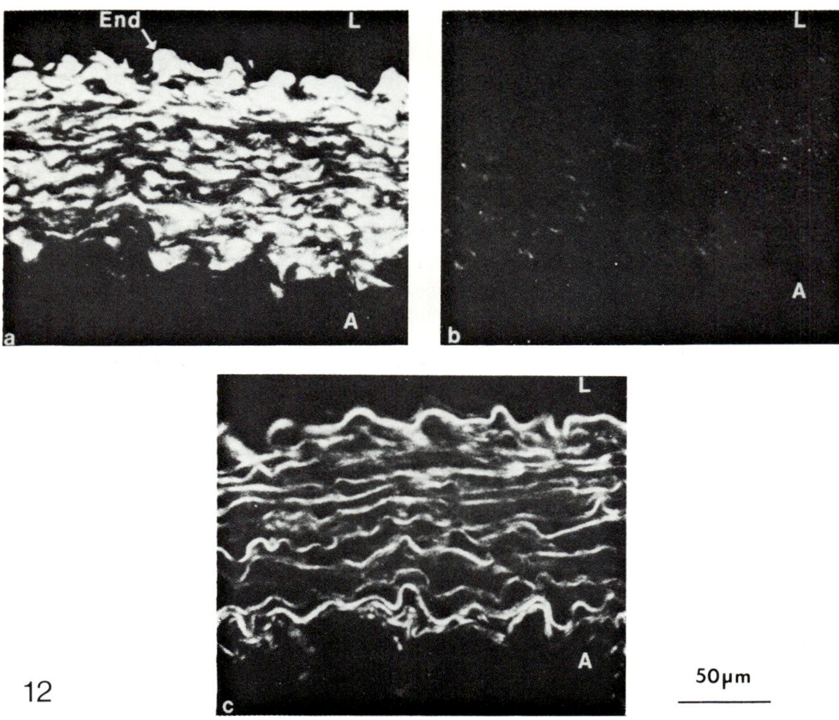

12

Fig. 12. Transverse sections of the abdominal aorta: (a) anti-hamster vimentin staining showing luminal (L) and adventitial (A) areas and endothelium (End); (b) nonspecific goat IgG staining; (c) autofluorescence of the internal elastic laminae. Bar, 50 µm. From ref. *92*.

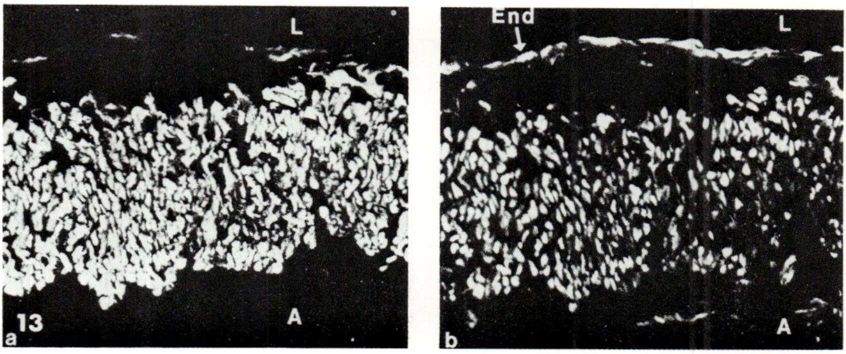

Fig. 13. Transverse sections of the PAMV; double immunofluorescence using the anti-chick desmin antibody and rhodamine (a) and monoclonal anti-vimentin and fluorescein (b) showing the luminal (L) and adventitial (A) areas with the endothelium (End). Modified from ref. *92*.

4. Innervation of Smooth Muscle

Autonomic nerves innervate most smooth muscle tissues (*14–16*). The density of nerve fibers and varicosities varies considerably between tissues. In some smooth muscle tissues, such as vas deferens, each cell is thought to be associated with a nerve terminal (*16*). Other smooth muscle tissues are more sparsely innervated and nerves end on only a small percentage of smooth muscle cells in the muscle bundle (*15*). The neurotransmitters released from the nerve varicosities are thought to modulate activity of the muscle cells either to stimulate or inhibit contractility. In tissues where each cell is not directly innervated, the inner cells are believed to be activated either by the diffusion of the neurotransmitter from the nerve or by the electrical coupling between the cells (*15*).

Nerve fibers with one to many axons usually are oriented in the direction of the muscle fibers, but in some cases form a network around the smooth muscle tissue. Large, preterminal nerve fibers, at considerable distances from smooth muscle cells, usually do not contain vesicles.

The neuromuscular junction in smooth muscle tissue differs greatly from the nerve–muscle contact in skeletal muscle. The nerves terminate in varicosities containing vesicles at distances from 20 nm to several microns (Fig. 16). There are no recognizable modifications in the smooth muscle cells in the region of the neuromuscular contact.

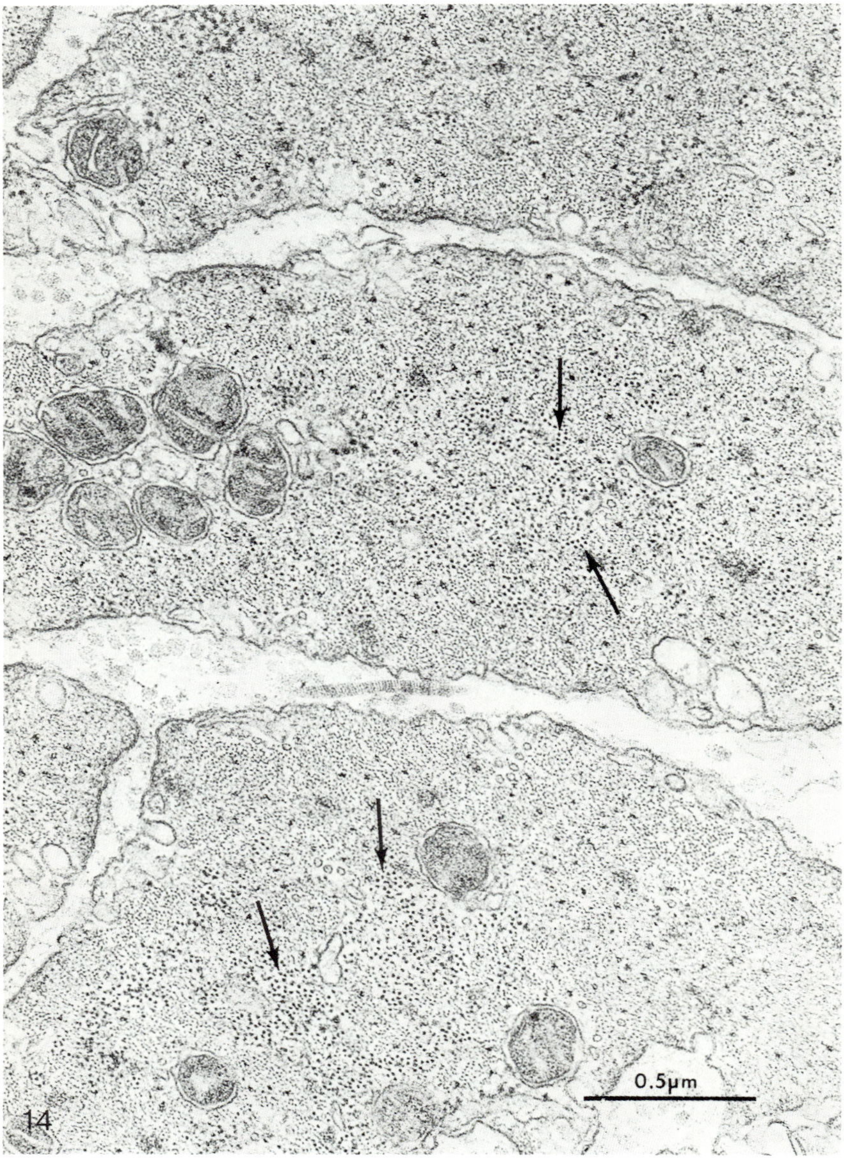

Fig. 14. Transverse section of a bundle of smooth muscle fibers from hypertrophied PAMV showing the predominantly central location of the abnormally increased number of intermediate filaments (arrows). Bar, 0.5 μm. From ref. *5*.

There are at least three different types of autonomic terminal varicosities in smooth muscle, recognized by the presence of vesicles and granules of characteristic size (*15*). Adrenergic terminals contain many small, granular vesicles about 30–60 nm in diameter with occasional large, dense vesicles of 60–150 nm. These terminals are thought to release noradrenaline or other catecholamines as the neurotransmitter. The vesicles in cholinergic varicosities, with acetylcholine as the transmitter, are approximately the same size (30–60 nm) as those seen in adrenergic fibers except the vesicles are agranular. Because the dense core of the vesicles is not readily preserved in all fixatives, adrenergic fibers may sometimes be mistaken for cholinergic fibers. The third type of varicosity contains large opaque vesicles and is thought to represent purinergic nerves with ATP as the transmitter (*15*).

4.1. Innervation of the Myometrium

The myometrium probably can be considered sparsely innervated, except during the later stages of pregnancy. Nerves innervating large muscular blood vessels and muscle bundles of the myometrium are present, except in late pregnancy, when they are thought to degenerate in response to hormones and stretch (*53, 59, 100, 102–104*). A large nerve fiber between the longitudinal and circular muscle layers of the rat uterus is shown in Fig. 15. Nerve varicosities are rare but when present they are in close proximity to bundles of myometrial cells (Fig. 16).

The exact functions of nerves in the uterus are not known. They may control contractility, blood flow, secretion, or sensation. It is generally assumed however, that nerves do not control myometrial contractility, as the muscle is thought to be myogenically controlled and its activity modulated by hormones and prostaglandins.

5. Structure of Cultured Cells

The ultrastructure of cultured vascular and intestinal smooth muscle cells has been described in detail (*18*). Studies of cultured uterine cells have also been described (*54*). Cultured smooth muscle cells are generally spindle-shaped with centrally located nuclei (Fig. 17). Many of the organelles in cultured smooth muscles such as mitochondria, endoplasmic reticulum, and Golgi are more prominent at the nuclear poles as in adult tissues. Mem-

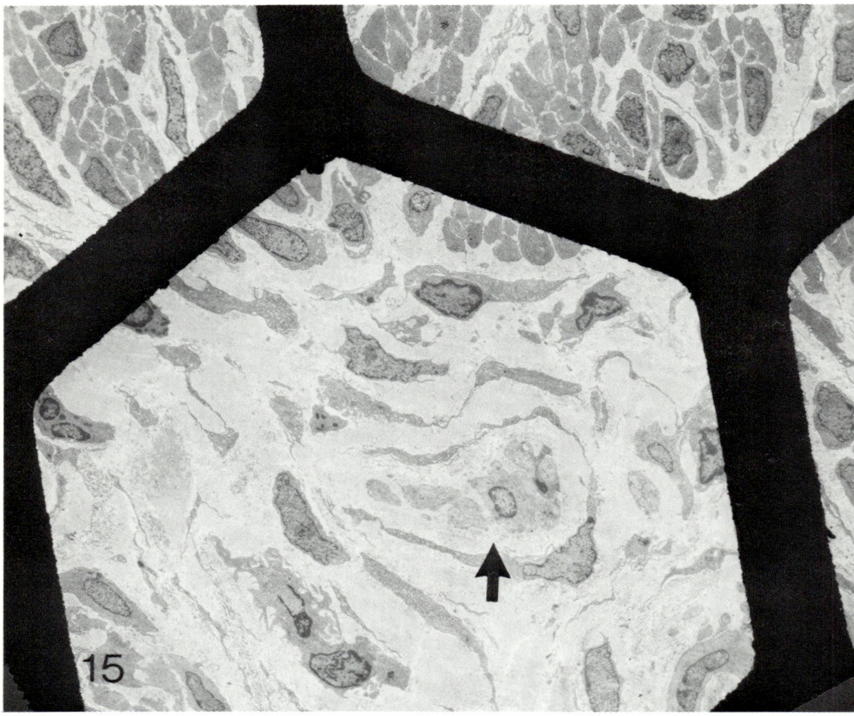

Fig. 15. Low magnification photomicrograph of uterine tissue from nonpregnant rat showing a section over several grid squares (black bars = grid). The longitudinal muscle layer cut in cross-section is at the top and the circular muscle layer in longitudinal orientation is to the left (lower corner). A large nerve fiber (arrow) is shown between the two muscle layers. ×1700.

brane vesicles occur along the plasma membrane. Filaments oriented along the long axis of the cell occupy much of the cytoplasm in confluent cultures. The sarcoplasmic reticulum, at least in certain phases of the cell cycle, is more extensive than in normal adult smooth muscles (*18*).

The abundance of myofilaments (6 nm diameter) as seen in the electron microscope (Fig. 17) has been used as one criterion of identity of vascular smooth muscle in culture (*66*). However, actin filaments have now been observed in a number of nonmuscle cells (*18*).The appearance of thick filaments in cultured smooth muscle seems to be related to the ability of the cells to contract. Myosin filaments and spontaneous contractility are both observed in primary cultures of taenia coli and vas deferens prepared by dissociation (Fig. 18).

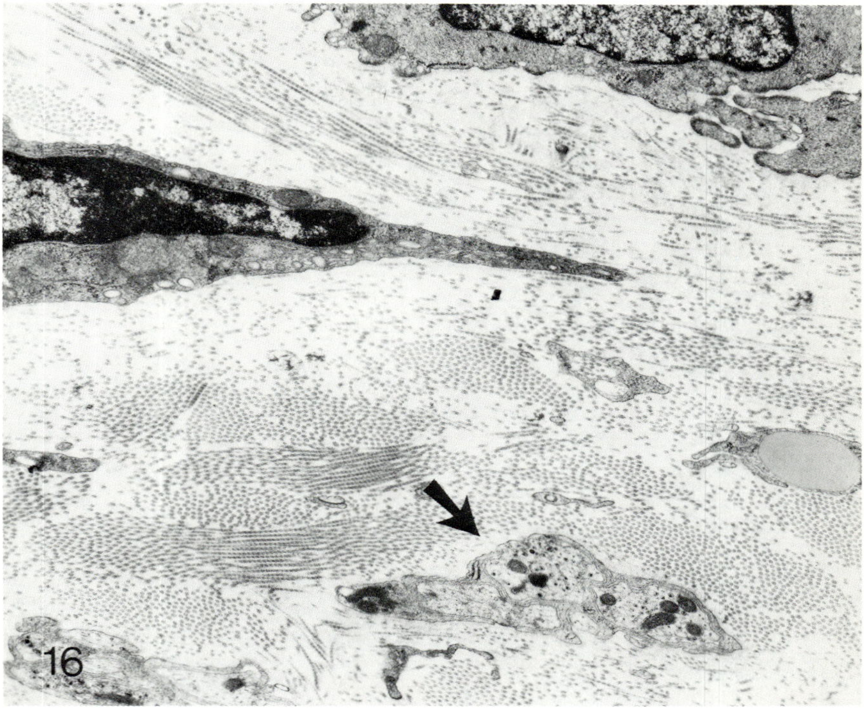

Fig. 16. Micrograph of nonpregnant rabbit uterus showing nerve terminal or varicosity (arrow) in close association with a bundle of smooth muscle cells of the longitudinal layer. The nerve is identified as adrenergic by the presence of small, dense-cored vesicles. ×28,500.

Several studies have suggested that smooth muscle cells of adult tissues are phagocytotic (*12, 18, 41, 45*). Phagocytosis by cultured aortic smooth muscle and by well-differentiated, spontaneously contracting vas deferens smooth muscle cells has also been demonstrated (*41, 44*). Cultured vascular smooth muscle cells very actively phagocytose latex spheres (Fig. 18). The phagocytotic pathway may be a route of nonreceptor-mediated entry of lipids and other large molecules into smooth muscle cells.

6. Localization of Ca^{2+} in Smooth Muscle

Techniques that can determine quantitatively and at high spatial resolution the distribution of elements in smooth muscle can provide essential information about the sources and sinks of ions

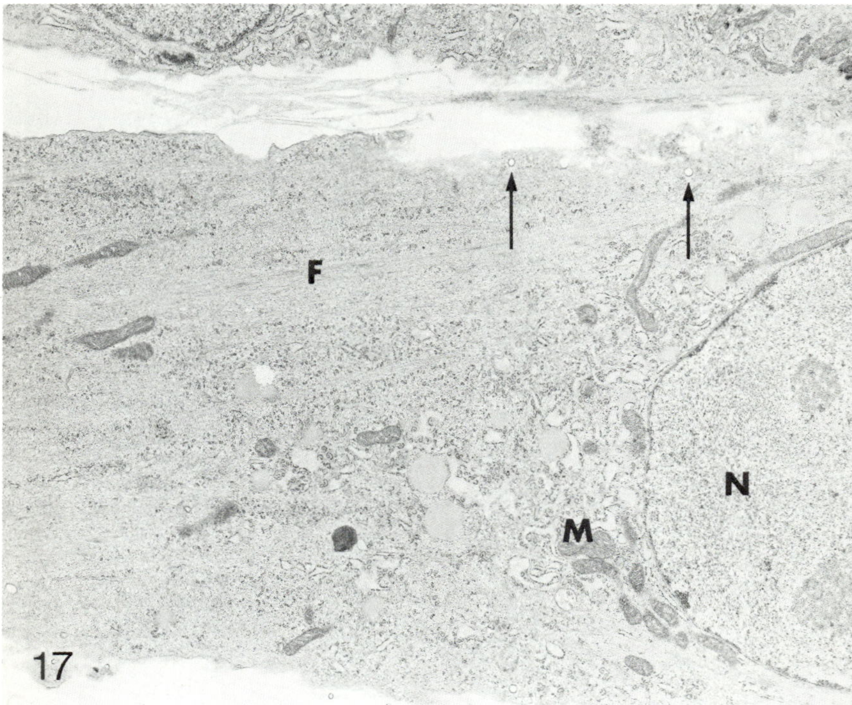

Fig. 17. Low magnification (X5000) electron micrograph of cultured smooth muscle cell from guinea pig aorta showing myofilaments (F) oriented in the long axis of the cell, mitochondria (M), nucleus (N), and vesicles (arrows).

controlling smooth muscle function and about the effects of drugs and neurotransmitters on the translocation of these ions, and particularly calcium. Several methodologies have been evolved to this end during the past decade but, unfortunately, the methods that are easy are not reliable, while those that are reliable are not easy.

6.1. Precipitating Agents, Dyes, and Autoradiography

Pyroantimonate and oxalate have been used in attempts to precipitate and localize Ca^{2+} in smooth muscle by viewing the deposits with the electron microscope (*23, 40, 60, 61*). The rationale behind these studies is that the precipitating agents will diffuse into the muscle cell during processing for electron microscopy and complex with Ca^{2+} to form an electron-opaque crystal at the ion's

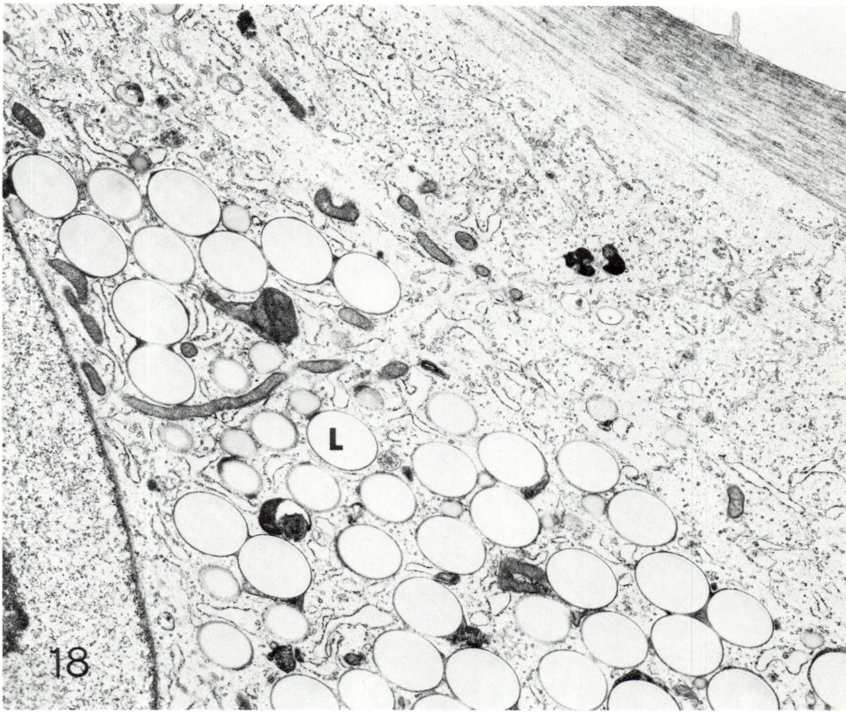

Fig. 18. Cultured vascular smooth muscle cell from guinea pig aorta after 18-h incubation with latex spheres to show phagocytosis. Note the abundance and location of the latex particles (L) within the smooth muscle cell. ×8200.

free or bound sites. Using these methods, electron dense precipitates have been observed in smooth muscle mitochondria, surface vesicles, sarcoplasmic reticulum, plasma membrane, and nuclei (*23, 61*). Unfortunately pyroantimonate *cannot provide valid information* about the localization of calcium in cells, because of the lack of specificity of the precipitating agent for calcium, the translocation of Ca^{2+} during the liquid fixation used for processing, and the failure of pyroantimonate to penetrate some of the intracellular compartments (*40*). Indeed, surface membrane-bound electron opaque deposits in smooth muscle treated with the pyroantimonate method contain primarily K, Os, and Sb, with only occasional and random traces of calcium (APS, unpublished observations). Precipitation with pyroantimonate suffers, in addition, from the fact that the X-ray lines of Sb and Ca overlap, and the quantitation of Ca in the presence of Sb requires relatively sophisticated computer fitting techniques. In view of its lack of specificity and the major translocations that can occur during preparatory procedures, the use of the pyroantimonate method should be discouraged as an exercise in futility. Oxalate is a more specific precipitating agent for calcium, but is useful only for the semiquantitative demonstrations of calcium in membrane-bound compartments (i.e., sarcoplasmic reticulum).

Autoradiography has been used in attempts to localize Ca^{2+} in smooth muscle (*55*), but neither the sensitivity nor the spatial resolution of the method, as used to date, has been sufficient to yield reliable results. In particular, photographic grains thought to represent calcium localized on the plasma membrane may be due to the deposition of extracellular Ca^{2+} in this region. The fact that some radioautographic studies suggest that mitochondria contain high concentrations of calcium, in contrast to the more reliable results obtained with electron probe analysis of cryosections, indicates that either translocation of Ca^{2+} occurred during preparatory steps for radioautography and/or that the spatial resolution and sensitivity of the technique is insufficient for these studies.

Electron optical techniques in current use detect total (free and bound) calcium at high spatial resolution, but provide no information about the concentration of free Ca^{2+}. Recent studies with a Ca-sensitive luminescent protein, aequorin, have demonstrated the rise in free Ca^{2+} in activated smooth muscle (*29, 56*). Although not as well resolved spatially as electron microscopic methods, these techniques can provide valuable complimentary information about transient fluctuations in free Ca^{2+} in the time domain. The potential and limitations of these techniques have been extensively reviewed by Blinks (*7*).

6.2. Electron Probe and Electron Energy Loss Analysis of Smooth Muscle

Energy dispersive electron probe X-ray microanalysis (EDS or EPMA) has been used extensively for the detection and quantitation of elements in biological materials (*11, 36, 38, 47, 60, 71, 72, 82, 86, 89, 90, 96*). Electron energy loss spectroscopy (EELS) is a second electron optical method suitable for obtaining compositional information at electron microscopic resolutions and is now being developed in several laboratories. Some preliminary results have been obtained with this technique on biological specimens (*73, 90*). EDS and EELS offer major advantages in that for the localization of elements in smooth muscle and other cells, these are the only methods currently available for determining the composition of cellular organelles *in situ*. EDS is based on the fact that irradiation of a microvolume with a focused electron beam, such as produced in an electron microscope, results in the production of X-rays that are characteristic of the elements present in the microvolume. These characteristic peaks are superimposed on a broad X-ray background (continuum X-rays) that, in ultrathin sections, is proportional to the mass of the microvolume irradiated (for review, *see* refs. *47, 71,* and *90*). The X-rays, from elements in the range of atomic numbers from that of Na and higher, are detected with an X-ray detector that is placed within the microscope column, as close to the specimen as possible. Spectra obtained by focusing the beam on organelles (e.g., mitochondria, sarcoplasmic reticulum, and nucleus) are collected and the concentrations in these regions obtained by using computerized methods for quantitation (*71*).

EDS has already produced considerable biologically significant information, while, as indicated above, biological EELS is still largely in the stage of development. In the long term, EELS offers the potential of providing information not attainable by EDS, such as molecular (rather than only atomic) species, greater sensitivity, and higher spatial resolution than practically attainable with EDS.

The tissue preparation techniques used for EDS or EELS are critical for preventing translocation artifacts and damage. To prevent the movement of diffusible elements the only reliable method involves rapid freezing of tissues. High spatial resolution requires the analysis of thin, freeze-dried cryosections. These preparation methods have been described in detail for smooth and striated muscle (*87, 88, 96*). Briefly these methods rely upon rapid freezing in liquid nitrogen, helium, or propane; cutting sections of the frozen tissue; and drying the frozen section under vacuum.

Studies using EDS on vascular smooth muscle and cardiac muscle revealed that mitochondria do not accumulate Ca^{2+} during contraction, contrary to what one might expect if they were important sites for regulation of the ion (*86, 87, 88*). Mitochondrial uptake of Ca^{2+} was observed only when damaged cells were examined. In other studies we attempted to localize Ca in cultured vascular smooth muscle cells (*36, 38*). Mitochondrial deposits were observed and analyzed by EDS. However, later studies showed that the deposits were associated with cell damage (*50*). These studies are particularly pertinent to the understanding of control of intracellular Ca, and provide *in situ* evidence that mitochondria do not play a physiological role in control of Ca^{2+}. In vascular smooth muscle cells made hyperpermeable with saponin to allow control of cytoplasmic-free Ca^{2+} with EGTA buffered solutions, the mitochondria did not accumulate Ca^{2+} at $10^{-6}M$ although they contained large deposits of Ca (granules) at $10^{-5}M$. Comparing these results with the absence of Ca^{2+} accumulation in maximally contracted normal (unskinned) smooth muscle, it was concluded that under normal conditions the free Ca^{2+} concentration in smooth muscle remains below $10^{-5}M$. In contrast, Ca accumulation in the SR at $10^{-6}M$ free Ca^{2+} was observed in the same experiments (Fig. 19).

In addition to information provided about the subcellular distribution of Ca^{2+}, electron probe analysis also showed that there were no major elemental gradients for K, Na, and Cl between the mitochondria and cytoplasm, and the high cellular Cl content in smooth muscle was not compartmentalized (*88*). The latter finding, also shown through the use of Cl selective electrodes (*1*), supported the concept that the high Cl content of smooth muscle (*17*), may be due to the operation of a Cl pump. EDS of rabbit portal anterior mesenteric vein also revealed cytoplasmic Na concentration higher than estimated from chemical measurements and showed, furthermore, that significant amounts of Na can be lost from cells at 2°C in Na-free (Li-substituted) solutions that had been thought to remove only extracellular Na^{+}. In view of the suggested role of Na^{+}/Ca^{2+} exchange in contractile regulation (e.g., ref. *6*), EDS may play an exceedingly valuable role for unequivocal measurement of cytoplasmic sodium in normal, abnormal, and drug-treated smooth muscles.

EDS has also been used to study ion content of sarcoplasmic reticulum in muscle (*87*). Studies of skeletal muscle by EDS have shown that the terminal cisternae of the sarcoplasmic reticulum stores (up to 120 mmol Ca^{2+}/kg dry wt) and releases Ca^{2+} during contraction (*87, 96*).These studies have confirmed that the sarcoplasmic reticulum in skeletal muscle plays the most important role in the regulation of cytoplasmic Ca^{2+}. Furthermore, these investigations revealed that (1) during tetanus more Ca^{2+} was released from the sarcoplasmic reticulum than previously believed, (2) there was an increase in K^{+} and Mg^{2+} content of the sarcoplasmic reticulum during Ca^{2+} release and (3) there was no large and/or significant electrical potential change between the cytoplasm–reticulum during tetanus.

More important to the discussion presented here is whether the sarcoplasmic reticulum (SR) in smooth muscle stores and re-

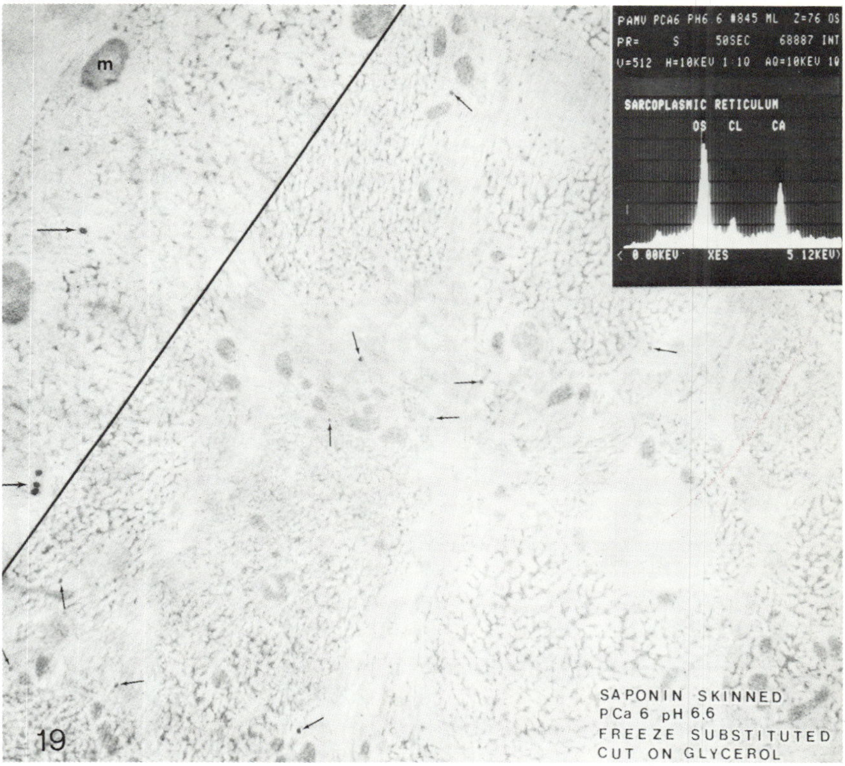

Fig. 19. Transverse section of saponin-skinned smooth muscle cells from rabbit portal vein. The skinned cells were exposed to a pCa 6, pH 6.6 EGTA buffer containing 2 m*M* free Mg for 30 min, were frozen, and were processed by freeze substitution. Deposits (arrows) of Ca associated with P in the SR are frequent at the periphery of the cells (shown at higher magnification at the upper left). Note the absence of mitochondrial loading. A typical spectrum from one of the SR deposits is shown. The osmium peak arises from the osmium included with the acetone to enhance contrast. Similar deposits were found in cryosections. M = mitochondrion. From ref. *91*.

leases Ca^{2+}. Studies of vascular smooth muscles with EDS have shown high concentrations of Ca^{2+} in tubules of the sarcoplasmic reticulum of normal cells (*86–88*). Furthermore, the release of identical stores of Ca^{2+} by norepinephrine and caffeine further suggest that the SR is the major source of activator Ca^{2+} in smooth muscle (*91*). The release of calcium from the junctional sarcoplasmic reticulum was demonstrated only recently, through electron probe analysis of the junctional (subplasma membrane) region of, respectively, relaxed and contracted guinea pig portal vein smooth muscle cells. In cryosections of rapidly frozen preparations, "hot spots" containing high concentrations of Ca^{2+}(and also P) were found with much higher frequency (approximately four per cell periphery) in relaxed cells than in cells frozen at the peak of a norepinephrine-induced contraction (*9*). These experiments were conducted in solutions containing lanthanum, but not Ca^{2+}, allowing the clear distinction of extracellular (La-containing) from intracellular regions. In parallel experiments on similar (50–80 μm diameter) smooth muscle bundles, repeated contractions could be observed in Ca-free solutions, when Ca^{2+} efflux was inhibited with La^{3+} in Na-free (high K^+) solutions and the stimulating drug (30 μ*M* norepinephrine) was removed at the peak of the contraction. The simultaneous demonstration of repeated contractions in Ca^{2+}free solutions and Ca^{2+} release from the junctional SR is consistent with recycling of Ca^{2+} between sarcoplasmic reticulum and cytoplasm during the contraction–relaxation cycle. In contrast to the high concentrations of Ca found in the region of the junctional SR, electron probe analysis of the plasma membrane itself revealed concentrations of Ca that, with the precision of the measurement, were not significantly different from cytoplasmic concentrations. In other experiments, electron probe analysis of rabbit portal anterior mesenteric vein showed that during a maintained maximal contraction cytoplasmic Ca^{2+} increased by 1.0 mmol/kg dry wt (*10*). If this calcium were to be released from the nonspecialized surface membrane, the Ca^{2+} concentration on the latter would have to be at least 30–40 mmol/kg dry membrane wt, very much in excess of the experimentally found values. These findings, although certainly not excluding the regulation of membrane permeability in smooth muscle by Ca^{2+}, argue against a significant role of the nonspecialized plasma membrane as a source or sink of activator Ca^{2+}.

The advantages of EELS over EDS have been noted and described in detail (*73, 90*). Briefly, they are (1) the ability to determine molecular as well as atomic information, (2) detectability of elements lighter than Na ($z < 11$), and (3) greater collection efficiency than EDS with a resolution of about 5 nm. EELS has been applied to the study of biological materials (*90*). The same preparation techniques used for EDS are required for EELS including careful freezing, drying, and handling. Unfortunately, relatively thin sections (30 nm thick) have to be prepared for analysis by EELS or methods used to correct for background from thin sections. Also, heavy metal stains must be avoided to prevent backscattering and without stain, imaging of cell organelles becomes limiting. It is anticipated that these problems will eventually be solved, but attention to correct preparatory techniques and an understanding of the physical processes of contrast involved in producing EELS images will require meticulous attention to obtaining and correctly interpreting biologically meaningful results. Use of EELS on isolated ferritin molecules has recently shown the iron core (7.5 nm), indicating the potential of this technique (*90*). In the long term, we anticipate that the investment required for developing EELS for biological analysis will justify itself in a capability for producing compositional information that is qualitatively more detailed (molecular, rather than merely atomic) and can be observed at higher spatial resolution than hereto attainable.

Acknowledgments

Supported by grants from the Medical Research Council of Canada to REG and by the National Institutes of Health (Grant No. HL 15835) to APS. REG is a Fellow of the Ontario Heart Foundation.

References

1. Aicken, C. C. and A. F. Brading. Measurement of intracellular chloride in guinea-pig vas deferens by ion analysis, ^{36}chloride efflux and micro-electrodes. *J. Physiol.* 326:139–154, 1980.
2. Ashton, F. T., A. V. Somlyo, and A. P. Somlyo. The contractile apparatus of vascular smooth muscle; intermediate high voltage stereo electron microscopy. *J. Mol. Biol.* 98:17–29, 1975.
3. Beams, H. W. and R. G. Kessel. The Golgi apparatus; Structure and function. *Int. Rev. Cytol.* 23:209–276, 1968.
4. Bergman, R. A. Uterine smooth muscle fibers in castrate and oestrogen treated rats. *J. Cell Biol.* 36:636–648, 1968.
5. Berner, P. F., A. V. Somlyo, and A. P. Somlyo. Hypertrophy-induced increase of intermediate filaments in vascular smooth muscle. *J. Cell Biol.* 88:96–101, 1981.
6. Blaustein, M. P. Sodium ions, calcium ions, blood pressure regulation and hypertension. A reassessment and a hypothesis. *Am. J. Physiol.* 232:C165–C173, 1977.
7. Blinks, J. R. Photoproteins as biological calcium indicators. *Pharmacol. Rev.* 28:1–93, 1976.

8. Bo, W. J., D. L. Odor, and M. Rothrock. The fine structure of uterine smooth muscle of the rat uterus at various time intervals following a single injection of estrogen. *Am. J. Anat.* 123:369–384, 1968.
9. Bond, M., T. Kitazawa, A. V. Somlyo, and A. P. Somlyo. Release and recycling of calcium by the sarcoplasmic reticulum in guinea pig portal vein smooth muscle. *J. Physiol.* (Lond.) 355:677–695, 1984.
10. Bond, M., H. Shuman, A. V. Somlyo, and A. P. Somlyo. Total cytoplasmic calcium in relaxed and maximally contracted rabbit portal vein smooth muscle. *J. Physiol.* (Lond.) 357:185–201, 1984.
11. Bond, M. and A. V. Somlyo. Dense bodies and actin polarity in vertebrate smooth muscle. *J. Cell Biol.* 95:403–413, 1982.
12. Brown, M. S. and J. L. Goldstein. Receptor-mediated control of cholesterol metabolism. *Science* 191:150–154, 1976.
13. Brandes, D. and E. Anton. Lysosomes in uterine involution: Intracytoplasmic degradation of myofilaments and collagen. *J. Gerontol.* 24:55–69, 1969.
14. Burnstock, G. Structure of smooth muscle and its innervation. In: *Smooth Muscle.* E. Bulbring, A. F. Brading, A. W. Jones and T. Tomita, Eds. Edward Arnold: London, 1–70, 1970.
15. Burnstock, G. Cholinergic and purinergic regulation of blood vessels. In: *Handbook of Physiology. The Cardiovascular System, Vol. II*: Vascular Smooth Muscle. D.F. Bohr, A.P. Somlyo, H.V. Sparks Jr., Eds. Bethesda, American Physiological Society, 567–612, 1980.
16. Burnstock, G. Development of smooth muscle and its innervation. In: *Smooth Muscle: An Assessment of Current Knowledge.* E. Bulbring, A.F. Brading, A.W. Jones and T. Tomita, Eds. London, UK: Edward Arnold, 431–457, 1981.
17. Casteels, R. The distribution of chloride ions in the smooth muscle cells of the guinea pig's taenia coli. *J. Physiol.* (London) 214:225–243, 1971.
18. Chamley-Campbell, J., G. R. Campbell, and R. Ross. The smooth muscle cell in culture. *Physiol. Rev.*59:1–61, 1979.
19. Chance, B. Energy-linked functions of mitochondria. Academic Press, New York, 1963.
20. Csapo, A. Actomyosin of uterus. *Am. J. Physiol.* 160:46-52, 1950.
21. Csapo, A. I. Smooth Muscle as a Contractile Unit. *Physiol. Rev. Suppl. 5.* 42:7–33, 1962.
22. Csapo, A. I. Force of labor. In: *Principles and Practice of Obstetrics and Perinatology.* L. Iffy, H.A. Kamientzky, Eds. New York, John Wiley and Sons Inc., 761–799, 1981.
23. Debbas, G., L. Hoffman, E. J. Landon, and L. Hurwitz. Electron microscopic localization of calcium in vascular smooth muscle. *Anat. Rec.* 182:447–471, 1975.
24. DeDuve, C. Lysosome concept. In: *Lysosomes.* A.V.S. DeReuck and M. P. Cameron, Eds. Little Brown Co., Boston, 1–35, 1963.
25. DeDuve, C. Lysosome in retrospect. In: *Lysosomes in Biology and Pathology, Vol. 1,* J.T. Dingle and H.B. Fell, Eds. North-Holland Publishing Co., Amsterdam, 3–40, 1969.

26. Devine, C. E., A. V. Somlyo, and A. P. Somlyo. Sarcoplasmic reticulum and excitation–contraction coupling in mammalian smooth muscle. *J.Cell Biol.* 52:690–718, 1972.
27. Devine, C. E., A. V. Somlyo, and A. P. Somlyo. Sarcoplasmic reticulum and mitochondria as cation accumulating sites in smooth muscle. *Philos. Trans. Roy. Soc.* 265:17–23, 1973.
28. Devine, C. E. and D. G. Rayns. Freeze-fracture studies of membrane systems in vertebrate muscle. II. Smooth Muscle. *J. Ultrastruc. Res.* 51:293–306, 1975.
29. Fay, F. S., H. H. Schlevin, W. C. Granger, Jr., and S. R. Taylor. Aequorin luminescence during activation of single isolated smooth muscle cells. *Nature* 280:506–508, 1979.
30. Finn, C. A. and D. G. Porter. *The Uterus*. Elek Science: London, 1975.
31. Gabella, G. Caveolae intracellulares and sarcoplasmic reticulum in smooth muscle. *J. Cell Sci.* 8:601–609, 1971.
32. Gabella, G. Hypertrophic smooth muscle. IV. Myofilaments, intermediate filaments and some mechanical properties. *Cell Tissue Res.* 201:27–28, 1979.
33. Gabella, G. Structure of smooth muscles. In: *Smooth Muscle*. E. Bulbring, A.F. Brading, A.W. Jones and T. Tomita, Eds. Edward Arnold: London, 1–46, 1981.
34. Garfield, R. E. and E. E. Daniel. Relation of membrane vesicles to volume control and Na^+ transport in smooth muscle. Studies on Na^+rich tissues. *J. Mechanochem. Cell Motility* 4:157–176, 1977.
35. Garfield, R. E. and E. E. Daniel. Relation of membrane vesicles to volume control and Na^+ transport in smooth muscle: Effect of metabolic and transport inhibition on fresh tissues. *J. Mechanochem. Cell Motility* 4:115–155, 1977.
36. Garfield, R. E. and A. P. Somlyo. Electron probe analysis and ultrastructure of cultured, freeze-dried vascular smooth muscle. In: *Proc. 33rd Ann. Meeting Electron Microscopy Society of America*. G.W. Bailey, Ed. Claitor's Publishing Division, Los Angeles. 558–559, 1975.
37. Garfield, R. E., and A. P. Somlyo. Golgi apparatus, lectin-binding sites and X537A action. *Proc. 1st Int. Congr. Cell Biol. J. Cell Biol.* 70:1012, 1976.
38. Garfield, R. E. and A. P. Somlyo. Ultrastructure, electron probe, and spectrophotometric microanalysis of cultured vascular smooth muscle. *Proc. 5th Int. Congr. Histochem. Cytochem.* 124, 1976.
39. Garfield, R. E., and A. P. Somlyo. Golgi apparatus and lectin-binding sites: Effects of *lasalocid* (X537A) *Exp. Cell Res.* 109:167–167, 1977.
40. Garfield, R. E., R. M. Henderson, and E. E. Daniel. Evaluation of the pyroantimonate technique for localization of tissue sodium. *Tissue and Cell* 4:575–589, 1972.
41. Garfield, R. E., S. Chacko, and S. Blose. Phagocytosis by muscle cells. *Lab. Invest.* 33:418–427, 1975.
42. Garfield, R. E. and A. P. Somlyo. Ultrastructural basis for vascular smooth muscle reactivity. In: *Factors Influencing Vascular Reactivity*. O. Carrier and S. Shibata, Eds. Igaku Shoin, Japan, 1–25, 1977.

43. Garfield, R. E., S. M. Sims, M. S. Kannan, and E. E. Daniel. The possible role of gap junctions in activation of the myometrium during parturition. *Am. J. Physiol.* 235:C168–C179, 1978.
44. Garfield, R. E. Smooth muscle regeneration: Ultrastructure and multipotential properties of smooth muscle. In: *Muscle Regeneration,* A Mauro, Ed. Raven Press, New York, 383–404, 1979.
45. Goldstein, J. L. and M. S. Brown. The low-density lipoprotein pathway and its relation to atherosclerosis. *Ann. Rev. Biochem.* 46:879–930, 1977.
46. Goodford, P. J. and M. W. Wolowyk. Localization of cation interactions in the smooth muscle of the guinea pig taenia coli. *J. Physiol.* 224:521–535, 1972.
47. Hall, T. A. The microprobe assay of chemical elements. In: *Physical Techniques of Biological Research,* G. Oster, Ed. New York, N.Y.: Academic Press, Inc. 151–267, 1971.
48. Ishikawa, H., R. Bischoff, and H. Holtzer. Formation of arrowhead complexes with HMM in a variety of cell types. *J. Cell Biol.* 7:281–308, 1969.
49. James-Kracke, M. R., B. F. Sloane, H. Shuman, and A. P. Somlyo. Lysosomal composition in cultured vascular smooth muscle cells: Electron probe analysis. *Proc. Natl. Acad. Sci.* 76:6461–6465, 1979.
50. James-Kracke, M. R., B. F. Sloane, H. Shuman, R. Karp. and A. P. Somlyo. Electron probe analysis of cultured vascular smooth muscle. *J. Cell Physiol.* 103:313–322, 1980.
51. Johansson, B. and A. P. Somlyo. Electrophysiology and excitation–contraction coupling. In: *The Handbook of Physiology: Vascular Smooth Muscle.* D.F. Bohr, A.P. Somlyo and H.V. Sparks, Eds. Baltimore, MD: Williams and Wilkins Co., 301–324, 1980.
51a. Junker, J. L., A. J. Wasserman, P. F. Berner, and A. P. Somlyo. Electron probe analysis of sodium and other elements in hypertrophied and Na-loaded smooth muscle. *Circ. Res.* 54:254–266, 1984.
52. Lehninger, A. L. Mitochondrion: Molecular basis for structure and function. Benjamin: New York, 1965.
53. Marshall, J. M. Effects of ovarian steroids and pregnancy on adrenergic nerves of uterus and oviduct. *Am. J. Physiol.* 240:C165-, 1981.
54. McCormack, S. A. and S. R. Glasser. Differential response of individual uterine cell types from immature rats treated with estradiol. *Endocrinol.* 106:1634–1649, 1980.
55. McGuffee, L. J. and R. M. Bagby. Ultrastructure, calcium accumulation, and contractile response in smooth muscle. *Am. J. Physiol.* 230:1217–1224, 1976.
56. Morgan, J. P. and K. G. Morgan. The first recorded Ca^{+} transients. *Pflugers' Arch.* 395:75–77, 1982.
57. Osa, T. and T. Katase. Physiological comparison of the longitudinal and circular muscles of the pregnant rat uterus. *Jap. J. Physiol.* 25:153–164, 1975.
58. Owens, G. K. and S. M. Schwartz. Vascular smooth muscle cell hypertrophy and hyperpoloidy in the Goldblatt hypertensive rat. *Circ. Res.* 53:491–501, 1983.

59. Owman, C., N. O. Sjoberg, and N. O. Sjostrand. Short adrenergic neurons, a peripheral neuroendocrine mechanism. In: *Amine Fluorescence Histochemistry.* M. Fujiwara and C. Tanaka, Eds. Igaku Shoin, Tokyo, 47–67, 1974.
60. Popescu, L. M. and I. Diculescu. Calcium in smooth muscle sarcoplasmic reticulum *in situ.* Conventional and X-ray analytical electron microscopy. *J. Cell Biol.* 67:911–918, 1975.
61. Popescu, L. M., I. Diculescu, U. Zeick, and L. Ionescu. Ultrastructural distribution of calcium in smooth muscle cells of guinea-pig taenia coli. *Cell Tissue Res.* 154:357–378, 1974.
62. Quinlan, R. A. and W. W. Franke. Heteropolymer filaments of vimentin and desmin in vascular smooth muscle tissue and cultured baby hamster kidney cells demonstrated by chemical crosslinking. *Proc. Natl. Acad. Sci.* U.S.A. Vol. 79:3452–3456, 1982.
63. Reynolds, S. R. M. *Physiology of the Uterus.* 2nd Edition. P.B. Hoeber Inc., New York, 1949.
64. Rhodin, J. A. G. Fine structure of vascular walls in mammals, with special reference to smooth muscle components. *Physiol. Rev.* 42(Suppl.5) 48–81, 1962.
65. Rhodin, J. A. G. Architecture of the vessel wall. In: *Handbook of Physiology. Section 2: The Cardiovascular System. Vol. II.* D.R. Bohr, A.P. Somlyo, H.V. Sparks Jr., Eds. American Physiol. Society, Bethesda, MD., 1–31, 1980.
66. Ross, R. The smooth muscle cell. II. Growth of smooth muscle in culture and formation of elastic fibers. *J. Cell Bio.* 50:172–186, 1971.
67. Ross, R. Connective tissue cells, cell proliferation and synthesis of extracellular matrix: A review. *Philos. Trans. Roy. Soc. London (Biol.)* 271:247–259, 1975.
68. Ross, R. and S. J. Klebanoff. The smooth muscle cell. I. *In vivo* synthesis of connective tissue proteins. *J. Cell Biol.* 50:159–171, 1971.
69. Shoenberg, C. F. An electron microscopic study of smooth muscle in pregnant uterus of the rabbit. *Biophys. Biochem. Cytol.* 4:609–614, 1958.
70. Shoenberg, C. F. and D. M. Needham. A study of the mechanism of contraction in vertebrate smooth muscle. *Biol. Rev.* 51:53–104, 1976.
71. Shuman, H., A. V. Somlyo, and A. P. Somylo. Quantitative electron probe microanalysis of biological thin sections: Methods and validity. *Ultramicroscopy* 1:317–339, 1976.
72. Shuman, H. and A. P. Somlyo. Energy filtered "conventional" transmission imaging with a magnetic sector spectrometer. In: *Analytical Electron Microscopy.* H.R. Geiger, Ed. San Francisco, CA., San Francisco Press, Inc. 202–204, 1981.
73. Shuman, H., P. Kruit, and A. P. Somlyo. Quantitative electron energy loss spectroscopy of low concentrations of calcium in carbon containing matrices. *Microbeam Analysis.* R. Gooley, Ed. San Francisco Press, San Francisco, CA., 247–251, 1983.
74. Singer, S. J. and G. L. Nicolson. The fluid mosaic model of the structure of cell membranes. *Science* 175:720, 1972.
75. Sloane, B. F., A. Scarpa, and A. P. Somlyo. Vascular smooth muscle mitochondrion: Magnesium content and transport. *Arch. Biochem. Biophys.* 189:409–416, 1978.

76. Sloane, B. F. Lysosomal apparatus in uterine muscle: Effects of estrogen and ovariectomy. *Biol. Reprod.* 23:867–876. 1980.
76a. Sloane, B. F. Isolated membranes and organelles from vascular smooth muscle. In: *The Handbook of Physiology: The Cardiovascular System.* Vol. II. Vascular Smooth Muscle. D. F. Bohr, A. P. Somlyo, and H. V. Sparks, Eds., Bethesda, MD, American Physiological Society, pp. 121–132, 1980.
77. Smith, R. C., W. Z. Cande, R. Craig, P. J. Tooth, J. M. Scholey, and J. Kendrik-Jones. Regulation of myosin filament assembly by light-chain phosphorylation. *Phil. Trans. R. Soc. Lond. B.* 302:73–82, 1983.
78. Somlyo, A. P. and A. V. Somlyo. Vascular smooth muscle. I. Normal structure, pathology, biochemistry and biophysics. *Pharmacol. Rev.* 22:197–272, 1968.
79. Somlyo, A. P. and A. V. Somlyo. Vascular smooth muscle. II. Pharmacology of normal and hypertensive vessels. *Pharmacol. Rev.* 22:249–253, 1970.
80. Somlyo, A. P., C. F. Devine, A. V. Somlyo, and S. R. North, Sarcoplasmic reticulum and the temperature-dependent contraction of smooth muscle in calcium-free solution. *J. Cell Biol.* 51:722–741, 1971.
81. Somlyo, A. P., C. E. Devine, A. V. Somlyo, and R. V. Rice. Filament organization in vertebrate smooth muscle. *Philos. Trans. Roy. Soc. Lond. (Biol.)* 265:223–229, 1973.
82. Somlyo, A. P., A. V. Somlyo, C. E. Devine, P. D. Peters, and T. A. Hall. Electron microscopy and electron probe analysis of mitochondrial cation accumulation in smooth muscle. *J. Cell Biol.* 61:723–742, 1974.
83. Somlyo, A. P., R. E. Garfield, S. Chacko, and A. V. Somlyo. Golgi organelle response to the antibiotic X537A. *J. Cell Biol.* 66:425–443, 1975.
84. Somlyo, A. P. Vascular smooth muscle. In: *Cellular Pharmacology of Excitable Tissues.* T. Harahashi, Ed. Charles C. Thomas, Springfield, Illinois, 360–407, 1975.
85. Somlyo, A. P., A. V. Somlyo, F. A. Ashton, and I. Vallieres. Vertebrate smooth muscle: Ultrastructure and function. *Cold Spring Harbor Conf. Cell Proliferation, Vol. 3, Cell Motility,* 165–183, 1976.
86. Somlyo, A. P., I. Vallieres, R. E. Garfield, H. Shuman, A. Scarpa, and A. V. Somlyo. Calcium compartmentalization in vascular smooth muscle: Electron probe analysis and studies on isolated mitochondria. In: *Biochemistry of Smooth Muscle.* N.L. Stephens, Ed. University Park Press, Baltimore, Maryland, 563–583, 1977.
87. Somlyo, A. P., A. V. Somlyo, H. Shuman, B. F. Sloane, and A. Scarpa. Electron probe analysis of calcium compartments in cryosections of smooth and striated muscles. In: *Ann. N.Y. Acad. Sci.* A. Scarpa and E. Carafoli, Eds. New York, N.Y. *New York Acad. Sci.* 307:523-544, 1978.
88. Somlyo, A. P., A. V. Somlyo, and H. Shuman. Electron probe analysis of vascular smooth muscle: Composition of mitochondria, nuclei and cytoplasm. *J. Cell Biol.* 81:316–335, 1979.
89. Somlyo, A. P. and A. V. Somlyo. Calcium storage sites and elemental distribution in smooth muscle. In: *Advances in Pharmacology and Thera-*

peutics. II. Vol. 3; Cardio-Renal and Cell Pharmacology. H. Yoshida, Y. Hagiwara and E. Ebashi, Eds. New York, N.Y. Pergamon Press, 55–60, 1982.

90. Somlyo, A. P., and H. Shuman. Electron probe and electron energy loss analysis in biology. *Ultramicroscopy* 8:219–234, 1982.
91. Somlyo, A. P., A. V. Somlyo, H. Shuman, and M. Endo. Calcium and monovalent ions in smooth muscle. *Fed. Proc.* 41:2883–2890, 1982.
92. Somlyo, A. P., A. V. Somlyo, T. Kitazawa, M. Bond, H. Shuman, and D. Kowarski. Ultrastructure, function, and composition of smooth muscle. *Ann. of Biomed. Engr.* (BMES). 11:579–588, 1983.
93. Somlyo, A. V., P. Vinall, and A. P. Somlyo. Excitation-contraction coupling and electrical events in two types of vascular smooth muscle. *Microvasc. Res.* 1:354–373, 1969.
94. Somlyo, A. V. and A. P. Somlyo. Strontium accumulation by sarcoplasmic reticulum and mitochondria in vascular smooth muscle. *Science* 174:955–958, 1971.
95. Somlyo, A. V. and A. P. Somlyo. Ultrastructure of smooth muscle. In: *Methods in Pharmacology.* Vol. 3. E. E. Daniel and D. M. Paton, Eds. Plenum Press, New York, 3–45, 1975.
96. Somlyo, A. V., H. Shuman, and A. P. Somlyo. Elemental distribution in striated muscle and effects of hypertonicity: Electron probe analysis of cryosections. *J. Cell Biol.* 74:828–857, 1977.
97. Somlyo, A. V. Bridging structures spanning the junctional gap at the triad of striated muscle. *J. Cell Biol.* 80:743–750, 1979.
98. Somlyo, A. V. Ultrastructure of vascular smooth muscle. In: *The Handbook of Physiology: Vascular Smooth Muscle.* D.F. Bohr, A. P. Somlyo, and H. V. Sparks, Eds. Baltimore, MD. Am. Physiol. Soc., Williams & Wilkins Co., 33–67, 1980.
99. Somlyo, A. V., T. M. Butler, M. Bond, and A. P. Somlyo. Myosin Filaments have nonphosphorylated light chains in relaxed smooth muscle. *Nature* 294:567–570, 1981.
100. Sporrong, B., P. Alm, C. Owman, N. O. Sjoberg, and G. Thorbert. Ultrastructural evidence for adrenergic nerve degeneration in the guinea pig uterus during pregnancy. *Cell Tiss. Res.* 195:189–193, 1978.
101. Suzuki, H., T. Kamata, H. Onishi, and S. Watanabe. ATP induced reversible change in the conformation of chicken gizzard myosin and HMM. *J. Biochem. Tokyo* 91:1699–1705, 1982.
102. Thorbert, B., P. Alm, C. Owman, and N. O. Sjoberg. Regional distribution of autonomic nerves in guinea pig uterus. *Am. J. Physiol.* 233:C25–C34, 1977.
103. Thorbert, G., P. Alm, C. Owman, and B. Sporrong. Regional changes in structural and functional integrity of myometrial adrenergic nerves in pregnant guinea pig, and their relationship to the localization of the conceptus. *Acta Physiol. Scand.* 103:120–131, 1977.
104. Thorbert, G. Regional changes in structure and function of adrenergic nerves in guinea pig uterus during pregnancy. *Acta Obstet. Gynecol. Scand. Suppl.* 79:5–39, 1979.
105. Todd, M.E., C.G. Laye, and D.N. Osborne. The dimensional characteristics of smooth muscle in rat blood vessels: A computer assisted analysis. *Circ. Res.* 53:319–331, 1983.

106. Trybus, K. M., T. W. Huiatt, and S. Lowey. A bent monomeric conformation of myosin from smooth muscle. *Proc. Natl. Acad. Sci. USA.* 158:6151–6155, 1982.
107. Vallieres, J., A. Scarpa, and A. P. Somlyo. Subcellular fractions of smooth muscle; Isolation, substrate utilization and Ca^{2+} transport by main pulmonary artery and mesenteric vein mitochondria. *Arch. Biochem. Biophys.* 170:659–669, 1975.
108. Wikstrom, M., P. Akonen, and T. Tuukkainen. The role of mitochondria in uterine contractions. *FEBS Letts.* 56:77–88, 1975.
109. Wootton, G. S. and P. J. Goodford. An association between mitochondria and vesicles in smooth muscle. *Cell Tissue Res.* 161:119–132, 1975.

Chapter 2

Calcium Antagonists and Ionophores

D. J. Triggle and R. A. Janis

1. Introduction

The delineation and quantitation of the sources and pathways of Ca^{2+} mobilized during excitation–contraction coupling in smooth muscle is a problem of both importance and complexity. It is now generally agreed that both intracellular and extracellular sources of Ca^{2+} are involved and that the relative extent of their involvement is dependent upon the smooth muscle, the stimulant and its concentration, and upon the component (phasic or tonic) of mechanical response measured.

The several pathways of Ca^{2+} mobilization in smooth muscle are discussed in detail elsewhere in this volume (Chapters 1, 3, 4) and are summarized in Fig. 1. Intracellular Ca^{2+} mobilization may occur from bound sources including sarcoplasmic reticulum and the internal sarcolemmal surface, whereas extracellular Ca^{2+} mobilization probably occurs through several distinct pathways including those that have been designated potential-dependent (PDC) channels and receptor-operated (ROC) channels as well as through a leak pathway (*10, 12, 76, 100*). It is likely that extracellular Ca^{2+} is derived both from free (nonbound) sources as well as from membrane-bound sources and that the latter may, in part, be associated with specific membrane receptor sites (*100*). It should be noted, as depicted in Fig. 1, that an influx

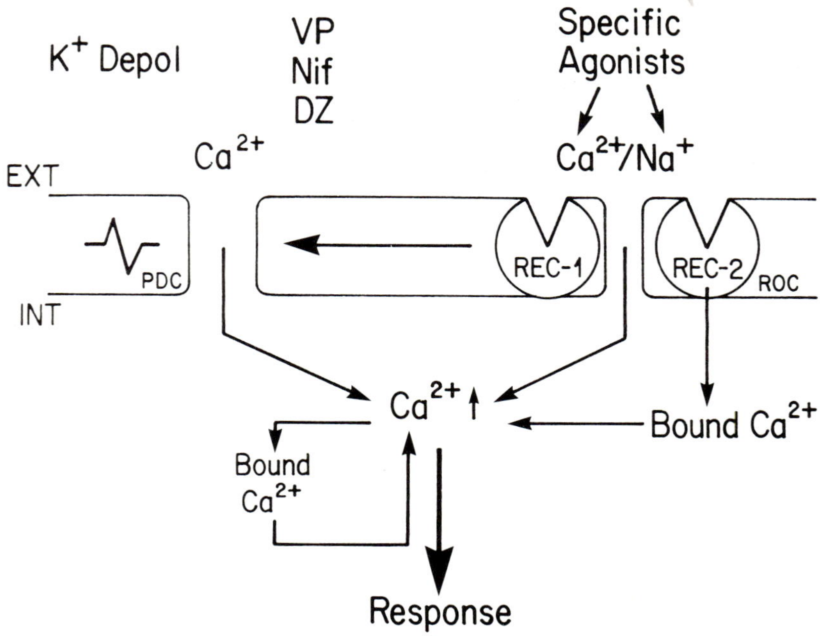

Fig. 1. Schematic representation of Ca^{2+} mobilization in smooth muscle. Ca^{2+} entry may occur through potential-dependent (PDC) and receptor-operated channels (ROC). Activation of receptors (REC-1, REC-2) may lead to mobilization of intracellular Ca^{2+} from plasma membrane and sarcoplasmic reticulum, and or in some types of smooth muscle may cause depolarization and subsequent activation of PDCs. VP, Nif, and DZ are verapamil, nifedipine, and diltiazem. Their location outside the membrane is not meant to represent the location of their binding site, which is probably within the channel (see text).

of Ca^{2+} may serve to release intracellular Ca^{2+} (induced release) thus indicating the importance of determining the structural and functional relationships existing between the several mobilizable Ca^{2+} pools in smooth muscle. The terms PDC and ROC are used here operationally; the possibility has not been excluded that there is only one basic type of PDC in mammalian smooth muscle, and that an association of neurotransmitter receptors or other regulating proteins with the PDC changes its voltage dependence and other properties (*14*).

Finally, as in other Ca^{2+}-dependent systems, the physiologic control of Ca^{2+} mobilization may be bypassed through the use of Ca^{2+}-selective ionophores, nobably A 23187 and ionomycin, that serve to elevate cytosolic Ca^{2+} by mobilization of

Ca^{2+} from both extracellular and intercellular sources (Fig. 1). These agents are of interest not only as pharmacologic tools with which to manipulate cytosolic Ca^{2+} levels, but also because they may provide clues to the action of endogenous ionophores that may mediate Ca^{2+} entry.

A number of approaches have been employed in attempts to analyze these several pathways of Ca^{2+} mobilization (*57, 76*). Within the past few years, increasing attention has been paid to the Ca^{2+} (channel) antagonists which appear to inhibit, with considerable affinity and selectivity, Ca^{2+} entry through the potential-dependent Ca^{2+} channels, and some receptor-operated Ca^{2+} channels.

2. Calcium Channel Antagonists

Although a diverse collection of molecules has been reported to antagonize Ca^{2+}-dependent events, in most instances their activity is secondary to other more potent and better defined activities (*69, 76, 78, 113*). However, the Ca^{2+} channel antagonists, which include verapamil, nifedipine, and diltiazem as the best known agents, although also a structurally heterogeneous group of molecules (Fig. 2), clearly exhibit substantial specificity and selectivity of action. Several recent reviews are available (*12, 14, 25, 38, 39*).

2.1. Pharmacology

The actions of the Ca^{2+} channel antagonists were first described for the inhibition of cardiac excitation–contraction coupling and the associated slow inward Ca^{2+} current (inter alia, *26, 33, 76, 109*). It was early recognized, however, that these agents were powerful relaxants of smooth muscle, both vascular and nonvascular (*10, 26, 76*). Although electrophysiologic characterization of Ca^{2+} currents in smooth muscle is sparse (*46*), several lines of evidence do indicate that selective inhibition of Ca^{2+} mobilization underlies the relaxant actions of these antagonists in smooth muscle.

Table 1 is a representative collection of data for inhibition of agonist- and K^+-depolarization-induced mechanical responses in smooth muscle. K^+-induced responses are usually very sensitive to the Ca^{2+} channel antagonists whereas the sensitivity of agonist-induced responses varies considerably (*12, 13*). Thus, in some studies an approximate equisensitivity is seen to the antago-

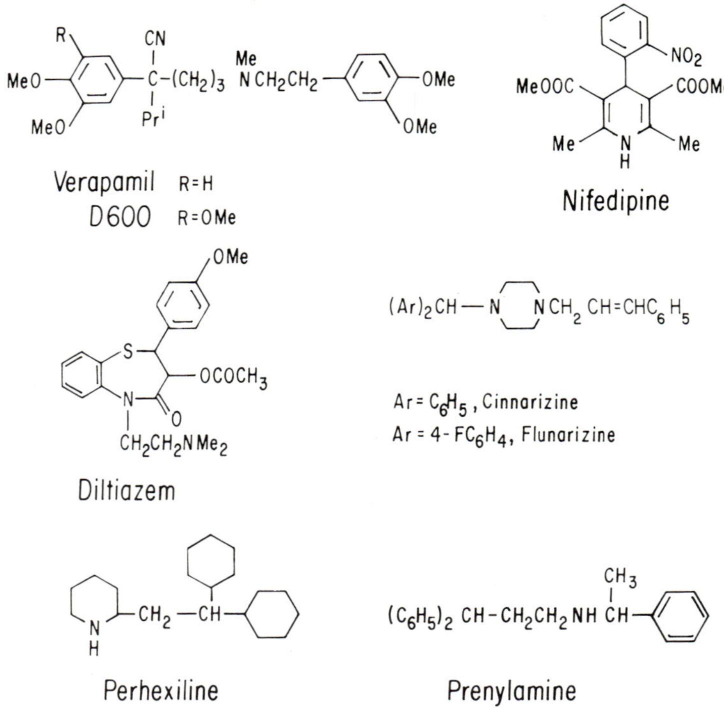

Fig. 2. Structural formulas of some Ca^{2+} channel antagonists.

nists with both agonist- and K^+-stimulation, and in other studies agonist-induced contractions were found to be more sensitive to these drugs than K^+-induced contractions (*7, 13, 50, 111*). In contrast, in many tissues, including rabbit aorta, regardless of the antagonist employed, agonist-induced responses are much less sensitive than are the corresponding K^+-induced responses.

The high sensitivity to the Ca^{2+} antagonists of K^+-induced responses in most smooth muscles is consistent with K^+-depolarization activating one or more populations of potential-dependent channels. In contrast, the widely varying sensitivity of agonist-induced responses is presumably a reflection of the operation of several modes of Ca^{2+} mobilization and pathways of excitation–contraction coupling. Furthermore, the sensitivity of agonist-induced responses to the Ca^{2+} channel antagonists may be very dependent both upon the concentration of agonist and upon the component of response measured. Several specific examples will serve to illustrate these complexities.

Table 1
Ca^{2+} Channel Antagonist Activities in Smooth Muscle

System		Antagonist	ID_{50}, *M*	Ref.
Rabbit aorta	NE	Verapamil	1.2×10^{-4}	*57,83*
	K^+	Verapamil	2.7×10^{-8}	
	NE	Diltiazem	10^{-4}	
	K^+	Diltiazem	5×10^{-7}	*106*
	NE	Nicardipine	2×10^{-4}	
	K^+	Nicardipine	2×10^{-9}	*90*
Guinea-pig ileum	ACh	D600	2.7×10^{-8}	*76*
	K^+	D600	3×10^{-8}	
	ACh	Nifedipine	5.1×10^{-9}	*76*
	K^+	Nifedipine	3.4×10^{-9}	
Rabbit ear artery	5-HT	Verapamil	3×10^{-7}	*50*
	K^+	Verapamil	4×10^{-9}	
	NE	Diltiazem	1×10^{-6}	*6*
	K^+	Diltiazem	2×10^{-6}	
Canine mesenteric	$PGF_{2\alpha}$	Nifedipine	3×10^{-8}	*111*
	K^+	Nifedipine	4×10^{-8}	*85*
Canine coronary	NE	D600	5×10^{-7}	*105*
	K^+	D600	2×10^{-7}	
Rabbit saphenous	5-HT	Nimodipine	$>2 \times 10^{-5}$	*94*
	K^+	Nimodipine	2.5×10^{-10}	
Rabbit basilar	5-HT	Nimodipine	3×10^{-10}	*94*
	K^+	Nimodipine	1.7×10^{-10}	

Responses of rabbit aorta to norepinephrine have been described as insensitive to Ca^{2+} channel antagonists (*12, 44, 57, 83, 90*; Table 1). However, responses to low concentrations of norepinephrine ($10^{-8}M$) are approximately equisensitive to inhibition by diltiazem as are responses to K^+ depolarization. Furthermore, under these conditions there is good agreement between the inhibition by diltiazem of mechanical responses and $^{45}Ca^{2+}$ uptake (*106*).

One explanation for this finding is that low concentrations of norepinephrine mobilizes Ca^{2+} through antagonist sensitive channels, but that higher concentrations mobilizes Ca^{2+} preferentially from intracellular sources. Similar observations have been made for norepinephrine responses in rabbit pulmonary artery. In canine trachea the mechanical responses to low concentrations of acetylcholine are more effectively antagonized by D600 than are the responses to high concentrations of acetylcholine (*23*). This may be associated with the observed correlation of tension and membrane potential at low acetylcholine concentrations and the continued development of tension in the absence of further depolarization with increasing acetylcholine concentration (*22*). Thus, the effects of agonist concentration upon Ca^{2+} antagonist sensitivity of mechanical responses may have their origin in the relative extents of electromechanical and pharmacomechanical coupling (*86*). However, the absence of adequate electrophysiologic data in most smooth muscle systems makes it difficult to evaluate the generality of this proposal (*14, 46, 97*). Furthermore, it is possible that the apparent transitions from electromechanical to pharmacomechanical coupling with increasing agonist concentration may reflect the activation of Ca^{2+}-activated outward currents as intracellular Ca^{2+} levels rise (*55*).

Transitions in Ca^{2+} mobilization mechanisms may also occur through the presence of distinct receptor subclasses within a given tissue. This is probable for adrenergic α-receptors where the presence of postsynaptic α_1 and α_2 receptors associated with contractile responses has been demonstrated *in vivo* and *in vitro* (*28, 65, 107–109*). In rat vascular tissues α_2-mediated responses are sensitive to the inhibitory effects of the Ca^{2+} channel antagonists whereas α_1-mediated responses are often insensitive (*28, 65, 108*). Thus, responses to an agonist such as norepinephrine which has both α_1 and α_2 activities may show differential sensitivity to the antagonists according to the concentration at which it is employed. The extent to which α_1- or α_2-receptors are responsible for Ca^{2+} influx appears to depend on species, tissue, and agonist concentration (*107*). Experimental design may also effect the apparent selectivity for a given receptor type, as may be expected from the effects of extracellular Ca^{2+} (*87*).

Vascular smooth muscle is the major therapeutic site of action of the Ca^{2+} channel antagonists as used as antianginal or antihypertensive agents (*2, 14, 25, 39, 44*). Recent evidence (*64*) indicates that γ-adrenergic receptors may be of great importance in mediating effects of norepinephrine that is released from the postganglionic sympathetic nerve ending and plays a major role in the normal regulation of blood pressure. According to this view, sympathetic neural regulation of vasoconstriction occurs via the very high concentrations of noradrenaline released from nerves that are close to the γ-adrenergic receptors. This norepinephrine may also activate extrajunctional α-receptors (*64*).

Activation of the γ-adrenergic receptors, which are insensitive to normal α-adrenergic receptor antagonists and to Ca^{2+} channel antagonists, may result in an action potential that is sensitive to the Ca^{2+} channel antagonists (*46*).

Consideration must also be given to the possibility that subclasses of Ca^{2+} channels exist in smooth muscle as has been demonstrated for certain other cell types (*31*). For receptor-operated channels, it is quite plausible that they are uniquely associated with specific smooth muscle receptors and that each may have distinguishing characteristics of ionic selectivity, antagonist sensitivity, and so forth. For potential-dependent channels the situation is more conjectural. Hurwitz et al. (*35*) have postulated the existence of two types of K^{+}-activated potential-dependent Ca^{2+} channel in guinea-pig ileal longitudinal smooth muscle that are associated with the development of the phasic (fast) and tonic (slow) components of mechanical response respectively and that can be distinguished by their La^{3+} sensitivity and apparent rates of inactivation. A similar subclassification has been proposed by Nachshen and Blaustein for Ca^{2+} channels in brain synaptosomes (*63*). It is of interest, therefore, that the phasic and tonic components of the K^{+}-induced mechanical responses in ileal smooth muscle, although both are dependent upon extracellular Ca^{2+}, do show differential sensitivity to inhibition by the Ca^{2+} antagonists, the tonic component being more sensitive than the phasic component (*75*). As mentioned above, it is possible that there is only one basic structure for the Ca^{2+} channel,the properties of which are modified when it is associated with ligand receptors, or is otherwise regulated, such as by phosphorylation, or membrane potential. It is well known that the characteristics of Ca^{2+} channels in other cell types are regulated and modified by many factors (*31, 104*). For example, the selective effect of Ca^{2+} channel antagonists on a given phase of contraction may also depend on factors such as the time that the Ca^{2+} channel spends in a given conformation state. Varying degrees of use-dependence has been reported for all Ca^{2+} channel antagonists studied (*47*). This is analogous to the variable sensitivity of Na^{+} channel to local anesthetics (*34*).

2.2. Structure–Activity Relationships

The evidence summarized in the previous section is clearly consistent with specificity and selectivity of action of the Ca^{2+} channel antagonists. This conclusion is substantiated by the existence of structure–activity relationships, including stereoselectivity, for the several structural classes of antagonists. Reviews of structure-activity relationships in these compounds are available (*39, 52, 58, 76, 101*). Structure–activity data in smooth muscle parallel data for cardiac preparations,

Stereoselectivity for verapamil, D600, and members of the 1,4-dihydropyridine series has been observed in several smooth muscle systems (*40, 84, 95, 96*). Of some interest is the absence of any stereoselectivity of the lower potency of D600 inhibition of muscarinic and α_1-adrenergic receptor binding, suggesting that stereoselectivity may be an important index of specificity of action.

Structure–activity relationships have been analyzed quite extensively in cardiac muscle for the verapamil, D600, and the 1,4-dihydropyridine series (*52*), but for smooth muscle data are available only for the dihydropyridines. The following generalizations may be made (*8, 39, 58, 76, 101*):

1. The presence of a 4-substituent, preferable substituted phenyl, in the 1,4-dihydropyridine ring increases activity.
2. Ester substituents in the 3- and 5-positions are optimal; —COMe and —CN substituents reduce activity. Ester substituents larger than —CO_2Me may increase activity.
3. The 1,4-dihydropyridine ring is essential; conversion to the pyridine derivative eliminates activity.
4. Substituents in the 4-phenyl ring modify activity: *ortho-* and *meta-*substitutions increase or decrease activity according to their electronic or steric effects, but *para-*substituents are invariably highly detrimental to activity.

2.3. Sites of Action

Although structure–activity data indicate specific sites of action they do not define these sites. However, a membrane locus of action is strongly indicated. Thus, the Ca^{2+} channel antagonists are effective as inhibitors of Ca^{2+} movements in isolated organelles or membrane vesicles only at concentrations that are higher than those necessary to inhibit tension responses and Ca^{2+} movements in cellular systems (*92*). Furthermore, the inactivity of most of these drugs in skinned preparations suggests a membrane locus of action (*12, 13, 18, 56, 79*). Electrophysiological

Table 2
Binding of [^{3}H]-1,4-Dihydropyridines to Smooth Muscle Membranes

Smooth muscle species	K_d, $\times 10^{-10} M$	B_{max}, fmol mg^{-1}	Reference
[^{3}H]Nitrendipine			
Ileum, guinea-pig	1.6	1100	*8*
Ileum, rat	2.6	25	*20*
Aorta, bovine	1.0	80	*81*
Aorta, bovine	21[a]	40–60	*115*
Aorta, canine	3.1	20	*98*
Coronary artery, pig	16[a]	35	*17*
Mesenteric artery, rat	1.0	18	*98*
Mesenteric artery, dog	2.5	25	*98*
Myometrium, rat	1.8	220	*103*
Stomach, rat	1.5	90	*103*
Vas deferens, rat	2	230	*103*
Bladder, rat	2	85	*103*
Bladder, guinea-pig	1.5	500	*103*
Lung, guinea-pig	1.8	40	*103*
[^{3}H]Nimodipine			
Ileum, guinea-pig	1.2	750	*37*

[a]Assay was carried out at high protein concentration and at 37°C. Other values were obtained at 25°C and at relatively low protein concentration.

studies on smooth muscle (*see* ref. *46* for review) as well as on cardiac muscle (*32, 47, 104*) also indicate that the sarcolemma is a major site of action of these agents.

This evidence for a sarcolemmal site of action of the Ca^{2+} channel antagonists has also been amply confirmed by several recent studies of [^{3}H]antagonist binding to crude and partially purified membrane preparations from a variety of smooth muscles (Table 2), although some controversy exists for cardiac membranes with regard to the localization of these binding sites (*114*).

2.3.1. Binding Affinities and Binding Densities of 1,4-Dihydropyridines for Smooth Muscle Membranes

For membranes from every smooth muscle where the equilibrium dissociation constant (K_d) was measured at 25°C and at low protein concentration, a similar high affinity was obtained, 1–3 $\times 10^{-10} M$ (Table 2). A lower binding affinity has been found at 37°C, as well as at high protein concentrations (*17, 115*). The amount of low affinity nonspecific binding is probably underestimated by the filtration technique when high protein concentrations are used and this may increase the apparent K_d.

Membrane preparations from several other cell types, including cardiac and neural, exhibit an identical high affinity binding (for example, *see* refs. *5, 9, 20, 53, 82*). A comparison of the ability of a series of nifedipine analogs to compete with [^{3}H]nitrendipine for sites in membranes from cardiac and smooth muscle shows that a marked similarity exists in the binding sites of these two tissues (*103*). In contrast, the potency of these compounds for inhibition of cardiac muscle contraction (*74*) and Ca^{2+} current (*47*) is over a hundredfold less than their potency for inhibition of K^+-induced contraction in guinea-pig ileal smooth muscle (*102*). However, the rank order of nifedipine analogs for inhibition of ileal smooth muscle and cardiac contraction is the same (*8, 9, 103*). The correct potency series for nifedipine analogs for inhibition of [^{3}H]nitrendipine binding has also been found for membranes from vascular smooth muscle (*81*). Many neurotransmitter receptor-active agents and other compounds have no effect on binding to membranes from smooth (*8*) and cardiac muscle (*103*), and brain (*27, 29*). Membranes from both guinea-pig ileal smooth muscle (*8*) and bovine aorta (*81*) exhibit stereoselective binding, consistent with known pharmacology (*95, 96*).

The dissociation constant for [^{3}H]nitrendipine determined by kinetic measurements is in good agreement with that estimated by Scatchard analysis (*8, 82, 103*). The results are consistent with the mass action association of ligand with a single population of high affinity sites. However, [^{3}H]nitrendipine binding to intact cultured cardiac cells exhibits both a high and low affinity site (*49*). It remains to be determined if a low affinity binding site can also be detected in intact smooth muscle cells. To date, the existence of low affinity sites corresponding to the low sensitivity of agonist-induced contraction of many smooth muscles (*12, 24*) has not yet been detected by studying [^{3}H]nitrendipine binding to isolated membranes. Nerve (*1*) and cardiac cells are usually less sensitive to the Ca^{2+} channel antagonists than are K^+-induced contractions of many smooth muscles, although other exceptions have been reported (*93*). Therefore, the relationship between the high affinity binding and the generally low sensitivity of these tissues requires further studies.

Only limited information can be obtained from a comparison between tissues of the apparent maximal number of binding sites (B_{max}) shown in Table 2, because the degree of purification of sarcolemmal membranes may differ by more than 10-fold for the preparations shown. Nevertheless, it should be noted that membranes from guinea-pig ileum were isolated by differential centrifugation (*8*), and therefore, they are probably not much richer in sarcolemmal than are those from aorta (*81, 98, 115*) and mesenteric artery (*98*) that were purified by a combination of differential and sucrose density gradient centrifugation. The low density of binding sites in blood vessels relative to synaptic rich areas in brain has been previously demonstrated by autoradiography (*61*). Therefore, unless there is a large loss of binding sites during density gradient centrifugation of vascular smooth muscle membranes, it can be concluded that guinea-pig ileum has a particularly high density of nitrendipine binding sites, relative to blood vessels. Table 3 illustrates that the ratio of [^{3}H]-nitrendipine binding sites to neurotransmitter sites varies widely among preparations.

2.3.2. Effect of Organic and Inorganic Ca^{2+} Channel Antagonists on [^{3}H]Nitrendipine Binding

Membranes from different tissues appear to differ in the effect of diltiazem on nitrendipine binding. Those from guinea-pig ileal smooth muscle (*8*) as well as skeletal muscle (*103*) show an apparent increase in the number of [^{3}H]nitrendipine binding sites. In contrast, membranes from pig coronary artery (*17*), rabbit ventricle, at 37°C (*103*), and brain membranes (*27, 60*) exhibit an increased affinity of nitrendipine binding with little or no change in the B_{max}. The increased binding affinity for nitrendipine in

Table 3
Ratio of Densities of Receptors to [^{3}H]Nitrendipine Binding Sites[a]

	Receptor type	Ratio
Guinea-pig ileum	Muscarinic cholinergic receptors	14:1
	Histamine	4:1
	Substance P	1:1
Guinea-pig bladder	Muscarinic cholinergic	3:1
Pig coronary arteries	β-Receptor	1:1
	Muscarinic cholinergic	2:1
Dog ventricle	β-Receptor	12:1
	Muscarinic cholinergic	91:1
	Ouabain	1600:1

[a]Data for ileum and bladder are from G. T. Bolger and D. J. Triggle, unpublished; those for arteries and ventricle are from ref. (*17*).

smooth muscle membranes produced by diltiazem is stereoselective (*17*).

In contrast to diltiazem, a decrease in the apparent binding affinity of nitrendipine to smooth muscle membranes is seen with verapamil, D600, and several other nondihydropyridines. The maximal decrease in specific binding, however, is less than complete and is dependent on the nitrendipine concentration (*8, 20, 60*). From these results it is concluded that diltiazem and verapamil are positive and negative heterotropic regulators, respectively, of the nitrendipine binding site. More recently, studies by Murphy et al. (*60*) demonstrated that D600 and diltiazem can reverse the inhibition of [^{3}H]nitrendipine binding caused by lidoflazine, tiapamil, and many related compounds. They concluded that all of the nondihydropyridine Ca^{2+} antagonists regulate the nitrendipine binding site in brain membranes by acting at a common allosteric site. Similar results are found with cardiac membranes (*103*).

Information on the relationship between cation binding sites and those for organic channel blockers has been obtained from ligand-binding studies. Treatment of smooth muscle membranes with EDTA decreases binding, as do La^{3+}, Cd^{2+}, and many other inorganic Ca^{2+} channel blockers (*8*). Similar results are found with membranes from brain and heart (*20, 27, 29, 53, 103*). Ca^{2+} restores binding to EDTA membranes by increasing the B_{max} (*8, 29, 103*). Other agents that carry current through this channel, such as Sr^{2+} and Ba^{2+}, also restore binding to EDTA-treated brain membranes.

2.3.3. Effects of Ca^{2+} Channel Antagonists at Other Sites

Most of the published results are consistent with the view that Ca^{2+} channel blockade alone can explain the main pharmacological effects of low concentration of this class of drugs. However, some compounds are extremely hydrophobic, so that high concentrations are achieved in the membrane phase (*41, 45*).

It is well established that the actions of Ca^{2+} channel antagonists, particularly the nondihydropyridines are not limited to the Ca^{2+} channel (*19, 39, 101*), and that the full pharmacological spectrum of these agents exhibits great diversity. Electrophysiological studies in cardiac cells demonstrated that D600 inhibits Ca^{2+}-independent K^+ currents (*19, 42*), an effect not shared by the dihydropyridine nisoldipine (*42*). High concentrations of verapamil and D600 also inhibit Na^+ channels (*4, 101*) and a variety of receptor types (*39, 41*); this inhibition is not stereoselective

(*4, 40*) in contrast to the inhibition of Ca^{2+} current and muscle contraction (*4, 8, 39*).

It has been proposed that diltiazem (*24*), nimodipine and nitrendipine (*see* ref. *39*) may produce Na^+ pump stimulation in smooth muscle. Although reduced intracellular Na^+ has been reported in vascular smooth muscle treated with diltiazem (*24*), this may be secondary to the blocking of Na^+ influx through the Ca^{2+} channel. Electrophysiological studies have resulted in the further suggestion that nitrendipine may stimulate an electrogenic Ca^{2+} pump (*32*). Studies measuring Ca^{2+} current and ion fluxes in smooth muscle will be required to determine if ion pump stimulation makes any contribution to smooth muscle relaxation.

The possibility that calmodulin inhibitors produce smooth muscle relaxation by inhibiting calmodulin-stimulated phosphodiesterase has been noted (*10, 21*); Ca^{2+} channel inhibitors in high concentrations also inhibit this enzyme (*21*), and therefore some of these drugs could also act, in part, by increasing the levels of cyclic nucleotides. It should be noted the phosphodiesterase inhibition in cardiac cells would oppose the effect of Ca^{2+} channel blockers (*104*) in contrast to smooth muscle where potentiation of their effects would be seen. Further studies are required to determine if the different pharmacologic spectrum of the different dihydropyridines (*44*) can be attributed in part to sites of action other than the Ca^{2+} channel.

3. Calcium Ionophores

Ionophores are agents that form lipid-soluble reversible complexes with ions and can thus serve as vehicles for their membrane translocation. Of particular interest to Ca^{2+}-demanding processes are divalent cation selective ionophores, particularly A 23187 and ionomycin (*68, 73, 76*) which form neutral membrane-permeant complexes.

A 23187, a monocarboxylic acid ionophore, forms 2:1 complexes (*67, 73*) and ionomycin 1:1 complexes with divalent cations (*43, 48*). Although both A 23187 and ionomycin serve as divalent cation ionophores, they differ in certain important respects. Ionomycin has a greater selectivity than does A 23187 for divalent cations and, moreover, exhibits a greater Ca^{2+}:Mg^{2+} selectivity. Thus, the actions of ionomycin are less likely to be complicated by Na^+, K^+, or Mg^{2+} transport.

The actions of A 23187 and, to a lesser extent, ionomycin, have been examined in a number of smooth muscle systems (Ta-

Table 4
Actions of A 23187 and Ionomycin in Smooth Muscle[a]

Ionophore	Tissue	Contraction	Ca^{2+} ext-dependence	Na^{+} ext-dependence	Ca^{2+} antagonist sensitivity	Ref.
A 23187	Toad stomach (cells)		Partial	—	—	*62*
	Toad stomach (strips)		—	—	—	
A 23187	Guinea-pig ileum					*77, 91*
						99
A 23187	Rat vas deferens	No	Partial	—	Partial	*36, 88*
	Guinea-pig stomach			—		*51*
	Guinea-pig taenia coli			—		
A 23187	Rat aorta, carotid	Small	—	—	—	*15*
	Rat jugular	Small	—	—	—	
A 23187	Dog coronary			—	Partial	*112*
	Rabbit aorta			—	No	*68, 112*
A 23187	Guinea-pig trachea			—	No	*11, 16*
A 23187	Cat nictitating membrane		—	—	—	*89*
Ionomycin	Guinea-pig ileum					*70*

[a](—) not reported.

ble 4). The data available suggest for smooth muscle, and other systems (*76*), that the actions of these ionophores may not be simply ascribable to translocation of extracellular Ca^{2+} across the plasmalemmal membrane. Thus, the actions may not be totally dependent on extracellular Ca^{2+} from intracellular stores. They may also act indirectly, as in the cat nictitating membrane, by release of norepinephrine (*89*). Additional complexities of action are indicated by observations in which ionophore responses are sensitive to extracellular Na^{+} and to organic calcium antagonists, including verapamil and D600 (*70, 77, 108*). The latter observations suggest that the ionophores may activate voltage-dependent

Ca^{2+} channels, perhaps by a Ca^{2+}-dependent Na^{+} entry leading to membrane depolarization (*70, 76, 77*).

Alternatively, ionophore-mediated Ca^{2+} entry may lead to activation of Ca^{2+}-dependent phospholipase action, to the production of arachidonic acid metabolites that subsequently initiate mechanical activity. The effects of inhibitors of arachidonic acid metabolism support this suggestion since A 23187 responses are blocked by indomethacin and aspirin in guinea-pig trachea (*11, 16*). Thus products of both lipoxygenase and cyclooxygenase pathways may be involved in the mechanical responses of smooth muscles to both A 23187 and ionomycin.

The actions of A 23187 and ionomycin as Ca^{2+}-selective ionophores in a wide variety of cellular systems raise the important question of whether there exist endogenous ionophores functioning to translocate Ca^{2+} across plasma and intracellular membranes. Several candidate molecules including phospholipids (*30*) and such phospholipid metabolites as arachidonic acid metabolites (*3, 54, 66*) and phosphatidic acid (*71, 72, 80*) have been proposed.

The role of phosphatidic acid is of particular interest since it is one of the products formed in receptor-stimulated phosphatidylinositol turnover in a number of secretory and smooth muscle systems and for which an association with processes of Ca^{2+} mobilization has long been advanced (*59, 71*). Phosphatidic acid has been demonstrated to serve as a Ca^{2+} ionophore in a number of systems including smooth muscle (*80, 110*). Thus, Vogt and Dakhil (*110*) showed that phosphatidic acid caused contractions of guinea-pig taenia coli in Ca^{2+}-ext-dependent fashion. Subsequently, Salmon and Honeyman (*80*) showed that carbachol stimulation of disaggregated toad smooth muscle cells provoked a rapid Ca^{2+}-ext-independent accumulation of phosphatidic acid. Furthermore, low concentrations of phosphatidic acid (10^{-8}–$10^{-6}M$) initiated rapid mechanical activity in these cells.

Although these findings are of great interest the role of phosphatidic acid (or other molecules) as endogenous Ca^{2+}-demanding systems has not been established. Phosphatidic acid or other carriers are not candidates for plasma membrane Ca^{2+} channels because their rates of Ca^{2+} translocation will be too slow (*104*). However, phosphatidic acid could be a signal originating at the plasma membrane as a consequence of an agonist–receptor stimulus, and serving to mobilize intracellular Ca^{2+}.

4. Summary

Both Ca^{2+} antagonists and ionophores exhibit a complex pharmacological spectrum. Nevertheless they have been useful tools to provide basic physiological and biochemical information about the regulation of smooth muscle contraction.

Several mechanisms are used by different smooth muscles under the influence of various stimuli to regulate intracellular Ca^{2+} levels and contraction, and this is undoubtedly reflected in a diversity of sensitivities of the contractile response to the Ca^{2+} channel antagonists. Those contractions that are believed to be dependent only on the release of intracellular Ca^{2+} are generally insensitive to these agents, whereas K^+-depolarization-induced contraction that is highly dependent on extracellular Ca^{2+} is usually very sensitive.

Structurally different Ca^{2+} channel antagonists have markedly different spectrums of action. Some of these differences can be attributed to the effects of these compounds at sites other than the Ca^{2+} channel, whereas other differences may be caused by preferential binding of different classes of these drugs to different states of the Ca^{2+} channel, or to pharmacokinetic differences. The two agents most commonly used in the past, verapamil and D600, are among the least specific of the Ca^{2+} channel antagonists and therefore are poor choices for specific Ca^{2+} channel inhibition.

The recent availability of radio-labeled Ca^{2+} channel antagonists promises to provide the tools necessary for the further characterization and isolation of Ca^{2+} channels from smooth muscle. Studies to date indicate that all smooth muscles contain a similar single population of high affinity binding sites for [^{3}H]nitrendipine. The characteristics of this binding site are consistent with the view that it represents the site mediating inhibition of smooth muscle contraction produced by nanomolar concentrations of dihydropyridines. Further studies are required to determine whether or not lower affinity binding sites can also be detected, particularly in intact smooth muscle cells. Nondihydropyridine Ca^{2+} channel antagonists appear to allosterically regulate the dihydropyridine binding site in a complex manner. More studies on the basic biochemistry and electrophysiology of smooth muscle are required before the pharmacology of Ca^{2+} antagonists and ionophores can be completely described.

Acknowledgments

Partial support for the preparation of this work was provided by a grant from the National Institutes of Health (HL 16003).

References

1. Akaike, N., A.M. Brown, K. Nishi, and Y. Tsuda. Actions of verapamil, diltiazem and other divalent cations on the calcium-current of *Helix* neurons. *Br. J. Pharmacol.* 74:87-95, 1981.
2. Antman, E.M., P.H. Stone, J.E. Muller, and E. Braunwald. Calcium channel blocking agents in the treatment of cardiovascular disorders. Pt. I. Basic and clinical electrophysiologic effects. *Ann. Int. Med.* 93:875–885, 1980.
3. Barritt, G.J. A proposal for the mechanism by which α-adrenergic agonists, vasopressin, angiotensin and cyclic AMP induce calcium release from intracellular stores in the liver cell: a possible role for metabolites of arachidonic acid. *Cell Calcium* 2:53–63, 1981.
4. Bayer, R., R. Klausche, R. Kaufmann, and R. Mannhold. Inotropic and electrophysiological actions of verapamil and D600 in mammalian myocardium. III. Effects of the optical isomers on transmembrane action potentials. *Naunyn-Schmiedeberg's Arch. Pharmacol.* 290:81–97, 1975.
5. Bellemann, P., D. Ferry, F. Lübbecke, and H. Glossmann. [^{3}H]-Nitrendipine, a potent calcium antagonist, binds with high affinity to cardiac membranes. *Arzneim. Forsch.* 31:2064–2067, 1981.
6. Bevan, J.A. Selective action of diltiazem on cerebral vascular smooth muscle in rabbit: antagonism of extrinsic but not intrinsic maintained tone. *Am. J. Cardiol.* 49:519–524, 1982.
7. Bevan, J.A. Diltiazem selectively inhibits cerebrovascular extrinsic but not intrinsic myogenic tone. *Circ. Res.* 52, (Supp.II):104–109, 1983.
8. Bolger, G.T., P. Gengo, R. Klockowski, E. Luchowski, H. Siegel, R.A. Janis, A.M. Triggle, and D.J. Triggle. Characterization of binding of the Ca^{2+}-channel antagonist, [^{3}H]-nitrendipine, to guinea pig ileal smooth muscle. *J. Pharmacol. Exp. Ther.* 225:291–309, 1983.
9. Bolger, G.T., P.T. Gengo, E.M. Luchowski, H. Siegel, D.J. Triggle, and R.A. Janis. High affinity binding of a calcium channel antagonist to smooth and cardiac muscle. *Biochem. Biophys. Res. Comm.* 104:1604–1609, 1982.
10. Bolton, T.B. Mechanisms of action of transmitters and other substances on smooth muscle. *Physiol. Revs.* 59:606–718, 1979.
11. Burke, J.F. and N.A.M. Paterson. A comparison of antigen-induced and calcium ionophore A 23187-induced contraction of isolated guinea pig trachea. *Can. J. Physiol. Pharmacol.* 59:1031–1038, 1981.
12. Cauvin, C., R. Loutzenhiser, and C. van Breemen. Mechanism of calcium antagonists-induced vasodilation. *Annu. Rev. Pharmacol. Toxicol.* 23:373–396, 1983.

13. Cauvin, C., K. Saida, and C. van Breemen. Effects of Ca antagonists on Ca fluxes in resistance vessels. *J. Cardiovasc. Pharmacol.*, 4:Supp 3, S287–S290, 1983.
14. Cohen, C.J., R.A. Janis, D.G. Taylor, and A. Scriabine. Where do Ca^{2+} antagonists act? In: *Calcium Antagonists in Cardiovascular Disease*, edited by L.H. Opie and R. Krebs. New York: Raven Press, pp. 151–163, 1984.
15. Cohen, M.L., K.S. Wiley, and R.H. Tust. Effect of the ionophore A 23187 on contraction and relaxation of rat arteries and veins. *Proc. Soc. Exp. Biol. Med.* 159:353–358, 1978.
16. Creese, B.R., and M.A. Denborough. The involvement of prostaglandins in the contractile and cyclic AMP responses of airways smooth muscle to two calcium ionophores. *Proc. Aust. Biochem. Soc.* 13:54, 1980.
17. DePover, A., M.A. Matlib, S.W. Lee, G.P. Dubé, I.L. Grupp, G. Grupp, and A. Schwartz. Specific binding of [^{3}H]nitrendipine to membranes from coronary arteries and heart in relation to pharmacological effects. Paradoxical stimulation by diltiazem, *Biochem. Biophys. Res. Commun.* 108:110–117, 1982.
18. Diecke, F.P.J. Studies with nitrendipine on chemically skinned muscle strips. In: *Nitrendipine*, edited by A. Scriabine, S. Vanov, and K. Deck. Baltimore: Urban & Schwarzenberg, in press, 1983.
19. Ehara, T., and R. Kaufmann. The voltage and time-dependence effects of (-)-verapamil on the slow inward current in isolated cat ventricular myocardium. *J. Pharmacol. Exp. Ther.* 207:49–55, 1978.
20. Ehlert, F.J., W.R. Roeske, E. Itoga, and H.I. Yamamura. The binding of [^{3}H]-nitrendipine to receptors for calcium channel antagonists in the heart, cerebral cortex and ileum of rats. *Life Sciences* 30:2191–2202, 1982.
21. Epstein, P.M. K. Fiss, R. Hachisu, and D,M. Adrenyak. Interaction of calcium antagonists with cyclic AMP phosphodiesterases and calmodulin. *Biochem. Biophys. Res. Commun..* 105:1142–1149, 1982.
22. Farley, J.M. and P.R. Miles. Role of depolarization in acetylcholine induced contractions of dog trachealis muscle. *J. Pharmacol. Exp. Ther.* 201:199–205, 1977.
23. Farley, J.M., and P.R. Miles. The sources of calcium for acetylcholine induced contractions of dog tracheal smooth muscle. *J. Pharmacol. Exp. Ther.* 207:340–346, 1978.
24. Flaim, S.F. Comparative pharmacology of calcium blockers based on studies with vascular smooth muscle. In: *Calcium Blockers. Mechanisms of Action and Clinical Applications*, edited by S.F. Flaim and R. Zelis. Baltimore: Urban & Schwarzenberg, p. 155-178, 1982.
25. Flaim, S.F. and R. Zelis (editors). *Calcium Blockers. Mechanisms of Action and Clinical Applications.* Baltimore: Urban & Schwarzenberg, 1982.
26. Fleckenstein, A. Specific pharmacology of calcium in myocardium, cardiac pacemakers and vascular smooth muscle. *Annu. Rev. Pharmacol. Toxicol.* 17:149–166, 1977.
27. Glossmann, H., D.R. Ferry, F. Lübbecke, R. Mewes, and F. Hofmann. Calcium channels: direct identification with radioligand binding studies. *Trends Pharmac. Sci.* 3:431–437, 1982.

28. Godfraind, T. and R.C. Miller. α_1 and α_2-adrenoceptor stimulation and Ca fluxes in isolated rat aorta. *Arch. Int. Pharmacodyn.* 256:171–172, 1982.
29. Gould, R.J., K.M.M. Murphy, and S.H. Snyder. [^{3}H]nitrendipine-labelled calcium channels discriminate inorganic calcium agonists and antagonists. *Proc. Natl. Acad. Sci.* 79:3656–3660, 1982.
30. Green, D.E., M. Fry, and G.A. Blondin. Phospholipids as the molecular instruments of ion and solute transport in biological membranes. *Proc. Natl. Acad. Sci.* 77:257–261, 1980.
31. Hagiwara, S., and L. Byerly. Calcium channel. *Annu. Rev. Neurosci.* 4:69–125, 1981.
32. Hermsmeyer, K. Nitrendipine and hyperpolarization in arterial smooth muscle. In: *Nitrendipine,* edited by A. Scriabine, S. Vanov and K. Deck. Baltimore: Urban & Schwarzenberg, pp. 161–168, 1983.
33. Hescheler, J., D. Pelzer, G. Trube, and W. Trautwein. Does the organic calcium channel blocker D600 act from inside or outside on the cardiac cell membrane. *Pfluegers Arch.* 393:287–291, 1982.
34. Hille, B. Local anesthetics: Hydrophilic and hydrophobic pathways for the drug–receptor reaction. *J. Gen. Physiol.* 69:497–515, 1977.
35. Hurwitz, L., L.J. McGuffee, S.A. Little, and H. Blumberg. Evidence for two distinct types of potassium-activated calcium channels in an intestinal smooth muscle. *J. Pharmacol. Exp. Ther.* 214:574–580, 1980.
36. Ishida, Y. and S. Shibata. Characteristic contractile responses to the calcium ionophore A 23187 in guinea pig vas deferens. *Br. J. Pharmacol.* 71:581–583, 1980.
37. Janis, R.A., S.C. Maurer, J.G. Sarmiento, G.T. Bolger, D.J. Triggle. Binding of [^{3}H]nimodipine to cardiac and smooth muscle membranes. *Eur. J. Pharmacol.* 82:191–194, 1982.
38. Janis, R.A., and A. Scriabine. Sites of action of Ca^{2+} channel inhibitors. *Biochem. Pharmacol.* 32:3499–3507, 1983.
39. Janis, R.A., and D.J. Triggle. New developments in Ca^{2+} channel antagonists. *J. Med. Chem.* 26:775–785, 1983.
40. Jim, K., A. Harris, L.B. Rosenberger, and D.J. Triggle. Stereoselective and nonstereoselective effects of D600 (methoxyverapamil) in smooth muscle preparations. *Eur. J. Pharmacol.* 76:67–72, 1981.
41. Karliner, J.S., H.J. Motulsky, J. Dunlap, J. Heller Brown, and A. Insel. Verapamil competitively inhibits α_1-adrenergic and muscarinic but not β-adrenergic receptors in rat myocardium. *J. Cardiovasc. Pharmacol.* 4:515–520, 1982.
42. Kass, R.S. Nisoldipine: A new, more selective calcium current blocker in cardiac purkinje fibers. *J. Pharmacol. Exp. Ther.* 223:446–456, 1982.
43. Kaufmann, R.F., R.W. Taylor, and D.R. Pfieffer. Cation transport and specificity of ionomycin: Comparison with ionophore A 23187 in rat liver mitochondria. *J. Biol. Chem.* 255:2735–2739, 1980.
44. Kazda, S., B. Garthoff, and A. Knorr. Nitrendipine and other calcium entry blockers (calcium antagonists) in hypertension. *Federation Proc.* 42, 1983.
45. Keefe, D.L. and R.E. Kates. Myocardial disposition and cardiac pharmacodynamics of verapamil in the dog. *J. Pharmacol. Exp. Ther.* 220:91–96, 1982.

46. Kuriyama, H., Y. Ito, H. Suzuki, K. Kitamura, and T. Ito. Factors modifying contraction–relaxation cycle in vascular smooth muscle. *Am. J. Physiol.* 243:H641–H662, 1982.
47. Lee, K.S. and R.W. Tsien. Mechanism of calcium channel block by verapamil, D600 diltiazem and nitrendipine in single dialyzed heart cells. *Nature, London* 302:790–794, 1983.
48. Liu, C.M. and T.E. Hermann. Characterization of ionomycin as a calcium ionophore. *J. Biol. Chem.* 253:5892–5894, 1978.
49. Loh, E., J.D. Marsh, D. LeChance, W.H. Barry, and T.W. Smith. Relationship of nitrendipine binding to contractility in intact cultured heart cells. *Circ. Res.* 53:539–543, 1983.
50. McCalden, T.A. and J.A. Bevan. Sources of activator Ca^{2+} in rabbit basilar artery. *Am. J. Physiol.* 241:H129–133, 1981.
51. Mandrek, K. and Golenhofen. Activation of gastro-intestinal smooth muscle induced by the calcium ionophore A 23187. *Pfluegers. Arch.* 371:119–124, 1977.
52. Mannhold, R., R. Rodenkirchen, and R. Bayer. Qualitative and quantitative structure–activity relationships of specific Ca antagonists. *Prog. Pharmacol.* 5:25–52, 1982.
53. Marangos, P.J., J. Patel, C. Miller, and A.M, Martino. Specific calcium antagonist binding in brain. *Life Sciences* 31:1575–1585, 1982.
54. Marshall, P.J., J.F. Dixon, and L.E. Hokin. Evidence for a role in stimulus-secretion coupling of prostaglandins derived from release of arachidonoyl residues as a result of phosphatidylinositol breakdown. *Proc. Nat. Acad. Sci.* 77:3292–3296, 1980.
55. Meech, R.W. Calcium-dependent potassium activation in nervous tissues. *Annu. Rev. Biophys. Bioeng.* 7:1–18, 1978.
56. Metzger, H., H.O. Stern, G. Pfitzer, and J.C. Ruegg. Calcium antagonists affect calmodulin-dependent contractility of a skinned smooth muscle. *Arzneim.-Forsch.* 32:1425–1427, 1982.
57. Meisheri, K.D., O. Hwang, and C. van Breemen. Evidence for two separate Ca^{2+} pathways in smooth muscle plasmalemma. *J. Mem. Biol.* 59:19–25, 1981.
58. Meyer, H. Antiangial and anti-ischemic agents. *Annu. Reports Med. Chem.* 17:71–77, 1982.
59. Michell, R.H., C.J. Kirk, L.M. Jones, C.P. Downes, and J.A. Creba. The stimulation of inositol lipid metabolism that accompanies calcium mobilization in stimulated cells: defined characteristics and unanswered questions. *Phil. Trans. Roy. Soc. Lond.* B. 296:123–137, 1981.
60. Murphy, K.M.M., R.J. Gould, B.L. Largent, and S.H. Snyder. A unitary mechanism of calcium antagonist drug action. *Proc. Nat. Acad. Sci.* 80:860–864, 1983.
61. Murphy, K.M.M., R.J. Gould, and S.H. Snyder. Autoradiographic visualization of [^{3}H]nitrendipine binding sites in rat brain: localization to synaptic zones. *Eur. J. Pharmacol.* 81:517–519, 1982.
62. Murray, J.J., P.W. Reed, and F.S. Fay. Contraction of isolated smooth muscle cells by ionophore A 23187. *Proc. Nat. Acad. Sci.* 72:4459–4463, 1975.

63. Nachshen, D.A. and M.P. Blaustein. Some properties of potassium-stimulated calcium influx in presynaptic nerve endings. *J. Gen. Physiol.* 76:709–728, 1980.
64. Neild, T.O. and E. Zelcer. Noradrenergic neuromuscular transmission with special reference to arterial smooth muscle. *Prog. Neurobiology* 19:141–158, 1982.
65. Nghiem, C., V.C. Swamy, and D.J. Triggle. Inhibition of D600 of norepinephrine- and clonidine-induced responses of the aortal from normotensive (WKY) and spontaneously hypertensive rats (SHR). *Life Sciences* 30:45-49, 1982.
66. Ohniski, S.T. and T.M. Devlin. Calcium ionophore activity of a prostaglandin B_1 derivative (PGB_x). *Biochem. Biophys. Res. Comm.* 89:240–245, 1979.
67. Pfeiffer, D.R. and H.A. Lardy. Ionophore A 23187: The effect of H^+ concentration on complex formation with divalent and monovalent cations and the demonstration of K^+ transport in mitochondria mediated by A 23187. *Biochemistry* 15:935–943, 1976.
68. Pressman, B.C. Properties of ionophores with broad range of cation selectivity. *Federation Proc.* 32:1698–1710, 1973.
69. Prozialeck, W.C. and B. Weiss. Inhibition of calmodulin by phenothiazines and related drugs: structure–activity relationships. *J. Pharmacol. Exp. Ther.* 222:509–516, 1982.
70. Putney, Jr., J.W. Stimulus-permeability coupling: role of calcium in the receptor regulation of membrane permeability. *Pharmacol. Rev.* 30:209–245, 1978.
71. Putney, J.W. Recent hypotheses regarding the phosphatidylinositol effect. *Life Sciences* 29:1183–1194, 1981.
72. Putney, J.W., S.J. Weiss, C.M. Van de Walle, and R.A. Haddas. Is phosphatidic acid a calcium ionophore under neurohumoral control? *Nature, London* 284–346, 1980.
73. Reed, P.W. Biochemical and biological effects of carboxylic acid ionohores. In: *Polyether Antibiotics: Carboxylic Ionophores.* edited by J.W. Westley, Vol. 1. New York: Dekker, 1980.
74. Rodenkirchen, R., R. Bayer, R. Steiner, F. Bossert, H. Meyer, and E. Möller. Structure–activity studies on nifedipine in isolated cardiac muscle. *Naunyn-Schmiedeberg's Arch. Pharmacol.* 310:69–78, 1978.
75. Rosenberger, L.B., M.K. Ticku, and D.J. Triggle. The effects Ca^{2+} antagonists on Ca^{2+} movements and mechanical responses in guinea-pig ileal longitudinal smooth muscle. *Can. J. Physiol, Pharmacol.* 57:333-347, 1979.
76. Rosenberger, L.B. and D.J. Triggle. Calcium, calcium translocation and specific calcium antagonists. In: *Calcium and Drug Action,* edited by G.B. Weiss. New York: Plenum Press, pp. 3–35, 1978.
77. Rosenberger, L.B. and D.J. Triggle. The mechanism of action of ionophore A 23187 on guinea pig intestinal smooth muscle. *Can. J. Physiol. Pharmacol.* 57:348–358, 1979.
78. Roufogalis, B.D. Specificity of trifluoperazine and related phenothiazines for Ca^{2+} binding proteins. In: *Calcium and Cell Function,* edited by W.Y. Cheung. London: Academic Press, Vol 3, 1983, p 129.

79. Saida, K. and C. van Breemen. Mechanism of Ca^{2+} antagonist induced vasodilation; intracellular actions. *Circ. Res.* 52:137–140, 1983.
80. Salmon, D.M. and T.W. Honeyman. Proposed mechanism of cholingeric action in smooth muscle. *Nature, London* 284:344–345, 1980.
81. Sarmiento, J.G., R.A. Janis, D. Jenkins, A.M. Katz, and D.J. Triggle. Characteristics of [^{3}H]nitrendipine binding to vascular smooth muscle membranes. In: *Nitrendipine,* edited by A. Scriabine, S. Vanov, and K. Deck. Baltimore MD: Urban & Schwarzenberg, pp 153–160, 1984.
82. Sarmiento, J.G., R.A. Janis, R.A. Colvin, F. Messineo, and A. Katz. Nitrendipine: Biochemical effects of nitrendipine. In: *Nitrendipine,* edited by A. Scriabine, S. Vanov, and K. Deck. Baltimore MD: Urban & Schwarzenberg, pp 53–68, 1984.
83. Schumann, H.J., B.D. Gorlitz, and J. Wagner. Influence of papaverine, D600 and nifedipine on the effects of noradrenaline and calcium on the isolated aorta and mesenteric artery of the rabbit. *Naunyn-Schmiedeberg's Arch. Pharmacol.* 289:409–417, 1975.
84. Shibanuma, T., M. Iwanani, K. Okida, K.T. Takenaka, and M. Murakami. Synthesis of optically active 2-(N-benzyl-N-methylamino) ethyl-methyl-2,6-dimethyl-4- (*m*-nitrophenyl) -1,4-dihydropyridine-3,5-dicarboxylate (Nicardipine). *Chem. Pharm. Bull.* 28:2809–2812, 1980.
85. Shimizu, K., T. Ohta, and N. Toda. Evidence for greater susceptibility of isolated dog cerebral arteries to Ca^{2+} antagonists than peripheral arteries *Stroke* 11:261–266, 1980.
86. Somylo, A.V. and A.P. Somlyo. Electromechanical and pharmacomechanical coupling in vascular smooth muscle. *J. Pharmacol. Exp. Ther.* 159:129–145, 1968.
87. Spedding, M. Assessment of "Ca^{2+} antagonist" effects of drugs in K^+-depolarized smooth muscle. Differentiation of antagonists subgroups. *Naunyn-Schmiedeberg's Arch. Pharmacol.* 318:234–240, 1982.
88. Swamy, V.C., M. Ticku, C.R. Triggle, and D.J. Triggle. The action of the ionophores X-537A and A 23187 on smooth muscle. *Can. J. Physiol. Pharmacol.* 53:1108–1114, 1975.
89. Tarizzo, V. and M.C. Rubio. Pharmacological effects of ionophore A 23187 on isolated guinea pig atria and isolated cat nictitating membrane. *Gen. Pharmacol.* 13:111–116, 1982.
90. Terai, M., T. Takenaka and H. Maeno. Inhibition of calcium influx in rabbit aorta by nicardipine hydrochloride (YC-93). *Biochem. Pharmacol.* 32:141–147, 1981.
91. Thomas, G. Inhibition of calcium ionophore A 23187-induced contractions of guinea-pig ileum by anti-inflammatory agents. *J. Pharm. Pharmacol.* 32:141–147, 1980.
92. Thorens, S. and G. Haeusler. Effect of some vasodilators on calcium translocation in intact and fractionated vascular smooth muscle. *Eur. J. Pharmacol.* 54:79–91, 1979.
93. Toll, L. Calcium Antagonists: High-affinity binding and inhibition of calcium transport in a clonal cell line. *J. Biol. Chem.* 257:13189–13192, 1982.

94. Towart, R. The selective inhibition of serotonin-induced contractions of rabbit cerebral vascular smooth muscle by calcium-antagonistic dihydropyridines. *Circ. Res.* 48:650–657, 1981.
95. Towart, R., E. Wehinger, and A. Meyer. Effects of unsymmetrical ester substituted 1,4-dihydropyridine derivatives and their optical isomers on smooth muscle. *Naunyn-Schmiedeberg's Arch. Pharmacol.* 317:183–185, 1981.
96. Towart, R., E. Wehinger, H. Meyer, and S. Kazda. The effects of nimodipine, its optical isomers and metabolites on isolated vascular smooth muscle. *Arzneim. Forsch.* 32:338–346, 1982.
97. Trapani, A., N. Matsuki, P. W. Abel, and K. Hermsmeyer. Norepinephrine produces tension through electromechanical coupling in rabbit ear artery. *Eur. J. Pharmacol.* 72:87–91, 1981.
98. Triggle, C.R., D.K. Agrawal, G.T. Bolger, E.E. Daniel, C.Y. Kwan, E.M. Luchowski, and D.J. Triggle. Calcium channel antagonist binding to isolated vascular smooth muscle membranes. *Can. J. Physiol. Pharmacol.* 60:1738–1740, 1982.
99. Triggle, C.R., W.F. Grant, and D.J. Triggle. Intestinal smooth muscle contraction and the effects of cadmium and A 23187. *J. Pharmacol. Exp. Ther.* 194:182–190, 1975.
100. Triggle, D.J. Effects of calcium on excitable membranes and neurotransmitter function. *Prog. Mem. Surface Sci.* 5:267–325, 1972.
101. Triggle, D.J. Biochemical pharmacology of calcium blockers. In: *Calcium Blockers. Mechanisms of Action and Clinical Applications,* edited by S. Flaim and R. Zelis. Baltimore, MD: Urban & Schwarzenberg, pp 125–134, 1982.
102. Triggle, D.J. Counting Ca^{2+} channels? *Trends Pharmacol. Sci.* 3:465–466, 1982.
103. Triggle, D.J. and R.A. Janis. Nitrendipine: Binding sites and mechanism of action. In: *Nitrendipine,* edited by A. Scriabine, S. Vanov and K. Deck. Baltimore, MD: Urban & Schwarzenberg, pp 33–52, 1984.
104. Tsien, R.W. Calcium channels in excitable cell membranes. *Annu. Rev. Physiol.* 45:341–358, 1983.
105. van Breemen, C. and B. Siegel. The mechanism of α-adrenergic activation of the dog coronary artery. *Circ. Res.* 46:426–429, 1980.
106. van Breemen, C., O. Hwang, and K.D. Meisheri. The mechanism of inhibitory action of diltiazem on vascular smooth muscle contractility. *J. Pharmacol. Exp. Ther.* 218:459–463, 1981.
107. Vanhoutte, P.M. and T.J. Rimele. Calcium and α-adrenoceptors in activation of vascular smooth muscle. *J. Cardiovas. Pharmacol.* 4:5280–5286, 1982.
108. van Meel, J.C.A., A. De Jonge, H.O. Kalkman, B. Wilffert, P.B.M.W.M. Timmermans, and P. A. Van Zwieten. Vascular smooth muscle contraction initiated by postsynaptic α_2-adrenoceptor activation is induced by an influx of extracellular calcium. *Eur. J. Pharmacol.* 69:205–208, 1981.
109. Van Zwieten, P.A. Calcium antagonists—recent developments. *Prog. Pharmacol.* 5:1–94, 1982.

110. Vogt, W. and Dakhil, T. Abhangigkeit der kontraktionswirkung von phosphatidsaure und acetylcholine auf die taenia coli vom extracellularenionenengehalt. *Naunyn-Schmiedeberg's Arch.* 251:344–354, 1965.
111. Walus, K.M., J.D. Fondacaro, and E.D. Jacobson. Effects of calcium and its antagonists on the canine mesenteric circulation. *Circ. Res.* 48:692–700, 1981.
112. Watson, E.J. Effects of ionophore A 23187 and X 537A on vascular smooth muscle activity. *Eur. J. Pharmacol.* 52:171–178, 1978.
113. Weiss, B., W.C. Prozialeck, and T.L. Wallace. Interaction of drugs with calmodulin. Biochemical, pharmacology and clinical implications. *Biochem. Pharmacol.* 31:2217–2226, 1982.
114. Williams, L.T. and L.R. Jones. Subcellular distribution of myocardial binding sites for a calcium antagonist, [^{3}H]nitrendipine. (abstract) *Circulation* 60:11–95, 1982.
115. Williams. L.T. and P. Tremble. Binding of a calcium antagonist, [^{3}H]nitrendipine, to high affinity sites in bovine aortic smooth muscle and canine cardiac membranes. *J. Clin. Invest.* 70:209–212, 1982.

Chapter 3

Calcium Compartments and Mobilization During Contraction of Smooth Muscle

Rodger Loutzenhiser, Paul Leyten, Kooichi Saida, and Cornelis van Breemen

1. Introduction

Although the details of the molecular events mediating smooth muscle contraction have not been fully resolved (*see* Chapters 7 and 8), there is little doubt that Ca^{2+} ions act as the second messenger regulating smooth muscle tension development. Accordingly, an elevation in the concentration of free Ca^{2+} in the myoplasm represents a final step linking receptor occupation or membrane depolarization to the activation of the contractile process. During smooth muscle stimulation, Ca^{2+} may be mobilized from both extracellular and intracellular sources. The rate at which extracellular Ca^{2+} enters the smooth muscle cell increases in response to receptor occupation or membrane depolarization. Alterations in extracellular Ca^{2+} bound to the cell surface may play a role in Ca^{2+} entry, either as a modulator of plasmalemmal Ca^{2+} permeability or as a potential source of activating Ca^{2+}. Furthermore, as mentioned in Chapters 1 and 11, various intracellular structures are capable of accumulating Ca^{2+}. The mobilization of such intracellularly bound Ca^{2+} pools can cause an increase in myoplasmic free Ca^{2+}, independent of Ca^{2+} entry. Of course, the steady-state level of myoplasmic free Ca^{2+} is determined not only by the rate of Ca^{2+} entry, but also by the rate of Ca^{2+} sequestration and Ca^{2+} extrusion. Thus, a complex cell Ca^{2+} cycle may be envisioned in which the concentration of Ca^{2+} ion in the myoplasm is regulated by modulation of the rates of Ca^{2+} entry, Ca^{2+} release, Ca^{2+} sequestration, and Ca^{2+} extrusion. The activation of smooth muscle in some cases may involve multiple actions at various sites of this cycle.

In this chapter, we discuss the transplasmalemmal and intracellular Ca^{2+} movements that occur during smooth muscle activation and relate these Ca^{2+} movements to tension development. Emphasis has been placed upon vascular and, to a lesser extent, intestinal smooth muscle. Though some generalizations are warranted, caution should be exercised when extrapolating concepts developed for one class of smooth muscle to another. The observed variations in the nature of the responses elicited in differing classes of smooth muscle by a particular agonist (e.g., norepinephrine or PGE_2) is an indication of the divergent activation characteristics of different types of smooth muscle.

2. Calcium Entry

The smaller diameter of smooth muscle cells (ca. 2 μm), and their slow rate of tension development allow for Ca^{2+} entry from the extracellular space to play a major role in the activation of this muscle type. In addition, a Ca^{2+} gradient of approximately 9300 cal/mol ($RT \ln \{[Ca]_o/[Ca]_i\} + 2E_mF$) is poised across the plasmalemma to provide a ready source of activator Ca^{2+}. There appear to be a number of Ca^{2+} entry pathways in smooth muscle plasmalemma. In addition to the intrinsic Ca^{2+} leak, separate agonist-activated and depolarization-activated Ca^{2+} entry processes have been found in smooth muscle cells.

2.1. Intrinsic Ca^{2+} Leak

All cell membranes are somewhat permeable to Ca^{2+} ions (*2*). The magnitude of the intrinsic Ca^{2+} leak into smooth muscle cells has been determined by means of ^{45}Ca flux measurements. Using the La^{3+} quenching technique, Casteels and van Breemen (*12*) found that when ATP production is inhibited, net Ca^{2+} content of guinea-pig taenia coli increases linearly by 300 μmol/kg tissue over the ensuing 60 min. The passive Ca^{2+} flux was calculated to be 0.02 pmol cm^{-2} min^{-1} in this tissue. Using the cold

EGTA quenching technique (see below), the unidirectional ^{45}Ca influx of resting rabbit aorta has been reported to be approximately 15–20 μmol/kg tissue/min or 0.03–0.04 pmol cm^{-2} min^{-1} (*48, 54*). These values were calculated assuming an extracellular volume of approximately 45%, and a surface area to volume ratio of 1.5 μm^{-1} (*40*). Without correcting for membrane potential, these fluxes correspond to a resting Ca^{2+} permeability (P_{Ca}) of 1.3–2.9 × 10^{-8} cm s^{-1}. Considering that the quenching techniques consistently underestimate cellular Ca^{2+} this P_{Ca} is very similar to that found in squid axon, 3.6 x 10^{-8} cm s^{-1} (*2*).

By definition, the intrinsic Ca^{2+} leak is not increased by smooth muscle stimulants, since any increase in the rate of Ca^{2+} entry would be considered a component of the stimulated Ca^{2+} influx pathway. However, the biophysical natures of the Ca^{2+} leak pathway and the Ca^{2+} influx processes activated by stimulants are, at present, undefined. The rate of ^{45}Ca entry into resting smooth muscle is not altered by concentrations of organic Ca^{2+} antagonists that completely inhibit the gain in ^{45}Ca occurring in response to high K^+ or low doses of norepinephrine (*13, 53, 68*). In contrast to organic Ca^{2+} entry blockers, inorganic cations such as Co^{2+}, Mn^{2+}, and La^{3+} seem to block Ca^{2+} movements in a less specific manner and inhibit both leak and stimulated ^{45}Ca entry. These observations indicate that the leak pathway must differ quantitatively from the Ca^{2+} influx processes which are activated by membrane depolarization or receptor occupation. In rabbit aorta, however, increasing concentrations of norepinephrine activate a Ca^{2+} influx process that becomes progressively insensitive to diltiazem and nisoldipine (*77*). The relationship between the Ca^{2+} entry pathway stimulated by high concentrations of agonists and the Ca^{2+} leak pathway remains to be determined.

The role of the intrinsic Ca^{2+} leak in the activation of smooth muscle has not been fully investigated. Since smooth muscle activation occurs when cytosolic Ca^{2+} rises above $10^{-7}M$, the amount of Ca^{2+} entering through the leak pathway (ca. 30 μmol/min/kg cell H_2O) is of sufficient magnitude to support a contraction. In order for tissues to remain in a relaxed state, the Ca^{2+} sequestering and Ca^{2+} extruding processes of the cell must constantly compensate for this large resting inward Ca^{2+} leak. However, the Ca^{2+} homeostatic processes of the cell are impaired, the leak pathway may play a direct role in activation (*51*). Furthermore, as elaborated upon below, during KCl- or agonist-induced stimulation ^{45}Ca influx increases only two- to fourfold. This observation suggests that a sizable fraction of the Ca^{2+} that enters the cell during stimulation enters through the Ca^{2+} leak pathway. The Ca^{2+} leak alone does not appear to cause an increase in smooth muscle tone of rabbit aorta under resting conditions. However, Bevan et al. (*3*) and Winquist and Baskin (*83*) have reported that the intrinsic myogenic tone of rabbit facial vein and basilar artery is insensitive to Ca^{2+} antagonists. In these vessels, the Ca^{2+} leak may be the Ca^{2+} entry pathway on which myogenic tone is dependent. It has also been suggested recently that the intrinsic Ca^{2+} leak may play a prominent role in the activation of arterial smooth muscle by agonists which are capable of releasing intracellular Ca^{2+} stores (*51*). This hypothesis is discussed in detail in the section of this chapter dealing with agonist-induced alterations in intracellular Ca^{2+} sequestration.

2.2. Ca^{2+} Entry Stimulated by Membrane Depolarization

In electrically active smooth muscle, depolarization of the plasma membrane below a threshold value will initiate action potentials. The inward current of smooth muscle action potentials appears to be carried primarily by Ca^{2+} and to a lesser extent by Na^+ ions [for review, *see* Tomita (*66*)]. These action potentials generate phase-linked, submaximal contractions. Both the action potentials and their accompanying contractions are blocked by lanthanides, divalent cations (*5*), and organic Ca^{2+} entry blockers (*13*). The concept thus evolved that the smooth muscle plasmalemma contains a Ca^{2+} entry pathway that is activated upon a decrease in the resting membrane potential. Furthermore, electrically quiescent smooth muscles (e.g., that present in large arteries) may generate Ca^{2+} dependent action potentials in the presence of tetraethylammonium (*9, 10*), and exhibit graded mechanical responses to progressive membrane depolarization (*38, 39*). Potential-operated Ca^{2+} channels thus appear to exist not only in electrically active smooth muscle, but also in smooth muscles which do not normally generate action potentials.

Potential-operated Ca^{2+} channels are also activated when membrane depolarization is induced by elevating extracellular K^+. Typically, when the concentration of K^+ in the media bathing smooth muscle cells is increased, tonic contractures are elicited. Early in the 1960s, Hinke et al. (*33, 34*), and Briggs (*7*) demonstrated that the contractions elicited in rabbit aorta by KCl are highly dependent upon the presence of extracellular Ca^{2+}. The interpretation of this observation was that KCl-induced contractions are supported by the entry of extracellular Ca^{2+} ions. Confirmation of this concept followed the advent of techniques that allowed the measurement of cellular ^{45}Ca fluxes. A major problem encountered with the determination of cellular ^{45}Ca in smooth muscle is that approximately 90% of the tissue tracer is associated with extracellular binding sites (*76*). A large fraction of this extracellularly bound ^{45}Ca exchanges at a rate indistinguishable from that of cellular ^{45}Ca; thereby obscuring the measurement of the much smaller cellular pool. The displacement of extracellular ^{45}Ca by La^{3+} was the basis of the first successful approach to the determination of alterations in cellular ^{45}Ca (*71, 76*). Reducing the temperature of the washing media (*79*) and evalating the external La^{3+} concentration (*31, 42*) have been reported to imporve the retention of cellular ^{45}Ca by retarding ^{45}Ca extrusion. It is possible, however, that La^{3+} ions might also displace a fraction of intracellular ^{45}Ca when plasmalemmal permeability is elevated (*79*). Accordingly, an additional technique was developed that employed an ice-cold wash in Ca^{2+}-free media containing the Ca^{2+} chelator EGTA (*54*).

Using both the La^{3+} method (*76, 31, 42*) and the EGTA method (*54*), it has been demonstrated that cellular Ca^{2+} content increases during activation of smooth muscle by KCl. Furthermore, preincubation of tissues in La^{3+} prevents both the contractile response to KCl and the stimulation of ^{45}Ca entry. When La^{3+} is applied at short intervals following the exposure to KCl, graded contractions and graded stimulations of ^{45}Ca uptake are observed (*73*). Using this approach, it has been determined that a minimal Ca^{2+} uptake of approximately 100 μmol/kg smooth muscle cells or 130 μmol/L cell H_2O is required for maximal activation of rabbit aorta by KCl (*73*). This is a lower limit since the quenching methods are known to underestimate cellular Ca^{2+}. Chemically, skinned smooth muscle fibers contract maximally at $10^{-5}M$ Ca^{2+} (*27, 37*). If one assumes that calmodulin is the activating site, and that cellular [calmodulin] is at most $10^{-5}M$, then during full actiation 40 μmol Ca/kg cell would be at the upper limit required to bind to the cytosolic calmodulin. Considering that at least 130 μmol Ca^{2+} are required for full activation, it seems that a large portion of the Ca^{2+} that enters the cell during KCl stimulation is sequestered by intracellular organelles or bound to cytosolic constituents. Furthermore, the activation curve relating KCl-stimulated ^{45}Ca uptake and tension development is "S-shaped", approximately 10–30 μmol Ca^{2+}/kg cell being required to elicit a threshold contractile response (*73*). This observation suggests that little intracellular Ca^{2+} release occurs in rabbit aorta during the steady-state phase of KCl activation, since Ca^{2+} release would cause a contraction to occur without siginificant ^{45}Ca uptake. As described below, however, it is possible that Ca^{2+}-induced Ca^{2+} release may play a role during the initial phase of depolarization induced activation.

The concept that KCl-induced smooth muscle contractions are mediated by the activation of a potential operated Ca^{2+} entry pathway has gained further support from studies examining the effect of KCl upon unidirectional ^{45}Ca influx and the actions of Ca^{2+} antagonists. Accordingly, when aortic tissues are exposed to ^{45}Ca for very short intervals, the rate of ^{45}Ca entry into tissues treated with KCl is 2–3 times that of control tissues (*54*). Furthermore, this KCl-stimulated ^{45}Ca influx is sensitive to inhibition by organic Ca^{2+} entry blockers. As illustrated in Fig. 1, the degree to which increasing concentrations of D600, diltiazem, and nisoldipine inhibit the contractile response to KCl closely approximates the degree to which these agents block the KCl-stimulated ^{45}Ca entry (*50*).

It thus follows that the depolarization-induced contractile response is proportional to Ca^{2+} influx. However, this does not rule out the possible presence of an amplifying mechanism. It has been calculated that during an action potential having a deflection of 60 mV, a maximum of 4 μmol Ca^{2+}/kg cell H_2O would enter the smooth muscle cell (*5*). This small amount of Ca^{2+} would be well below the threshold (ca. 10–30 μmol Ca/kg cell) required to elicit a contractile response. It has, therefore, been suggested that a depolarization- or Ca^{2+}-induced release of intracellular Ca^{2+} may contribute to the contractile response accompanying action potentials (*5*). However, since KCl cannot induce contractions in the absence of extracellular Ca^{2+} (*71*), a potential-induced release of intracellular Ca^{2+} seems unlikely. As stated above, the relationship between ^{45}Ca uptake and KCl-induced tension development suggests that Ca^{2+}-induced Ca^{2+} release does not contribute to the steady-state contractile response to KCl. A resolution to this apparent paradox may follow from an hypothesis advanced by Fabiato and Fabiato (*26*). According to this theory, Ca^{2+}-induced Ca^{2+} release occurs as a function of the rate of change in cytosolic Ca^{2+} concentration, and would occur only during the initial phase of KCl-induced activation. As suggested below (Role of Surface Bound Ca^{2+}), the initial rate of KCl-stimulated ^{45}Ca influx may greatly exceed the two- to fourfold increase measured under steady-state conditions (*47, 52*). Also, if one assumes a rate of membrane potential deflection of 5 V/s and a membrane capaci-

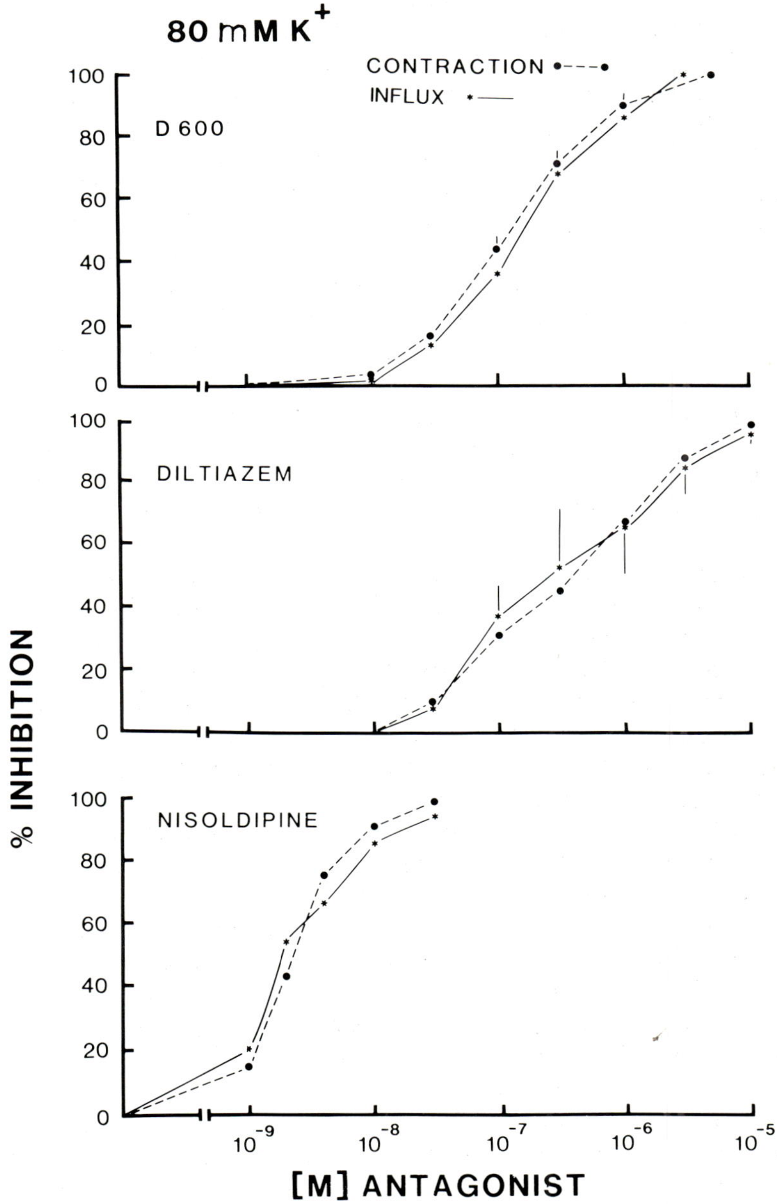

Fig. 1. Effects of D600, diltiazem and nisoldipine on contraction and ^{45}Ca influx stimulated by 80 m*M* K^+ in rabbit aorta. (From R. Loutzenhiser and C. van Breemen. In G. F. Merrill and H. R. Weiss, eds., *Calcium Entry Blockers, Adenosine, and Neurohumors,* Baltimore: Urban and Schwarzenberg, 1983, pp. 73–89, by permission.)

tance of 10^{-6} F/cm (*44*), 2 × 10^{-11} mol of Ca^{2+} would cross each cm^2 of cell surface during the upstroke of the action potential, assuming that the current is carried exclusively by Ca^{2+} ions. This change in Ca^{2+} flux corresponds to a transient, 1000-fold increase in P_{Ca} during the initial phase of the action potential. As suggested by the Fabiatos to occur in the heart (*25, 26*), a rapid increase in cytosolic Ca^{2+} may trigger the release of intracellulary bound Ca^{2+}, contributing to the initial phase of KCl-induced activation and to the contractile response obtained during smooth muscle action potentials. During the steady-state phase of the KCl-induced contraction, Ca^{2+} release would not occur and intracellular sites would resume sequestering Ca^{2+}.

2.3. Receptor-Stimulated Ca^{2+} Entry

Arterial smooth muscle contractions induced by norepinephrine (*20, 31, 78*). prostaglandins (*49*), and histamine (van Breemen, unpublished observations), and visceral smooth muscle contractions induced by cholinergic agonist (*68*) are all accompanied by an increase in the rate of ^{45}Ca uptake. Theoretically, an agonist could stimulate Ca^{2+} influx directly (i.e., pharmacomechanical coupling), or indirectly via membrane depolarization (*63* and Chapter 5). Casteels and coworkers have reported that the contractions elicited by submaximal doses of histamine and norepinephrine in the rabbit ear artery (*22*), and pulmonary artery (*9*) are not accompanied by depolarization. In contrast, Hermsmeyer *et al.* found that norepinephrine causes dose-dependent depolarizations in rat tail and rabbit ear artery over the entire concentration range employed (*32*). At present, it is unclear whether the depolarization induced by norepinephrine is an obligate transduction step or rather is a manifestation of the agonist-induced increase in the plasmalemmal P_{Ca} and other modifications in the membrane properties. Organic Ca^{2+} entry blockers have been reported to block the norepinephrine-induced depolarization (*13, 64*) suggesting that the latter possibility may, in fact, be the case.

In 1979, Bolton (*5*) and van Breemen et al. (*74*), independently, suggested that separate receptor-operated and potential-operated Ca^{2+} influx pathways exist in smooth muscle. This concept received direct confirmation in a study by Meisheri and van Breemen (*54*). As illustrated in Fig. 2, the ^{45}Ca influx stimulated by maximally effective doses of KCl (80 m*M*) and norepinephrine ($10^{-5}M$) exhibit additivity when the two agents are simultaneously administered to isolated rabbit aorta. Furthermore, as illustrated in Fig. 3 (from the same study), a concentration of D600 ($10^{-6}M$) sufficient to completely block the KCl-stimulated ^{45}Ca

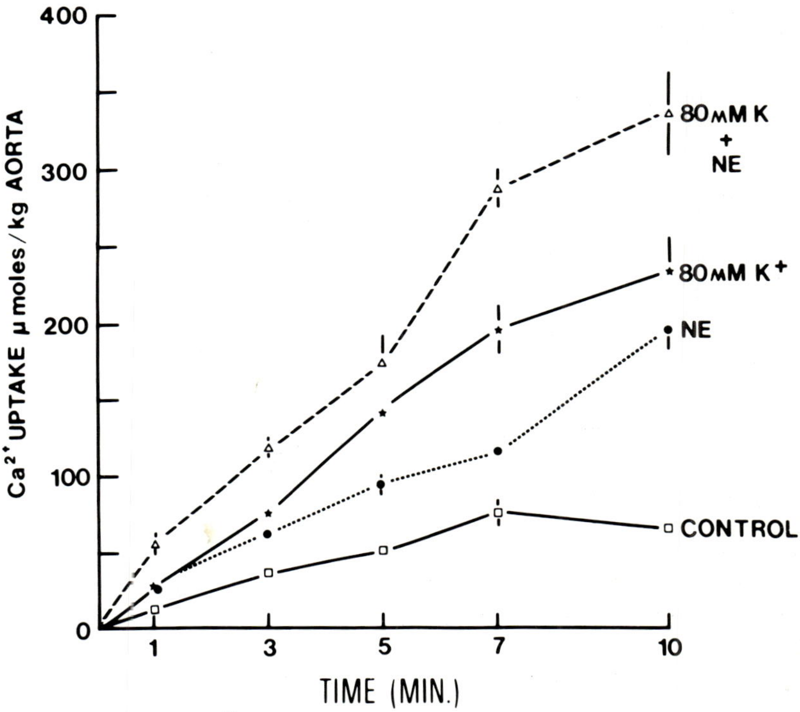

Fig. 2. Initial ^{45}Ca uptakes in rabbit aorta under control conditions and during stimulation with $10^{-5}M$ norepinephrine, 80 mM K^+, and the two stimulants combined. (From K. Meisheri, O. Hwang, and C. van Breemen, *J. Membr. Biol. 59*:19–25, 1981, by permission.)

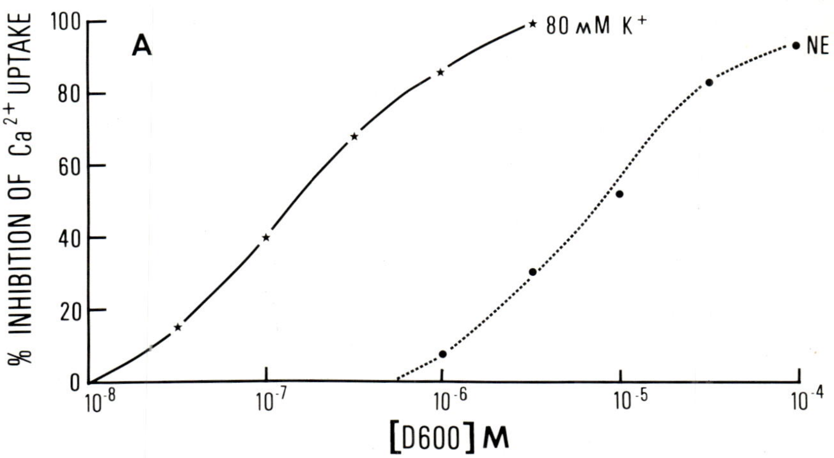

Fig. 3. The effect of D600 on ^{45}Ca uptake by rabbit aorta induced by 80 mM K^+ and by $10^{-5}M$ norepinephrine. (From K. Meisheri, O. Hwang, and C. van Breemen, *J. Membr. Biol. 59*:19–25, 1981, by permission.)

influx does not affect the ^{45}Ca influx pathway activated by norepinephrine. Thus, based upon the classical criteria of additivity and selectivity, two separate Ca^{2+} influx pathways appear to be operative in this tissue. Other distinctions have been found between the Ca^{2+} influx pathway stimulated by norepinephrine and that activated by KCl. Beta-adrenergic stimulation produces a c-AMP-dependent, partial blockade of the KCl-stimulated ^{45}Ca influx, but has no effect upon the norepinephrine- or methoxamine-stimulated ^{45}Ca influx in rabbit aorta (*55*). Furthermore, the norepinephrine-stimulated ^{45}Ca entry pathway exhibits unique characteristics in regard to its sensitivity to blockade by organic Ca^{2+} antagonists, as described below.

At present, it has not been determined whether different agonists activate different classes of receptor-operated channels, or if a particular agonist might operate different classes of Ca^{2+} channels in different blood vessels. These possibilities have been suggested, however, based upon differences in the sensitivity of norepinephrine- (*13*) and serotonin-induced (*67*) contractions to Ca^{2+} antagonists in different blood vessels. A comparison of the diltiazem sensitivity of the norepinephrine-induced ^{45}Ca influx in isolated rabbit aorta to that of isolated rabbit mesenteric resistance vessels (*14*) suggests that norepinephrine may activate a separate, more sensitive Ca^{2+} entry pathway in resistance vessels. In contrast, the Ca^{2+} entry pathway activated upon KCl-induced depolarization exhibits comparable sensitivity to diltiazem and other Ca^{2+} antagonists in almost all vessels in which this has been examined (*13, 14*).

Another striking difference between the receptor-operated Ca^{2+} influx pathway and the potential-operated Ca^{2+} influx pathway concerns the relationship between the sensitivity to Ca^{2+} antagonists and the degree of activation. van Breemen et al. (*78*) originally reported that the sensitivity of the norepinephrine-stimulated ^{45}Ca influx in rabbit aorta to diltiazem decreased as the concentration of norepinephrine used to activate the tissue was increased. This inverse relationship between the degree of activation by norepinephrine and sensitivity to Ca^{2+} influx blockade was subsequently observed with other classes of organic Ca^{2+} entry blockers (*77*), indicating that the phenomenon reflects an intrinsic property of the norepinephrine-activated Ca^{2+} influx pathway. In contrast, as shown in Fig. 4, the sensitivity of the KCl-stimulated ^{45}Ca influx to Ca^{2+} antagonists does not demonstrate this property. In this respect, the Ca^{2+} influx

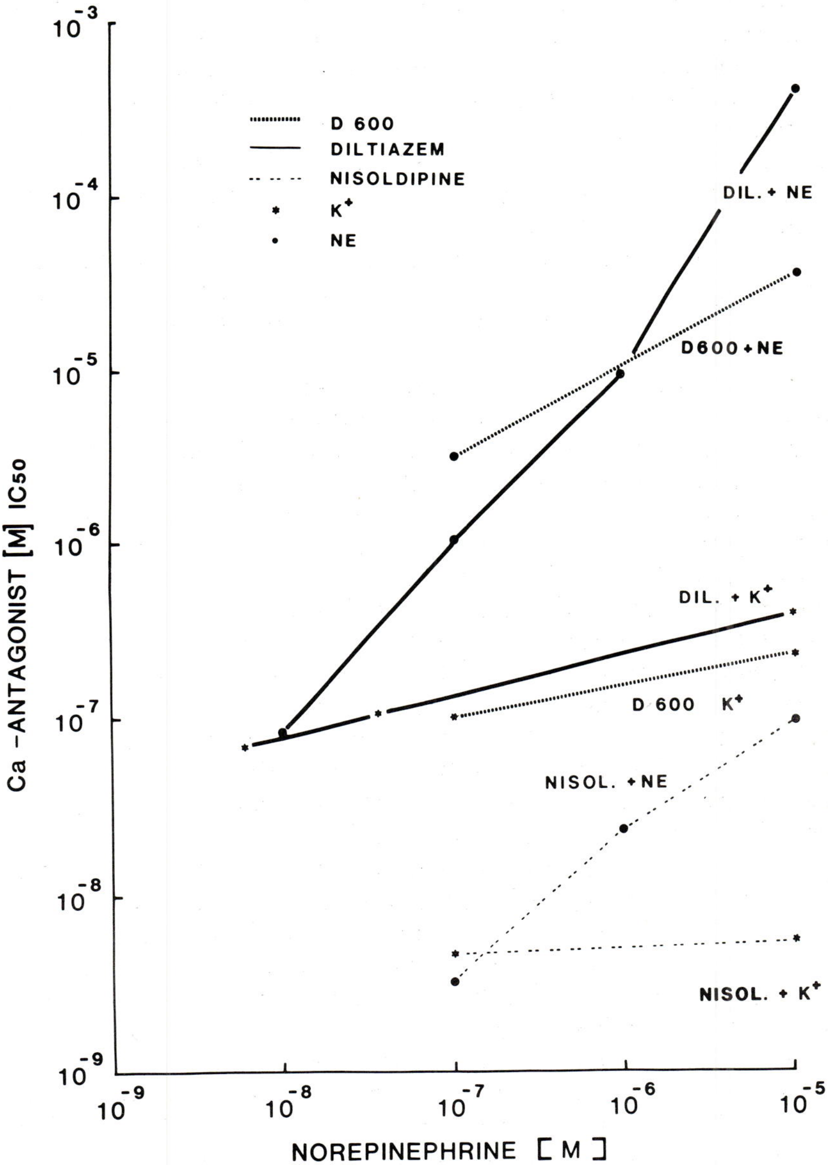

Fig. 4. Effects of norepinephrine concentration and equipotent K^+ concentrations on the half maximal inhibitory concentrations (IC_{50}s) of D600, diltiazem, and nisoldipine. The dependent variable was inhibition of stimulated ^{45}Ca influx in rabbit aorta.

pathway activated by KCl behaves in a manner consistent with the concept of a Ca^{2+} entry channel, possessing specific binding sites for the Ca^{2+} antagonists (*58, 68*). The receptor-operated Ca^{2+} influx pathway, however, appears to have some characteristics that are not incorporated into such a simplistic model. It is possible that high concentrations of norepinephrine may alter membrane fluidity or destabilize the membrane lipids, thereby stimulating the entry of Ca^{2+} ions through a pathway that is not sensitive to agents that block proteinic Ca^{2+} channels

2.4. Role of Surface Bound Ca^{2+}

Most of the extracellular Ca^{2+} in smooth muscle exists in the bound state. Potential Ca^{2+} binding sites include negatively charged membrane proteins and phospholipids, sialic acid residues of the glycocalyx, and the negatively charged sites associated with collagen, elastin, and the ground substance comprising the extracellular matrix (*82*). The possible role of Ca^{2+} bound to the external surface of the cell membrane as a source of activating Ca^{2+} or as a factor regulating plasmalemmal P_{Ca} has been the subject of much speculation.

Hurwitz and Sutra (*36*) observed that the relationship between the magnitude of KCl-induced contractions in ileal smooth muscle and the concentration of Ca^{2+} in the bathing media is hyperbolic. These authors suggested that a loosely bound Ca^{2+} pool which is mobilized by a saturable transport mechanism may be involved in the activation of this tissue. In the same preparation, Chang and Triggle (*15*) found evidence that the different components of the contractile response to muscarinic agonists could be preferentially altered by pretreating the tissue with different Ca^{2+} concentrations. The magnitude of the initial, phasic component of the contraction was dependent upon the concentration of Ca^{2+} the tissues were exposed to prior to stimulation, whereas the magnitude of the sustained component was more dependent upon the concentration of Ca^{2+} present during stimulation. Although such findings suggest that activating Ca^{2+} derived from a bound pool mediates the phasic component, they do little towards defining the location of such a pool.

It is difficult to determine whether a nonionic Ca^{2+} source is extracellular or intracellular. The early observation that the contractile response of vascular smooth muscle to agonists is lost slowly following the removal of extracellular Ca^{2+} was originally interpreted as an indication that extracellulary bound Ca^{2+} might be used as a source of activating Ca^{2+} (*34*). As elaborated upon below, however, subsequent findings suggest that the major component of this residual, agonist-induced contractile response observed in Ca^{2+}-free media is due to the release of intracellular Ca^{2+}. Unlike the response to agonists, the contractile response to KCl is lost very rapidly in Ca^{2+}-free media (*34*), and is completely dependent upon Ca^{2+} entry (*71, 76*). DeFelice and Joiner (*17*) found that the magnitude of the contractile response of rat aorta to KCl was proportional to the amount of Ca^{2+} contained in a loosely bound Ca^{2+} pool. These authors found that aortic strips from male rats accumulated more Ca^{2+} when bathed in Ca^{2+}-rich buffer, and responded more readily to KCl than did aortic strips from females. The extra Ca^{2+} was found to be held in a rapidly exchanging pool ($t_{1/2}$ = 2.5–5 min). Uchida (*69*) used the fluorescence of chlorotetracycline as an indicator of nonionic or membrane-associated Ca^{2+} in smooth muscle microsomes. Uchida found that histamine caused a decrease in membrane-bound Ca^{2+} in the presence of microsomes from intestinal smooth muscle that was sensitive to histamine, but not in the presence of microsomes from smooth muscle insensitive to histamine. Unfortunately, the design of this study did not allow a clear distinction between the release of extracellularly bound Ca^{2+} or intracellularly bound Ca^{2+} (*11, 15*).

A more direct demonstration of the involvement of extracellularly bound Ca^{2+} during activation of smooth muscle has been obtained using ^{45}Ca flux measurements. The rationale behind this approach is based upon the observation that intracellular ^{45}Ca is much more resistant to removal by washing in ice-cold buffer containing EGTA than is extracellular free ^{45}Ca or extracellularly bound ^{45}Ca. Thus, by measuring the uptake of ^{45}Ca into the EGTA-resistant or intracellular compartment, one can directly monitor the movements of ^{45}Ca into the cell rather than relying upon contractile data alone. Furthermore, by examining the kinetic behavior of the ^{45}Ca influx response, inferences can be made concerning the nature of the extracellular Ca^{2+} pools that are used as a source of activating Ca^{2+}. Using this approach it was determined that Ca^{2+} derived from a slowly exchanging extracellular pool enters the smooth muscle cells upon activation of rabbit aorta by the prostaglandin H_2 (PGH_2) analog U-44069 (*52*). The slow exchange rate of this Ca^{2+} source could be demonstrated by the observation that when the PGH_2 analog and ^{45}Ca were administered to tissues simultaneously, a 2.5 min delay in the stimulation of ^{45}Ca uptake was observed, whereas no such delay was observed when tissues were pretreated with tracer before the application of the prostaglandin.

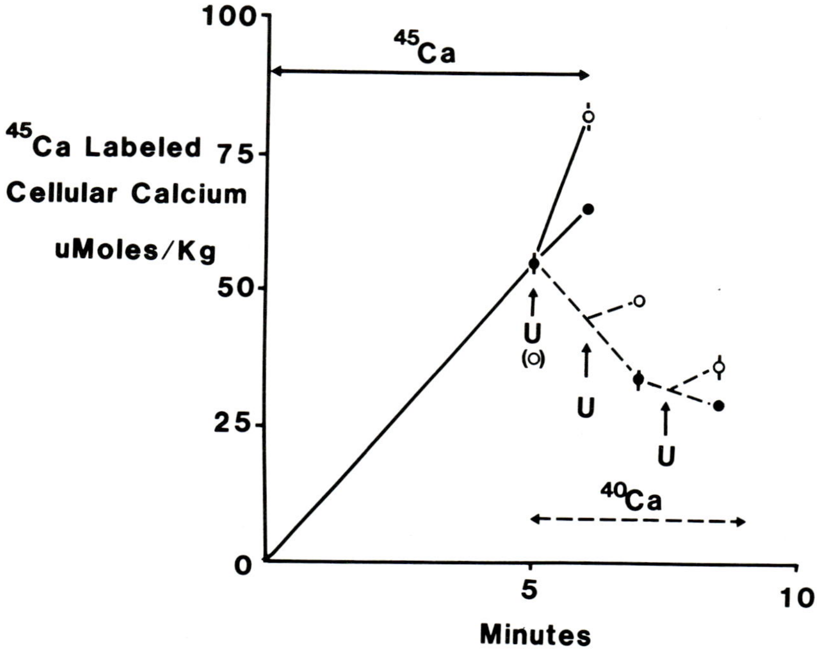

Fig. 5. U-44069-stimulated cellular uptake of ^{45}Ca during the washout of extracellular ^{45}Ca. Solid line represents control tissues (●) and tissues treated with U-44069 (○) in ^{45}Ca-PSS. Dashed line represents control tissues (●) and tissues treated with U-44069 (○) during washout of tracer with ^{40}Ca-PSS. (From R. Loutzenhiser and C. van Breemen, *Blood Vessels*, 20:295–305, 1983, by permission.)

The slow rate of exchange of this external, activating Ca^{2+} pool is also evident from the study illustrated in Fig. 5. In this experiment, the extracellular Ca^{2+} pools were prelabeled by placing tissues in ^{45}Ca buffer for 5 min. The extracellular ^{45}Ca was then washed out by returning the tissues to buffer containing only ^{40}Ca. The addition of the prostaglandin during the washout of extracellular ^{45}Ca resulted in a cellular uptake of residual tracer. By determining the rate of decay of this residual ^{45}Ca uptake, the rate of exchange labeling of the extracellular ^{45}Ca source was estimated to be 0.33 min^{-1} ($t_{1/2}$ = 2 min). In the same study, the rates of exchange of interstitial free ^{45}Ca and two extracellularly bound ^{45}Ca pools were determined to be 3.24, 0.31, and 0.026 min^{-1}, respectively. The kinetic behavior of the ^{45}Ca pool utilized for the initial phase of the U-44069-stimulated ^{45}Ca influx, thus, corresponded to that of an extracellularly bound ^{45}Ca pool, rather than to that of the extracellular free ^{45}Ca ions. When tissues were prelabeled with tracer and the temporal profile of the U-44069-stimulated ^{45}Ca influx examined, two phases were observed—an initial rapid phase that was transient and a slower, secondary phase that was more sustained. Essentially, the same type of evidence has been obtained during activation of rabbit aorta by norepinephrine and KCl. Lodge and van Breemen (*47*) demonstrated that norepinephrine and KCl also stimulate ^{45}Ca influx in a biphasic manner. The initial, transient rapid phase of Ca^{2+} influx is hyperbolically dependent upon extracellular free Ca^{2+} and the magnitude of this component depends upon prelabeling the extracellular Ca^{2+} binding sites. Thus, in rabbit aorta, during the initial rapid phase of activation by the prostaglandins, norepinephrine, and KCl, Ca^{2+} derived from an extracellularly bound pool may enter the smooth muscle cell.

The relative importance of surface bound Ca^{2+} in the activation of smooth muscle may differ depending upon the mode of activation, species, and gender. In the studies sited above, DeFelice and Joiner (*17*) found that the contractile response of rat aorta to KCl was dependent upon slowly exchanging Ca^{2+} and thus augmented in the male. The magnitude of the norepinephrine-induced contractions of rat aorta did not exhibit such gender-related differences. Karanian et al. (*43*), however, reported similar gender differences in the responsiveness of isolated rat aorta to prostaglandins. These authors reported that the maximal contractile responses to PGE_2, PGD_2, PGF_2, and a stable PGH_2 analog were greater in the male than female, whereas no differences in the responses to norepinephrine and serotonin were observed. Although these studies and those presented above, suggest that agonist, species, and gender-related differences in the involvement of extracellularly bound Ca^{2+} may play an important role in determining vascular responsiveness, this question will require substantial further investigation. Furthermore, the underlying mechanisms mediating the involvement of extracellularly bound Ca^{2+} remain to be determined.

The simple concept of a plasmalemmal Ca^{2+} channel that, upon activation, allows the free passage of Ca^{2+} down its electrochemical gradient, does not predict the observed dependency of ^{45}Ca influx upon the labeling of extracellularly bound Ca^{2+}. Also, such a model does not explain the observation that in stimulated tissues (prelabeled with ^{45}Ca), the initial rate of ^{45}Ca entry is much more rapid than the steady-state rate of ^{45}Ca entry. Both would occur, however, if upon activation of the influx pathway, Ca^{2+} were released from extracellular binding sites located on the external surface of the plasma membrane. Such a model of receptor-stimulated Ca^{2+} uptake has recently been proposed by Tan and Tashjian (*65*) for pituitary cells. These authors found that treatment with thyrotropin-releasing hormone (TRH) caused an increase in cellular ^{45}Ca and a simultaneous decrease in extracellularly bound ^{45}Ca. The release of surface-bound Ca^{2+} pool may play two roles in the activation of smooth muscle: (1) as originally proposed by Langer (*45, 46*) for the heart, Ca^{2+} that is released near the cell surface may increase the Ca^{2+} gradient, thereby facilitating Ca^{2+} influx; (2) Ca^{2+} released from stabilizing sites on the plasmalemma may also constitute an early transduction step in the link between receptor occupation and the increase in plasmalemmal Ca^{2+} permeability.

3. Release of Intracellularly Bound Ca^{2+}

The Ca^{2+} content of the smooth muscle cells of rabbit aorta, at rest, is approximately 3–5 × 10^{-4} mol Ca/kg cell H_2O (*31, 54*). Since tension development is initiated at free Ca^{2+} concentrations above 10^{-7}, almost all of this Ca^{2+} is bound. This large pool of intracellularly bound Ca^{2+} is a potential source of activating Ca^{2+}. Furthermore, since the Ca^{2+} released from intracellular sites would have to diffuse only a minimal distance to reach the myofilaments, this source of activating Ca^{2+} may effect very rapid increases in smooth muscle tension.

3.1. Agonist-Induced Release of Ca_i

In 1963 Bohr (*4*) discovered that the initial phase of the norepinephrine-induced aortic contraction was less dependent on extracellular Ca^{2+} than the tonic contraction. Hinke (*33*) and Hudgins and Weiss (*35*) confirmed this observation showing that the norepinephrine-induced contraction washed out only slowly in 0 Ca^{2+} EDTA solutions. They also showed that in contrast the high K^+ contraction disappeared rapidly upon Ca^{2+} removal. It was thus clear that norepinephrine released a bound Ca^{2+} source to supply activator Ca^{2+} during the initial phase of contraction. Initially there was some disagreement on the location of this bound Ca^{2+} source since Hinke (*34*) thought that it was bound tightly to the outer membrane surface, whereas Daniel (*16*) located it within the cell membrane. It was subsequently demonstrated that a transient contractile response to norepinephrine could be obtained in the presence of La^{3+} (*71, 76*). La^{3+} has been shown to be very effective in blocking Ca^{2+} fluxes across biological and artificial membranes and to displace Ca^{2+} from anionic binding sites (*70, 71, 76*). La^{3+} also completely blocked the contractile response to KCl. It was therefore argued that the residual norepinephrine-induced contraction observed in the presence of La^{3+} indicated that this agonist was causing the release of an intracellularly bound Ca^{2+} store (*71, 76*).

The release of intracellular Ca^{2+} by norepinephrine can be visualized, more directly, by measuring ^{45}Ca efflux (*20, 75*). When norepinephrine is administered to ^{45}Ca-labeled, aorta, the ^{45}Ca associated with the agonist-sensitive intracellular store is released into the myoplasm. This release of tracer increases the rate of ^{45}Ca extrusion, visualized as stimulation of ^{45}Ca efflux (*20*). This phenomenon is not to be confused with the stimulation of ^{45}Ca efflux observed when Ca^{2+} influx is stimulated, for example, by KCl (*81*). During KCl stimulation ^{45}Ca efflux occurs only in the presence of ^{45}Ca and is prevented by Ca^{2+} entry blockers. Therefore, KCl stimulation of ^{45}Ca efflux depends upon a displacement of intracellularly bound tracer by ^{40}Ca (*81*). On the other hand, the agonist-induced stimulation of ^{45}Ca efflux may be observed in the presence of EGTA, and Ca^{2+} antagonists (*77*), and thus represents a true agonist-induced release of intracellularly bound ^{45}Ca.

The observation that the norepinephrine-sensitive intracellular Ca^{2+} pool is limited in size and is refilled from the outside (*76*) opened the possibility for studying its relationship to other agonist-sensitive Ca^{2+} pools. The contractile responses of large arteries to angiotensin II, histamine, serotonin, vasopressin, and prostaglandins (*19, 49*) are also mediated, in part, by an agonist-induced release of intracellular Ca^{2+}. Deth and van Breemen (*19*) originally recognized that norepinephrine, histamine, and angiotensin II each release Ca^{2+} from a common, agonist-sensitive, intracellular pool in rabbit aorta. Furthermore, these authors noted that the ability of an agonist to deplete the Ca^{2+} pool utilized by another agonist was related to its efficacy in Ca^{2+}-free (EGTA) buffer. It was subsequently determined that prostaglandin H_2 (PGH) releases Ca^{2+} from the same pool (*49*). The Ca^{2+} pool released by the prostaglandin was completely depleted by prior treatment with norepinephrine or histamine, but was relatively unaffected by angiotensin II. Angiotensin II and PGH_2, thus, release Ca^{2+} from separate sites, even though each releases Ca^{2+} from a pool sensitive to depletion by norepinephrine and histamine. It was suggested (*49*) that agonists may release intracellular Ca^{2+} from sites located within specific receptor-associated domains. The efficacy of agonist-induced Ca^{2+} release would be a function of receptor density, or the area limited by the combined domains of its plasmalemmal receptors. The interrelationship between the releasable Ca^{2+} pools of various agonists would be a function of the spatial distribution of these domains. The probability of overlap between the domains of two lesser efficacious agonists would be relatively small, even though each agonist might share a portion of the releasable Ca^{2+} pool belonging to the larger domain of a third agonist. The inherent implication of this hypothesis is that occupation of a receptor initiates a process resulting in the release of Ca^{2+} only from those sensitive sites located within a limited distance of the receptor. This would place the location of the agonist-releasable Ca^{2+} pool very near the cell surface.

The precise morphological location of the agonist-releasable Ca^{2+} has not been determined. However, since only a single, agonist-induced phasic contractile response can be elicited in the absence of external Ca^{2+}, or in the presence of La^{3+} (*19, 20, 76*), it is thought that the pool will refill only from the extracellular space. This observation also suggests that the possible morphological candidates should be restricted to those cellular structures in the immediate vicinity of the plasmalemma. Two possible locations appear quite plausible: a portion of the sarcoplasmic reticulum that is closely associated with the plasmalemma (*6, 21*), and the Ca^{2+} binding sites located on the inner surface of the plasmalemma (*76*). Mitochondria have been demonstrated to accumulate Ca^{2+} and have been implicated as a possible source of releasable Ca^{2+}. However, the intracellular ^{45}Ca pool released by dinitrophenol (DNP) exchanges at a slower rate than that released by norepinephrine (*18*). Recently, Grover et al. (*29*) discovered that the aortic NE-sensitive Ca fraction has a pH sensitivity that is similar to that of high affinity Ca^{2+} binding to partially purified smooth muscle sarcolemma and suggested that these Ca fractions overlap. A more complete understanding of agonist-induced Ca^{2+} release awaits future work delineating the morphologioal identity of the agonist-sensitive Ca^{2+} stores and the biochemical processes responsible for the release of Ca^{2+} from these sites.

The degree to which agonist-induced release of intracellular Ca^{2+} is involved in activation of vascular smooth muscle varies in different species and in different arteries within the same species. Agonist-induced Ca^{2+} release plays a major role in the activation of rabbit aorta (*71*), rat aorta (*30*), human coronary artery (R. Ginsburg, private communication), and human renal artery (Loutzenhiser, unpublished results), but does not appear to occur in dog coronary artery (*80*) or basilar artery (*67*). Within the mesenteric vasculature of the rabbit, activation by norepinephrine involves progressively less release of intracellular Ca^{2+} as one descends from the superior mesenteric artery to the resistance segment of the vascular bed. Norepinephrine-induced activation of the terminal resistance segment of this arterial tree involves Ca^{2+} influx alone (*14*). It remains to be determined whether the lack of agonist-induced release of Ca^{2+} observed in mesenteric resistance vessels is representative of all resistance vessels, or whether it reflects a specific characteristic of the mesenteric vascular bed.

3.2. Caffeine-Induced Release of Ca_i

In skeletal muscle caffeine can release Ca^{2+} from the sarcoplasmic reticulum by enhancing the Ca^{2+}-induced Ca^{2+} release mechanism (*23*).

The ability of caffeine to release intracellular Ca^{2+} from vascular smooth muscle cells is strongly suggested by following experimental data:

(1) Caffeine can still produce a contraction in the absence of extracellular Ca^{2+} (*18*).
(2) The caffeine sensitive internal Ca^{2+} store is at least partially preserved after skinning procedures (*61*).

Since caffeine does not increase control ^{45}Ca influx (*72*), its response appears to arise solely from release of intracellular Ca^{2+}. The quantity of released Ca^{2+} as measured by the caffeine-induced contraction varies with different vascular smooth muscle types. In the intact rabbit mesenteric artery the magnitude of contraction induced by a maximal dose of caffeine is similar to that induced by a maximal norepinephrine dose (*44*). In the intact cells of the porcine coronary artery a maximal caffeine dose gives a contractile response that is 50% of the acetylcholine-induced contraction (*37*). In rat aorta and bovine besilar artery caffeine produces a contraction which is 20% of the agonist-induced one (*l4*). In rabbit aorta (*18*), rabbit ear artery (*8*), and rabbit pulmonary artery (*9, 10*) the caffeine induced contraction is less than 15% of the one induced by norepinephrine. It is not clear if this variability in contractility by caffeine is dependent on the size of function of the specific blood vessel, although there is a tendency toward an increasing agonist/caffeine contraction ratio in smaller vessels. Another possibility is that differences in the thickness of the preparations affect the final contractile response induced by caffeine. In the thicker peparations strong inhibitory effects of caffeine (induced in more superficial muscle layers) could prevent a further increase in tension development in deeper muscle layers. A third reason for variations in force development induced by caffeine could be that the amount of Ca^{2+} available in the caffeine-sensitive Ca^{2+} store is different in different types of vascular smooth muscle. If, as in skeletal muscle, the sarcoplasmic reticulum represents the caffeine-releasable Ca^{2+} store, one is tempted to assume that the size of the caffeine-induced contraction is proportional to the volume of SR. Although there may be a good correlation between the size of SR and caffeine-induced ^{45}Ca efflux, this does not seem to hold true for the size of contraction. The volume of SR in rabbit aorta is three times greater than that in rabbit mesenteric artery (RA = 5% vs RMA = 1.8%, ref. *21*), whereas the caffeine-induced contraction is about six times smaller (RA: norepinephrine/caffeine ratio = 6 vs RMA: norepinephrine/caffeine ratio = 1).

It is intriguing that in the rabbit aorta the amount of ^{45}Ca released by caffeine is approximately the same as that released by norepinephrine, whereas the size of the norepinephrine-induced contraction is six times greater. From the work of Deth et al. (*18*) it was concluded that in rabbit aorta norepinephrine and caffeine mobilize a common cellular Ca^{2+} source. This sharing of a common Ca^{2+} store by agonists and caffeine has been confirmed for other types of vascular smooth muscle. However, in rabbit mesenteric artery (*62*) and rabbit aorta (Leyten, unpublished results), caffeine seems to be more potent than norepinephrine in releasing intracellular Ca^{2+}, and in the porcine coronary artery acetylcholine has a greater potential for releasing Ca^{2+} than caffeine (*37*). Estimates of the size of caffeine-sensitive Ca^{2+} stores in the rabbit aorta range from 40–60 μmol Ca^{2+} (*18*). There appear to be two procedures by which the amount of Ca^{2+} in the caffeine-sensitive store can be enhanced: (1) the application of 1 m*M* procaine during the loading period of the caffeine store increases the subsequent amplitude of the caffeine contraction in 0 Ca^{2+} (*39*) and (2) after previous high K^+ depolarization the contractile response to caffeine is enhanced (*18*).

3.3. Ca^{2+}-Induced Release of Ca_i

In 1970, Endo et al. (*24*) and Ford and Podolsky (*28*) first reported, independently, that Ca^{2+} can induce a release of Ca^{2+} from the sarcoplasmic reticulum (SR) of skinned skeletal muscle fibers. Because the conditions necessary to elicit a significant Ca^{2+}-induced Ca^{2+} release (e.g., addition of caffeine, low Mg^{2+} or high Ca^{2+}) were not physiological, the role of this process in skeletal muscle contraction remained controversial. Using mechanically skinned cardiac fibers, Fabiato and Fabiato (*25*) demonstrated that Ca^{2+}-induced Ca^{2+} release can be more readily evoked at physiologically relevant Ca^{2+} concentrations in cardiac muscle. Fabiato also suggested that the rate of change of Ca^{2+} may be a determining factor regulating Ca^{2+}-induced Ca^{2+} release (*26*).

In visceral smooth muscle, Saida (*59, 60*) demonstrated Ca^{2+}-induced Ca^{2+} release from the SR of saponin-treated skinned smooth muscle cells at 3×10^{-6} *M* Ca^{2+} without additional conditions facilitating the release mechanism. Saida, thus, concluded that Ca^{2+}-induced Ca^{2+} release may be of physiological relevance in the activation of visceral smooth muscle. A possible role of Ca^{2+}-induced Ca^{2+} release in vascular smooth muscle has been proposed by Itoh et al. (*39*).

Saida and van Breemen (*62*) have also obtained evidence that Ca^{2+}-induced Ca^{2+} release may be coupled to agonist-induced Ca^{2+} release in vascular smooth muscle. A direct comparison between the kinetics of the norepinephrine-induced Ca^{2+} release and caffeine-induced Ca^{2+} release in isolated mesenteric artery at room temperature indicated that these two agents initially release Ca^{2+} from separate stores. This conclusion was based upon the observation that the norepinephrine-sensitive, intracellular Ca^{2+} store refilled after bathing the muscle in 1.5 m*M* Ca^{2+} buffer for 3 min, whereas the caffeine-sensitive, intracellular Ca^{2+} store refilled only after bathing the muscle for 6.5 min. However, a more complex interrelationship between these two Ca^{2+} pools was indicated by subsequent studies. It was found that when the SR was depleted by prior treatment with caffeine, the rate of recovery of the norepinephrine response becomes as slow as that of the caffeine response. Furthermore, the magnitude of the norepinephrine-induced contraction elicited in Ca^{2+}-free media (2 m*M* EGTA) was found to be directly correlated with the amount of Ca^{2+} present in the caffeine-sensitive pool (SR). It was, therefore, postulated that during norepinephrine-induced activation, Ca^{2+} is released from an agonist-sensitive Ca^{2+} pool of limited size. The released Ca^{2+} causes a rapid, transient increase in the Ca^{2+} concentration near the SR. This rapid change in Ca^{2+} then, in turn, would initiate a Ca^{2+}-induced Ca^{2+} release from the SR. This model combines the concept that the Ca pools which are directly releasable by norepinephrine and caffeine are kinetically distinct with the observation that the contractions evoked by the release of Ca^{2+} from the norepinephrine-sensitive pool are dependent upon the amount of Ca^{2+} present in the caffeine-sensitive pool. As stated in a previous section, indirect evidence also suggests that Ca^{2+}-induced Ca^{2+} release may occur during smooth muscle action potentials.

4. Modulation of Intracellular Ca^{2+} Sequestration

As described above (*see* Section 2.2, Ca^{2+} Entry Stimulated by Membrane Depolarization), full activation of the calmodulin system of smooth muscle would require a maximum of 40 μmol Ca^{2+}/kg cell. However, in order to elicit a maximal contractile response to KCl a minimum Ca^{2+} uptake of 100 μmol Ca^{2+}/kg cell is required (*73*). Clearly, only a fraction of the Ca^{2+} ions which enter the cell can be used to activate the contractile machinery. The remaining Ca^{2+} ions are distributed between passive Ca^{2+} buffers and ATP-dependent Ca^{2+} sequestering sites. An alteration in the rate of Ca^{2+} sequestration represents an additional step whereby fine control may be exerted over tension development.

4.1. Ca^{2+} Sequestration During Relaxation

The sequestration of Ca^{2+} into intracellular sites may mediate the relaxation induced by some direct-acting vasodilators. A stimulation of Ca^{2+} uptake by sarcoplasmic reticulum in response to an elevation in cAMP has been suggested to be the mechanism of action of phosphodiesterase inhibitors, beta-adrenergic agonists (*1*). This concept is supported by the observation that such agents induce relaxation without causing a decrease in smooth muscle ^{45}Ca content (*56*). However, a direct and reproducible *in vitro* demonstration of cAMP stimulated Ca^{2+} uptake by smooth muscle microsomes has not been forthcoming. Furthermore, recent evidence has supported an alternate theory, involving a cAMP-induced decrease in the Ca^{2+} sensitivity of the contractile system (*see* Chapters 7 and 8). Either theory could explain how cAMP mediated relaxation might occur without a decrease in ^{45}Ca content. Improved membrane isolation techniques that allow a greater purification of SR vesicles, and skinned smooth muscle preparations in which the Ca^{2+} sequestering capacity of the SR is preserved may lead to a resolution of this controversy.

Figure 6 illustrates another observation which implies that intracellular Ca^{2+} sequestration may be intimately involved in the relaxation of smooth muscle. As indicated, when isolated rabbit aorta is treated with norepinephrine, a gain in ^{45}Ca content accompanies the development of tension. When the agonist is removed, tension decreases rapidly. The cellular ^{45}Ca content, however, remains elevated for a considerable period of time (ca. 10 min) following the relaxation. A possible involvement of the Ca^{2+} uptake into the SR in this type of relaxation is suggested by

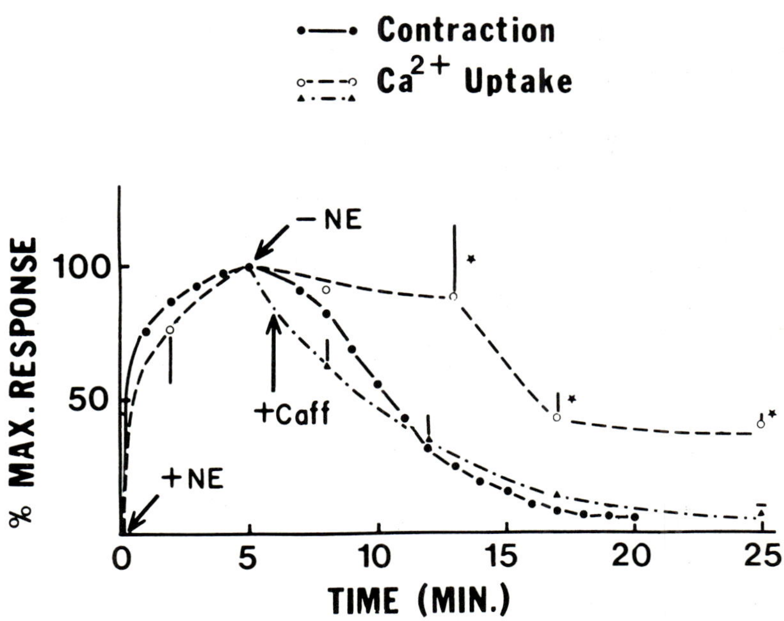

Fig. 6. NE-induced force development and net Ca^{2+} uptake in the rabbit aorta are expressed as % of the maximal response to 10^{-5} *M* NE. The decrease in cellular Ca^{2+} upon NE removal (−NE) is followed in the presence (lower dashed curve) and absence of 10^{-2}*M* caffeine. Bars equal SEM; lack of bars indicate SEM < 5%. Stars indicate significant differences between the time courses of Ca^{2+} loss and reduction in developed force.

the finding that the cellular ^{45}Ca content decreases much more rapidly if caffeine is added upon norepinephrine removal.

4.2. Agonist-Induced Alterations in Intracellular Ca^{2+} Sequestration

As mentioned in the above section on agonist-induced Ca^{2+} release, the activation of conduit arteries by agonists involves the mobilization of an intracellularly bound Ca^{2+} store. The release of Ca^{2+} from this store contributes to the initial rapid phase of the contractile response obtained in physiological media, and supports the transient contraction that is observed when tissues are stimulated by agonists in Ca^{2+}-free buffers. The release of intracellularly bound Ca^{2+} was first demonstrated to occur during activation by norepinephrine, however, other agonists (i,e., histamine, AII, prostaglandins, 5-HT) release Ca^{2+} from the same intracellular site. Recent findings indicate that the intracellular site containing this agonist-releasable Ca^{2+} store (ARC) may also function as an intracellular Ca^{2+} buffer (*48, 51*). The release of Ca^{2+} from this site during stimulation by agonists may be but one manifestation of an agonist-induced reduction in the ability of this site to sequester and contain Ca^{2+}. The other consequence of this process would be that during agonist-induced stimulation a greater portion of the Ca^{2+} that enters the cell would be available for activation.

The first indication that the norepinephrine-releasable intracellular Ca^{2+} is sequestered in a site that may act as an intracellular Ca-buffer was the observation by Deth and Casteels (*18*) that the norepinephrine-induced contractions in EGTA are potentiated by pretreatment with KCl. Karaki et al. (*41*) also found that the pretreatment with high K^+ increased the magnitude of the subsequent Ca^{2+}-free norepinephrine-induced contractions. This effect was prevented by the Ca^{2+} entry blocker D600. Casteels and Droogmans (*8*) reported that the Ca^{2+} entry blocker nicardipine actually reduced the size of the norepinephrine-releasable Ca^{2+} pool below resting levels in the presence of elevated K^+, but not in low K^+ buffer. Finally, Loutzenhiser and van Breemen (*51*) found that high K^+ buffer also increased and D600 decreased the rate of refilling of the agonist-releasable Ca^{2+} pool. The D600 treatment also slowed the rate of $^{45}Ca^{2+}$ uptake to the same extent that it slowed refilling, suggesting that the refilling of the agonist-releasable Ca^{2+} pool was closely linked to Ca^{2+} entry.

These observations led Loutzenhiser and van Breemen to conclude that the intracellular Ca^{2+} pool that is released by agonists refills using Ca^{2+} ions that have just crossed the plasma membrane. The refilling of this pool would thus be altered by a variety of manipulations that modify the rate of Ca^{2+} entry. High K^+ stimulates Ca^{2+} influx, thereby increasing the rate of refilling and the size of the pool. Since, Ca^{2+} antagonists block the K^+-stimulated Ca^{2+} influx, D600 and nicardipine would be expected to prevent these effects. Indeed, when the KCl-induced Ca^{2+} influx is prevented, the net effect of K^+-induced depolarization would be to decrease the electrical component of the Ca^{2+} gradient. This latter effect could cause a decrease in the resting Ca^{2+} leak and may explain the nicardipine-induced decrease in the size of the release pool in the presence of KCl (*8*). Since the resting Ca^{2-} permeability is not decreased by organic Ca^{2+} entry blockers, the lack of effect of D600 and nicardipine in low K^+ buffers would be predicted by this model. Finally, La^{3+} and EGTA would prevent refilling of this pool by abolishing the resting Ca^{2+} permeability and eliminating the Ca^{2+} gradient, respectively. However, in contrast, Casteels and Droogmans (*8*) have suggested that the release pool refills via a direct communication with the extracellular space. In support of this latter hypothesis, these authors found that in rabbit ear artery, the magnitude of the norepinephrine-induced contraction in Ca^{2+}-free buffer was greater following loading in 10 m*M* Ca^{2+} low K^+ buffer than in 0.2 m*M* Ca^{2+} high K^+ buffer, whereas the tissue remained relaxed in low K^+ buffer and contracted in high K^+, low Ca^{2+} buffer. One possible complication that may confound the interpretation of such studies is the stimulatory effect of Ca^{2+} depletion in EGTA upon smooth muscle Ca^{2+} permeability (Loutzenhiser and Cauvin, unpublished results). Thus, after prolonged exposure to Ca^{2+}-free buffer containing 2 m*M* EGTA, the initial rate of ^{45}Ca entry into rabbit aorta is much greater in 10 m*M* Ca^{2+}, 5 m*M* K^+ than in 0.2 m*M* Ca^{2+}, 80 m*M* K^+ (Loutzenhiser and van Breemen, unpublished observations).

Another indication that the site containing the agonist-releasable Ca^{2+} pool may act as an intracellular Ca^{2+} buffer is the apparent ability of this pool to modulate the contractile response of smooth muscle to Ca^{2+} entry. Casteels and Droogmans first noted that when the NE-releasable Ca^{2+} pool was depleted, the contractile response of rabbit ear artery to high K^+ was slowed (*8*). Loutzenhiser and van Breemen (*51*) also reported that depletion of the agonist-releasable Ca^{2+} pool slowed the response of rabbit aorta to 80 m*M* KCl, but did not effect the KCl-induced stimulation of ^{45}Ca entry. When the relationship between tension development in response to KCl treatment and the refilling of the agonist-releasable Ca pool was examined in detail, it was found that no contractile response to KCl was elicited until the releasable Ca pool had refilled to at least 90%. This relationship was maintained as extracellular Ca^{2+} concentration was varied between 0.2 and 1.5 m*M* (*48*). These observations suggest that following the depletion of the agonist-releasable Ca^{2+} pool, the Ca^{2+} ions that enter the cell in response to KC1-included stimulation are sequestered into this site at an initial rate sufficient to prevent Ca^{2+} ions from reaching the myofilaments. Subsequently, after the pool has refilled to near resting levels, a portion of the Ca^{2+} ions escape sequestration and initiate the contractile response, During this phase, however, the agonist-releasable pool continues to sequester incoming Ca^{2+} ions thereby modulating the rate of tension development.

In contrast, prior depletion of the agonist-releasable Ca^{2+} pool does not cause a delay in the contractile response to agonists (*51*), although such treatment, of course, eliminates the initial rapid, phasic component of the contraction. The presence of the agonist apparently prevents the sequestration of Ca^{2+} ions into the releasable Ca^{2+} pool. Karaki et al. (*41*) demonstrated that the presence of NE prevents refilling by 60–70%. This agonist-induced decrease in the sequestering ability of the releasable Ca^{2+} pool may increase the availability of Ca^{2+} that enters the cell during agonist-induced activation. In support of this concept, the presence of NE appears to accelerate the rate of tension development in response to KCl (*51*). This mechanism may also allow Ca^{2+} that enters the cell via the leak or nonstimulated pathway to contribute to the contractile response to agonists. Accordingly, the sites containing the agonist-releasable Ca^{2+} pool may play a role in cell Ca^{2+} homeostasis. The Ca^{2+} that leaks into the cell under resting conditions may be sequestered into these sites before being extruded from the cell. During stimulation by agonists, the Ca^{2+} ions that enter the cell via the leak pathway may bypass the agonist-releasable Ca^{2+} sequestering sites and directly contribute to the activation of the tissue. Such a mechanism could explain the observation that a Ca^{2+}-dependent component of the contractile response of rabbit aorta to agonists may remain in the presence of complete blockade of "stimulated" Ca^{2+} influx by organic Ca^{2+} entry blockers, since the Ca^{2+} leak pathway is not blocked by these agents (*51*).

5. Summary

In this chapter we have reviewed evidence for Ca^{2+} movements that accompany and are causative of the contraction–relaxation cycle in smooth muscle. Membrane excitation increases the influx of Ca^{2+} ions above the passive leak by the opening of receptor-operated and/or potential-operated Ca^{2+} channels. Simultaneously, bound-Ca^{2+} sources on both sides of the plasmalemma are released to varying degrees depending on the mode of activation and the type and size of artery. Sarcoplasmic reticular Ca^{2+} is also released and indications are that this event may be mediated by Ca^{2+}-induced Ca^{2+} release.

Intracellular Ca^{2+} fractions, in addition to supplying Ca^{2+} during the initial phase of activation, function as Ca^{2+} buffers or sinks during the maintained phases of depolarization-induced contractions and during relaxation. Removal of Ca^{2+} ions from the cytoplasm during relaxation is mediated by active extrusion, sarcoplasmic reticular pumping and rebinding to the inner plasmalemmal surface. The relative contributions of these three processes again depend on the type of artery and mode of prior activation.

Acknowledgment

P. L. was supported by grant No.82014 of the Nederlandse Hartstichting.

References

1. Anderson, R., and K. Nilsson. Cyclic AMP and calcium in relaxation of intestinal smooth muscle. *Nature New Biol.* 238:119, 1972.
2. Baker, P. F. Transport and metabolism of calcium ions in nerve. *Progr. Biophys. Mol. Biol.*24:177–223, 1972.
3. Bevan, J. A. Diltiazem selectively inhibits cerebrovascular extrinsic but not intrinsic myogenic tone. *Circ. Res.* 52(Suppl I): I104–I109, 1983.
4. Bohr, D. F. Vascular smooth muscle: dual effect of calcium. *Science* 139:597–599, 1963.
5. Bolton, T. B. Mechanisms of action of transmitter and other substances on smooth muscle. *Physiol. Revs.* 59:606–718, 1979.
6. Brading, A. F., M. Burnett, and F. Sneddon. The effect of sodium removal on the contractile response of the guinea pig taenia coli to carbachol. *J. Physiol* 306:411–429, 1980.
7. Briggs, A. H. Calcium movement during potassium contracture in isolated rabbit aortic strips. *Am. J. Physiol.* 203:849–852, 1962.

8. Casteels, R., and G. Droogmans. Exchange characteristics of the noradrenaline-sensitive calcium store in vascular smooth muscle cells of rabbit ear artery. *J. Physiol.* 317:263–279, 1981.
9. Casteels, R., K. Kitamura, H. Kuriyama, and H. Suzuki. The membrane properties of the smooth muscle cells of the rabbit main pulmonary artery. *J. Physiol. London* 271:41–61, 1977.
10. Casteels, R., K. Kitamura, H. Kuriyama, and H. Suzuki. Excitation–contraction coupling in the smooth muscle cells of the rabbit main pulmonary artery. *J. Physiol. London* 271:63–79, 1977.
11. Casteels, R., and L. Raeymaekers. The action of acetylcholine and catecholamines on an intracellular Ca store in the smooth muscle of the guinea pig taenia coli. *J. Physiol.* 294:51–68, 1979.
12. Casteels, R., and C. van Breemen. Active and passive Ca^{2+} fluxes across cell membranes of the guinea pig taenia coli, *Pflugers Arch.* 359:197–207, 1975.
13. Cauvin, C. R. Loutzenhiser, and C. van Breemen. Mechanisms of calcium antagonist-induced vasodilation. *Ann. Rev. Pharmacol. Toxicol.* 23:373–396, 1983.
14. Cauvin, C., K. Saida, and C. van Breemen. Extracellular Ca dependence and diltiazem inhibition of contraction in rabbit conduit arteries and mesenteric resistance vessels. *Blood Vessels* 21:23–31, 1984.
15. Chang, K., and D. Triggle. Quantitative aspects of drug-receptor interactions. *J. Theor. Biol.* 40:125–154, 1973.
16. Daniel, E. E. Attempted synthesis of data regarding divalent ions in muscle function. In: *Muscle,* edited by W. M. Paul, E. E. Daniel, C. M. Kay, and G. Monckton. Oxford: Pergamon, 1965, p 295–313.
17. Defelice, A., and P. Joiner. Comparison of aortic calcium and contractility in male, female and lactating female rats. *J. Pharmacol. and Exp. Ther.* 194:191–204, 1975.
18. Deth, R., and R. Casteels. A study of releasable Ca fractions in smooth muscle cells of the rabbit aorta. *J. Gen. Physiol.* 69:401–416, 1977.
19. Deth, R., and C. van Breemen. Relative contributions of Ca^{2+} influx and cellular Ca^{2+} release during drug induced activation of the rabbit aorta. *Pluegers Arch.* 348:13–22, 1974.
20. Deth, R., and C. van Breemen. Agonist induced $^{45}Ca^{2+}$ release from smooth muscle cells of the rabbit aorta, *J. Memb. Biol.* 30:363–380, 1977.
21. Devine, C. E., A. V. Somlyo, and A. P. Somlyo. Sarcoplasmic reticulum and excitation-contraction coupling in mammalian smooth muscles. *J. Cell Biol.* 52:690–718, 1972.
22. Droogmans, G., L. Raeymaekers, and R. Casteels. Electro- and pharmacomechanical coupling in the smooth muscle cells of rabbit ear arteries. *J. Gen. Physiol.* 70:129–148, 1977.
23. Endo, M. Calcium release from the sarcoplasmic reticulum. *Physiol. Rev.* 57:71–108, 1977.
24. Endo, M., M. Tanaka, and Y. Ogawa. Calcium induced release of calcium from the sarcoplasmic reticulum of skinned skeletal muscle fibers. *Nature (Lond.)* 228:34–36, 1970.
25. Fabiato, A., and F. Fabiato. Excitation–contraction coupling of isolated cardiac fibers with disrupted or closed sarcolemma. Calcium-dependent cyclic and tonic contractions. *Circ. Res.* 31:293–301, 1972.

26. Fabiato, A., and F. Fabiato. Calcium and cardiac excitation-contraction coupling. *Ann. Rev. Physiol.* 41:473–483, 1979.
27. Filo, R. S., D. F. Bohr, and J. C. Ruegg. Glycerinated skeletal and smooth muscle; calcium and magnesium dependence. *Science* 147:1581–1583, 1965.
28. Ford, L. E., and R. J. Podolsky. Regenerative calcium release within muscle cells *Science* 167:58–59, 1970.
29. Grover, A. K., J. Crankshaw, C. R. Triggle, and E. E. Daniel. Nature of norepinephrine-sensitive Ca-pool in rabbit aortic smooth muscle: effect of pH. *Life Sciences* 32:1553–1558, 1983.
30. Godfraind, T., and A. Kaba. The role of calcium in the actions of drugs on vascular smooth muscle. *Arch. Int. Pharmacodyn. Ther.* 196: (Supp): 35–49, 1972.
31. Godfraind, T. Calcium exchange in vascular smooth muscle, action of noradrenaline and lanthanum. *J. Physiol. London* 260:21–35, 1976.
32. Hermsmeyer, K., A. Trapani, and P. W. Abel. Membrane potential-dependent tension in vascular smooth muscle. In: *Vasodilatation,* edited by P. Vanhoutte and I. Leusen. New York: Raven, 1981, pp 273–284.
33. Hinke, J. A. M. Calcium requirements for noradrenaline and high potassium ion contraction in arterial smooth muscle. In: *Muscle,* edited by W. M. Paul, E. E. Daniel, C. M. Kay, and G. Monckton, Oxford: Pergamon, 1965, p 269–284.
34. Hinke, J. A. M., M. L. Wilson, and H. R. Burnham. Calcium and the contractility of arterial smooth muscle. *Am. J. Physiol.* 206:211–217, 1964.
35. Hudgins, P., and G. B. Weiss. Differential effects of calcium removal upon vascular smooth muscle contraction induced by norepinephrine, histamine, and potassium. *J. Pharmacol Exp Ther.* 159:91–97, 1968.
36. Hurwitz, L. and A. Suria. The link between agonist action and response in smooth muscle. *Ann. Rev. Pharm.* 11:303–326, 1971.
37. Itoh, T., M. Kajiwara, K. Kitamura, and H. Kuriyama. Roles of stored calcium on the mechanical response evoked in smooth muscle cells of the porcine coronary artery. *J. Physiol. London* 322:107–125, 1982.
38. Ito, Y., and H. Kuriyama. Membrane properties of the smooth muscle fibers of the guinea-pig portal vein. *J. Physiol. London* 214:427–441, 1971.
39. Ito, T., H. Kuriyama, and H. Suzuki. Excitation-contraction coupling in smooth muscle cells of the guinea-pig mesenteric artery. *J. Physiol. (London)* 321:513–535, 1981.
40. Jones, A. W. Vascular smooth muscle and alterations during hypertension. In: *Smooth Muscle: An Assessment of Current Knowledge,* edited by E. Bulbring, A. S. Brading, A. W. Jones and T. Tomita. E. Arnold, 1981, pp 397–429.
41. Karaki, H., H. Kuboto, and N. Urakawa. Mobilization of stored calcium for phasic contractions induced by norepinephrine in rabbit aorta. *Eur. J. Pharmaool.* 56:237–245, 1979.
42. Karaki, H., and G. B. Weiss. Alterations in high and low affinity binding of ^{45}Ca in rabbit aortic smooth muscle by norepinephrine and potassium after exposure to lanthanum and low temperature. *J. Pharmacol. Exp. Ther.* 211:86–92, 1979.

43. Karanian, J., F. Moran, E. Ramey, and P. Ramwell. Gender differences in prostaglandin recepators of rat aorta. *Br. J. Pharm.* 72:10–12, 1981.
44. Kuriyama, H., Y. Ito, H. Suzuki, K. Kitamura, and T. Itoh. Factors modifying contraction-relaxation cycle in vascular smooth muscle. *Am. J. Physiol.* 243:H641–H662, 1982.
45. Langer, G. Ionic movements and the control of contraction. In *The Mammalian Myocardium,* edited by G. A. Langer and A. J. Brady. New York: Wiley Interscience, 1974, pp 193-217.
46. Langer, G. The role of calcium in the control of myocardial contractility: an update. *J. Mol. Cell. Cardiol.* 12:231–239, 1980.
47. Lodge, N., and C. van Breemen. Initial and 'steady state' rates of calcium influx during norepinephrine (Ne) and high potassium (high K) activation of rabbit aortic smooth muscle. (abs) *Fed. Proc.* 42:314, 1983.
48. Loutzenhiser, R., C. Cauvin, and C. van Breemen. Studies on the refilling of the norepinephrine-releasable intracellular calcium pool of rabbit aorta. (abs) *Fed. Proc.* 42:314, 1983.
49. Loutzenhiser, R., and C. van Breemen. The mechanisms of activation of isolated rabbit aorta by the stable PGH_2 analogue, U-44069. *Am. J. Physiol.* 241:C243–C249, 1981.
50. Loutzenhiser, R., and C. van Breemen. Mechanisms of stimulated Ca influx and consequences of Ca influx inhibition. In: *Calcium Entry Blockers, Adenosine, and Neurohumors,* edited by G. F. Merrill and H. R. Weiss. Baltimore: Urban and Schwarzenberg, 1983, pp 73–89.
51. Loutzenhiser, R., and C. van Breemen. The influence of receptor occupation on Ca^{++} influx-mediated vascular smooth contraction. *Circ. Res.* 52(Suppl. I): 97–103, 1983.
52. Loutzenhiser, R., and C. van Breemen. Involvement of extracellularly bound calcium in the activation of arterial smooth muscle. *Blood Vessels* 20:295–305, 1983.
53. Mayer, C. J., C. van Breemen, and R. Casteels. The action of lanthanum and D600 on the calcium exchange in smooth muscle cells of the guinea pig taenia coli. *Pflugers Arch.* 337:333–350, 1972.
54. Meisheri, K., O. Hwang, and C. van Breemen. Evidence for two separate Ca^{2+} influx pathways in smooth muscle plasmalemma. *J. Membr. Biol.* 59:19–25, 1981.
55. Meisheri, K., and van Breemen. Effects of α-adrenergic stimulation on Ca^{2+} movements in rabbit aortic smooth muscle: relationship with cyclic AMP. *J. Physiol.* (London) 331:429–441, 1982.
56. Mueller, E., and C. van Breemen. Role of intracellular Ca^{2+} sequestration in β-adrenergic relaxation of a smooth muscle. *Nature* 281:682–683, 1979.
57. Ohashi, H., T. Takewaki, and T. Okada. Calcium and the contractile effect of carbachol in the depolarized guinea-pig taenia caecum. *Jap. J. Pharmac.* 24:601–611, 1974.
58. Rosenberger, L. B., M. K. Ticku, and D. J. Triggle. The effects of calcium antagonists on mechanical responses and Ca^{2+} movements in guinea pig ileal longitudinal smooth muscle. *Can. J. Physiol. Pharmacol.* 57:333–347, 1979.

59. Saida, K. Ca^{2+} and "depolarization"-induced Ca^{2+} release in skinned smooth muscle fibers. *Biomedical Res.* 2849:453-455, 1981.
60. Saida, K. Intracellular Ca release in skinned smooth muscle. *J. Gen. Physiol.* 80:191–202, 1982.
61. Saida, K. and C. van Breemen. A possible Ca^{2+}induced Ca^{2+} release mechanism mediated by norepinephrine in vascular smooth muscle. *Pflugers Arch.* 397:166–167, 1983.
62. Saida, K. and C. van Breemen. Characteristics of the norepinephrine-sensitive Ca store in vascular smooth muscle. *Blood Vessels* 21:43–52. 1984.
63. Somlyo, A. V., and A. P. Somlyo. Electromechanical and pharmacomechanical coupling in vascular smooth muscle. *J. Pharmacol. Exp. Ther.* 159:129–145, 1968.
64. Stekiel, W. J., S. J. Contney, J. H. Lombard, and D. R. Harder. Response of venous membrane potentials to calcium blockade in spontaneous hypertension. *Hypertension* 5(Suppl. I):I64–I69, 1983.
65. Tan, N., and A. Tashjian. Receptor-mediated release of plasma membrane-associated calcium and stimulation of calcium uptake by thyrotropin-releasing hormone in pituitary cells in culture. *J. Biol. Chem.* 256:8994–9002, 1981.
66. Tomita, T. Electrophysiology of mammalian smooth muscle. *Prog. Biophys. Mol. Biol.* 30:185–203, 1975.
67. Towart, R. The selective inhibition of serotonin-induced contractions of rabbit cerebral vascular smooth muscle by calcium-antagonistic dihydropyridines. *Circ. Res.* 48:650–657, 1981.
68. Triggle, D. J. Biochemical pharmacology of calcium blockers. In: *Calcium Blockers, Mechanisms of Action and Clinical Application,* edited by S. Flaim and R. Zelis. Baltimore: Urban & Schwarzenberg, 1982, pp 125–134.
69. Uchida, M. Histamine-induced decrease of membrane-bound calcium ions in the membrane fraction of rabbit taenia coli, *Eur. J. Pharmacol.* 64:357–360, 1980.
70. van Breemen, C. Permselectivity of a porous phospholipid cholesterol artificial membrane. Calcium and lanthanum effects. *Biochem. Biophys. Res. Comm.* 32:977–983, 1968.
71. van Breemen, C. Blockade of membrane calcium fluxes by lanthanum in relation to vascular smooth muscle contractility. *Int. Arch. Physiol. et Biochem.* 77:710–717, 1969.
72. van Breemen, C. Transmembrane calcium transport in vascular smooth muscle. In: *Second International Symposium on Vascular Neuroeffector Mechanisms,* edited by J. A. Bevan *et al.* Basel: S. Karger, 1975, pp 67–79.
73. van Breemen, C. Ca^{2+} requirement for activation of intact aortic smooth muscle. *J. Physiol. Lond.* 272:317–329, 1977.
74. van Breemen, C., P. Aaronson, and R. Loutzenhiser. Na–Ca interactions in mammalian smooth muscle. *Pharmacol, Rev.* 30:167–208, 1979.
75. van Breemen, C. and R. Deth. La^{3+} and excitation–contraction coupling in vascular smooth muscle. In: *Symposium on the Role of Ions in Transmission of Signals from Tissue and Blood to the Vascular Smooth Muscle Cells,* edited by E. Betz. Basel: S. Karger, 1976, pp 26–31.

76. van Breemen, C., B. R. Farinas, P. Gerba, and E. D. McNaughton. Excitation–contraction coupling in rabbit aorta studied by the lanthanum method for measuring cellular calcium influx. *Circ. Res.* 30:44–54, 1972.
77. van Breemen, C., O. Hwang, and C. Cauvin. Ca antagonist inhibition of norepinephrine stimulated Ca influx in vascular smooth muscle. In: *Calcium Modulators,* edited by T. Godfraind and A. Albertini. Amsterdam: Elsevier Biomedical Press, 1982, pp 185–198.
78. van Breemen, C., O. Hwang, and K. D. Meisheri. The mechanism of inhibitory action of diltiazem on vascular smooth muscle contractility. *J.Pharmacol. Exp. Ther.* 218:459–463, 1981.
79. van Breemen, C., O. Hwang, and B. Siegel. The lanthanum method. In: *Excitation–Contraction Coupling in Smooth Muscle,* edited by R. Casteels, T. Godfraind, and J. C. Ruegg. Amsterdam: Elsevier, 1977, pp 243–252.
80. van Breemen, C., and B. Siegel. The mechanism of α-adrenergic activation of the dog coronary artery. *Circ. Res.* 46:426–429, 1980.
81. van Breemen, C., F. Wuytack, and R. Casteels. Stimulation of ^{45}Ca efflux from smooth muscle cells by metabolic inhibition and high K depolarization. *Pflugers Arch.* 359:183–196, 1975.
82. Villamil, M., V. Rettori, and N. Yeyatil. Calcium exchange and in the arterial wall. *Am. J. Physiol.* 224:1314–1319, 1973.
83. Winquist, R. J. and E. P. Baskin. Calcium channels resistant to organic calcium entry blockers in a rabbit vein. *Am. J. Physiol.* 245:H1024–1030, 1983.

Chapter 4

Mechanisms of Smooth Muscle Relaxation

T. Godfraind and R. C. Miller

1. Introduction

Normally, contractions of smooth muscle, which follow activation of the ATPase properties of myosin by phosphorylation in the presence of calmodulin and Ca (Chapter 2), are terminated by withdrawal of the initiating stimulus followed by a fall in the free cytoplasmic Ca concentration, mediated by the physiological systems that usually maintain the cation at a low level. Relaxation (or inhibition of contraction) in the presence of a contractile stimulus, can be achieved by a reduction in free cytoplasmic Ca levels or by reducing the rate of myosin phosphorylation in the presence of Ca-calmodulin through phosphorylation of myosin light chain kinase by a cyclic AMP-dependent phosphokinase (Chapter 2).

The Ca responsible for the increase in free cytoplasmic levels is potentially available from the extracellular space and also from several compartments having various affinities for Ca (*88*). In many cells, particularly smooth muscle cells, there is a pool of Ca releasable by agonists such as noradrenaline, but not by depolarization (see below). The precise location of this Ca is unknown, but it is often assumed to be the sarcoplasmic reticulum because isolated vesicles that probably originate from the sarcoplasmic reticulum are able to take up calcium (*64, 70, 91*). In tissues such as liver, mitochondrial Ca may be involved (*66, 82*). The cell nucleus contains significant amounts of Ca and there are also numerous caveoli associated with the plasma membrane of many cells that may also be a source of Ca for contraction that is extracellular, but inaccessible to chelating agents such as EGTA (*45*). Most cells are surrounded by a glycocalix to which calcium is bound, perhaps providing a relatively high concentration of calcium close to the external surface of the plasma membrane that might be utilized by stimuli such as depolarization.

Intracellular free Ca levels could therefore be reduced by inhibiting the influx of extracellular Ca or the release of intracellular Ca, or by enhancing the activity of the normal Ca removing mechanisms (Chapter 2). A further possibility is that the catalytic ability of myosin light chain kinase might be reduced by agents that interfere with the binding of Ca to calmodulin or by phosphorylation by a cAMP dependent mechanism.

These possible mechanisms of relaxation, with reference to pharmacological agents thought to affect them are examined in this chapter.

2. Relaxation Mediated by Direct and Indirect Actions on Smooth Muscle

It is possible to abolish (by local anesthetics, adrenergic neurone blockers, or ganglion blockers) (*80*) the amount of excitatory neurotransmitters released from nerve terminals, and the interaction of excitatory transmitters or circulating hormones with their specific receptors on smooth muscle cells can be selectively antagonized (for example, alpha-adrenoceptor antagonists). Such possible means of relaxing smooth muscle will not be dealt with here.

2.1. Actions at the Cell Membrane

Physiological antagonism of contraction can occur via the interaction of a compound (endogenous or exogenous) with a specific membranal receptor that mediates relaxation of the smooth muscle in question (for example, by the stimulation of vascular or pulmonary beta-adrenoceptors, vascular histamine or bradykinin receptors, and so forth). In many cases such membranal interactions are correlated with changes in cellular levels of cyclic nucleotides that are thought to be involved in the regulation of intracellular Ca levels (see below). The effect of endogenous compounds interacting with specific rceptors to produce relaxation can be enhanced by reducing the effectiveness of the physiological systems that remove them from the vicinity of the receptor. For example, the removal of adenosine by uptake into smooth muscle cells can be inhibited selectively by dipyridamole, and the uptake of epinephrine and norepinephrine into long paranchyma by hydrocortisone. The movement of extracellular Ca, essential for contraction, through the sarcolemma of smooth muscle cells can be inhibited by compounds collectively called Ca entry blockers (for example cinnarizine, flunarizine, and nifedipine).

2.2. Intracellular Actions

Compounds may act intracellularly to change the rate of production or destruction of cyclic nucleotides by interacting with the enzymes involved in these processes. Production is stimulated by nitroglycerin, nitroprusside, and other nitroso-containing molecules, and destruction is inhibited by the xanthines. Another possibility is that compounds might compete with Ca intracellularly for binding sites on calmodulin. Such compounds might properly be called Ca antagonists, but are usually referred to as calmodulin inhibitors (e.g., chlorpromazine; see below).

2.3. Involvement of Endothelium in Relaxation of Vascular Tissue

It has recently been shown that isolated precontracted arterial preparations respond to acetylcholine with a relaxation only when the intimal cells are present (*31*). The identity of the chemical mediator, endothelial relaxing factor, released from the endothelial cells is unknown, but seems not to be a product of cyclo-oxygenase activity since relaxation is unaffected by indomethacin or aspirin (*86*). The relaxant effect of acetylcholine is however sensitive to ETYA (*5,8,11,14*-eicosatetraynoic acid), that inhibits both cyclo-oxygenase and lipoxygenase pathways of arachidonic acid metabolism (*29*), and to mepacrine, an inhibitor of phospholipase A_2 (*38*) which catalyses the liberation of arachidonic acid.

It is now apparent that methacholine, substance P, and the Ca ionophore A 23187 are also dependent on the integrity of the endothelium to produce arteriodilatation (*15, 84, 92, 93*). However, bradykinin has been reported to produce relaxation of vascular tissue dependent on the endothelium and insensitive to indomethacin in arteries of some species but not others when it stimulates prostaglandin synthesis and also produces relaxation sensitive to indomethacin (*1, 14, 15, 83*). Other mediators of vascular relaxation such as ATP and ADP are partly dependent on the endothelium and vasoactive intestinal peptide, glycerol trinitrate, sodium nitrite, sodium azide, AMP, adenosine, and isoprenaline exert a direct relaxant effect on the smooth muscle in vessels so far examined (*15, 19, 22, 32*).

The role of the endothelium as a mediator of relaxation is therefore complex and must be investigated for each agonist in a variety of vascular tissues from a number of species before any definite picture is likely to emerge.

Some vascular smooth muscles contract in the presence of cyclo-oxygenase inhibitors (*44*) or can be relaxed by arachidonic acid, the relaxation being reversed by indomethacin (*17*). Isolated coronary arteries sometimes display rhythmic contractile activity that may be a prelude to coronary spasm (*43, 44*). Such activity in contracted sheep coronary vessels can be inhibited by indomethacin (*17*), whereas spontaneous contractile activity of human coronary arteries is depressed by arachidonic acid (*44, 56, 57*). These observations indicate the production of an important endogenous vasodilatator modulating vascular reactivity. The major metabolite of arachidonic acid metabolism in vascular tissue has been identified as prostacyclin (*74*), which is a potent inhibitor of contraction of smooth muscle of many, but not all, origins (*63*). Prostaglandin synthetase activity is associated with the endothelium and subendothelium of blood vessels (*60, 62, 78, 89*), and as well as synthesizing prostacyclin from endogenous arachidonic acid, endothelial cells may be able to convert into prostacyclin the short-lived prostaglandin endoperoxides liberated during platelet aggregation. The relaxant effect of prostacyclin and other prostaglandins on smooth muscle probably occurs via a specific receptor mediating stimulation of cAMP production (*63*).

3. Mechanisms of Relaxation

This section surveys in more detail the interaction of drugs with particular cellular processes that have the potential to mediate relaxation. In many cases direct convincing evidence of the involvement of particular processes is lacking, and no known drugs have been suggested to exert a relaxant effect by the stimulation of the physiological processes removing Ca from the cell. Some emphasis is therefore placed on the Ca entry blockers that have a clearly defined mode of action.

3.1. Relaxation Through Interaction with Cyclic Nucleotides

Much evidence is available to support the idea that many cell processes are regulated by cAMP-dependent phosphorylation of proteins and that cellular levels of cAMP can be influenced by the interaction of various drugs and hormones with membranal receptors (*49, 50, 54, 67*). However, much confusion has arisen over the sometimes apparent lack of correlation between the effect of different agonists on cellular cAMP levels in smooth muscle and the corresponding state of contraction of the tissue (*20, 21*). These discrepancies may possibly reflect a compartmentation of cAMP in the cell (*71*).

The enzyme adenylate cyclase, mostly located in association with plasma membranes, catalyzes the production of cAMP from Mg-ATP or Mn-ATP. An agonist-sensitive adenylate cyclase system is composed of at least three separable components: the agonist receptor, adenylate cyclase, and a guanine-nucleotide-binding protein that regulates the rate of enzyme hydrolysis of ATP to cAMP (*5, 72*). However, there are two types of guanine–nucleotide-binding proteins or nucleotide regulatory units that, when activated, exert a stimulatory (N_s) or inhibitory (N_i) effect on the enzyme adenylate cyclase. It is probable that the separate regulatory units are associated with separate membranal receptors (*72*). Each regulatory unit in its basal, nonactivated state binds GDP, and the associated adenylate cyclase converts ATP to cAMP at some basal rate. Receptor activation increases the affinity of the appropriate nucleotide regulatory unit for GTP, which when bound, exerts an inhibitory effect (GTP-N_i) or a stimulatory effect (GTP-N_s) on the associated adenylate cyclase, thus decreasing or increasing the rate of production of cAMP (*6, 72*). If cAMP levels rise, there may be a direct inhibitory effect on contraction (Chapter 2).

Cyclic AMP might also induce relaxation by stimulating the ATP-dependent uptake of Ca into cell structures such as sarcoplasmic reticulum (*2, 7, 25*). In some tissues such a correlation is not always evident (*3, 16, 75*), and cGMP may be the mediator (*68*).

Agonists such as acetylcholine, histamine, norepinephrine, serotonin, and bradykinin can increase cGMP levels in various tissues (*48*), but it is not known if such changes in cGMP in vascular smooth muscle are a consequence of the same receptor interactions that promote the stimulation or inhibition of adenylate cyclase activity or not, and may indeed be secondary to the effects of these agents on endothelial cells (see above). These agonists increase cGMP levels through a mechanism that may depend on Ca (*48, 77*), though other agonists such as nitroglycerin and sodium nitroprusside also increase cGMP levels (*8, 65*), but by a Ca-independent mechanism, possibly involving oxidation of guanylate cyclase. An oxidation mechanism of activating guanylate cyclase may be an important step leading to muscle relaxation mediated by some oxidants that can generate hydroxyl-free radicals (*65*) and perhaps by lipid peroxidation products from fatty acids (*50*).

The criteria that must be fulfilled before cyclic nucleotides can be implicated in the regulation of smooth muscle contraction have been reviewed by Hardman (*50*) and, in general, conclusive evidence of a correlation is lacking. A correspondence between receptor stimulation, increased cAMP production and smooth muscle relaxation exists for adrenoceptor-mediated relaxation of rat vascular smooth muscle. In this tissue, reponsiveness to isoproterenol decreases with age (*27, 34, 35*) and this has been correlated with a decline in beta-adrenoceptor number (*28*) and a decrease in isoproterenol-stimulated cAMP production (*23, 76*). Similarly, in hypertensive rats arterial responsiveness to isoproterenol and the associated increase in cAMP levels are diminished (*12*). However these membranal changes do not appear to alter the inhibitory activity of such agonists as nitroglycerin, sodium nitrite, adenosine, papaverine and PGE_1 (*28, 51, 84*) and the relaxant effect of adenosine has been linked to stimulation of cAMP via specific receptors in coronary arteries (*55*). It has recently been reported that isoproterenol-induced relaxations are correlated with increased cAMP levels and an inhibition of Ca influx produced by K^+ depolarization (*61*), which may imply a direct effect of cAMP on the mechanism mediating the increase in membrane permeability to Ca.

The cyclic nucleotides are degraded intracellularly by cyclic nucleotide phosphodiesterases of which several molecular forms with different substrate specificities have been described (*81, 90*). These enzymes also exhibit varying sensitivities to activation by calmodulin and may be compartmentalized within the cell (*71*). The xanthines, perhaps the most utilized phosphodiesterase inhibitors, have other actions such as inhibition of adenosine uptake and of extraneuronal uptake of catecholamines (*18, 55*). It is therefore difficult to predict the effects of phosphodiesterase inhibitors on contracted smooth muscle and as yet no compound can be said with certainty to produce relaxation by this mechanism alone.

3.2. Relaxation by Interaction with Calmodulin Entry

Many neuroleptics, such as chlorpromazine, haloperidol, sulpiride, desipramine, and trifluoperazine, some of which also produce vascular relaxation (*36*), have been shown to bind to calmodulin and to inhibit calmodulin-dependent enzymes such as adenylate cyclase, as has the coronary dilatator prenylamine (*24, 52, 53, 58, 59*). However, the IC_{50} concentrations for inhibition are in the 10–100 μ*M* concentration range (*59*) and the activity of the neuroleptics as inhibitors of calmodulin does not seem to be related to their clinical efficacity (*69*). The ability of chlorpromazine and prenylamine to inhibit vascular contractions depends on the contractile agent (*4*) and trifluoperazine does not inhibit portal vein contractions, indicating that an interaction with calmodulin is unlikely to be their primary mode of action.

In intact smooth muscle a calmodulin antagonist would be expected to produce relaxation correlated with an increased cellular Ca^{2+} content since the sarcolemmal Ca–Mg ATPase which removes Ca from the cell (Chapter 2) would also be inhibited. To date no such observation has been reported.

It has been suggested that felodipine, a structural analog of the Ca entry blocker nifedipine (*37*), does not reduce Ca entry into rat portal vein but does bind to calmodulin, inhibition of which may be a likely mechanism of action (*10*). However, the published data show that felodipine inhibits the spontaneous activity of the portal vein and the associated changes in membrane polarity that are caused by an increase in membrane Ca permeability (*9*). Also the Ca influx measurements were made in unstimulated tissue, which would not be expected to exhibit a sensitivity to Ca entry blockers that only inhibit stimulated Ca influx (see below). Furthermore, the IC_{50} value of felodipine against spontaneous activity and norepinephrine-induced contractions is about 1 n*M* in this tissue, but felodipine reacted with calmodulin at much higher concentrations, in the millimolar range (the cellular concentration of calmodulin is thought to be in the μ*M* range). It has also been shown that neither cinnarizine, flunarizine, nor nifedipine inhibit the norepinephrine-stimulated efflux of Ca from rat aorta (*37, 38, 43, 45*) that is dependent on the activity of a calmodulin-activated (Mg–Ca)-ATPase. A recent report (*79*) shows that 100 μ*M* concentrations of the Ca entry

Table 1
Concentrations of Calcium Entry Blockers Producing a 50% Inhibition of Contractions of Rat Aorta

	Cinnarizine	Flunarizine	Nifedipine
IC_{50} (*M*)	2.7×10^{-8}	1.9×10^{-8}	2.5×10^{-9}
RC_{50} (*M*)	2.5×10^{-8}	2.2×10^{-8}	1.4×10^{-9}

[a]Induced by 35 min exposure to 100 m*M* K^+ after a preincubation of 90 min with the blocker (IC_{50}) and the concentrations producing a 50% relaxation of depolarization contracted arteries (RC_{50})

blockers, cinnarizine, flunarizine, and fendilin, will inhibit contraction of skinned smooth muscle fibers to varying extents, but verapamil, diltiazem, and nifedipine do not. The IC_{50} values for such Ca entry blockers range from about 2–7 n*M* (*26, 37*) for nifedipine to about 30–50 n*M* for cinnarizine and flunarizine (Table 1) (*38, 41, 43, 45*). It is therefore unlikely that inhibition of calmodulin is an important site of action of Ca antagonists either.

3.3. Relaxation Through Blockade of Calcium Entry

The term "Ca entry blocker" has been recently introduced in order to specify among the group of Ca antagonists those whose pharmacological actions result from the blockade of Ca entry through membrane channels opened during cell excitation (*36*). The concept of Ca^{2+} antagonism emerged from early studies with the antianginal drug, lidoflazine, and with cinnarizine, which was introduced as an antihistaminic (*40, 47*). Ca entry blockers have a broad spectrum of clinical use mostly in the cardiovascular field.

3.3.1. Actions of Drugs on Muscle Contraction and Inhibition of Calcium Fluxes

A comparison of the action of putative Ca entry blockers on the response to norepinephrine in the presence and absence of external Ca and on the response to Ca in depolarized muscle gives valuable indications as to the mode of action of the blockers. This is illustrated by the study of Broekaert and Godfraind (*11*), who compared the inhibitory effect of cinnarizine and papaverine on norepinephrine and Ca-induced contractions of rabbit thoracic aorta and mesenteric artery. They observed that the cinnarizine concentration inhibiting contractions evoked by Ca^{2+} in depolarized arteries was 200-fold lower in mesenteric artery than in aorta. Papaverine produced the same action, but was less potent than cinnarizine and it showed no tissue selectivity. These actions of cinnarizine can be attributed to blockade of Ca entry since the drugs block depolarization-dependent ^{45}Ca entry into the muscle cell as measured by ^{45}Ca influx into the compartment resistant to dixplacement by lanthanum.

In a Ca-free medium, 10 μM norepinephrine evokes a fast nonsustained contraction of rabbit arteries; after the readmission of Ca, a slow sustained contraction develops. The action of papaverine and cinnarizine on the two components of the contractile response was studied in the mesenteric artery using concentrations of inhibitors that reduced by 50% the contraction evoked by norepinephrine. Cinnarizine selectively reduced the extracellular Ca-dependent response, whereas papaverine inhibited the initial fast response more than the extracellular Ca-dependent one. This action of papaverine was confirmed in rabbit aorta, where cinnarizine is ineffective as an inhibitor of norepinephrine-induced contractions. Such results indicate that cinnarizine and papaverine antagonize norepinephrine action by different mechanisms. Cinnarizine seems to act only by blockade of Ca entry (see below). Papaverine inhibits contraction not only by an action similar to that of cinnarizine, but also by an additional action. Inhibition of phosphodiesterase is the most likely intracellular action and may be responsible for a rise in cellular cAMP content. An increase in cAMP may enhance the accumulation of Ca in endoplasmic reticulum (Chapter 2) and allow the phosphorylation of myosin light chain kinase. These two effects will reduce the amplitude of contraction. More recently, studying the action of the Ca-antagonist diltiazem, Saida and van Breemen have observed that at concentrations higher than those required to block Ca entry, the drug also inhibits the contraction evoked by intracellular calcium release (*73*). Since it is unlikely that this action is related to cyclic nucleotides, it could result from inhibition of the release of intracellular calcium.

Rat aorta is another preparation that can be used to characterize the membranal or intracellular site of action of such drugs. In this tissue, norepinephrine-evoked sustained contractions are dependent on the Ca concentration of the perfusion fluid. Ca entry blockers inhibit that part of the contraction dependent on extracellular Ca, but do not alter the contraction evoked by norepinephrine in Ca-free medium (*36–43*).

Receptor–response coupling might occur through the opening of membrane Ca channels and/or the release of intracellularly stored Ca. The relative importance of these two processes may be estimated by the ratio between that part of the response sensitive to Ca entry blockade and that part of the response resistant to Ca entry blockers.

In rat aorta, studies with several Ca entry blockers have shown that maximal blockade of norepinephrine effects has the following features (*43*): it does not attain the reduction observed in Ca-free solution; there is an incomplete blockade of norepinephrine-dependent Ca entry; and there is no marked reduction of intracellular Ca release even in Ca-free solution (Fig. 1). The type of alpha-adrenoceptor responsible for the effects of norepinephrine is not known with certainty. However, it has been observed that the alpha$_2$-agonists, oxymetazoline and clonidine, evoked contractions that were only supported by Ca entry and were completely blocked by cinnarizine (*41, 45*). The refilling of intracellular Ca stores is likely to occur through leak channels weakly sensitive to Ca entry blockers (*38*) and this could explain how the smooth muscle can maintain a sustained contraction when norepinephrine-dependent Ca influx is blocked. That blockade of Ca entry is responsible for inhibition of contraction is shown by superimposition of dose–inhibition curves of contraction and ^{45}Ca entry. This has been reported for cinnarizine (*33, 45*), flunarizine (*38*), diltiazem (*85*), and nifedipine (*37, 43*) with noradrenaline, PGF$_2\alpha$ and K-depolarization evoked contractions.

3.3.2. Relaxant Effect of Calcium Entry Blockers and Calcium Antagonists

Ca entry blockers not only inhibit contraction of smooth muscle as shown in the preceding section, but they also relax the muscle once the contraction is extablished. This observation was made by Godfraind and Kaba with cinnarizine (*39*), the first drug reported as being a Ca antagonist (*47*). In arterial smooth muscle contracted by depolarization, cinnarizine decreased the contractile tension and this relaxant effect increased with time and with increasing cinnarizine concentration. The residual muscle tension remaining when relaxation reached a plateau with a given concentration of cinnarizine, was approximately equal to that achieved by antagonism of the induction of contraction after preincubation cinnarizine for a time sufficient to achieve equilibrium (*39, 43*).

The rate of relaxation for a given drug may depend on the concentration used. When different drugs are compared, it also depends on the compound itself. This is illustrated by a comparison made in aorta between nifedipine and flunarizine, the latter showing a much slower rate of action (Fig. 2). Such experiments

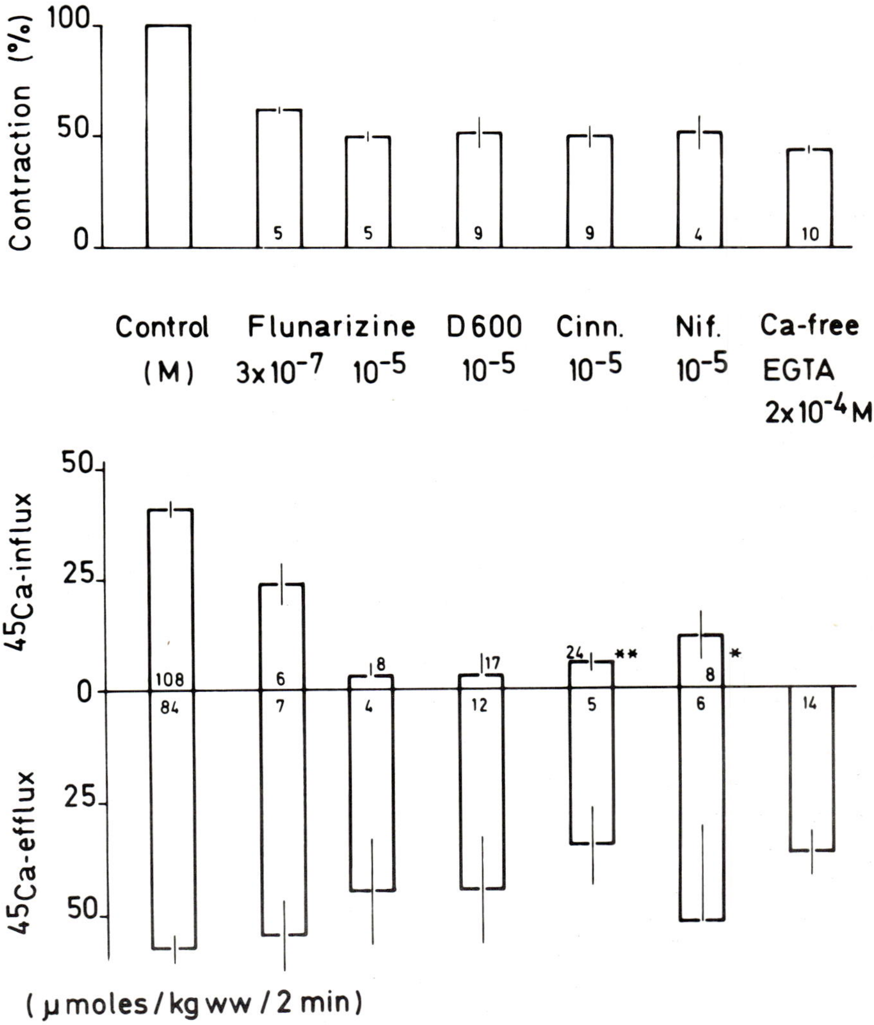

Fig. 1. Effect of calcium-free solution and of calcium entry blocking drugs, on contraction (upper panel) and ^{45}Ca influx and efflux (lower panel) produced by norepinephrine 10^{-5} *M* in rat aorta. Tissues were incubated with Ca entry blockers for 90 min before contractions were induced with norepinephrine. After washing the aorta in Ca^{2+}-free solution for 5 min, the tissue does not contract in response to depolarization (100 m*M* K^+). All contractions are expressed as % of the maximal contraction induced by norepinephrine in the absence of blocker. Numbers in columns equal number of determinations. Vertical bars represent SEM. (Modified from ref. *54*.)

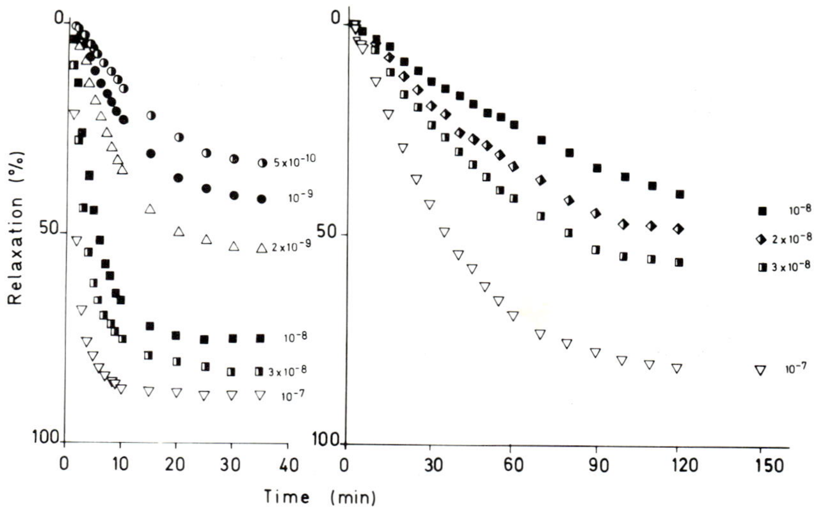

Fig. 2. Time-dependent relaxation of rat aorta produced by flunarizine (right) and nifedipine (left). Aortic segments maximally contracted by depolarization (100 m*M* KCl) were exposed at time 0 to flunarizine or nifedipine at the indicated final concentrations and relaxation was followed for the next 150 min. Points at each concentration are the means of at least three determinations; the concentration of flunarizine and nifedipine are indicated on the graph (modified from ref. *54*).

allow a study to be made of the reversal of relaxant effect as shown in Fig. 3 for nifedipine action in rat aorta. There is other quantitative evidence showing that relaxation of contraction and inhibition of contraction are caused by the same mechanism. The most convincing is a comparison of the concentration required to inhibit by 50% an induced contraction, to that producing a 50% relaxation of a contraction (Table 1). The relaxant effect of Ca entry blockers appears therefore to result from the association of two processes: blockade of Ca entry and a normal functioning of the Ca extrusion pump, both together producing a reduction of calcium available to the contractile machinery. Because the Ca pump is not affected by specific Ca entry blockers, the rate of relaxation is dependent on the rate of interaction of the drug with Ca channels. As shown in Table 1, RC_{50} values are similar to IC_{50} values measured after completion of the use-dependency, the Ca channel activated by depolarization being the most sensitive to the blocker.

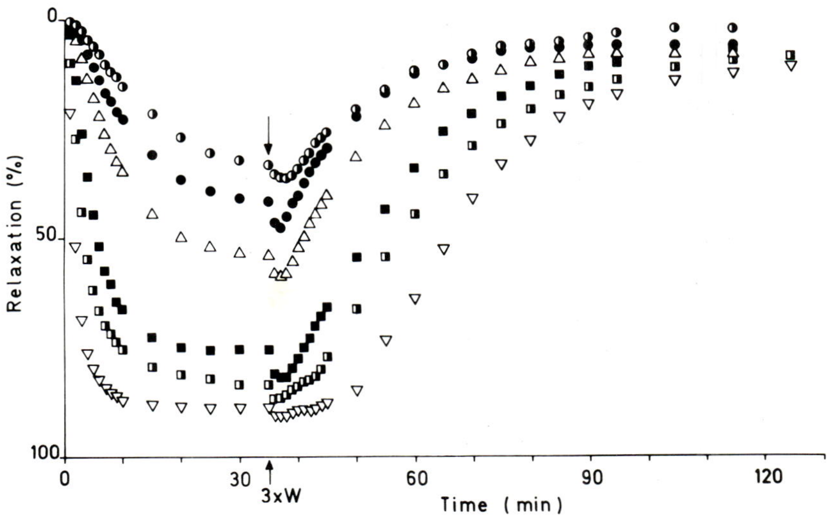

Fig. 3. Reversibility of the relaxing effect of nifedipine on rat aorta contracted by depolarization (100 m*M* KCl). Nifedipine was added to the bath at time 0 at the indicated concentrations when the depolarization-induced contraction was maximal. After 35 min, when the nifedipine-induced relaxation was maximal, the tissues were washed with depolarizing solution and the resulting contractions followed for the next 95 min. Each point represents the mean of at least five determinations for nifedipine at the following concentrations:
5×10^{-10} *M* (◑); 10^{-9} *M* (●); 2×10^{-9} *M* (△); 10^{-8} *M* (■); 3×10^{-8} *M* (◨); 10^{-7} *M* (▽).

3.3.3. Ca Channel Specificity

It has been proposed (*6*) that Ca channels could be divided into leak, potential-operated, and receptor-operated, but attempts to label Ca channels with radioactive calcium blockers have not yet differentiated a binding heterogeneity (*46*). In a given artery such as the rat aorta, the calcium entry blockers, nifedipine, lidoflazine, and flunarizine, are more potent as inhibitors of calcium influx and contraction stimulated by depolarization than they are as inhibitors of norepinephrine-stimulated responses when measured at equilibrium (*38, 43*). Such observations indicate that even if there is only one macromolecular structure for Ca channels, it is possible to distinguish, functionally and pharmacologically, between different Ca gating mechanisms. The mechanism sensitive to membrane polarity is differently affected from that activated by norepinephrine, and that activated by $PGF_2\alpha$ appears to be different from both of the others. Similar differences can be detected for other vessels and agonists (*42, 43*).

A feature of the inhibitory effect of Ca entry blockers on depolarization-induced contractions of vascular tissue is use dependency. That is, the time course of the inhibitory effect of Ca entry blockers is markedly different when contractions are induced by depolarization or by agonists. In the former case the degree of inhibition of the response increases with the duration of depolarization to reach a plateau, but this effect is not seen when contractions are induced by agonist (Fig. 4) (*38*). Different Ca entry blockers exhibit different degrees of use dependency as is illustrated for cinnarizine, flunarizine, D600, and nifedipine (Fig. 5). The degree of contraction measured at different time intervals is variable, the maximal change being most evident with flunarizine and least evident with nifedipine (*43*). This depression of contraction is correlated with a depression of Ca influx measured at the same time intervals (*45*) (Fig. 6). These observations also suggest that depolarization and receptor stimulation activate different Ca gating mechanisms, having different sensitivities to

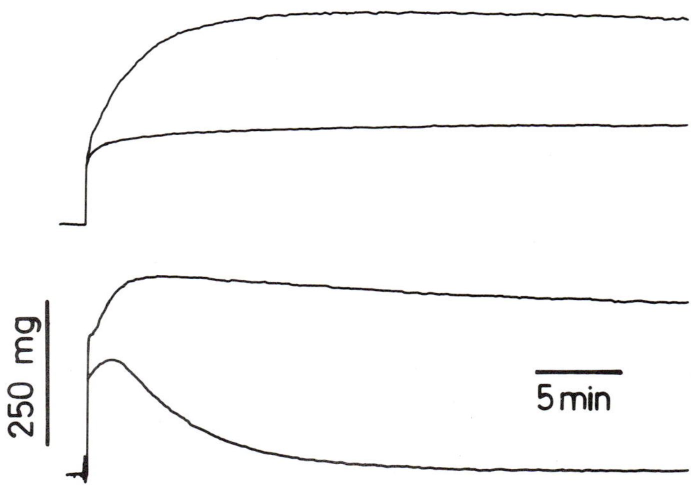

Fig. 4. Representative traces showing the depressant effect of 0.03 μ flunarizine on contractions of rat mesenteric artery stimulated by norepinephrine 10 μ*M* (upper traces) and depolarization in 100 m*M* KCl containing physiological solution (lower traces). Note the absence of use-dependency in the inhibitory effect when the artery is contracted by norepinephrine, and its presence in the response of depolarization. (Modified from ref. *52*.)

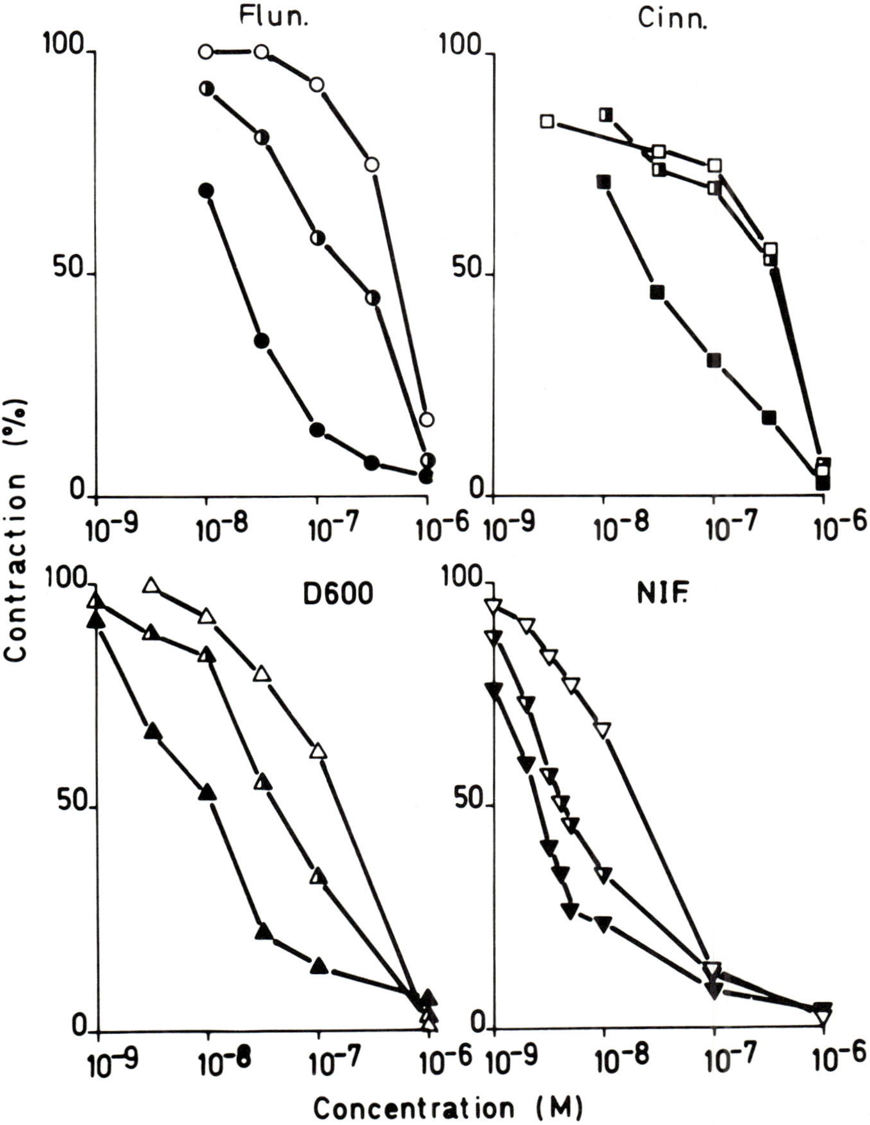

Fig. 5. Inhibition by Ca entry blockers of maximal depolarization (100 m*M* K^+) induced contractions in rat aorta. Tissues were incubated with the drugs for 90 min before contractions were induced. The degree of contraction in the presence of the drugs was measured after 30 (open symbols), 2 min (black and white) and 35 min (black), and expressed as a percentage of the maximal contraction obtained in the absence of Ca entry blockers after the same time intervals. Each point is the mean of at least three determinations. Flun = flunarizine; Cin = cinnarizine; Nif = nifedipine (modified from ref. *54*).

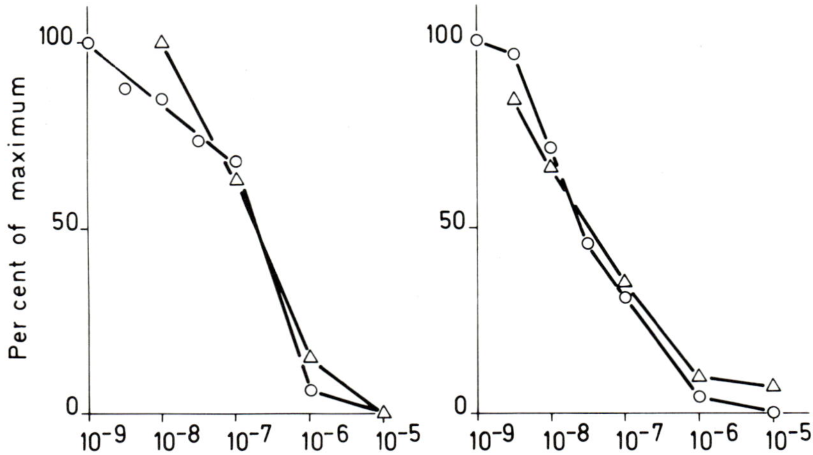

Fig. 6. Correlation of depression of contraction and depression of Ca^{2+} influx. The depressant effect of cinnarizine on KCl depolarization-induced contractions (○) and KCl-dependent uptake of ^{45}Ca (△) in the rat aorta. Responses were measured after 2 min (left) and after 35 min (right) exposure to the depolarizing solution. The preparations were preincubated for 90 min in the presence of cinnarizine. Each point is the mean of at least 5 observations. The standard error at any point did not exceed 10% of the calculated mean (modified from ref. 56).

the blockers, or that separate Ca channels, having differing characteristics, are activated.

3.3.4. Tissue Selectivity

Results obtained with flunarizine on rat aorta and mesenteric vessels show that the sensitivity of the contractile response to the blocker is much higher in mesenteric artery than in aorta (*38*). This observation established the concept of a difference in sensitivity according to the anatomical origin of the vessel, a concept not only documented within the group of diphenylpiperazines, but also within the group of dihydropyridines. In terms of Ca channels, this means that Ca channel properties could be different according to the smooth muscle studied. Thus, tissue selectivity means that a given blocker exhibits different potencies against the same stimulus in different tissues (*38, 40*). For example, noradrenaline contracts both rat aorta and rat mesenteric arteries but flunarizine is about 10-fold more potent as a blocker in the mesenteric artery. Differences are also apparent in the sensitivity of the responses induced by depolarization (Table 2) (*38*). Nifedipine has a similar potency in each vessel as an inhibitor of contractions induced by a particular agonist.

Table 2
Concentrations (M) of Cinnarizine, Flunarizine and Nifedipine Producing 50% of the Maximal Reduction of the Contractile Response (IC_{50})[a]

	Rat aorta	Mesenteric artery	Ratio, aorta/ mesenteric artery
		Cinnarizine	
Depolarization	2.7×10^{-8}	9.1×10^{-8}	0.3
NOR	1.7×10^{-7}	6.0×10^{-8}	2.8
		Flunarizine	
Depolarization	1.9×10^{-8}	2.4×10^{-9}	9.5
NOR	2.0×10^{-7}	2.4×10^{-8}	8.3
		Nifedipine	
Depolarization	2.5×10^{-9}	3.8×10^{-9}	0.7
NOR	2.0×10^{-8}	1.3×10^{-8}	1.5

[a]Induced by depolarization (100 mM K^+), and norepinephrine (NOR) (10 μM) in rat aorta and mesenteric arteries. The ratio of IC_{50} values in the two tissues indicates the degree of selectivity of the blockers.

A further determinant of the overall sensitivity of smooth muscle to Ca entry blockers is the degree to which the contraction is dependent on the entry into the cytoplasm of extracellular Ca and as far as vascular smooth muscle is concerned very large differences have been described. Further, this degree of dependence on extracellular Ca depends not only on the origin of the vessel concerned, but also on the agonist. In the rat aorta about 40% of the contraction induced by norepinephrine is independent of extracellular Ca and in the mesenteric artery only about 15%, whereas in the rabbit aorta about 80% of the norepinephrine-induced contraction is resistant to cinnarizine (*39*, Table 3). This pool of Ca releasable by noradrenaline in rat aorta is not accessible to other alpha-adrenergic agonists such as clonidine or

Table 3
Approximate Percentage of Agonist-Induced Contraction Independent of Extracellular Ca in a Variety of Isolated Arteries.[a]

	NOR	5HT	$PGF_{2\alpha}$	Clonidine
Rat aorta	40	40	40	0
Rat mesenteric artery	10–15		0	0
Rabbit aorta	80–90			22
Human coronary artery	0	0	0	

[a]NOR = norepinephrine; 5HT = serotonin.

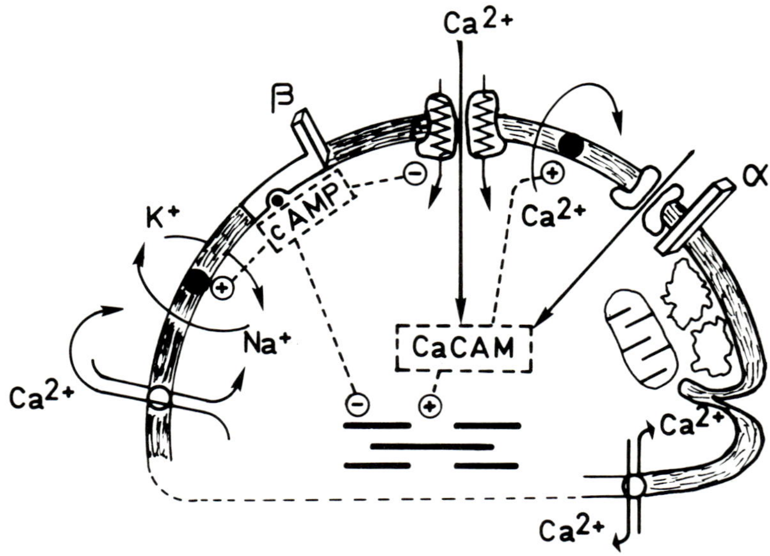

Fig. 7. Schematic representation of the cellular factors involved in smooth muscle relaxation.

From the left to the right:

1. Na-Ca exchange mechanism
2. Na-K pump
3. Beta-adrenoceptor coupled to adenylate cyclase responsible for the cyclization of ATP into cyclic AMP; the latter has a stimulatory action on the Na–K pump, a negative action on the contractile machinery and has been proposed to impair the opening of the potential operated channels.
4. Potential operated calcium channels; their opening is dependent upon the level of the membrane potential; once they are open, calcium entry occurs as a function of the calcium electrochemical gradient.
   Inside the cell, calcium forms a complex with calmodulin (CaCAM); this complex activates the contractile machinery and the calcium extrusion pumps.
5. ATP-dependent Ca pump stimulated by calmodulin.
6. Alpha-adrenoceptor associated with receptor-operated channel and with intracellular calcium stores, likely the endoplasmic reticulum. Once the channel is open, calcium entry is also passive (*see* 4).
7. The association formed by mitochondria and endoplasmic reticulum, responsible for intracellular calcium buffering.
8. Ca–Ca exchange mechanism occurring through a leak channel insensitive to calcium entry blockers.

Other explanations are given in the text.

oxymetazoline (*45*) and is less accessible to $PGF_{2}\alpha$ or serotonin (unpublished observations). Contractions of human coronary arteries also have a high dependence on extracellular Ca (Table 3) (*44*).

4. Concluding Remarks

The experimental data summarized in this chapter and schematized in Fig. 7 show that smooth muscle relaxation can be achieved by different mechanisms. To study the biochemical regulation of the processes involved in the cycle of contraction and relaxation, drugs having specific mechanisms of action are very useful. Among the various compounds so far examined, calmodulin antagonists, phosphodiesterase inhibitors, cyclic nucleotide promoters and Ca entry blockers, only the latter appear to interact with their target at concentrations much lower than those with which they interact with any other cellular process. However, as far as the general pharmacodynamic or therapeutic aspects are concerned, tissue specificity is a major determinant factor. Within the group of Ca entry blockers, it is now evident that some drugs show a higher selectivity for the Ca channels of a given group of smooth muscles, being much less active on other varieties. This is the case with cinnarizine and flunarizine, whose clinical uses are different from those of other Ca entry blockers such as nifedipine or verapamil. However, even for those drugs, the concentrations required to act on the heart are higher than those for reducing contraction of vessels (*87*). It may be anticipated that selectivity will also be found outside the cardiovascular system and be an important feature of pulmonary or gastrointestinal pharmacotherapy. It is now widely claimed that Ca entry blockers (Ca antagonists) are specific for potential operated channels, being less potent blockers of receptor operated channels (*13*). However, this is not true, neither for every drug nor for every vessel, and only more extensive studies will allow clarification of this question. It is most likely that important physiological and pharmacological information will be obtained by further studies on calcium antagonists.

References

1. Aiken, J. W. Inhibitors of prostaglandin synthesis specifically antagonize bradykinin and angiotensin-induced relaxations of the isolated celiac artery from the rabbit. *Pharmacologist* 16:295, 1974.

2. Alexander, R. W., J. A. Fontana, and W. Lovenberg. cAMP dependent protein kinase in rabbit aortic microsomes and its relation to Ca^{2+} uptake. *Federation Proc.* 32:711, 1973.
3. Allen, C. J. Ca^{2+}-binding properties of canine aortic microsomes: lack of effect of c-AMP. *Blood Vessels* 14:91–104, 1977.
4. Asano, M., Y. Suzyki, and H. Hidaka. Effects of various calmodulin antagonists on contraction of rabbit aortic strips. *J. Pharmacol. Exp. Ther.* 220:191–196, 1982.
5. Aurbach, G. D. Polypeptide and amine hormone regulation of adenylate cyclase. *Ann. Rev. Physiol.* 44:653–666, 1982.
6. Berridge, M. J. Receptors and calcium signalling. *TIPS* 1:419–424, 1980.
7. Bhalla, R. C., R. C. Webb, D. Singh, and T. Brock. Role of cyclic AMP in aortic microsomal phosphorylation and calcium uptake. *Am. J. Physiol.* 234:H508–H514, 1978.
8. Bohme, E., H. Graf, and G. Schultz. Effects of sodium nitroprusside and other smooth muscle-relaxants on cyclic GMP formation in smooth muscle and platelets. *Adv. Cyclic Nucleotide Res.* 9:131–143, 1978.
9. Bolton, T. B. Mechanisms of action of transmitters and other substances on smooth muscle. *Physiol. Rev.* 59:606–718, 1979.
10. Boström, S. L., B. Ljung, S. Mardh, S. Forsen, and E. Thulin. Interaction of the antihypertensive drug felodipine with calmodulin. *Nature* 292:777–778, 1981.
11. Broekaert, A., and T. Godraind. A comparison of the inhibition by cinnarizine and papaverine of the noradrenaline evoked contraction of isolated rabbit vessels in presence and absence of extracellular calcium. *Eur. J. Pharmacol.* 53:281–288, 1979.
12. Bucher, B., C. Heitz, and J. C. Stoclet. *In vivo* decrease of adenylate cyclase responsiveness to isoproterenol in the spontaneously hypertensive rat. *Biochem. Pharmacol.* 30:2583–2506, 1981.
13. Cauvin, C., R. Loutzenhiser, and C. van Breemen. Mechanisms of calcium antagonist-induced vasodilation. *Ann. Rev. Pharmacol. Toxicol* 23:373–396, 1983.
14. Cherry, P. D., R. F. Furchgott, and J. V. Zawadzki. The indirect nature of bradykinin relaxation of isolated arteries: endothelial dependent and independent components. *Federation Proc.* 40:689, 1980.
15. Cherry, P. D., R. F. Furchgott, J. V. Zawadzki, and D. Jothianandan. Role of endothelial cells in relaxation of isolated arteries by bradykinin. *Proc. Natl. Acad. Sci.* 79:2106–2110, 1982.
16. Clyman, R. I., V. C. Manganiello, C. J. Lovell-Smith, and M. Vaughan. Calcium uptake by subcellular fractions of human umbilical artery. *Am. J. Physiol.* 231:1074–1081, 1976.
17. Cornish, E. J., R. G. Goldie, E. V. Krstew, and R. C. Miller. Effect of indomethacin on responses of sheep isolated coronary artery to arachidonic acid. *Clin. Exp. Pharmacol. Physiol.* 10:171–175, 1983.
18. Cornish, E. J., Goldie, and R. C. Miller. Catecholamine uptake by isolated coronary arteries and atria of the kitten. *Br. J. Pharmacol.* 63:445–456, 1978.

19. De Mey, J. G., and P. M. Vanhoutte. Role of the intima in cholinergic and purinergic relaxation of isolated canine femoral arteries. *J. Physiol., London* 316:347–355, 1981.
20. Diamond, J. Evidence for dissociation between cyclic nucleotide levels and tension in smooth muscle. In: *The Biochemistry of Smooth Muscle,* edited by N. L. Stephens. Baltimore: University Park Press, pp 343–360, 1977.
21. Diamond, J. Role of cyclic nucleotides in control of smooth muscle contraction. In: *Advances in Cyclic Nucleotide Research,* edited by W. G. George and L. J. Ignarro: New York: Raven Press, Vol 9, p 327, 1978.
22. Duckles, S. P., and S. I. Said. Vasoactive intestinal peptide as a neurotransmitter in the cerebral circulation. *Eur. J. Pharmacol.* 78:371–374, 1982.
23. Ericsson, E. E. Adrenergic beta-receptor activity and cyclic AMP-metabolism in vascular smooth muscle, variations with age. *Scand. J. Clin. Lab. Invest.* 34:Suppl 141, 63, 1974.
24. Filburn, C. R., F. T. Colpo, and B. Sacktor. Mechanism of phenothiazine inhibition of Ca^{2+}-dependent guanosine 3′,5′-(cyclic) monophosphate phosphodiesterase of brain. *Mol. Pharmacol.* 15:257–262, 1979.
25. Fitzpatrick, D. F., and A. Szentivanyi. Stimulation of calcium uptake into aortic microsomes by cyclic AMP and cyclic AMP dependent protein kinase. *Naunyn-Schmiedeberg's Arch. Pharmacol.* 298:255–257, 1977.
26. Fleckenstein, A. Specific pharmacology of calcium in myocardium, cardiac pacemakers, and vascular smooth muscle. *Ann. Rev. Pharmacol. Toxicol.* 17:149–166, 1977.
27. Fleisch, J. H. Further studies on the effect of aging on beta-adrenoceptor activity of rat aorta. *Br. J. Pharmacol.* 42:311–313, 1971.
28. Fleisch, J. H. Age-related changes in the sensitivity of blood vessels to drugs. *Pharmacol, Ther.* 8:477–487, 1980.
29. Flower, R. J. Drugs which inhibit prostaglandin biosynthesis, Pharmacol. Rev. 26:33–67, 1974.
30. Flower, R. J. and C. J. Blackwell. The importance of phospholipase-A_2 in prostaglandin biosynthesis. *Biochem. Pharmacol.* 25:285–291, 1976.
31. Furchgott, R. F., and J. V. Zawadzki. The obligatory role of endothelial cells in the relaxation of arterial smooth muscle by acetylcholine. *Nature* 288:373–376, 1980.
32. Furchgott, R. F., and J. V. Zawadzki. ATP relaxes rabbit aortic smooth muscle by both an indirect action via endothelial cells and a direct action. *Pharmacologist* 22:271, 1980.
33. Godfraind, T. The action of cinnarizine and of phentolamine on the noradrenaline-depressant calcium influx in vascular smooth muscle. *Br. J. Pharmacol.* 52:120P, 1974.
34. Godfraind, T. Isoprenaline relaxation in vascular smooth muscle of aged rats: effect of flunarizine. *Arch. int. Pharmacodyn. Ther.* 230:331, 1977.
35. Godfraind, T. Alternative mechanisms for the potentiation of the relaxation evoked by isoprenaline in aortae from young and aged rats. *Eur. J. Pharmacol.* 53:273–279, 1979.

36. Godfraind, T. Mechanism of action of calcium entry blockers. *Federation Proc.* 40:2866–2871, 1982.
37. Godfraind, T. Actions of nifedipine on Ca fluxes and contraction in isolated arteries. *J. Pharmacol. Exp. Ther.* 224:443–450, 1983.
38. Godfraind, T., and D. Dieu. The inhibition by flunarizine of the norepinephrine-evoked contraction and calcium influx in rat aorta and mesenteric arteries. *J. Pharmacol. Exp. Ther.* 217:510–515, 1981.
39. Godfraind, T., and A. Kaba. Blockade or reversal of the contraction induced by calcium and adrenaline in depolarized arterial smooth muscle. *Br. J. Pharmacol.* 36:549–560, 1969.
40. Godfraind, T., A. Kaba, and P. Polster. Differences in sensitivity of arterial smooth muscles to inhibition of their contractile response to depolarization by potassium. *Arch. int. Pharmacodyn. Ther.* 172:235–239, 1968.
41. Godfraind, T., and R. C. Miller. Actions of $PGF_{2\alpha}$ and noradrenaline on Ca exchange and contraction in rat mesenteric arteries and their sensitivity to calcium entry blockers. *Br. J. Pharmacol.* 75:229–236, 1982.
42. Godfraind, T., and R. C. Miller. The study of receptors and Ca movements in smooth muscle as a model for presynaptic events. In: *Neurotransmitter Interactions and Compartmentation,* edited by H. F. Bradford: New York, London: pp 121–139, 1982.
43. Godfraind, T., and R. C. Miller. Specificity of action of Ca entry blockers, a comparison of their actions in rat arteries and in human coronary arteries. *Circulation Res.* 52:Suppl. 1, 181–191, 1983.
44. Godfraind, T., and R. C. Miller. Pharmacology of coronary arteries. In: *Factors Influencing the Course of Myocardial Ischemia,* edited by Gotto and DeBakey: Amsterdam–New York–Oxford, Elsevier Biomedical Press, pp 101–119, 1983.
45. Godfraind, T., R. C. Miller, and J. Socrates Lima. Selective $alpha_1$-and $alpha_2$-adrenoceptor agonist-induced contractions and ^{45}Ca fluxes in the rat isolated aorta. *Br. J. Pharmacol.* 77:597–604, 1982.
46. Godfraind, T., and N. Morel. Identification of the specific binding of flunarizine to rat aorta. *Br. J. Pharmacol.* 75:517P, 1981.
47. Godfraind, T., and P. Polster. Etude comparative de médicaments inhibant la réponse contractile de vaisseaux isolés d'origine humaine ou animale. *Thérapie* 23:1209–1220, 1968.
48. Goldberg, N. D., and M. K. Haddox. Cyclic GMP metabolism and involvement in biological regulation. *Ann. Rev. Biochem.* 46:823–896, 1977.
49. Greengard, P. Cyclic nucleotides, phosphorylated proteins and neuronal function. *Distinguished lecture series of the Society of General Physiologists,* Vol 1, New York: Raven Press, 1978.
50. Hardman, J. G. Cyclic nucleotides and smooth muscle contractions: some conceptual and experimental considerations. In: *Smooth Muscle,* edited by E. Bulbring, A. F. Brading, A. W. Jones, and T. Tomita. London: Edward Arnold, pp 249–262, 1981.
51. Hayashi, S., and N. Toda. Age related changes in the response of rabbit isolated aortae to vasoactive agents. *Br. J. Pharmacol.* 64:229–237, 1978.

52. Hidaka, H., T. Yamaki, M. Naka, T. Tanaka, H. Hayashi, and R. Kobayashi. Calcium-regulated modulator protein interacting agents inhibit smooth muscle calcium-stimulated protein kinase and ATPase. *Mol. Pharmac.* 17:66–72, 1980.
53. Hidaka, H., T. Yamaki, T. Totsuka, and M. Asano. Selective inhibitors of Ca^{2+}-binding modulators of phosphodiesterase produce vascular relaxation and inhibit actinomyosin interaction. *Mol. Pharmacol.* 15:49–59, 1979.
54. Krebs, E. G., and J. A. Beavo. Phosphorylation-dephosphorylation of enzymes. *Ann. Rev. Biochem.* 48:923–959, 1979.
55. Kukovetz, W. R., A. Wurm, I. Rinner, S. Holzmann, and G. Poch. Stimulation of adenylcyclase in coronary smooth muscle by adenosine. In: *Excitation-contraction Coupling in Smooth Muscle,* edited by R. Casteels, T. Godfraind, and J. C. Rüegg. Amsterdam: Elseviar Biomedical Press, pp 399–406, 1977.
56. Kulkarni, P. S., R. Roberts, and P. Needleman. Endogenous synthesis of a labile coronary dilating substance from arachidonic acid. *Federation Proc.* 35:298, 1976.
57. Kulkarni, P. S., R. Roberts, and P. Needleman. Paradoxical endogenous synthesis of a coronary dilating substance from arachidonate. *Prostaglandins* 12:337–353, 1976.
58. Levin, R. M., and B. Weiss. Binding of trifluoperazine to the calcium-dependent activator of cyclic nucleotide phosphodiesterase. *Mol. Pharmacol.* 13:690–697, 1977.
59. Levin, R. M., and B. Weiss. Selective binding of antipsychotics and other psychoactive agents to the calcium-dependent activator of cyclic nucleotide phosphodiesterase. *J. Pharmacol. Exp. Ther.* 208:454–459, 1979.
60. MacIntyre, D. E., J. D. Pearson, and J. L. Gordon. Localization and stimulation of prostacyclin production in vascular cells. *Nature* 271:549–551, 1978.
61. Meisheri, K. D., and C. van Breemen. Effects of beta-adrenergic stimulation on calcium movements in rabbit aortic smooth muscle: relationship with cyclic AMP. *J. Physiol. London* 331:429–441, 1982.
62. Moncada, S., A. G. Herman, E. A. Higgs, and J. R. Vane. Differential formation of prostacyclin (PGX or PGI_2) by layers of the arterial wall. An explanation for the anti-thrombotic properties of vascular endothelium. *Thromb. Res.* 11:323–344, 1977.
63. Moncada, S., and J. R. Vane. Arachidonic acid metabolites in the cardiovascular system. In: *Cell Membrane in Function and Dysfunction of Vascular Tissue,* edited by T. Godfraind, and P. Meyer. Holland: Elsevier Biomedical Press, pp 105–126, 1981.
64. Morel, N., M. Wibo, and T. Godfraind. A calmodulin-stimulated Ca^{2+} pump in rat aorta plasma membranes. *Biochim. Biophys. Acta* 644:82–88, 1981.
65. Murad, F., C. K. Mittal, W. P. Arnold, and J. M. Braughler. Effect of nitrocompound smooth muscle relaxants and other materials on cyclic GMP metabolism. In: *Advances in Pharmacology and Therapeutics,* Vol 3, Ions-Cyclic Nucleotides-Cholinergy, edited by J. C. Stoclet. Oxford: Pergamon Press, pp 125–132, 1978.

66. Murphy, E., K. Coll, T. L. Rich, and J. R. Williams. Hormonal effects on calcium homeostasis in isolated hepatocytes. *J. Biol. Chem.* 255:6600–6608, 1980.
67. Nimmo, H. G., and P. Cohen. Hormonal control of protein phosphorylation. *Adv. Cyclic Nucleotide Res.* 8:145–266, 1977.
68. Nishikori, K., T. Takenaka, and H. Maeno. Stimulation of microsomal calcium uptake and protein phosphorylation by adenosine cyclic 3′, 5′-monophosphate in rat uterus. *Mol. Pharmacol.* 13:671–678, 1977.
69. Norman, J. A., A. H. Drummond, and P. Moser. Inhibition of calcium-dependent regulator-stimulated phosphodiesterase activity by neuroleptic drugs is unrelated to their clinical efficacy. *Mol. Pharmacol* 16:1089–1094, 1979.
70. Popescu, L. M. Cytochemical study of the intracellular calcium distribution in smooth muscle. In: *Excitation–Contraction Coupling in Smooth Muscle,* edited by R. Casteels, T. Godfraind, and J. C. Rüegg. Holland: Elsevier Biomedical Press, pp 13–23, 1977.
71. Rapoport, R. M., M. B. Draznin, and F. Murad. Sodium nitroprusside-induced protein phosphorylation in intact rat aorta is mimicked by 8-bromo cyclic GMP. *Proc. Natl. Acad. Sci.* 79:6470–6474, 1982.
72. Rodbell, M. The role of hormone receptors and GTP-regulatory proteins in membrane transduction. *Nature* 284:17–22, 1980.
73. Saida, K., and C. van Breemen. Mechanisms of Ca^{2+} antagonist-induced vasodilation. Intracellular actions. *Circulation Res.* 52:137–142, 1983.
74. Salmon, J. A., D. R. Smith, R. J. Flower, S. Moncada, and J. R. Vane. Further studies on the conversion of prostaglandin endoperoxide into prostacyclin by porcine aorta microsomes. *Biochim. biophys. Acta* 523:250–262, 1978.
75. Sands, H., J. Mascali, and E. Paietta. Determination of calcium transport and phosphoprotein phosphatase activity in microsomes from respiratory and vascular smooth muscle. *Biochim. Biophys. Acta* 500:223–234, 1977.
76. Schoeffter, P., and J. C. Stoclet. Age-related decrease of *in vitro* isoproterenol-induced cyclic AMP accumulation in rat aorta. *Eur. J. Pharmacol.* 77:183–186, 1982.
77. Schultz, G., and J. G. Hardman. Regulation of cyclic GMP levels in the ductus deferens of the rat. *Adv. Cyclic. Nucleotide Res.* 5:339–351, 1975.
78. Silberbauer, K., H. Sinzinger, M. Winter, W. Fiegl, and F. Ring. Prostacyclin-like activity of endothelium and subendothelium—important for atherosclerosis? *Experientia* 34:1471–1472, 1978.
79. Spedding, M. Comparison of "Ca^{2+}-antagonists" and trifluoperazine in skinned smooth muscle fibres. *Br. J. Pharmacol.* 75:25P, 1982.
80. Starke, K. Regulation of noradrenaline release by presynaptic receptor systems. *Rev. Physiol. Biochem. Pharmacol.* 77:1–124, 1977.
81. Stoclet, J. C., and C. Lugnier. Inhibitors of cyclic nucleotide phosphodiesterases. In: *Medicinal Chemistry Advances,* edited by F. G. De Las Heras, and S. Vega. Oxford and New York: Pergamon Press, pp 435–445, 1981.

82. Taylor, W. M., V. Prpić, J. H. Exton, and F. L. Bygrave. Stable changes to calcium fluxes in mitochondria isolated from rat livers perfused with alpha-adrenergic agonists and with glucagon. *Biochem. J.* 188:443–450, 1980.
83. Terragno, D. A., K. Crowshaw, N. A. Terragno, and J. C. McGiff. Prostaglandin synthesis by bovine mesenteric arteries and veins. *Circulation Res.* 36(Suppl.1), 176–180, 1975.
84. Toda, N., and S. Hayashi. Age-dependent alteration in the response of isolated rabbit basilar arteries to vasoactive agents. *J. Pharmacol. Exp. Ther.* 211:716–721, 1979.
85. van Breemen, C., O. Hwang, and K. D. Meisheri. The mechanism of inhibitory action of diltiazem on vascular smooth muscle contractility. *J. Pharmacol. Exp. Ther.* 218:459–463, 1981.
86. Vane, J. R. Inhibition of prostaglandin synthesis as a mechanism of action for aspirin-like drugs. *Nature* 231:232–235, 1971.
87. Van Nueten, J. M. Selectivity of calcium entry blockers. In: *Calcium Modulators,* edited by T. Godfraind, A. Albertini, and R. Paoletti. Holland: Elsevier Biomedical Press, pp 199–208, 1982.
88. Weiss, G. B. Sites sensitive to different types of calcium antagonist. In: *Calcium Modulators,* edited by T. Godfraind, A. Albertini, and R. Paoletti. Holland: Elsevier Biomedical Press, pp 209–220, 1982.
89. Weksler, B. B., A. J. Marcus, and E. A. Jaffe. Synthesis of prostaglandin I_2 (prostacyclin) by cultured human and bovine endothelial cells. *Proc. Natl. Acad. Sci.* 74:3922–3926, 1977.
90. Wells, J. N., and J. G. Hardman. Cyclic nucleotide phosphodiesterases. *Adv. Cyclic Nucleotide Res.* 8:119–143, 1977.
91. Wibo, M., N. Morel, and T. Godfraind. Differentiation of Ca^{2+} pumps linked to plasma membrane and endoplasmic reticulum in the microsomal fraction from intestinal smooth muscle. *Biochim. Biophys. Acta* 649:651–660, 1981.
92. Zawadzki, J. V., P. B. Cherry, and R. F. Furchgott. Comparison of endothelial-dependent relaxation of rabbit aorta by A23187 and by acetylcholine. *Pharmacologist* 22:271, 1980.
93. Zawadzki, J. V., R. F. Furchgott, and P. Cherry. The obligatory role of endothelial cells in the relaxation of arterial smooth muscle by substance P. *Federation Proc.* 40:689, 1981.

Chapter 5

Smooth Muscle Relaxants

Charles V. Jackson and John H. McNeill

1. Introduction

The term smooth muscle relaxant encompasses a broad spectrum of agents that relax both vascular and non vascular smooth muscle. Therapeutically these relaxants are used for the treatment of bronchospastic disorders, premature labor, and a variety of cardiovascular disorders (e.g., hypertension, angina pectoris, peripheral vascular disease, arrhythmias, and congestive heart failure). A few of the relaxants have proven to be useful tools for investigational purposes, for example, xanthines (phosphodiesterase inhibition and the role of cyclic 3′,5′-monophosphate in mediating cellular events) and calcium-channel blockers (elucidating the role of calcium in normal cell function and drug action). This chapter attempts to discuss current pharmacodynamics, pharmacokinetics, and clinical pharmacology of smooth muscle relaxants and to give the researcher, clinician, graduate and medical student basic information on these compounds.

2. Beta-Adrenoceptor Agonists

2.1. Isoproterenol

2.1.1. Pharmacological Actions

Isoproterenol is a nonspecific beta-adrenoceptor agonist that stimulates both $beta_1$- and $beta_2$-adrenoceptors, with little or no alpha-adrenoceptor stimulation. Its main actions are the relaxation of almost all types of smooth muscle ($beta_2$) and excitation of the heart ($beta_1$).

The mechanism of action responsible for the relaxation of smooth muscle induced by isoproterenol is thought to involve the adenylate cyclase system (*1, 48*). $Beta_2$-adrenoceptor activation by isoproterenol stimulates adenylate cyclase, which in turn generates cyclic 3′,5′-monophosphate (cAMP). As discussed in earlier chapters cAMP then mediates a decrease in the intracellular calcium level leading to relaxation.

2.1.2. Pharmacokinetics

Isoproterenol is readily absorbed by the systemic circulation upon parenteral administration or by inhalation. However, it is rapidly metabolized by the enzyme, catechol-*O*-methyl transferase (COMT), found in the liver and extraneuronal tissues. Enteral administration of isoproterenol provides unreliable absorption from the gastrointestinal tract, since there is extensive degradation in the gut. Therefore, isoproterenol has a relatively short duration of action in comparison with the noncatecholamine $beta_2$-specific adrenoceptor agonists (Table 1).

2.1.3. Toxicities and Side Effects

The majority of the side effects observed upon isoproterenol administration result from stimulation of the heart and vasodilation of the peripheral circulation. Stimulation of $beta_1$-adrenoceptors

Table 1
Beta-Adrenoceptor Agonists: Route of Administration and Duration of Action

Agonists	Route of administration	Duration of action, h
Isoproterenol	Inhalation	<1 (*53*)[a]
Salbutamol	Inhalation	3 (*43, 47*)
	Oral	6 (*84*)
Terbutaline	Inhalation	>3 (*47*)
	Oral	6 (*84*)
Metaproterenol	Inhalation	>3 (*6*)
	Oral	<2 (*24*)
Fenoterol	Inhalation	>8 (*41*)
	Oral	>6 (*24*)

[a]Number in parenthesis denotes reference number.

results in tachycardia and increased cardiac work. These two conditions at high doses of isoproterenol could result in angina pectoris, because of the increase in oxygen demand on the heart. Vasodilation of peripheral blood vessels results in flushing of the skin, and at high doses a fall in mean blood pressure. Less frequent side effects are nausea, dizziness, and skeletal muscle tremors. Cardiotoxicity may lead to death in individuals predisposed to cardiac anomalies. Toxic doses of isoproterenol have been shown to induce cardiac hypertrophy in rats, which may be followed by cardiac infarction and congestive heart failure (*78, 58*).

2.1.4. Therapeutic Uses

Isoproterenol's major therapeutic use is as a bronchodilator in respiratory disorders, especially for treatment of asthma. The bronchodilator effect produced by isoproterenol in asthmatics may also be due in part to inhibition of antigen-induced release of histamine, a potent bronchoconstrictor (*2, 72*). Isoproterenol's use as a bronchodilator is contraindicated in patients with cardiac tachycardias.

2.2. Selective Beta$_2$-Adrenoceptor Agonists

More selective noncatecholamine beta$_2$-adrenoceptor agonists have been developed in response to the clinician's need for a bronchodilator that has less cardiac side effects than isoproterenol. Over the past several years a large number of these agonists have been developed mainly for the treatment of respiratory disorders. Metaproterenol (orciprenaline) and terbutaline are approved for clinical use in the US and Canada, and salbutamol (albuterol) and fenoterol in Canada. Ritodrine has been approved for use in premature labor in the US. A large number of other beta$_2$-adrenoceptor agonists are still undergoing investigational trials. The beta$_2$-adrenoceptor agonists possess similar pharmacological actions, pharmacokinetics, and toxicological properties as does isoproterenol, and differ mainly in their potency and duration. Some, salbutamol and fenoterol for example, are active when taken by the oral route. However, these differences give the selective beta$_2$-adrenoceptor agonists a therapeutic advantage over isoproterenol.

2.2.1. Pharmacological Actions

The beta$_2$-adrenoceptor agonists relax smooth muscle of the bronchi, uterus, and vascular supply to skeletal muscle in the same manner as does isoproterenol. This relaxation is thought to involve the adenylate cyclase–cAMP system. They have little direct effect on the heart (some beta$_1$ stimulation at high doses), but may evoke sinus tachycardia, presumably as a reflex compensation in response to a fall in blood pressure from vasodilation (*68*). All of these agonists when used as bronchodilators produce an increase in 1-s forced expiratory volume (FEV$_1$), in the maximal rate of forced expiratory flow (FEF), and a decrease in airway resistance. Brittain, et al. (*8*) demonstrated that in the anesthetized guinea pig, relative potencies of intravenous beta$_2$-adrenoceptor agonists producing a 50% decrease in intratracheal pressure when compared to isoproterenol were Isoproterenol > Fenoterol ≥ Salbutamol > Terabutaline > Metaproterenol (relative potencies of 1, 4.3, 5.6, 15, and 56, respectively). Terbutaline has been shown to be more potent than isoproterenol in the relaxation of pregnant rabbit myometrium (*77*).

2.2.2. Pharmacokinetics

The noncatecholamine chemical structure of the beta$_2$-adrenoceptor agonists makes these drugs more resistant to degradation by sulfatase in the gastrointestinal tract, as well as by tissue COMT. This property allows for oral administration of these compounds and provides for a longer duration of action than isoproterenol (Table 1). Approximately 40–60% of these compounds is absorbed by the gastrointestinal tract after oral administration.

Smaller doses of the beta$_2$-adrenoceptor agonists are needed to produce equivalent bronchodilation when inhaled (*17*). Inhalation provides for a rapid onset of action, a potent bronchodilation, and wide patient acceptability. This type of administration is very effective in the treatment of intermittent symptoms or during acute irritations of chronic respiratory disease. A large portion of an inhaled dose is actually swallowed and absorbed by the gastrointestinal tract. Buchelt and Rominger (*10*) demonstrated that inhaled fenoterol possessed a two-staged process of absorption, the first being dose-independent and the second similar to absorption after oral administration. Table 1 also shows that the inhaled drugs, salbutamol and terbutaline, possess a shorter duration of action than when they are taken orally.

The beta$_2$-adrenoceptor agonists are resistant to degradation by COMT and are metabolized mainly in the liver and kidneys. The majority of the metabolites are in conjugated forms and are excreted in the urine. A small percentage of the agonists are excreted in the urine in the nonmetabolized form.

2.2.3. Toxicities and Side Effects

Side effects observed with these compounds are those you would expect upon beta-adrenoceptor stimulation. They include tachycardia, hypotension, palpitations, nausea, vomiting, and skeletal muscle tremor. These side effects are observed when high therapeutic doses are administered either orally or intravenously. Myocardial dilation and lesions have been observed in rabbits and rats receiving high doses of fenoterol for one to six months, orally or intravenously (*30*). Tolerance to beta$_2$-adrenoceptor bronchodilators has been demonstrated both in the laboratory (*7*) and in clinical trials (*65*).

2.2.4. Therapeutic Uses

These compounds are used mainly in the therapy of asthma and other bronchospastic diseases. In asthmatics, these compounds may also inhibit the antigen-induced release of histamine (*11, 77*). These compounds are also used in the treatment of premature labor. Careful monitoring of both maternal and fetal heart rates is important during this therapy. The beta$_2$-adrenoceptor agonists should be used with caution in individuals with congestive heart failure and preexisting cardiac arrhythmias.

3. The Xanthines

Caffeine, theophylline, and theobromine are the three naturally occurring xanthine alkaloids. These xanthines are found in coffee, tea, and cocoa, or chocolate. The xanthines are smooth muscle relaxants, CNS stimulants, cardiac stimulants, and diuretics. Therapeutically, theophylline is the most widely used xanthine for its bronchodilator effects, although caffeine is used in over-the-counter analgesic preparations. The pharmacology of these compounds will be briefly discussed in this section; for a more detailed discussion see Goodman and Gilman (*26*).

3.1. Pharmacological Actions

3.1.1. Smooth Muscle

The xanthines relax nonvascular smooth muscle (bronchial, uterine, intestinal) as well as vascular smooth muscle. These compounds are thought to produce relaxation of smooth muscle by inhibition of the enzyme, cAMP-phosphodiesterase, which hydrolyzes cellular cAMP. Inhibition of the phosphodiesterase allows for the accumulation of cAMP, which in turn mediates subsequent cellular events that lead to relaxation. However, with theophylline, the doses used clinically for treatment of asthma produce little if any inhibition of phosphodiesterase. A lack of correlation between increased levels of cAMP and the degree of relaxation produced by xanthines has also been reported in vascular smooth muscle (*15*) and in the spontaneously contracting rat uterus (*66*).

3.2. Pharmacokinetics

The xanthines are rapidly and completely absorbed after oral, rectal or parenteral administration. After oral administration, theophylline reaches peak plasma levels in approximately 2 h, whereas caffeine is more rapidly absorbed and reaches peak plasma levels within 1 h (*45*). Theophylline, the most widely used xanthine, appears to distribute rapidly throughout extracellular and, to a lesser extent, intracellular body water (*44, 85*).

Metabolism of the xanthines occurs primarily in the liver. In adults, theophylline has a plasma half-life of about 8–9 h and caffeine has a plasma half-life of about 3.5 h (*31*). Hepatic biotransformation of theophylline produces the urinary metabolites, 1,3-dimethyluric acid, 1-methyluric acid, and 3-methylxanthine (*9*). Approximately 10% of theophylline and 1% of caffeine are excreted via the urine unchanged.

3.3. Toxicities and Side Effects

The major side effects produced by the xanthines are nausea, vomiting and headache. These usually occur when serum concentrations are higher than 20 μg/mL (*35*) and are thought to occur as a result of CNS stimulation. Other adverse effects observed are diarrhea, irritability, and insomnia. Side effects appear to be more related to serum levels rather than patient idiosyncrasy. They can be eliminated by reducing the dose into the lower therapeutic range.

3.4. Therapeutic Uses

Caffeine is used in a variety of nonprescription drugs, such as aspirin analgesics, for its CNS stimulant effect to provide a feeling of well-being. In the doses used it is unlikely to produce any effect and its use in such preparations is not justified. Vasoconstriction of CNS blood vessels, a reported effect of xanthines, would also not occur with the doses used in nonprescription medications. Theophylline is used mainly as a bronchodilator because of its smooth muscle relaxant effect. It is used in the treatment of asthma and chronic pulmonary obstructive disease. Theophylline is also used in the treatment of *status asthmaticus* refractory to the administration of adrenergic agonists. In recent years, premature infants suffering from prolonged apnea have been treated successfully with theophylline (*16*).

4. Alpha-Adrenoceptor Antagonists

The alpha-adrenoceptor antagonists are agents that bind to the alpha-adrenoceptors (alpha$_1$ and/or alpha$_2$), and in turn inhibit the actions of sympathetic stimulation and the actions of sympathomimetic amines at these receptors (*see* Chapter 13). Phenoxybenzamine and phentolamine along with their structural analogs, dibenamine and tolazoline, have been the classical alpha-adrenoceptor antagonists in both investigational and clinical studies. The pharmacology of these compounds has been well characterized (*26*) and will not be discussed in this section. However, researchers in the last ten years have developed new and more selective alpha-adrenoceptor antagonists. One of the new antagonists is prazosin, a selective alpha$_1$-adrenoceptor antagonist, possessing vasodilator properties and antihypertensive activity.

4.1. Prazosin

4.1.1. Pharmacological Actions

Prazosin differs from the classical alpha-adrenoceptor antagonists in that it is selective only for the alpha$_1$-adrenoceptor, with no effect on alpha$_2$-adrenoceptors (*28, 71*). This selectivity for the alpha$_1$-adrenoceptor, allows the negative feedback mechanism to remain intact and norepinephrine can inhibit its own release from the nerve terminals. Therefore, one observes a decline in blood pressure without an accompanying tachycardia (*52*). Prazosin has also been shown to be more effective in relaxing veins than either phentolamine or sodium nitroprusside (*3*). Cohen and coworkers (*14*) have demonstrated that prazosin relaxes both arteries and veins from rats as a result of alpha$_1$-adrenoceptor inhibition and not by direct smooth muscle relaxation. Therefore, prazosin reduces peripheral vascular resistance and increases venous capacitance.

4.1.2. Pharmacokinetics

Prazosin, at the present time, is available only in the oral form. It is readily absorbed after oral administration; peak plasma levels are reached in about 1–2 h with a plasma half-life of approximately 3–4 h (*5*). Approximately 97% of circulating prazosin is bound to plasma albumin. Prazosin is metabolized in the liver, with 99% being metabolized and first-pass metabolism being significant.

4.1.3. Toxicities and Side Effects

Drowsiness, dizziness, headaches, and palpitations are usual side effects observed after prazosin administration. These side effects are usually well tolerated. One potential side effect that may occur with the first dose of prazosin is postural hypotension, which may produce syncope. Therefore, the patient should be given small doses initially and the patient should also be in the supine position. Packer and coworkers (*60*) and other (*19*) have evaluated this phenomenon and observed that although the first dose always produces beneficial hemodynamic effects, there appears to be a tolerance to subsequent prazosin therapy. Spironolactone has been shown to help prevent the tolerance observed in long-term prazosin therapy in patients with congestive heart failure (*74*). Other side effects that may be observed are dry mouth, blurred vision, constipation, and nasal decongestion.

4.1.4. Therapeutic Uses

Prazosin is currently available as an antihypertensive drug and is highly effective in forms of hypertension because of increased stimulation of alpha-adrenoceptors by catecholamines. Prazosin is also undergoing clinical trials as a therapeutic agent in the treatment of congestive heart failure.

5. Calcium-Channel Antagonists

Calcium antagonist is a term given to any compound that inhibits calcium and its action. There is a class of antagonists, within this category, that specifically inhibits the movement of calcium into a cell by entry through the slow channel during depolarization of a cell membrane (*22*). These agents are called the slow calcium-channel antagonists. These antagonists have been in clinical use in Europe for the past 20 yr and in recent years have gained clinical use in the US and Canada. A large number of these compounds are available for investigational purposes, especially for use as tools in calcium research. The calcium-channel antagonists currently marketed for clinical use (cardiovascular disorders) in the US and Canada are verapamil, diltiazem, and nifedipine. Perhexiline is currently undergoing clinical trials. For a thorough review of the mechanisms of action of these antagonists, refer to Chapter 7.

5.1. Verapamil

5.1.1. Pharmacological Actions

Verapamil is a synthetic derivative of papaverine, a nonnarcotic constituent of crude opium. It was the first calcium-channel antagonist to be marketed for clinical use in the US and Canada. Verapamil's major pharmacological actions are prolongation of electrical conduction in the heart, a negative inotropic effect in the heart and relaxation of vascular smooth muscle (Table 2).

5.1.2. Pharmacokinetics

Approximately 90% of an orally administered dose of verapamil is absorbed by the gastrointestinal tract, with a large first-pass metabolism in the liver (*32*). Verapamil is 90% bound by plasma proteins, with peak plasma levels occurring 1–2 h after oral administration (*18*). Schomerus and coworkers (*76*) have demonstrated that verapamil has a distribution half-life of 18–35 min and a slower elimination half-life of 3–7 h. The majority of the drug is excreted in the urine as conjugated metabolites (70%) and about 3–4% excreted as unchanged drug (*18*). The *N*-demethylated metabolite (norverapamil) has been shown to possess about 20% of the pharmacological activity as a coronary vasodilator of the par-

Table 2
Electrophysiological and Hemodynamic Effects of Calcium Channel Antagonists

Antagonist	SANC[a]	AVNC[b]	CI[c]	VSMC[d]
Verapamil	↓/↑	↓↓	↓↓	↓↓
Nifedipine	↓/↑	0	0/↑	↓↓↓
Diltiazem	↓	↓	↓	↓
Perhexiline	↓	↓	↓	↓

[a]Denotes sino-atrial node conduction.
[b]Denotes atrio-ventricular node conduction.
[c]Denotes cardiac inotropic response.
[d]Denotes vascular smooth muscle contractile response.

ent compound in the rat and dog (*81*). Verapamil has also been shown to accumulate in the myocardium (*37*).

5.1.3. Toxicities and Side Effects

Side effects produced by verapamil can be divided into acute and chronic adverse effects. Acute side effects are results of hemodynamic changes, i.e., bradycardia and hypotension. These side effects are usually temporary and mild. Withdrawal of the drug will lessen the hypotension and bradycardia, but it may be necessary to use atropine, calcium gluconate, or vasopressors as temporary support (*50*). Chronic verapamil therapy may cause a variety of nonspecific side-effects, i.e., nausea, constipation, headache, dizziness, hepatotoxicity, galactorrhea, edema, and aching limbs. The major toxic effects of verapamil are hypotension, sinus bradycardia, and AV intranodal block. Therefore, verapamil is contraindicated in individuals with severe heart failure, second or third degree AV block, sick sinus syndrome, and cardiogenic shock. Verapamil-induced heart failure may be reversed by administering intravenous calcium or beta-adrenoceptor agonists (*59*).

5.1.4. Therapeutic Uses

At the present time, verapamil is marketed for the therapeutic treatment of supraventricular tachycardia, and exercise-induced and variant angina. Clinical trials have also shown an effectiveness of verapamil in the treatment of atrial flutter and fibrillation. However, extreme caution should be used in individuals with severe left ventricular dysfunction (*79*). Verapamil has also shown therapeutic effectiveness in clinical trials in the treatment of hypertrophic cardiomyopathy.

5.2. Nifedipine

5.2.1. Pharmacological Actions

Vater and coworkers (*82*) were the first investigators to review the pharmacology of nifedipine, a dihydropyridine derivative. Nifedipine is one of the most potent calcium antagonists available. However, it differs from verapamil in that it affects the peripheral vasculature more than the heart (*21*). Table 2 illustrates that nifedipine has essentially no effect on AV nodal conduction, but a pronounced relaxation of vascular smooth muscle. The therapeutic effectiveness of nifedipine in treatment of coronary artery disease involves two mechanisms: (1) increased oxygen supply to myocardium because of its coronary artery vasodilator properties and (2) decreased myocardial oxygen demand, indirectly, by reduction in afterload as a consequence of decreased arterial pressure from peripheral vasodilation (*49*). Table 2 also illustrates that nifedipine may increase heart rate and force of contraction of the heart; however, this is an indirect reflex sympathetic effect produced by marked reduction in arterial pressure.

5.2.2. Pharmacokinetics

Nifedipine is rapidly and almost completely absorbed after oral or sublingual administration. Hepatic first-pass metabolism is not high and about 65% remains bioavailable in the circulation (*34*). Approximately 90% of nifedipine that remains in the circulation is protein bound, and has an elimination half-life of approximately five hours.

5.2.3. Toxicities and Side Effects

Nifedipine is well tolerated and exhibits minor side effects. The major side effect is hypotension, which causes headache, flushing, dizziness, palpitations, and edema. Nausea and vomiting are observed in a small population of patients and there appears to be no renal or hepatic toxicities (*49*). No toxic effects have been observed on AV node conduction characteristics.

5.2.4. Therapeutic Uses

At the present time, nifedipine is available for treatment of variant and exercise-induced angina. A beta-adrenoceptor antagonist may have to be used along with nifedipine to inhibit the reflex tachycardia that may result from its hypotensive effect. Nifedipine is undergoing investigation to determine its usefulness in the treatment of congestive heart failure and as an antihypertensive, especially in the treatment of hypertensive emergencies.

5.3. Diltiazem

5.3.1. Pharmacological Actions

Diltiazem is the most recently approved channel antagonist. This agent is a benzothiazepine derivative (*75*) that exhibits cardiovascular actions ranging in effectiveness between verapamil and nifedipine (Table 2). Diltiazem, like nifedipine, appears to preferentially affect vascular smooth muscle contraction; however, it is less potent than nifedipine (*39*).

5.3.2. Pharmacokinetics

At the present time, there are limited data on the absorption, distribution, metabolism, and elimination of diltiazem. Studies in dogs and normal individuals have indicated that after oral administration, diltiazem appears in the plasma in approximately 30–60 min and has a plasma half-life of 4 h (*20, 38*). Approximately 80% of the drug was bound to plasma proteins. Deacetylation of diltiazem appears to be the primary hepatic metabolism, with 35% of an administered dose excreted in the urine and 60% excreted in the feces.

5.3.3. Toxicities and Side Effects

Diltiazem exhibits side effects similar to those observed with verapamil and nifedipine; however, they are less severe and the drug is well tolerated. Major side effects are bradycardia, hypotension, headache, and dizziness. Diltiazem may antagonize preexisting sick sinus syndrome and AV node block.

5.3.4. Therapeutic Uses

At the present time, clinical trials indicate a therapeutic usefulness for diltiazem in the treatment of various forms of angina (*87*)

5.4. Perhexiline

Perhexiline is a calcium-channel antagonist that is used in Europe and elsewhere as an antianginal agent. It is currently undergoing investigational trials in the US. Perhexiline is very similar to diltiazem in its pharmacological actions (Table 2), and it is the least potent of the compounds illustrated (*63*).

This antagonist is rapidly absorbed from the gastrointestinal tract after oral administration. The half-life in the plasma is rather long, 2–6 d. Perhexiline appears to be metabolized to hydroxylated metabolites. Perhexiline may not gain wide clinical acceptance because of serious side effects and toxicities. Minor side effects include dizziness, anorexia, insomnia, nausea, and vomiting. More serious side effects include hypoglycemia, polyuria, hepatotoxicity, and peripheral neuropathy.

6. Nonspecific Vasodilators

In previous sections of this chapter, specific classes of smooth muscle relaxants have been discussed. However, there remain several drugs that are widely used as vasodilators in the treatment of angina pectoris and hypertension. These agents can not be unified as a class according to similar chemical structures or mechanisms of action, only as direct, nonspecific smooth muscle relaxants. Nonspecific relaxation is defined as relaxation of smooth muscle that has been contracted by a variety of drugs or nerve stimulation. These agents relax arteries and veins as well as nonvascular smooth muscle (*see* Chapter 3). Such vasodilators are the organic nitrates and nitrites, hydralazine, diazoxide, sodium nitroprusside, and minoxidil.

6.1. Organic Nitrates and Nitrites

6.1.1. Pharmacological Actions

The organic nitrates and nitrites are represented by nitroglycerin (glyceryl trinitrate), isosorbide dinitrate, amyl nitrite, and others. Low doses of these vasodilators relax veins to a greater degree than arteries (Table 3). Nitroglycerin-induced relaxation of veins causes a reduction in venous return to the heart, thereby decreasing preload on the heart and decreasing the oxyen consumption of the heart (*23, 86*). Currently, this is thought to be the predominant therapeutic effect in the treatment of angina pectoris. Nitrates also cause the relaxation of arteries, both resistance (*25, 54*) and large coronary arteries (*4, 13*). The relaxation of resistance vessels decreases peripheral vascular resistance and cardiac afterload, whereas relaxation of the coronary arteries increases myocardial perfusion (especially during coronary vasospasm). This results in a decreased oxygen demand and increased oxygen supply to the heart, respectively.

6.1.2. Pharmacokinetics

The organic nitrates and nitrites are readily absorbed by the systemic circulation upon oral, sublingual, intravenous, or topical

Table 3
Sites of Action of Nonspecific Vasodilators

Vasodilator	Site of action[a]
Nitrates	Veins
Hydralazine	Arteries
Diazoxide	Arteries
Nitroprusside	Arteries/veins
Papaverine	Arteries
Minoxidil	Arteries

[a]Denotes predominant site of action.

administration. However, they are rapidly metabolized by the hepatic enzyme, glutathione organic nitrate reductase. The metabolites produced by the reduction process are denitrated metabolites and the nitrite ion (*56*). After sublingual administration, nitroglycerin reaches peak plasma levels in approximately 5 min and it has a plasma half-life of 1–4 min. Dinitrate metabolites, which are less potent vasodilators, have a half-life of approximately 2 h. Oral administration of nitroglycerin increases the plasma half-life to 1–2 h (*57*). The major metabolite of isosorbide dinitrate is isosorbide glucuronide formed by enzymatic denitration in the liver. Sublingual administration of isosorbide produces peak plasma levels in 5–6 min with a plasma half-life of 30 min. These agents are almost completely metabolized, nitroglycerin—99%, and isosorbide dinitrate—100% and excreted via the urine in metabolic form.

6.1.3. Toxicities and Side Effects

The most common side effect observed from nitrate administration is moderate to severe headache. Headaches can be controlled by decreasing the dose and they usually disappear in a few days. Other side effects observed are transient dizziness and weakness associated with the postural hypotension that results from vasodilation causing cerebral ischemia.

6.1.4. Therapeutic Uses

These vasodilators have been used for over 100 yr for the treatment of angina pectoris (typical and variant). The relaxation of both arteries and veins decreases oxygen demand by the heart and also increases oxygen supply to the heart, alleviating the pain and preventing further myocardial ischemia. These agents can be used prophylactically just prior to exercise or strenuous work. Nitrates are also used in the treatment of congestive heart failure since they can reduce peripheral arterial resistance and allow the heart to eject blood more efficiently.

6.2. Hydralazine

6.2.1. Pharmacological Actions

Hydralazine directly relaxes vascular smooth muscle and has a predominant effect on resistance vessels (Table 3). Relaxation of the resistance vessels produces a decrease in peripheral vascular resistance with a resulting decrease in blood pressure. Hydralazine also increases heart rate, stroke volume, and cardiac output (*62, 83*). The effects of hydralazine on the heart are indirect and are produced by the reflex stimulation of the sympathetic nervous system which is activated by the hypotensive response. The reduction in peripheral resistance decreases afterload on the heart and allows the heart to empty more efficiently, thereby increasing cardiac output. Hydralazine can cause sodium and water retention and decrease urine volume. It may also increase plasma renin activity caused by the reflex sympathetic discharge to the juxtaglomerular cells of the kidney.

6.2.2. Pharmacokinetics

Hydralazine is readily absorbed from the gastrointestinal tract, with peak plasma levels being reached between 30 min and 2 hours. The plasma half-life averages about 3 h. (*80*). In patients with congestive heart failure it takes approximately 24 h to reach peak hemodynamic effectiveness. Simultaneous ingestion of food and hydralazine can increase the bioavailability of the drug (*51*).

Metabolism of hydralazine is primarily by *N*-acetylation in the liver, with a large first-pass metabolism (*69*). Individuals also absorb acetylate hydralazine at different rates that will affect its bioavailability. Less than 15% of a parenteral dose is excreted unchanged in the urine.

6.2.3. Toxicities and Side Effects

Hydralazine therapy is plagued by a wide variety of side effects. Sweating, dizziness, headache, palpitation, anorexia, and nausea are common side effects. Side effects that occur less frequently are flushing, nasal congestion, lacrimation, conjunctivitis, edema, tremors, and muscle cramps. By increasing the dose of hydralazine slowly, side effects are less frequent and less severe. If reflex sympathetic discharge stimulates the heart to a great degree, anginal attacks may be produced. Therefore, a beta-adrenoceptor antagonist may be required as adjunct therapy. A major toxic effect of hydralazine observed with long term therapy is a lupuslike syndrome that is reversible (*64*). This syndrome occurs almost exclusively in Caucasians who are slow acetylators.

6.2.4. Therapeutic Uses

The most common therapeutic use of hydralazine is as an antihypertensive, and its effectiveness is optimized by the simultaneous use of a beta-adrenoceptor antagonist. Smaller doses of hydralazine are required when a thiazide diuretic is used along with beta-adrenoceptor blocker. This combination therapy promotes longer therapy with less limiting side effects. Hydralazine has also demonstrated a therapeutic use in the treatment of chronic congestive heart failure, because of its effect on cardiac output.

6.3. Diazoxide

6.3.1. Pharmacological Actions

Diazoxide is a vasodilator that is chemically similar to the thiazide diuretics. It has a greater specificity for arterial relaxation than relaxation of capacitance vessels (Table 3). It also is a potent relaxant of nonvascular, uterine smooth muscle (*42*). Diazoxide is less potent in relaxing vessels precontracted by a variety of agonists, especially norepinephrine and angiotensin II. This agent may produce its relaxation by inhibition of calcium (*70*). The hemodynamic effects of diazoxide are similar to those of hydralazine. Unlike thiazide diuretics, diazoxide causes retention of sodium and water and may lead to edema formation unless the individual is treated with a thiazide diuretic.

6.3.2. Pharmacokinetics

Diazoxide is readily absorbed after oral administration. Approximately 90% of an intravenous dose or the amount absorbed after oral administration is bound by plasma proteins. This agent has a relatively long plasma half-life of 24–36 h (*67*). Approximately 33% of diazoxide is excreted unchanged in urine.

6.3.3. Toxicities and Side Effects

Excessive hypotension and reflex sympathetic stimulation are the two major side effects of diazoxide. Intense reflex stimulation leading to prolonged tachycardia could produce anginal pain and myocardial ischemia. Diazoxide also inhibits release of insulin from the pancreas and produces hyperglycemia. Therefore, it is contraindicated in diabetic individuals.

6.3.4. Therapeutic Uses

Diazoxide is most commonly used in the treatment of hypertensive crisis. It does not interfere with CNS function or the subsequent administration of a slower acting antihypertensive agent. Diazoxide is used occasionally in the treatment of hypoglycemia due to hyperinsulinemia.

6.4. Sodium Nitroprusside

6.4.1. Pharmacological Actions

Sodium nitroprusside is a very potent vasodilator, affecting both arteries and veins to the same degree (Table 3). This vasodilator has little relaxant effect on nonvascular smooth muscle. Therefore, the major effects are a decrease in both peripheral vascular resistance (afterload) and venous return (preload) on the heart (*29, 73*).

Sodium nitroprusside is thought to produce its vasodilator effect by altering calcium mobilization. Investigators have shown that sodium nitroprusside decreases calcium efflux from vascular smooth muscle and appears to increase the binding of calcium to intracellular sites (*33, 40*). This agent has also been shown to stimulate guanylate cyclase and increase cGMP levels without altering cAMP levels; however, no cause and effect relationship has been demonstrated in relaxation (*36*).

6.4.2. Pharmacokinetics

Sodium nitroprusside is administered only as an intravenous infusion. Peak plasma levels are reached in minutes, with a plasma half-life of minutes. Metabolism of sodium nitroprusside is rapid, producing cyanide as a metabolic intermediate and thiocyanate as the major end product. Rapid metabolism allows titration of the blood pressure to the desired level by varying the rate of infusion. Thiocyanate has a half-life of about 4 d in normal individuals and is prolonged in individuals with renal failure.

6.4.3. Toxicities and Side Effects

The major side effects observed with sodium nitroprusside are nausea, vomiting, sweating, palpitations, and headache, all secondary to the hypotension. These effects are reduced and disappear upon drug withdrawal. Because of the dynamic hypotensive effect, individuals should be monitored closely and given the drug slowly. Cyanide toxicity is rare, but thiocyanate accumulation during prolonged therapy can lead to weakness, hypoxia, muscle spasms and psychosis.

6.4.4. Therapeutic Uses

Sodium nitroprusside is very useful in hypertensive emergencies because of its short-lasting but rapid hypotensive effect.Sodium nitroprusside has also gained therapeutic use in acute myocardial infarction and acute congestive heart failure.

6.5. Minoxidil

Pharmacologically, minoxidil produces a hypotensive action through direct relaxation of resistance vessels (Table 3). Minoxidil affects cardiovascular hemodynamics secondarily to the hypotensive response, as does hydralazine. However, it is more potent than hydralazine. It also has the ability to increase cardiac output and lower pulmonary wedge pressure in congestive heart failure patients (*12*). Minoxidil is well absorbed from the gastrointestinal tract after oral administration. Peak plasma levels are reached within 1 h and it has a plasma half-life of about 4 h (*27*). Metabolites of minoxidil have not as yet been identified. Hepatic biotransformation is responsible for metabolism of minoxidil and less than 10% is excreted unchanged in the urine.

Minoxidil possesses side effects similar to hydralazine and diazoxide; these involve the reflex activation of the sympathetic nervous system. Fluid and sodium retention are side effects that can produce problems, especially in individuals with renal failure (*46*). Adjunct therapy with diuretics can alleviate this problem. Minoxidil may also produce the development of pericardial effusion and excessive growth of body hair.

Recently, minoxidil has been released in the United States as a treatment for moderate to severe hypertension. The compound is an especially effective antihypertensive where conventional therapy has failed. Minoxidil has also been used in clinical trials in the treatment of acute and chronic congestive heart failure (*55, 61*). Patients do not appear to develop tolerance to minoxidil during chronic therapy as they do with hydralazine.

References

1. Andersson, R., and K. Nilsson. Cyclic AMP and calcium in relaxation in intestinal smooth muscle. *Nature New Biol.* 238:119–120, 1972.
2. Assen, E. S. K., and H. O. Schild. Inhibition by sympathomimetic amines of histamine release induced by antigen in passively sensitized human lung. *Nature* 224:1028–1029, 1969.
3. Awan, N. A., R. R. Miller, and D. T. Mason. Comparison of effects of nitroprusside and prazosin on left ventricular function and peripheral circulation in chronic heart failure. *Circulation* 57:152–159, 1978.
4. Bache, R. J., and B. A. Tockman. Effect of nitroglycerin and nifedipine on subendocardial perfusion in the presence of a flow limiting coronary stenosis in the awake dog. *Circ. Res.* 50:678–687, 1982.
5. Bateman, D. N., D. C. Hobbs, T. M. Twomey, E. A. Stevens, and M.D.Rawlins. Prazosin, pharmacokinetics and concentration effect. *Eur. J. Clin. Pharmacol.* 16:177–181, 1979.
6. Beardshaw, J., L. McLean, and M. Chan–Yeung. Comparison of the bronchodilator and cardiac effects of hydroxyphenylorciprenaline and orciprenaline. *Chest* 65:507–511, 1974.
7. Benoy, G. G., M. S. El–Fellah, R. Schneider, and 0.L. Wade. Tolerance to sympathomimetic bronchodilators in guinea pig isolated lungs following chronic administration *in vivo*. *Br. J. Pharmacol.* 55:547–554, 1975.
8. Brittain, R. T., C. M. Dean, and D. Jack. Sympathomimetic bronchodilator drugs. *Pharmacol. Ther.* 2:423–469, 1976.

9. Brodie, B. B., J. Axelrod, and J. Reichenthal. Metabolism of theophylline (1,3–methylxanthine) in man. *J. Biol. Chem.* 194:215–222, 1952.
10. Buchelt, L., and K. L. Rominger. The pharmacokinetics and metabolism of berotec (fenoterol) in man. *Med. Proc.* 18:15, 1972.
11. Casterline, C. L., R. Evans III, and G. W. Ward, Jr. The effect of atropine and albuterol aerosols on the human bronchial response to histamine. *J. Allergy Clin. Immunol.* 58:607–613, 1976.
12. Chatterjee, K., D. Drew, W. W. Parmley, J. C. Klausner, J. Polansky, and B. Zacherle. Combination vasodilator therapy for severe chronic congestive heart failure. *Ann. Intern. Med.* 85:467–470, 1976.
13. Chiariello, M., H. Gold, R. Leinbach, M. Davis, and P. R. Moroko. Comparison between the effects of nitroprusside and nitroglycerin on ischemic injury during acute myocardial infarction. *Circulation* 54:766–773, 1976.
14. Cohen, M. L., K. S. Wiley, and I. H. Slater. *In vitro* relaxation of arteries and veins by prazosin alpha–adrenergic blockade with no direct vasodilation. *Blood Vessels* 16:144–154, 1979.
15. Collins, G. A., and M. C. Sutter. Quantitative aspects of cyclic AMP and relaxation in the rat anterior mesenteric portal vein. *Can. J. Physiol. Pharmacol.* 53:989–997, 1976.
16. Dietrich, J., A. N. Krauss, M. Reidenberg, D. E. Drayer, and P.A.M. Auld. Alterations in state in apneic preterm infants receiving theophylline. *Clin. Pharmacol. Ther.* 24:474–478, 1978.
17. Dulfano, M., and P. Glass, Evaluation of a new β–adrenergic receptor stimulant terbutaline, in bronchial asthma. II. Oral comparison with ephedrine. *Curr. Ther. Res.* 15:150–157, 1973.
18. Eichelbaum, M., M. Ense, G. Remberg, M. Schomerus, and H.J. Dengler. The metabolism of ^{14}C–verapamil in man. *Drug Metab. Dispos.* 7:145–148, 1979.
19. Elkayam, U., T. Lejemtel, M, Mathur, H. S. Ribner, W. H. Frishman, J. Strem, and E. H. Sonnenblick. Marked early attenuation of hemodynamic effects of oral prazosin therapy in chronic congestive heart failure. *Am. J. Cardiol.* 44:540–545, 1979.
20. Flaim, S. F., and R. Zelis. Clinical use of calcium entry blockers. *Fed. Proc.* 40:2877–2881, 1981.
21. Fleckenstein, A. On the basic pharmacological mechanisms of nifedipine and its relation to therapeutic efficacy. In *Third International Adalat Symposium*. Edited by A.D. Jatere and P. R. Lichtlen. Amsterdam, Excerpta Medica, 1976, pp, 1–13.
22. Fleckenstein, A. Specific pharmacology of calcium in myocardium, cardiac pacemakers and vascular smooth muscle. *Ann. Rev. Pharmacol. Toxicol.* 17:149–166, 1977.
23. Gerry, J. L., Jr., H. V. Schaff, C. H. Kallman, and J. T. Flaherty. Effects of nitroglycerin on regional myocardial ischemia induced by atrial pacing in dogs. *Circ. Res.* 48:569–576, 1981.
24. Geumei, A. M., W. F. Miller, J. Miller, and L. R. Gast. Bronchodilator effect of a new oral β–adrenoceptor stimulant, TH1165a. A comparison with metaproterenol sulfate. *Chest* 70:460–465, 1976.
25. Gold, H. K., R. C. Leinback, and C. A, Sanders. Use of sublingual nitroglycerin in congestive failure following acute myocardial infarction. *Circulation* 46:839–845, 1972.

26. Goodman, L. S., A. G. Gilman, and A. Gilman (eds.). *The Pharmacological Basis of Therapeutics*. Sixth Edition, New York, MacMillan Publishing Co., 1980.
27. Gottlieb, T. B., R. C. Thomas, and C. A. Chidsey. Pharmacokinetic studies of minoxidil. *Clin. Pharmacol.* Ther. 13:436–441, 1972.
28. Graham, R. M., H. F. Oates, L. M. Stoker, and G. S. Stokes. Alpha–blocking action of the antihypertensive agent, prazosin. *J. Pharmacol. Exp. Ther.* 201:747–752, 1977.
29. Greenway, C. V. Effects of sodium nitroprusside, isosorbide dinitrate, isoproterenol, phentolamine, and prazosin on hepatic venous responses to sympathetic nerve stimulation in the cat. *J. Pharmacol. Exp. Ther.* 209:56–61, 1979.
30. Heel, R. C., R. N. Brogden, T. M. Speight, and G. S. Avery, Fenoterol: a review of its pharmacological properties and therapeutic efficacy in asthma. *Drugs* 15:3–32, 1978.
31. Hendeles, L.. M. Weinberger, and L. Bighley. Disposition of theophylline after a single intravenous infusion of aminophylline. *Am. Rev. Respir. Disease.* 118:97–103, 1978.
32. Henry, P. D. Comparative pharmacology of calcium antagonists: nifedipine, verapamil, and diltiazem. *Am. J. Cardiol.* 46:1047–1058, 1980.
33. Hester, R. K., G. B. Weiss, and W. J. Fry. Differing actions of nitroprusside and D600 on tension and ^{45}-Ca fluxes in canine renal arteries. *J. Pharmacol. Exp. Ther.* 208:155–160, 1979.
34. Horster, F. A. Pharmacokinetics of nifedipine–^{14}C in man. In: *Second International Adalat Symposium,* edited by W. Lochner, W. Braasch, and G. Kroneberg. New York, Springer–Verlag, 1975, pp 124–127.
35. Jacobs, M. H., R. M. Senior, and G. Kessler. Clinical experience with theophylline relationships between dosage, serum concentration, and toxicity. *J. Am. Med. Assoc.* 235:1983–1986, 1976.
36. Katsuki, S., W. P. Arnold, and F. Murad. Effects of sodium nitro–-prusside, nitroglycerin, and sodium azide on levels of cyclic nucleotides and mechanical activity of various tissues. *J. Cycl. Nucleotide Res.*3:239–247, 1977.
37. Keefe, D. L., S. R. Harapat, and R. E. Kates. Relationship between pharmacodynamics and myocardial concentration of verapamil. *Am. J. Cardiol.* 47:406, 1981.
38. Kohno, K., Y. Takeuchi, A. Etoh, and K. Noda. Pharmacokinetics and biovailability of diltiazem (CRD–401) in dog. *Arzneim, Forsch.* 27:1424–1428, 1977.
39. Kondo, K., H. Suzuki, T. Okuno, M. Suda, and T. Saruta. Effects of nifedipine, diltiazem and verapamil on the vasoconstrictor responses to norepinephrine and potassium ions in the rat mesenteric artery. *Arch. Int. Pharmacodyn.* 245:211–217, 1980.
40. Kreye, V. A. W., G. D. Baron, J. B. Luth, and H. Schmidt–Gayk. Mode of action of sodium nitroprusside on vascular smooth muscle. *Naunyn–Schmiedeberg's Arch. Pharmacol.* 288:381–402, 1975.
41. Krieger, E. Bronchial antispasmodic effect in obstructive airway disorders. Double–blind comparison of aludrin, alupent, berotec and placebo. *Int. J. Clin. Pharmacol. Ther. Toxicol.* 4 (Suppl):102–103, 1971.

42. Landesman, R., E. M. Coutinho, K. H. Wilson, and A. C. Vieira–Lopes. The relaxant effect of diazoxide on nongravid human myometrium *in vivo. Am. J. Obstet. Gynecol.* 102:1080–1084, 1968.
43. LeBlanc, K. Dosis–Wirkungs–Vergleich von berotec und salbutamoldosier–aerosol. *Int. J. Clin. Pharmacol. Ther. Toxicol.* 4 (Suppl.):104–105, 1971.
44. Levy, G., and Koysooko, R. Pharmacokinetic analysis of the effect of theophylline on pulmonary function in asthmatic children. *J.Pediatr.* 86:789–793, 1975.
45. Lillehei, J. P. Aminophylline. Oral vs rectal administration. *J. Am. Med. Assoc.* 205:530–533, 1968.
46. Limas, C. J., and E. D. Freis. Minoxidil in severe hypertension with renal failure. Effects of its addition to conventional antihypertensive drugs. *Am. J. Cardiol.* 31:355–361, 1973.
47. Manicatide, M. A., C. L. Racoveanu, V. V. Nicolaescu, and D. B. Teculescu. Comparative assessment of the efficiency of three new bronchodilators (salbutamol, terbutaline, Th1165a) in the management of patients with chronic obstructive lung disease. *Rev. Roum. Med. Intern.* 10:49–56, 1973.
48. Marshall, J. M., and E. A. Kroeger. Adrenergic influences on uterine smooth muscle. *Phil. Trans. R. Soc.* B265:135–148, 1973.
49. McAllister, A. G., Jr. Clinical pharmacology of slow channel blocking agents. *Prog. Cardiovasc. Diseases.* 25:83–102, 1982.
50. McGoon, M. D., R. E. Vlietstra, D. R. Holmes, and J. E. Osborn. The clinical use of verapamil. *Mayo Clin. Proc.* 57:495–510, 1982.
51. Melander, A., K. Danielson, A. Hanson, B. Rudell, B. Schersten, T. Tholin, and E. Whalin. Enhancement of hydralazine bioavailability by food. *Clin. Pharmacol. Ther.* 22:104–107, 1977.
52. Miller, R. R., W. H. Fennell, J. B. Young, A. R. Palomo, and M. A. Quinones. Differential systemic arterial and venous actions and consequent cardiac effects of vasodilator drugs, *Prog. Cardiovasc. Diseases.* 24:353–374, 1982.
53. Minette, A. Spirometry of the bronchodilator effects of hydroxyphenylorciprenaline (TH1165a) in various forms in a group of 124 coal miners suffering from bronchospasm. *Respiration* 27:276–315, 1970.
54. Naggar, C. Z., K. Stanley, and L. L. Downing. Effects of amyl nitrite on left ventricular relaxation in patients with borderline hypertension. *Am. J. Cardiol* 50:979-984, 1982.
55. Nathan, M., S. A. Rubin, D. Siemienczuk, and H. J. C. Swan. Effects of acute and chronic minoxidil administration on rest and exercise hemodynamics and clinical status in patients with severe, chronic heart failure. *Am. J. Cardiol.* 50:960–965, 1982.
56. Needleman, P., and E. M. Johnson, Jr. The pharmacologic and biochemical interaction of organic nitrates with sulfhydryls: possible correlation with the mechanism for tolerance development, vasodilation, and mitochondrial reactions. In: *Organic Nitrates,* edited by P. Needleman. New York, Springer–Verlag, 1975, pp 57–114.
57. Neurath, G. B., and N. Dunger. Blood levels of the metabolites of glyceryl trinitrate and pentaerythritol tetranitrate after administration of a two–step preparation. *Arzneim. Forsch.* 27:416–519, 1977.

58. Nomura, Y., H. Kajiyama, and T. Segawa. Alteration in sensitivity to isoproterenol and acetylcholine in the rat heart after repeated administration of isoproterenol. *J. Pharmacol. Exp. Ther.* 220: 411–416, 1982.
59. Opie, L. H. Calcium antagonists. *Lancet* 1:806–810, 1980.
60. Packer, M., J. Meller, R. Gorlin, and M. V. Herman. Hemodynamic and clinical tachyphylaxis to prazosin–mediated afterload reduction in severe chronic congestive heart failure. *Circulation.* 59:531–539, 1979.
61. Packer, M., J. Meller, N. Medina, and M. Yushak. Sustained effectiveness of minoxidil in heart failure after development of tolerance to other vasodilator drugs. *Am. J. Cardiol.* 48:375–379, 1981.
62. Parmley, W. W., J. L. Rouleau, and K. Chatterjee. Vasodilators in heart failure secondary to coronary artery disease. *Am. Heart J.* 103:625–632, 1982.
63. Perez, J. E., L. Borda, R. Schuchlieb, and P. D. Henry, Inotropic and chronotropic effects of vasodilators. *J. Pharmacol. Exp. Ther.* 221:609–613, 1982.
64. Perry, H. M. Late toxicity to hydralazine resembling systemic lupus erythematosus or rheumatoid arthritis. *Am. J. Med.* 54:58–72, 1973.
65. Plummer, A. L. The development of drug tolerance to β2-adrenergic agents. *Chest* 73:949–957, 1978.
66. Polacek, I., J. Bolan, and E. E. Daniel. Accumulation of adenosine 3′,5′–monophosphate and relaxation in the rat uterus *in vitro. Can. J. Physiol. Pharmacol.* 49:999–1004, 1971.
67. Rane, A., and J. T. Wilson. Clinical pharmacokinetics in infants and children. *Clin. Pharmacokinet.* 1:2–24, 1976.
68. Reed, C. E. Physiology and pharmacology of β2–adrenergic agents. *Chest* 73:914–917, 1978.
69. Reidenberg, M. M., D. Drayer, A. Demarco, and C. T. Bello. Hydralazine elimination in man. *Clin. Pharmacol. Ther.* 14:970–977, 1973.
70. Rhodes, H. J., and M. C. Sutter. The action of diazoxide on isolated vascular smooth muscle. Electrophysiology and contraction. *Can. J. Physiol. Pharmacol.* 49:276–287, 1971.
71. Roach, A. G., F. LeFevre, and I. Cavero. Effect of prazosin and phentolamine on cardiac presynaptic α-adrenoceptors in the cat, dog and rat. *Clin. Exp. Hypertension.* 1:87–101, 1978.
72. Rosenthal, R. R, Effects of β-adrenergic agents on mediator release. *Chest.* 73 (Suppl.):972–973, 1978.
73. Ross, G. and P. V. Cole. Cardiovascular actions of sodium nitroprusside in dogs. *Anaesthesia* 28:400–406, 1973.
74. Rouleau, J. -L., J.W. Warnica, and J. H. Burgess. Prazosin and congestive heart failure: short and long–term therapy. *Am. J. Med.* 71:147–152, 1981.
75. Sato, M., T. Nagao, I. Yymaguchi, H. Nakijima, and A. Kiyomoto. Pharmacological studies on a new 1,5–benzothiazepine derivative (CRD,401-diltiazem). *Arzneim. Forsch.* 21:1338–1343, 1971.
76. Schomerus, M., B. Spiegelhalder, B. Stieren, and M. Eichelbaum. Physiological disposition of verapamil in man. *Cardiovasc. Res.* 10:605–612, 1976.
77. Schulman, H. The comparative actions of uterine inhibiting drugs. *Am. J. Obstet. Gynecol.* 130:684–688, 1978.

78. Stanton, H. C., and E. D. Mayfield. Studies on isoproterenol–induced cardiomegaly in rats. *Am. Heart J.* 77:72–80,1969.

79. Talano, J. V., and C. Tommaso. Slow channel calcium antagonists in the treatment of supraventricular tachycardia. *Prog. Cardiovasc. Diseases.* 25:141–156, 1982.

80. Talseth, T. Clinical pharmacokinetics of hydralazine. *Clinical Pharmacokinet.* 2:317–329, 1977.

81. Todd, G. D., and R. G. McAllister. Distribution of verapamil and its active metabolite, norverapamil, in rat and dog. *Pharmacologist* 23:167, 1981.

82. Vater, W., G. Kroneberg, F. Hoffmeister, H. Kaller, K. Meng, A. Oberdorf, W. Puls, K. Schlossman, and K. Stoepel. Zur Pharmacologie von 4–(2′nitrophenyl)–2,6–dimethyl–1,4–dihydropyridin–3,5–dipyridin–3, edin ethylester (nifedipine, Bay a 1040). *Arzneim Forsch.* 22:1–14, 1972.

83. Weber, K. T., V. Andrews, G. T. Kinasewitz, J. S. Janicki, and A. Fishman. Vasodilator and inotropic agents in treatment of chronic cardiac failure: clinical experience and response in exercise performance. *Am. Heart J.* 102:569–577, 1981.

84. Weinberger, M., and L. Hendeles. Pharmacotherapy of asthma. *Am. J. Hosp. Pharm.* 33:1071–1080, 1976.

85. Weinberger, M., L. Hendeles, and R. Ahrens. Clinical pharmacology of drugs used for asthma. *Pediatr. Clin. N.A.* 28:47–75, 1981.

86. Williams, D. O., E. A. Amsterdam, and D. T. Mason. Hemodynamic effects of nitroglycerin in acute myocardial infarction. Decreases in ventricular preload at the expense of cardiac output. *Circulation* 51:421–427, 1975.

87. Yasue, H., S. Omote, A. Takizawa, M. Nagao, K. Miwa, and S. Tanaka. Exertional angina pectoris caused by coronary artery spasm: effects of various drugs. *Am. J. Cardiol.* 43:647–652, 1979.

Chapter 6

Cell-to-Cell Communication in Smooth Muscle

R. E. Garfield

1. Introduction

During the last two decades many structural studies have focused on cell-to-cell contacts and intercellular communication. These studies have been complemented in some cases with functional studies. The rationale behind these inquiries is obvious since cell-to-cell communication may have a profound influence on cell and tissue function.

It has long been recognized that cells communicate with one another by long-range (humoral) or short-range (synaptic or neuromuscular) communication systems through chemical messengers. It is only recently that direct cell-to-cell contacts have been visualized and the concept of direct communication between the interiors of cells has been recognized.

It is now acknowledged in most types of cells that specialized types of cell-to-cell contacts, termed gap junctions, may form the structural foundation for intercellular communication. Gap junctions are now regarded as the structural correlate for direct ionic (electrical) and metabolic exchange between most types of cells. Recent studies have shown that the junctions may be dynamic structures between some cells and they may be regulated in a variety of ways.

Investigations of various smooth muscles have shown that the presence of gap junctions does not always correlate with electrical coupling between cells. These results have led to debates about whether gap junctions are necessary for cell-to-cell coupling and about the function of gap junctions between smooth muscle cells.

This chapter will deal mainly with the structure and function of gap junctions and the concept of the existence of channels between cells. These subjects have been reviewed extensively for other cell systems (*1, 27, 28, 32, 37, 39, 45, 57*) and also form parts of reviews of the structure and physiology of smooth muscle (*4, 15, 31, 51*). Particular attention will be devoted to recent studies of gap junctions in the myometrium and their presence and regulation during parturition (*16–26*).

2. Cell-to-Cell Junctions

2.1. Structural Studies

The cells in any smooth muscle tissue are usually separated from each other by wide spaces that are occupied by extracellular material. The smooth muscle cells make contact at selected points and form junctions between the plasma membranes. Cell-to-cell junctions have been observed by thin section electron microscopy in all types of smooth muscle tissues. There are at least four different types of junctions between smooth muscle cells, but the number and type of junction may vary from one tissue to another (*15*). The types of contacts have been termed (1) intermediate contact, (2) simple apposition or close contact, (3) interdigitation and (4) gap junction or nexus (*31*).

Intermediate contacts are junctions characterized by an increased density of the cytoplasm in the apposing cell membranes as seen in thin section (Fig.1). These junctions resemble the zona adherens (desmosomes) seen between epithelial cells and for this reason they have been referred to as desmosome-like in smooth muscle (*15, 30, 31*). The intercellular space of the intermediate contact is wider (from 20 to 100 nm) than the other types of cell contacts in smooth muscle. The structure of the intermediate junctions has not been defined in freeze-fracture replicates and the function of these contacts is not known. It has been suggested that the structure of the intermediate contacts is similar to the intercalated disc found between cardiac muscle cells and that they therefore may serve as a mechanical support to fasten cells together (*15*).

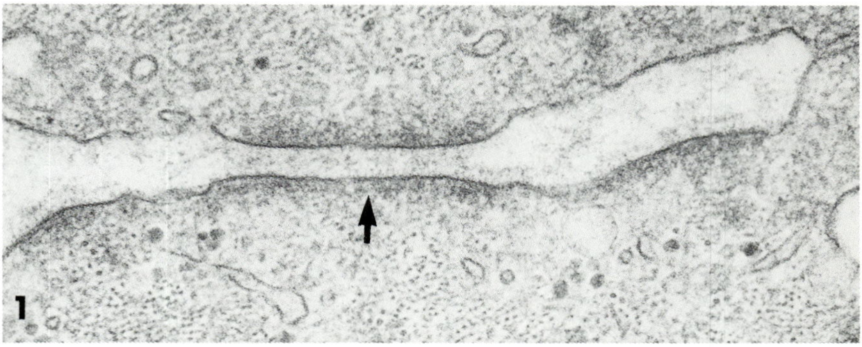

Fig. 1. Electron micrograph of an intermediate contact between two smooth muscle cells from rat uterus as seen in thin section. Note the increased density of the cytoplasm on each side of the contact as well as between the membranes in the extracellular space. ×81,000.

Areas of very close approximation of cell membranes (10 nm or less) without any other specialized features have been termed simple apposition or close contact (*15, 31*). Like the intermediate contacts, the close contact has not been attributed to any structures seen in freeze-fracture replicates of the plasma membrane and the function of these contacts, if any, is also not known. The structure of these contacts does not appear to offer any advantages for anchoring the cells together. It has been proposed that the simple apposition type contact may be involved in electrical coupling between smooth muscle cells which lack gap junctions (*15*). The simple apposition may be analogous in some cases to a very small gap junction or point contact but because they are small they cannot fulfill the criteria of a gap junction (i.e., 7-lines, *see below*). Because these contacts may occupy only a limited region of the cell surface, even if numerous, the probability of observing them in sections would be extremely low (*see* Fig. 2). These contacts also may be sites of gap junction formation in some types of smooth muscle tissues.

A third type of contact found between smooth muscle cells are the interdigitations (*15, 16, 31*). These contacts are identified by the minute protrusion of one cell into an indentation of another cell. The membranes of the interdigitation are usually accompanied by a close apposition type contact. Gap junctions have also been found on the indentation of one cell into another. The function of the interdigitation-type contact is unknown; however, these types of contacts are more frequently found in contracted tissues where the cells have a more contorted surface (*15, 16, 31*).

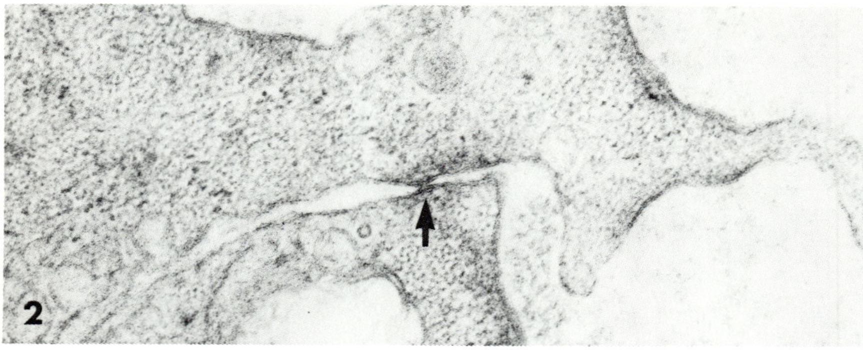

Fig. 2. Close or point contact (arrow) between two myometrial cells from a sheep. ×81,000.

A gap junction is a specialized type of cell-to-cell contact that in thin section usually appears as two symmetrical portions of the plasma membranes from two apposing cells (Fig. 3). Often, but not always, gap junctions are located on protrusions between cells or on protrusions of one cell into another (interdigitation). Gap junctions do not necessarily involve components from two different cells since they have been observed between plasma membranes of the same cell on cytoplasmic extensions (reflexive gap junction).

Formerly gap junctions were referred to as nexuses (meaning connections or bonds) because the plasma membranes look as if fused together into a five-layered structure (*9*). It is now recognized that after *en bloc* staining a 2–3 nm gap is present between the cell membranes and these comprise a seven-membered struc-

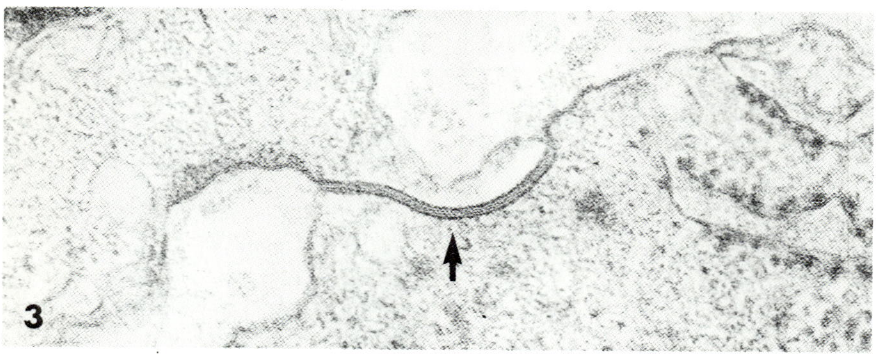

Fig. 3. Gap junction (arrow) between two uterine muscle cells from pregnant rabbit fixed during parturition. ×81,000.

4

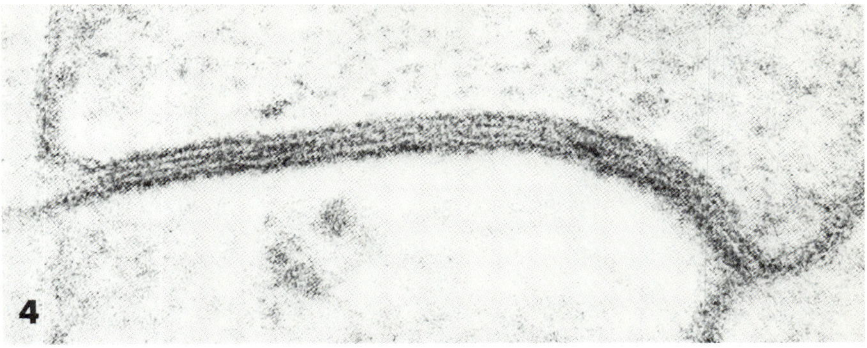

Fig. 4. High magnification micrograph of gap junction from pregnant rat uterus at parturition. Note 7-lined appearance. ×273,000.

ture. The septilaminar structure of the gap junction (Fig. 4) is thought to represent the two plasma membranes (outer and inner aspects) and the gap between the cells (i.e., four electron-opaque and three electron-translucent lines). The 5- or 7-lined appearance has been used as criteria for identification of the gap junction.

Gap junctions have been extensively studied in other tissues. A number of excellent reviews on the structure and function of gap junctions have been published within the last few years (*1, 28, 32, 39, 45*). Intramembraneous particles (proteins), which protrude through each cell membrane, are thought to align and bridge the gap between the membranes (*39, 45*). Colloidal lanthanum or other tracers and substances penetrate the extracellular spaces and outline the arrangement of the intramembraneous particles (*45*). When a gap junction exposed to lanthanum has been cut *en face,* the lanthanum outlines an array of hexagonal structures that represent the membrane particles or proteins (*45*). Figure 5 shows a gap junction between myometrial cells after fixation in glutaraldehyde with 8% tannic acid. The junction has been sectioned tangentially and the tannic acid reveals a regular periodicity between the membranes corresponding to the intramembraneous particles spanning the membranes.

Gap junctions appear as an aggregation of membrane particles or an array of pits in freeze-fracture replicates (Fig. 6) (*1, 28, 41*). The particles represent partially exposed membrane proteins and the pits are the impression left by the particles when the membrane is cleaved (*1, 28*). In a fracture face through a region of a gap junction, the P face (protoplasmic portion of the membrane) of the fracture usually contains a circular array of protruding particles about 7 nm in diameter with a 7–14 nm center-to-center

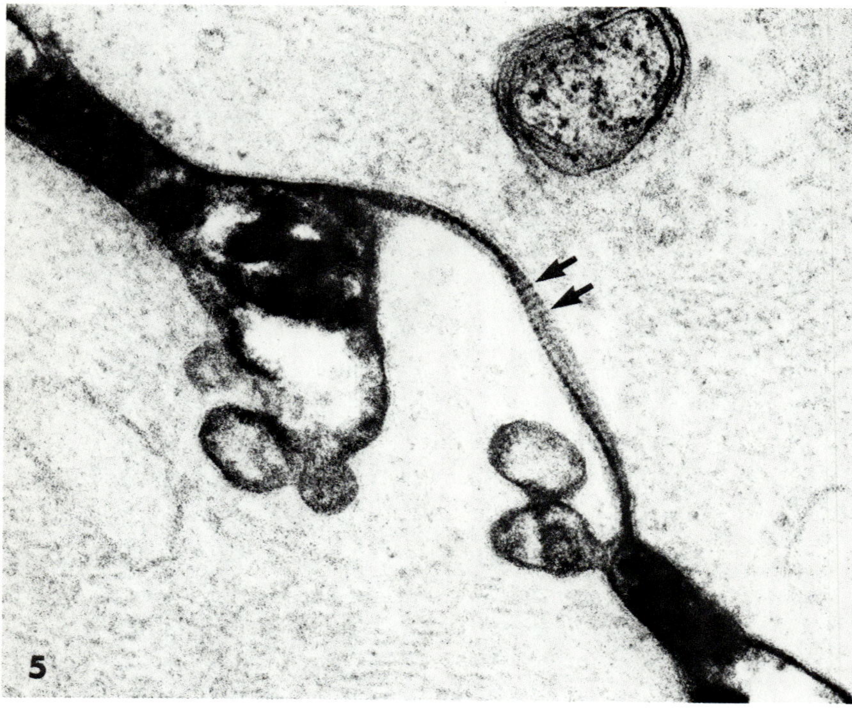

Fig. 5. Gap junction sectioned tangentially after fixation in glutaraldehyde and 8% tannic acid. Note that the electron dense tannic acid penetrates into the exterior portion of the junction and delineates a periodicity (arrows) that corresponds to the external part of the membrane particles in alignment between the two cell membranes. ×173,000.

spacing of particles (*45*). This array of particles is usually regular, but may vary depending upon the functional state of the contact. The E fracture face (extracellular face of membrane) contains the pits corresponding to the attachment sites of the proteins. In smooth muscle the particles more frequently adhere to the P face than to the E face of the membrane. Membrane protein aggregates of only a few to hundreds of particles have been observed in replicates of smooth muscle and these are thought to represent gap junctions of various sizes (*15, 18*). It has been argued that an array of particles should not be considered a gap junction unless the fracture plane steps from one cell membrane to the other cell in the region of the junction because particle arrays may be present in the membrane and not associated with another symmetri-

Fig. 6. Electron micrograph of a freeze-fracture replicate showing portions of an oval-shaped gap junction between muscle cells of the rat uterus. The fracture plane has passed between cell membranes within the region of the gap junction (larger arrow) to reveal the P- (P) and E-face (E). Note the pits in the E-face of junction and the aggregation of particles (proteins) in the P-face. Small arrows indicate caveolae on the surface of the cell. ×83,000.

cal region on an opposing cell membrane (*45*). However, replicates stepping between membranes have not always been used to identify gap junctions and replicates stepping between small gap junctions composed of only a few particles would be difficult or impossible to interpret. In this context it is interesting to note that clusters of membrane particles have been found associated with membrane receptors for hormones or other substances that have no relationship to cell-to-cell contact (*45*). It should also be recognized that exposure of membranes to low temperatures or fixation may result in particle aggregation that may resemble gap junctions. Another disadvantage to freeze-fracture may be the inability to discriminate between cell types in replicates of tissues composed of a mixture of cells (i.e., fibroblasts vs smooth muscle).

Gap junctions are present between most types of cells at least during some phase of life of the cell (*28, 45*). Though present between cells throughout the animal kingdom, the number or frequency of gap junctions varies considerably depending upon the tissue or cell type (*45*). As examples, gap junctions comprise 3.7% of the plasma membrane area in mammalian heart, 1.5% in liver, 0.05% in fibroblasts, and 80% in lens cells (*28, 45*). The only cell types known to be incapable of intercellular communication are ones that either do not undergo cell division, such as skeletal muscle (gap junctions are present in developing muscle, i.e., between myoblasts) or cells that are not part of an organized tissue, i.e,. blood cells. It is apparent that there are many configurations of gap junctions based upon the wide variety seen in published micrographs. Interestingly, gap junctions have been shown to form between a variety of cultured cells from different species and different cell types (*39, 45*).

Gap junctions have been observed in most but not all types of smooth muscle. Henderson (*31*) has made a very thorough review of the early studies of smooth muscle. There are at least five smooth muscle tissues where the presence of gap junctions has been questioned. These tissues include the longitudinal muscle of the intestine and taenia coli, vas deferens, oviduct, and myometrium (*15, 31*). In the myometrium, it is now recognized that gap junctions are present during periods when the muscle is active at parturition. The failure to find gap junctions in other smooth muscles may arise in part from the lack of appropriate criteria to identify the junctions and to incomplete quantification. Quantitative estimations based upon cell junctions per number of smooth muscle cells or similar measurements are inadequate.

2.2. Methods for Quantification of Gap Junctions

Two ultrastructural methods are used to demonstrate the presence of gap junctions between cells in any tissue, although functional studies of the presence of electrical or dye coupling has come to imply the existence of gap junctions between cells. One method is to determine the presence of groups or aggregates of particles or pits in freeze-fracture replicates of the membrane. The other method is the occurrence of a 5- or 7-lined structure of the membranes as seen by thin section microscopy.

It has been suggested that for gap junction demonstration and analysis the freeze-fracture method may be more dependable or accurate than the thin section method for smooth muscle (*15*) and other cell types (*28*). These claims are made even though there are very few quantitative studies of gap junctions using freeze-fracture methods and no studies comparing the two methods.

We have shown by analysis of thin sections that gap junctions develop between myometrial cells during parturition (*17–26*). We have confirmed these results using freeze-fracture methods (*18*). More recently, these studies have been duplicated in a number of laboratories and similar results obtained (*3, 42, 60*). Quantitative morphometric analyses of thin sections were used in our studies to determine the number and area of gap junctions. It should be pointed out that the estimates on thin sections do not include measurements of the membrane vesicles or the caveolae and therefore the final proportion is slightly overestimated. In addition, the quantity of junctions is underestimated because of the inability to identify all the junctions in sections cut at angles that do not reveal a 5- or 7-lined structure at least at some portion of the junction. Another disadvantage to analysis of thin sections is the inability to detect small gap junctions (ca. 30 nm in diameter or from 1 to 3 channels).

We have used a morphometric analysis system based upon principles described by others (*61*) to quantify gap junction area. This system is convenient, very rapid, simple, and inexpensive except for costs of photographic paper. A transparent grid is superimposed over photomicrographs of smooth muscle cells. The total length of the cell membrane in each photograph is determined by the formula:

$$B = \frac{\pi}{2} \times \frac{I}{L} \times A \times CF$$

B is the cell boundary or length of plasma membrane; *I* is the average number of intersects determined from two estimates taken at right angles to each other; *L* is the total length of the probe lines of the grid; *A* is the area of the grid; and *CF* is the magnification correction factor. Using these methods we expressed our data as number of gap junctions per unit length of membrane (*17–21*). The validity of this method has been confirmed in other studies of uterine muscle (*60*).

More recently, we have expressed our data as the proportional area of gap junctional membrane per area of plasma membrane as a percentage, by using the following formula (*23, 24*):

$$\left(\frac{A_j}{A_{pm}}\right) \times 100 = \left(\frac{L_j \times 2}{L_{pm} + L_j \times 2}\right) \times 100$$

where A_j and L_j are the area of gap junction membrane and the total cross-sectional length of gap junction membrane (sum of all gap junction lengths) and A_{pm} and L_{pm} are the area and length of all membranes (gap junction + nongap junction). The cross-sectional length of gap junctions (L_j) is corrected (× 2) to obtain the area of gap junctional membrane (A_j) because gap junctions comprise components of two cells, and in determination of gap junctional area (A_j) both components must be considered.

2.3. Functional Studies of Gap Junctions

There is convincing evidence that gap junctions form a pathway between cells to provide for the passage of current (electrical coupling or ionic coupling); i.e., gap junctions are low resistance or low impedance contacts between cells (*1, 9, 27, 39*). Whether junctions are necessary for electrical coupling is a question considered by all investigators. However, it is generally believed that the junctions are sufficient for electrical coupling. There are numerous studies that show that when current pulses are injected into one cell with a microelectrode, a voltage change occurs in an adjacent cell impaled with a microelectrode if the cells are structurally coupled by gap junctions (*1, 39, 45*). Most of these studies have been accomplished on cultured cells or simple cell systems where the geometry is not as complex as the arrangement of cells in intact smooth muscle. There have been no similar studies of the relationship of electrical coupling and gap junctions in cultured smooth muscle. Figure 7 shows a simplified model of how ionic current may spread between cells coupled by gap junctions.

Electrical coupling among cell aggregates has been observed in the absence of ultrastructurally defined gap junctions (*6*). These investigators propose that organized gap junctions are unnecessary for electrotonic coupling and that isolated channels scattered throughout the membrane area of closely apposed cells may provide low resistance cell-to-cell channels. A similar arrangement or organization of channels may be present in some types of smooth muscle tissues where organized gap junctions have not been observed. There is little evidence that electrical coupling can occur in the absence of cell-to-cell channels.

It has been proposed that cell-to-cell propagation of current between myocardial cells does not require the presence of low resistance junctions between cells (*48*). This hypothesis is based upon a model for electrical transfer of excitation from one cell to

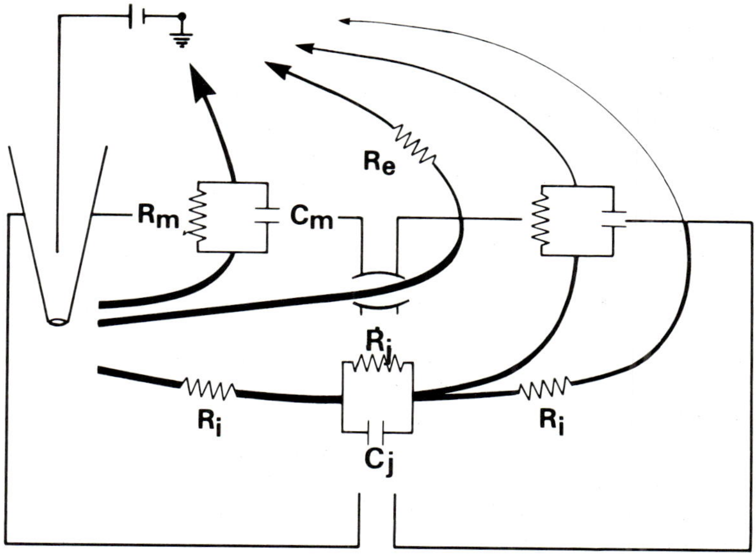

Fig. 7. Simplified model of current spread between two cells. When pulses (either hyperpolarizing or depolarizing) are injected into one cell with a microelectrode (upper left) the current will flow by the movement of charged ions in a circular motion back to the ground. Since the resistances in the intracellular (R_i) and extracellular (R_e) spaces are low because of the presence of electrolytes, little attenuation of current will occur in these areas. However, because the membrane resistance (R_m) is much greater than the junctional resistance (R_j) current will follow the pathway of least obstruction and flow between cells.

the next that utilizes the electrical potential generated in the junction cleft between cells. Unfortunately this model will be very difficult to verify experimentally and may be unnecessary if all cells are coupled by gap junctions in one form or another.

Gap junctions also provide a pathway for the direct exchange of metabolites (metabolic coupling) between cells via permeable intercellular channels (*32, 33, 45, 46*). It seems reasonable to conclude that gap junctions or multiple single dispersed channels are required for metabolic cooperation between cells. Without these pathways, large molecules must enter the extracellular space through the plasma membrane, which they do not normally penetrate, and exchange occurs between cells. This mechanism appears unlikely. Studies using radioactive tracers or fluorescent tracers have shown that charged and uncharged molecules can diffuse from cell to cell through channels approximately 2 nm in diameter (*39, 45*).

It has become apparent that cells connected by gap junctions may not necessarily always show a free exchange of dyes, isotopes, and current (*39, 45*). Electrophysiological evidence has shown that gap junctions can rapidly switch from a low resistance to a high resistance state (*39, 45*). It is now recognized that the channels or pores created by the gap junction may not always be in the open state (*1, 27, 32, 29, 44, 45, 53, 59*). There may be times when the channels are open and times when they are closed. This uncoupling mechanism may be a safety device to uncouple injured cells or to uncouple cells when metabolically or electrically desirable (*39*). Functional uncoupling has been achieved by changes in internal and external Ca^{2+} and pH (*1, 27, 32, 39, 44, 45, 53, 59*). Peracchia (*45*) has proposed that Ca^{2+} or pairs of hydrogen ions may bind to negative sites on the gap junction proteins normally occupied by K^+ and produce conformational changes to close the channels.

The available evidence indicates that gap junctions are sufficient for electrical and metabolic communication between cells when present in the open state. Electrical and metabolic cooperation between cells may control a number of physiological processes of cells. Lowenstein (*39*) has suggested that the most basic function of the gap junction is to maintain homeostasis and to buffer individual variations in channel-permeant molecules. In coupled cell systems, concentration variations, including changes in nutrients, would be rapidly distributed between all cells. Other roles considered for the junctions are: (1) intercellular movement of regulatory molecules such as cyclic nucleotides, (2) transmission of electrical signals that effect voltage sensitive processes in cells, (3) control of cell growth, proliferation, and differentiation, (4) propagation and synchronization of electrical activity in heart, smooth muscle, and nerve cells (*39, 45*).

2.4. Regulation of Gap Junctions

Gap junctions appear to be dynamic structures in most cell systems but not all. The formation of gap junctions has been studied primarily in vitro where cells are separated and allowed to reaggregate and form communicating junctions (*35, 37, 46*). The formation of gap junctions has been evaluated structurally and functionally by electrophysiological measurements, passage of dyes or isotopes, or by synchronous contractility in the case of cardiac cells (*19, 27, 28, 32, 33, 35, 37, 39, 45, 46, 59*). The formation of gap junctions in some cells, but not all, appears to involve the synthesis of new proteins (*28*).

In cell systems where gap junctions are believed to be dynamic structures, gap junction degradation is thought to occur by either dispersal of gap junction particles within the plasma membrane or internalization of the entire gap junction within one cell. The latter mechanism has been suggested by many because internalized gap junctions (annular) have been observed in many types of cells (*28*). Changes in gap junction numbers and sizes have been noted in response to a number of hormones (*37*).

3. Myometrial Gap Junctions

3.1. Presence of Myometrial Gap Junctions

Gap junctions have only recently been demonstrated satisfactorily in uterine muscle (*17–26*). These studies began several years ago following reports that described the presence of gap junctions in other smooth muscle tissues (*8, 9, 13, 30, 58*), including uterine smooth muscle (2). We were unable to find gap junctions in the myometrium in nonpregnant animals injected with estrogen or in tissues from pregnant animals prior to term (*16*). However, when we prepared myometrial tissues from animals that were in the process of delivering their young, we found numerous gap junctions between the cells (*17*).

Later we found that gap junctions were present between myometrial cells in increased number and sizes only in tissues from pregnant rats that were at term, during delivery and immediately postpartum (*18*). We proposed that gap junctions may be essential for normal labor and delivery for synchronous contraction of the myometrium. A model was proposed for the control of labor and parturition (*18*). The model (Fig. 8) suggested that some hormonal event (Step I) brings about the formation of gap junctions between the smooth muscle cells (Step II). The presence of gap junctions then leads to labor (Step III) and delivery of the fetus. The formation of the gap junctions (Step II) may or may not lead automatically to labor. A trigger mechanism from a stimulant such as oxytocin, prostaglandins, or stimulation from pacemaker cells may be necessary to excite the muscle once gap junctions have formed (Step III); i.e., there needs to be a source of current as well as a low resistance pathway for it. Agents used to induce

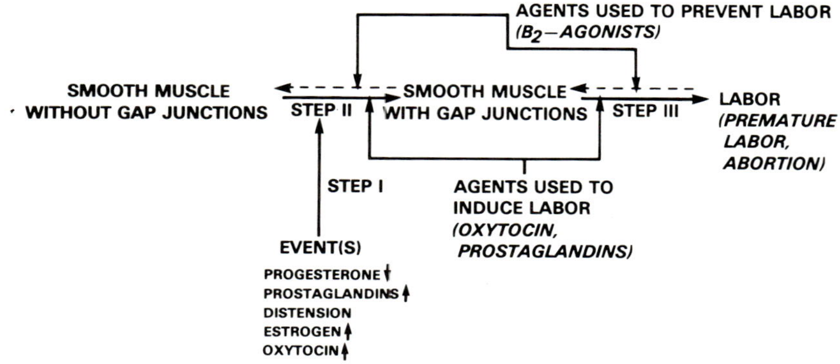

Fig. 8. Model of role and regulation of gap junctions by myometrium during parturition. Some event (Step I, change in hormones or distention) brings about the formation of the junctions between smooth muscle cells (Step II). The development of the junctions between cells leads to the conversion of the muscle to an active and reactive tissue and labor occurs (Step III). Agents used to stimulate labor (oxytocin and prostaglandins) may either promote the formation of gap junctions (Step II) or stimulate the muscle once the junctions have formed (Step III). Drugs used to inhibit labor (β_2-agonists) may either inhibit gap junction formation, reverse this process (Step II) or inhibit the muscle cells directly. (From Garfield et al., 1978.)

labor may act either to stimulate gap junction formation (Step II) or stimulate the muscle directly once gap junctions have formed (Step III). Drugs used to inhibit labor may either prevent gap junction formation or reverse this process (Step II), or may inhibit the muscle cells directly once the contacts have formed (Step III). The consequences of drug actions at different sites are likely to differ in important ways; thus inhibition of gap junction development is likely to have more permanent effects than inhibition of the contractile responses of the muscle cells directly.

Further studies of other species have shown that gap junctions occupy a significant portion of the cell membrane only during a limited period, during parturition, as in the rat (*19, 20, 26, 47*). The area of gap junctions, as a percentage of myometrial plasma membrane area, during pregnancy, parturition, and postpartum in various animals is shown in Fig. 9. These studies show that regardless of the species: (1) The myometrial gap junctions

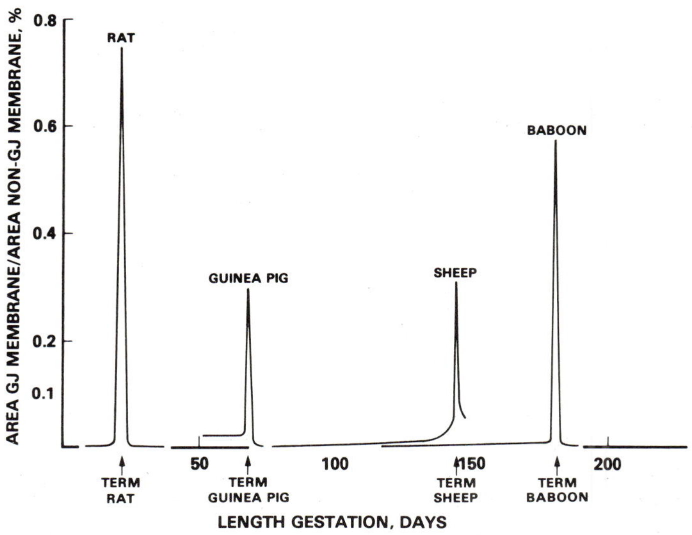

Fig. 9. Relative area of myometrial gap junction (GJ) membrane (percentage area of GJ membrane per area of plasma membrane) versus time during pregnancy, delivery, and postpartum for rats, guinea pigs, sheep, and baboons. For purposes of illustration, tissues from animals either delivering or postpartum were referenced to the term point regardless of their actual day of gestation.

are absent or present in low frequency throughout pregnancy. (2) At term, gap junction frequency increases. (3) The junctions are present in high frequency and increased sizes during delivery of the fetuses. (4) The gap junctions revert back to the predelivery state, low frequency and small sizes, following delivery. The above studies demonstrate that during parturition there are approximately 1000 gap junctions on each smooth muscle cell coupling that cell to surrounding cells. These estimates are based upon the relative mean area of gap junction membrane to plasma membrane area (0.2%, *see* ref. *22*), the mean area of a myometrial gap junction (0.06 μm^2), and the average surface area of a smooth muscle cell (23,000 μm^2 minus the caveolae).

The junctions also appear in human myometrium during term and preterm labor (*24*). These studies show that in pregnancy a good correlation exists between the presence of myometrial gap junctions and increased frequency of uterine contractions or increased cervical dilation. However, there are many difficulties in obtaining human specimens all from the same uterine region and from a homogenous population of women. Furthermore, all human tissues for these studies were obtained during cesarian section which may not appropriately represent spontaneous labor and delivery.

Gap junctions are also present in myometrial tissues from animals induced to deliver prematurely (*25, 26*) and in high frequency in human myometrium during premature labor (*24*). These studies further demonstrate the importance of the presence of gap junctions. Gap junctions are present in all conditions studied that lead to parturition (ovariectomy and prostaglandin treatment) and absent in tissues from control animals or animals treated similarly, but prevented hormonally from delivery (progesterone treatment). Thus, myometrial gap junctions appear to be necessary for term or preterm parturition.

Myometrial tissues incubated in physiological solutions develop large numbers of gap junctions and this system has been used to study control mechanisms (see below). Tissues from midterm pregnant animals, where gap junctions are rare, begin to develop gap junctions in significant numbers after about two hours incubation (*18, 21, 22*). It is then obvious that electrophysiological studies which were done in vitro (*36*) must be interpreted with caution when extrapolated to the muscle in vivo. It is clear, however, that tissues in vitro behave differently than those in vivo in regard to gap junctions; they develop more (2 or 3 times the in vivo value at 48 h incubation) and once they develop they fail to decrease as they do in vivo following delivery (*17, 22*).

3.2. Function of Myometrial Gap Junctions

The presence of gap junctions between uterine smooth muscle cells exclusively limited to the period just before, during, and immediately following normal and premature parturition has significant implications for the maintenance and termination of pregnancy. The absence of gap junctions between smooth muscle cells throughout gestation may maintain pregnancy by limiting electrical or metabolic communication between cells, thereby preventing coordinated contractions of the uterus. The rapid formation and appearance of gap junctions may initiate or allow initiation of parturition and terminate pregnancy by providing large areas of low resistance between cells, allowing spread of electrical information. The spread of current over the myometrium may synchronize uterine contractility and terminate pregnancy by allowing effective expulsive forces by the wall of the uterus. That large gap junctions are necessary for effective labor, regardless of their function, is seen in the many observations that indicate their presence during term and preterm labor and their absence or reduction in tissues from similar animals prevented experimentally from labor (*3, 17–26, 29, 42, 47, 49, 60*).

The conversion of uterine muscle from inactive to active at the end of pregnancy is evident in the development of synchronous electrical and mechanical activity within different regions of the uterus (*7*). In all species, parturition is accompanied by rapidly propagating trains of electrical discharge that synchronously activate billions of individual muscle cells that compose a major portion of the uterine wall (*7, 10–12, 56*). The ability of current to spread between muscle cells and induce synchronous activity is usually taken to imply the existence of low resistance contacts between the cells. The development of large areas of gap junction contacts between the muscle cells may be the structural basis for the synchronized activity of the tissue during labor. When gap junctions are present current may then spread down the surface of the smooth muscle cells by a local circuit mechanism and propagate between cells through the gap junction.

Two independent methods have recently been used to evaluate the relationship between electrical coupling in the myometrium and the development of increased gap junctional area (*49*). Cable properties of the myometrium were determined in myometrial tissues before and during labor and delivery. These studies reveal that the space constant (λ) was significantly increased in tissues from delivering animals (3.7 ± 1.0 mm) as compared to similar animals that were not delivering (2.6 ± 0.8 mm). Measurements of electrical impedance in myometrium also showed lower junctional resistivity in tissues from animals during delivery (139 Ωcm) compared to junction resistence immediately prior to delivery (375 Ωcm) and 1.5–2 d postpartum (1451 Ωcm). Therefore, these two in vitro studies provide electrophysiological evidence for better coupling of myometrial cells during delivery when gap junctions are present.

The fact that relatively good electrical coupling is observed between cells prior to the in vivo development of the junctions may mean that the contacts develop in vitro before or during recording or that small gap junctions are sufficient for coupling. There have been no studies of the relationship between gap junctions and electrical measurements of the myometrium in vivo. However, electrical and mechanical activity of the myometrium of various animals, as recorded by EMG (electromyography) and intrauterine pressure devices, increase at a time when the gap junctions are present (*7, 10–12, 63*).

Recordings of EMG activity at various sites on the myometrium have shown that bursts of action potentials are asynchronous prior to labor and become synchronous (phase-locked) during labor (*7, 63*). The possible explanations for this phenomenon are (1) increased cell-to-cell coupling and propagation of electrical events from region to region, or (2) coordinated release or response to a neurotransmitter or hormone. The latter has not been completely eliminated as a possible source of initiating simultaneous activity in different regions, but the former does correlate with the development of gap junctions.

Possibly another important role for gap junctions in the myometrium during labor is to connect the cells for metabolic cooperation (*32, 33, 45, 56*). If so, then the cells of the myometrium may be metabolically synchronized during labor. Studies of metabolite transfer between myometrial or other smooth muscle cells have not been reported. It should be remembered that gap junctions are present in nonexcitable cells in which electrical coupling may not be essential. Perhaps in these cells gap junctions synchronize or coordinate metabolism and they may do so in smooth muscle as well.

Another important role of gap junctions may be to control responses of myometrial tissues to drugs and receptors. A myometrial tissue with gap junctions would be expected to be more sensitive to exposure to a given agonist than a tissue without gap junctions. Perhaps this is one important mechanism responsible for increased reactivity and sensitivity of the myometrium to oxytocin or other agonists at term (*50*). Figure 10 shows how an increase in gap junctions could increase sensitivity of the myometrium to a drug released from a blood vessel. A drug must occupy the receptors on most of the muscle cells to produce a response when gap junctions are absent. When gap junctions are present only receptors on some cells need be occupied as the response is propagated from cell to cell. Increased sensitivity would shift the dose-response curve to the left (Fig. 10). An increase in reactivity of the muscle to a drug would be manifested in an elevation in maximum response, but may not shift the dose–response curve to the left.

Similarly the presence of gap junctions in a tissue could play a major role in controlling muscle responses to release of transmitters from nerves. Burnstock (*4*) has proposed that gap junc-

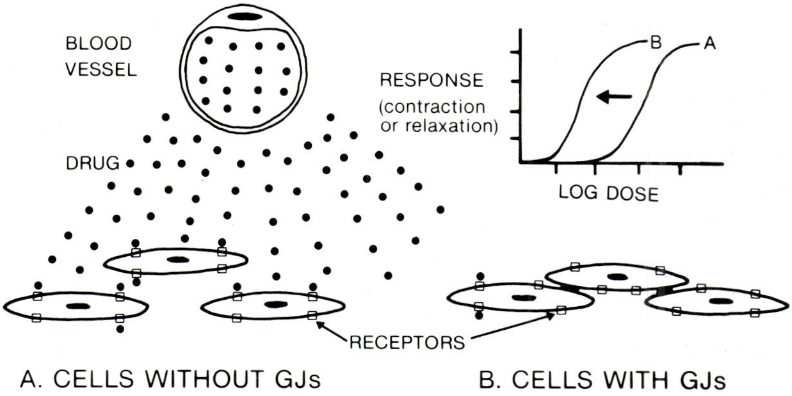

Fig. 10. Model for the increased responsiveness of smooth muscle tissues when gap junctions (GJs) are present between cells. A higher concentration of a drug is required to produce a response (contraction or relaxation) in a tissue without gap junctions (A) than in a tissue with junctions (B). In A the response is proportional to the occupation of receptors on all cells and in B the response is dependent upon occupation of receptors in cells close to blood vessels as the reaction is propagated from cell to cell.

tions are necessary for coupling of smooth muscle cells in sparsely innervated tissues. Nerves that terminate only on a few smooth muscle cells (key cells) in a muscle bundle, generate an electrical response in these cells as a result of release of transmitter and interaction with membrane receptors. The electrical response is transmitted from cell to cell via the gap junctions. The presence of nerves, transmitter, receptors, and gap junctions are all necessary for this mechanism to function. Figure 11 shows the relationship between a nerve varicosity and smooth muscle cells in the circular muscle of the intestine. Nerves in the uterus are thought to degenerate during the late stages of pregnancy (*40, 52, 54, 55*) when gap junctions may be forming. A similar inverse relationship between the presence of nerves and gap junctions has been described for other smooth muscle tissues. For example, denervation of rat vas deferens and decreased innervation of intestinal muscle result in increased gap junctions (*14, 62*).

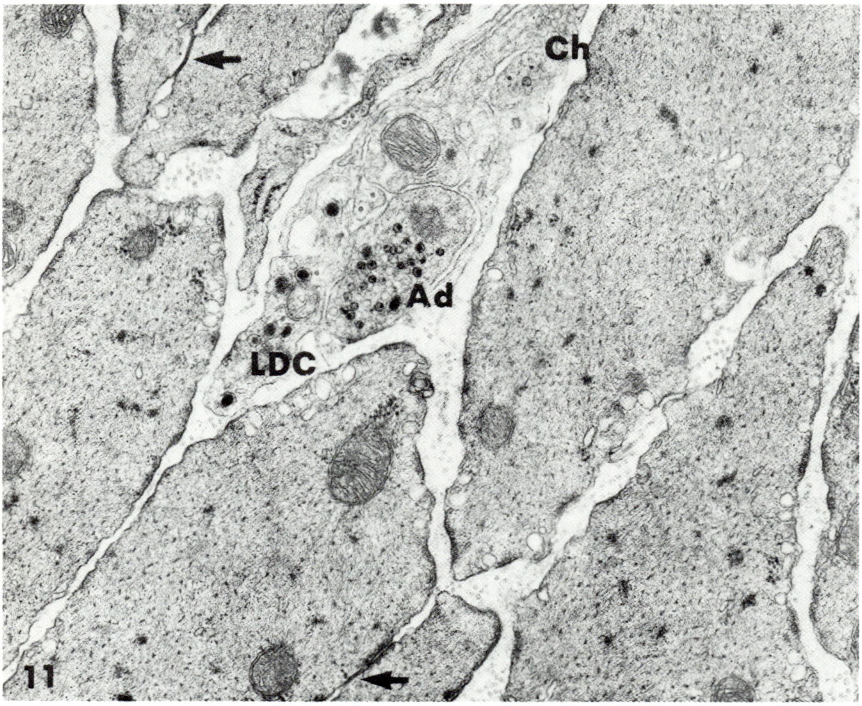

Fig. 11. Electron micrograph of nerve fiber from circular muscle of rat intestine showing a nerve fiber surrounded by smooth muscle cells. Three varicosities are present in the nerve and are characterized by small dense-cored vesicles (Ad: adrenergic), small agranular vesicles (Ch: cholinergic) and large dense-cored vesicles (LDC: identity unknown). Arrows indicate gap junctions between cells. ×27,000.

It has been suggested that the presence, lack of gap junctions, or their alteration from normal may be responsible for the normal or abnormal growth of tissue including cancers, some of which may be hormonally dependent (*39*). It is therefore interesting and perhaps significant that in the myometrium, which grows tremendously during pregnancy and in response to the steroid hormones, gap junctions are absent or present in low frequency. The cessation of growth of the myometrium either at term or before is accompanied by hormonal changes and the development of gap junctions (*17–26*). It is apparent then that the levels of hormones throughout pregnancy allow the expression of growth and proliferation of myometrial cells while suppressing the formation of gap junctions, and when the hormonal balance is changed at the end of pregnancy, gap junctions form between cells and growth ceases. Growth of intestinal muscle cells after injury has also been shown to be associated with increased gap junctions (*14*), but whether this increase occurs prior to or after the growth, has not been established.

We have recently (*22*) demonstrated electron-dense deposits associated with the gap junction following treatment of tissues with noradrenergic agents that affect adenylate cyclase and result in relaxation of the muscle (Fig. 12). These deposits may represent a reaction product of an enzyme associated with the gap junction. Similar deposits have been observed in other tissues, in which they are thought to represent calcium phosphate crystals formed during hydrolysis of adenosine triphosphate by an enzyme linked with gap junction proteins (*37*).

It has been proposed that gap junctions may be aggregates of hormone receptors or aggregates of adenylate cyclase (*37*). There

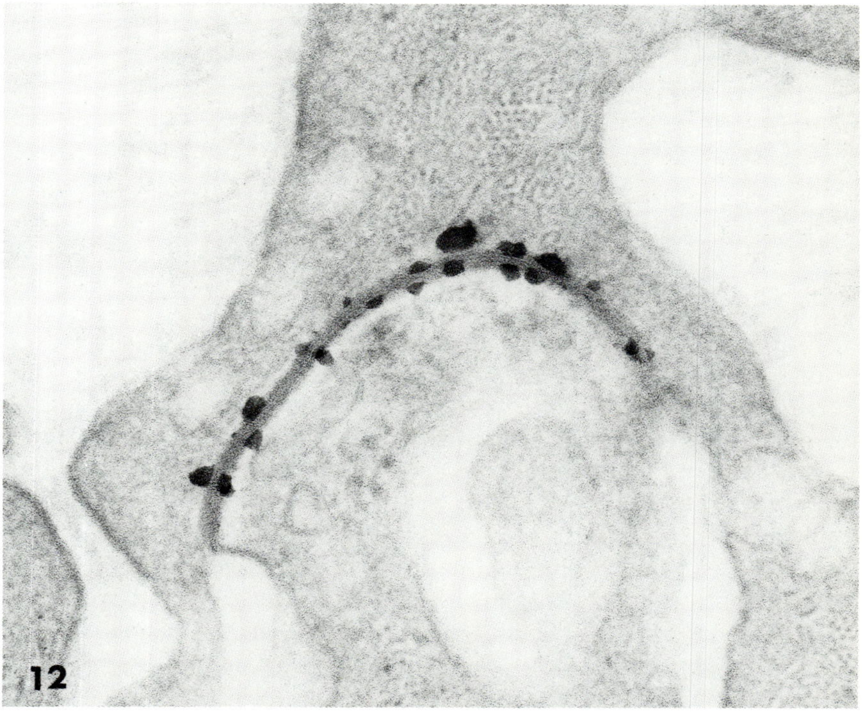

Fig. 12. A gap junction between smooth muscle cells of rat uterus after 48 h incubation with isoxsuprine. Note electron-opaque deposits associated with the junction. No lead citrate stain. ×143,000. (From Garfield et al., *Am. J. Physiol.* 138:569, 1980.)

is good correlation between binding of some hormones, the activation of adenylate cyclase, and the presence of gap junctions in a variety of tissues (*37*). Further studies are needed to define this relationship, if it exists, in the myometrium and other tissues.

3.3. Regulation of Gap Junctions in the Myometrium

Normal or premature labor in most species of animals, is thought to be controlled by changes in plasma and myometrial levels of steroid hormones and prostaglandins (*7, 38, 43, 47, 56*). The development of gap junctions in the myometrium is probably regulated by the changes in these hormones. An understanding of the mechanisms regulating gap junctions and their function may lead to effective procedures to initiate or inhibit labor.

3.3.1. Control by Steroid Hormones

Our studies suggest (*40–50*) that changes in the steroid hormones that precede labor initiate gap junction formation in the myometrium. Evidence from studies of rat tissues in vivo and in vitro indicates that progesterone may inhibit, whereas estrogen may stimulate, gap junction formation. In sheep and rats there is good correlation between gap junction formation and an increase in estrogen, a decrease in progesterone, and an increase in ratios of the two hormones measured in maternal and fetal blood and uterine tissues.

Figure 13 shows the blood and tissue levels of progesterone, estradiol, and prostaglandin F_2 (PGF), and the area of myometrial gap junctions during pregnancy and parturition in the rat. The hormone changes that occur in the rat include a decline in progesterone levels on day 19 of pregnancy, followed by increases in levels of estradiol and PGF after day 20. Moreover, an increase in the area of myometrial gap junctions was apparent on day 21, increasing until labor and delivery on day 22 of pregnancy.

Treatment of animals with progesterone, starting on day 19 of gestation, prevented the increase in gap junctions and delivery of the fetuses (*21*). Other studies have shown that ovariectomy of pregnant rats on day 16 resulted in decline in progesterone, development of gap junctions, and premature labor and delivery (*25*). Treatment of similar ovariectomized animals with progesterone prevented the fall in progesterone, the formation of gap junctions, and delivery (*25*). Furthermore, studies of myometrial tissues in vitro indicate that progesterone inhibits and estrogens stimulate the formation of gap junctions (*21, 22*). These results are consistent with the hypothesis (*17–26, 47*) that changes in the lev-

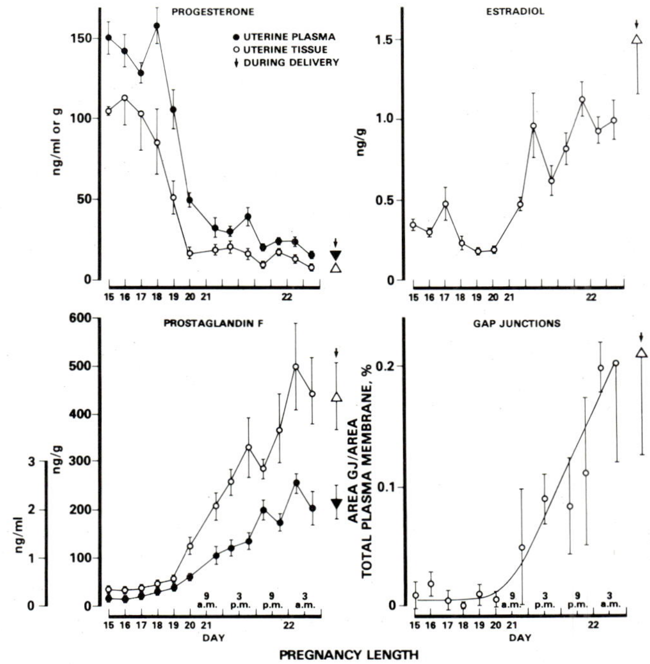

Fig. 13. Changes (mean ± SEM) in the levels of progesterone, estradiol, PGF, in uterine vein plasma (closed circles) and uterine tissues (open circles) and the myometrial gap junctional area during the latter days of pregnancy and during delivery (triangles) in rats. Note the axis of the time scale is expanded from day 21 onwards to delivery. (From Puri and Garfield, 1983.)

els of hormones that precede (progesterone withdrawal) or accompany (increase in estrogens and prostaglandins) the development of gap junctions are responsible for the regulation of their appearance.

The changes in the levels of steroid hormones may initiate synthesis of proteins that comprise gap junctions. It is well known that estrogen and progesterone bind to specific cytoplasmic receptors, move to the nucleus, and regulate receptor and protein synthesis. In the myometrium, estrogen stimulation is required to achieve the full development of gap junctions as well as to permit progesterone to inhibit their formation (*21*). This may be the result of estrogen stimulating the formation of receptors for progesterone, which then allows it to enter the cell and express an inhibitory effect on protein synthesis. Cycloheximide treatment prevents the development of gap junctions in myometrial tissues in vitro, confirming that protein synthesis is essential for gap junction formation (*22*).

3.3.2. Regulation by Prostaglandins

Our studies also suggest that prostaglandins directly influence gap junction formation in the myometrium. Prostaglandin synthesis increases before and during labor (Fig. 13) in various animals (*5, 7, 11, 12, 38, 43, 47, 56, 63*) and prostaglandins, like oxytocin, directly stimulate the myometrium to contract (*5, 7, 43, 56*) and indirectly affect the myometrium by altering steroid synthesis (*34*). The administration of prostaglandins to humans and animals results in termination of pregnancy after a period of time (*11, 43, 56*).

Indomethacin, a prostaglandin synthesis inhibitor, prevents gap junction formation in myometrial tissues in vitro (*21*). Also, both an antagonist and an inhibitor of thromboxane A_2 synthesis, as well as a prostacyclin analog (carbacyclin) prevent gap junction development (*22*). These studies demonstrate that there may be prostaglandins (thromboxanes and possibly endoperoxides) that stimulate gap junction development, and other prostaglandins (prostacyclin) that inhibit the process. It is also possible that the products of the lipoxygenase pathway, leukotrienes or slow reacting substance (SRSA), are involved in gap junction regulation either in uterine or other types of smooth muscle. We have proposed that the prostaglandins may control the aggregation of the gap junction proteins by effecting crosslinking reactions or by changing the fluidity of the membrane (22). Another possibility is that prostaglandins may regulate the steroid receptors.

The administration of PGE_1, PGE_2, and PGF_2 as abortifacients may alter steroid synthesis and effect changes in hormone profiles similar to those that occur at term; thus, prostaglandins may indirectly regulate synthesis of gap junction proteins, in addition to acting directly on the process of protein aggregation. The termination of pregnancy in rats after PGF_2 treatment has been related to changes in progesterone (*12*).

3.3.3. Gap Junction Permeability

We have implied in our studies that the presence of gap junctions in myometrial tissues during periods of intense uterine contractility indicated that the muscle cells were functionally coupled and the pores in the junctions were open. However, it is most likely that pores in the myometrial gap junction can exist in either an open or closed state, as they do in other cell systems (*1, 27, 44, 45, 59*). It is tempting to speculate that gap junctions may be receptor sites for oxytocin binding, as these sites increase at the same time as gap junctions during parturition and both are thought to be regulated by the same mechanisms (*50*). If oxytocin does bind to gap junction proteins that are partially exteriorized, the binding may result in a conformational change in the proteins and alter the permeability of the gap channels.

3.3.4. Degradation of Gap Junctions

Gap junctions, once formed in the myometrium during labor, decrease to preterm levels following parturition (*17, 18, 26*). The exact mechanism for gap junctional degradation is unclear. The protein components may detach any connection they may have between adjacent cells and the proteins become dispersed within the cell membrane (*28*). Another possibility is that gap junctions become interiorized within one of the cells connected by the junction and are digested by an endocytotic-lysosomal mechanism (*28*). Our studies support the latter mechanism (*37*). A large gap junction which may be in the initial stage of internalization is shown in Fig. 14. A possible interiorized or annular gap junction in rat myometrium is shown in Fig. 15.

3.3.5. Model for Control of Myometrial Gap Junctions

Evidence from studies of the myometrium demonstrates that gap junctions are dynamic structures regulated by a synthetic and a degradative pathway. These processes could be modulated by steroid hormones, prostaglandins, and calcium. A diagrammatic model for regulation of gap junctions is illustrated in Fig. 16.

It is proposed that the steroid hormones regulate gap junctions in the myometrium by controlling protein synthesis, as described. Once the proteins (connexons) are synthesized, they are inserted into the plasma membrane to interact with others via disulfide bridges or by crosslinking by PGs to form aggregates. The aggregates may be functional proteins associated with adenylate cyclase, hormone binding, or a calcium transport or binding mechanism. The intercellular channels created by the functional proteins may be either open or closed. When the aggregates have grown to sufficient size or become superfluous, they are incorporated into one of the connected cells by an endocytotic mechanism where they are digested by a lysosomal process. Thus, this dynamic system has at least four separate sites for control of the junctions and possible control of labor.

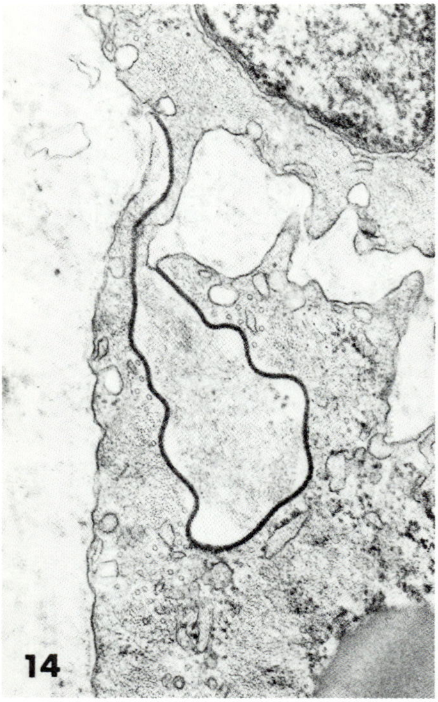

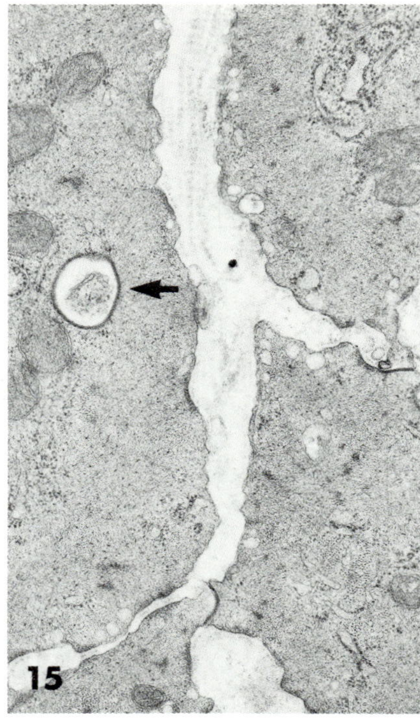

Fig. 14. Electron micrograph from rat myometrium showing large gap junction present between invagination of one cell into another. ×38,000. (From Garfield et al., *Am. J. Physiol.* 138:569, 1980.)

Fig. 15. Transverse section of rat myometrium showing interiorized or annular gap junction (arrow). ×22,000. (From Garfield et al., *Am. J. Physiol.* 138:569, 1980.)

4. Summary

In this review I have briefly outlined some of the the important details of the concept of direct communication between cells. Structural and functional evidence for the presence of channels between cells was presented. It appears that gap junctions are the structural correlate of direct cell-to-cell communication; however, evidence for this hypothesis in smooth muscle has been less than adequate. I have given considerable attention to studies of gap junctions between smooth muscle cells of the uterus. These studies support the view that gap junctions are channels between cells and show that the junctions are dynamic structures, regulated by hormones, that couple the cells together to improve and

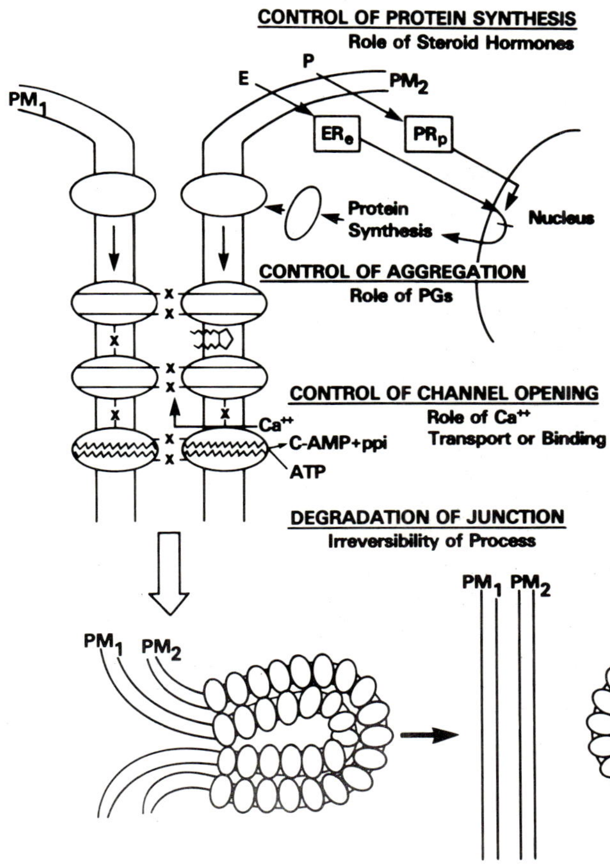

Fig. 16. Schematic representation of possible events controlling formation, functional coupling and destruction of gap junctions between plasma membranes (PM_1 and PM_2) of smooth muscle cells of the uterus. Shown are possible roles for (1) estrogen (E) and progesterone (P) interacting with their receptors (ER_e and PR_p) to control protein (connexon) synthesis; (2) prostaglandins in controlling crosslinking or connexon–connexon aggregation (X); (3) Ca^{2+} in regulating channel opening and possible sites for transport or binding Ca^{2+}; and (4) an irreversible endocytotic pathway for degradation of the junction.

coordinate contractility. The development of myometrial gap junctions during parturition may be necessary for activation of the muscle to produce labor contractions.

Acknowledgments

Supported by grants from the Medical Research Council of Canada. R. E. Garfield is a senior fellow of the Ontario Heart Foundation.

References

1. Bennett, M. V. L., and D. A. Goodenough. Gap junctions, electronic coupling, and intercellular communication. *Neurosci. Res. Prog. Bull.* 16:373–486, 1978.
2. Bergman, R. A. Uterine smooth muscle fibers in castrate and oestrogen treated rats. *J. Cell Biol.* 36:639–648, 1968.
3. Burden, H. W., M. L. Capps, and I. E. Lawrence. Gap junctions in the myometrium of pelvic-neurectomized ras with blocked parturition. *Am. J. Anat.* 156:105–111, 1979.
4. Burnstock, G. Cholinergic and purinergic regulation of blood vessels. In: *Handbook of Physiology. The Cardiovascular System, Volume II: Vascular Smooth Muscle.* Eds.: D. F. Bohr, A. P. Somlyo, H. V. Sparks, Jr. Bethesda, American Physiological Society, 1980, pp 567–612.
5. Challis, J. R. G., and B. F. Mitchell. Hormonal control of preterm and term parturition. *Semin. Perinatol.* 5:192, 1981.
6. Clapham, D. E., A. Shrier, and R. L. Dehann. Junctional resistance and action potential delay between embryonic heart cell aggregates. *J. Gen. Physiol.* 75:633–654, 1980.
7. Csapo, A. I. Force of labor. In: *Principles and Practice of Obstetrics and Perinatology.* Eds.: L. Iffy, H. A. Kaminetzky, New York, John Wiley & Sons Inc. 1981, pp 761–799.
8. Daniel, E. E., V. P. Daniel, G. Duchon, R. E. Garfield, M. Nichols, S. K. Malhotra, and M. Oki. Is the nexus necessary for cell-to-cell coupling of smooth muscle? *J. Membrane Biol.* 28:207–239, 176.
9. Dewey, M. M. and L. Barr. Intercellular connection between smooth muscle cells: the nexus. *Science* (Wash. D.C.) 137:670–672, 1962.
10. Fuchs, A. R. Uterine activity in late pregnancy and during parturition. *Biol. Reprod.* 1:344–353, 1969.
11. Fuchs, A. R. Parturition in rabbits and rats. In: *Endocrine Factors in Labor.* Eds.: A. Klopper and J. Gardner, Cambridge Press. Mem. Soc. Endocrinol. 20:163–185, 1973.
12. Fuchs, A. R. Hormonal control of myometrial function during pregnancy and parturition. *Acta Endocrinol.* (Suppl) (Copenh) 221:1, 1978.

13. Gabella, G. Intercellular junctions between circular and longitudinal intestinal muscle layers. *Z. Zellforsch. Mikroskop. Anat.* 125:191–199, 1972.
14. Gabella, G. Hypertrophic smooth muscle. III. Increase in number and size of gap junctions. *Cell Tissue Res.* 201:263–276, 1979.
15. Gabella, G. Structure of smooth muscles. In: *Smooth Muscle.* Eds.: E. Bulbring, A. F. Brading, A. W. Jones, and T. Tomita. Edward Arnold: London, 1–46, 1981.
16. Garfield, R. E. and E. E. Daniel. The structural basis of electrical coupling (cell-to-cell contacts) in rat myometrium. *Gynecol. Invest.* 5:284–300, 1974.
17. Garfield, R. E., S. Sims, and E. E. Daniel. Gap junctions: their presence and necessity in myometrium during parturition. *Science,* 198:958–960, 1977.
18. Garfield, R. E., S. M. Sims, M. S. Kannan, and E. E. Daniel. The possible role of gap junctions in activation of the myometrium during parturition. *Am. J. Physiol.* 235: *Cell Physiology* 4, C168–C179, 1978.
19. Garfield, R. E., S. Rabideau, J. R. G. Challis, and E. E. Daniel. Ultrastructural basis for maintenance and termination of pregnancy. *Am. J. Obstetrics and Gynecol.* 133:308–315, 1979.
20. Garfield, R. E., S. Rabideau, J. R. G. Challis, and E. E. Daniel. Hormonal control of gap junctions in sheep myometrium. *Biol. Reprod.* 21:999–1007, 1979.
21. Garfield, R. E., M. S. Kannan, and E. E. Daniel. Gap junction formation in myometrium. Control of estrogens, progesterone and prostaglandins. *Am. J. Physiol.* 238: Cell Physiology 7, C81–C89, 1980.
22. Garfield, R. E., D. Merrett, and A. K. Grover. Studies on gap junctions in the myometrium of women during various stages of menstruation. *Am. J. Obstet. Gynecol.* 138:569–574, 1980.
23. Garfield, R. E. and R. H. Hayashi. Presence of gap junctions in the myometrium of women during various stages of menstruation. *Am. J. Obstet. Gynecol.* 138:569–574, 1980.
24. Garfield, R. E. and R. H. Hayashi. Appearance of gap junctions in the myometrium of women during labor. *Am. J. Obstet. Gynecol.* 140:254–260, 1981.
25. Garfield, R. E., C. P. Puri, and A. I. Csapo. Endocrine, structural, and functional changes in the uterus during premature labor. *Am. J. Obstet. Gynecol.* 142:21–27, 1982.
26. Garfield, R. E., E. E. Daniel, M. Dukes, and J. D. Fitzgerald. Changes in gap junctions in myometrium of guinea pig at parturition and abortion. *Can. J. Physiol. Pharmacol.* 60:335–341, 1982.
27. Gilula, N. B. and M. L. Epstein. Cell-to-cell communication, gap junctions and calcium. *Symp. Soc. Exp. Biol.* 30:257–272, 1976.
28. Griepp, E. B. and J. P. Revel. Gap junctions in development. In: *Intercellular Communication.* Ed.: W. C. DeMello. New York, Plenum Publishing Corp. 1977, 1–32.
29. Hayashi, R. H., R. E. Garfield, and T. J. Kuehl. Gap junction formation in the myometrium of pregnant baboon. A Possible Model. Submitted to *Am. J. Obstet. and Gynecol.*

30. Henderson, R. M., G. Duchon, and E. E. Daniel. Cell contacts in duodenal smooth muscle layers. *Am. J. Physiol.* 221:564–574, 1971.
31. Henderson, R. Cell-to-cell contacts. In: *Methods in Pharmacology. Vol. 3. Smooth Muscle.* Eds.: E. E. Daniel and D. M. Paton. Plenum Press, New York, 47–77, 1975.
32. Hertzberg, E. L., T. S. Lawrence, and N. B. Gilula. Gap junctional communication. *Ann. Rev. Physiol.* 43:479–491, 1981.
33. Hooper, M. L., and J. H. Subak-Sharpe. Metabolic cooperation between cells. *Int. Rev. Cytol.* 69:45–104, 1981.
34. Horton, E. W. and N. L. Poyser. Uterine Luteolytic hormone: a physiological role for prostaglandin F_2. *Physiol. Rev.* 56:595–651, 1976.
35. Johnson, R., M. Hammer, J. Sheridan, and J. P. Revel. Gap junction formation between reaggregated Novikoff hepatoma cells. *Proc. Natl. Acad. Sci.* 71:4536–4540, 1974.
36. Kuriyama, H. and H. Suzuki. Changes in electrical properties during gestation and following hormonal treatments. *J. Physiol. (Lond.)* 260:315–333, 1971.
37. Larsen, W. J. Gap junctions and hormone action. In: *Transport of Ions and Water in Epithelia.* Eds.: B. J. Wall, J. L. Oschman, B. Moreton *et al.* Academic Press, London, 333–361, 1977.
38. Liggins, G. C. Initiation of parturition. *Br. Med. Bull.* 35:145–150, 1979.
39. Loewenstein, W. R. Junctional intercellular communication: The cell-to-cell membrane channel. *Physiol. Rev.* 61:829–913, 1981.
40. Marshall, J. M. Effects of ovarian steroids and pregnancy on adrenergic nerves of uterus and oviduct. *Am. J. Physiol.* 240:C165–C174, 1981.
41. McNutt, N. and R. S. Weinstein. Membrane ultrastructure at mammalian intercellular junctions. *Prog. Biophys. Mol. Biol.* 26:45–101, 1973.
42. Merk, F. B., P. W. L. Kwan, and I. Leav. Gap junctions in the myometrium of hypophysectomized estrogen-treated rats. *Cell Biol. Intl. Rep.* 4:287–294, 1980.
43. Nathanielsz, P. W. Endocrine mechanisms of parturition. *Ann. Rev. Physiol.* 40:411–437, 1978.
44. Peracchia, C. Gap junctions. Structural changes after uncoupling procedures. *J. Cell Biol.* 72:628–641, 1977.
45. Peracchia, C. Structural correlates of gap junction permeation. *Intl. Rev. Cytol.* 66:81, 1980.
46. Pitts, J. D. Direct communication between animal cells. In: *Intl. Cell Biol.* Eds.: B. R. Brinkley and K. R. Porter. Rockefeller Univ. Press, New York, 43–49, 1977.
47. Puri, C. and R. E. Garfield. Changes in hormone levels and gap junction during pregnancy, labor and parturition in the rat. *Biol. Reprod.* 27:967–975, 1982.
48. Ruffner, J. A., N. Sperelakis, and J. E. Mann Jr. Application of the Hodgkin-Huxley equations between excitable cells. *J. Theor. Biol. 87L:129–152, 1980.*
49. Sims, S. M., E. E. Daniel, and R. E. Garfield. Improved electrical coupling in uterine smooth muscle is associated with increased numbers of gap junctions at parturition. *J. Gen. Physiol.* 80:353–375, 1982.

50. Soloff, M. S., M. Alexondrova, and M. J. Fernstrom. Oxytocin receptors: triggers for parturition and lactation. *Science* 204:1313, 1979.
51. Somlyo, A. V. Ultrastructure of vascular smooth muscle. In: *Handbook of Physiology. Section 2: The Cardiovascular System. Vol. II. Vascular Smooth Muscle.* Eds.: D. F. Bohr, A. P. Somlyo, H. V. Sparks. *Am. Physiol. Soc.,* Bethesda, 33–67, 1980.
52. Sporrong, B., P. Alm, C. Owman, N. O. Sjoberg, and G. Thorbert. Ultrasturctural evidence for adrenergic nerve degeneration in guinea pig uterus during pregnancy. *Cell Tiss. Res.* 195:189–193, 1978.
53. Spray, D. C., A. L. Harris, and M. V. L. Bennett. Gap junctional conductance is a simple and sensitive function of intracellular pH. *Science,* 211:712–715, 1981.
54. Thorbert, G., P. Alm, C. Owman, and N. O. Sjoberg. Regional distribution of autonomic nerves in guinea pig uterus. *Am. J. Physiol.* 233:C25–C34, 1977.
55. Thorbert, G. Regional changes in structure and function of adrenergic nerves in guinea pig uterus during pregnancy. *Acta. Obstet. Gynecol. Scand. Suppl.* 79:5–39, 1979.
56. Thorburn, G. D., and J. R. G. Challis. Endocrine control of parturition. *Physiol. Rev.* 59:863, 1979.
57. Staehelin, L. A. Structure and function of intercellular junctions. *Int. Rev. Cytol.* 39:191–283, 1974.
58. Uehara, Y., and G. Burnstock. Demonstration of "gap junctions" between smooth muscle cells. *J. Cell Biol.* 44:215–217, 1970.
59. Unwin, P. N. T. and G. Zampighi. Structure of the junction between communicating cells. *Nature (London),* 283:454–549, 1980.
60. Wathes, D. C. and D. G. Porter. Effect of uterine distension and oestrogen treatment on gap junction formation in the myometrium of the rat. *J. Reprod. Fert.* 65:497–505, 1982.
61. Weibel, E. R. Stereological techniques for electron microscope morphometry. In: *Principles and techniques of electron microscopy: Biological Applications.* Ed.: M. A. Hayat. Van Nostrand-Reinhold, New York, Vol. 3, 237–297, 1973.
62. Westfall, D. P., L. L. Millecchia, T. J. F. Lee, S. P. Corey, D. Smith, and W. W. Fleming. Effects of denervation and reserpine on nexuses in the rat vas deferens. *Eur. J. Pharmacol.* 41:239–242, 1977.
63. Wolfs, G. M. J. A. and M. VanLeeuwen. Electromyographic observations on the human uterus during labor. *Acta Obstet. Gynecol. Scand.* (Suppl) 90:1–61, 1979.

Chapter 7

Calcium Regulation of Smooth Muscle Actomyosin

Samuel Chacko, Arline Rosenfeld, and George Thomas

1. Introduction

The contractile process in muscle is manifested by an increased cross-bridge cycling and the development of tension or, in biochemical terms, by an activation of Mg^{2+}-ATPase activity of myosin by actin. It is generally accepted that this process is regulated by free calcium ion concentration in the muscle cell. The factors and the cell components that regulate the intracellular calcium concentration in smooth muscle is the topic of other chapters (Chapters 3–7) in this book. This chapter focuses on the contractile proteins in smooth muscles, particularly the regulation of actin-activated ATP hydrolysis of smooth muscle myosin.

2. Components of the Contractile Apparatus

The contractile apparatus in smooth muscle consists of thick and thin filaments. The organization of these filaments, which are the components essential for the sliding filament model for contraction, in smooth muscle is different from that of striated muscle (for review on the organization of contractile apparatus, *see* Chapter 1 in this book and ref. *1*).

2.1. Thick Filament

The thick filament is made up of myosin, a hexameric molecule composed of two high molecular weight (200,000 daltons) heavy chains and four low molecular weight (two 20,000 and two 16,000 daltons) light chains. The molecular structure of the heavy chains of smooth muscle myosin is thought to be basically the same as that of other myosins (*2*). The heavy chains form alpha helices that wind together in the same direction to build a left-handed coiled rod of about 130 nm in length. These chains unwind and fold separately to form the two globular head portions close to the N-terminus. The head portion of the myosin molecule contains the active sites for ATP hydrolysis.

The light chains (one 16,000 and one 20,000 daltons) are located on each head. The 16,000-dalton light chain is analogous to the essential light chains of vertebrate and invertebrate striated muscle myosin. Removal of this light chain is associated with the loss of ATPase activity. The removal of the 20,000-dalton light chain (20KLC) on the other hand is not associated with a loss of ATPase activity. However, it has a role in the regulation of actin-activated ATP hydrolysis. Hence, it appears to be analogous to the regulatory light chain of the invertebrate striated muscle. The mode of regulation of ATPase by the regulatory light chain of invertebrate muscle is different from that of smooth muscle, however (*3*). The 20KLC serves as the substrate for a specific calcium-dependent kinase. This kinase, named myosin light chain kinase (MLCK), is reported to be present in muscle and nonmuscle cells (*4–6*). In the presence of Ca^{2-} and Mg^{2+}-ATP, the MLCK transfers the terminal phosphate from ATP to the serine residue at the C-terminus of the light chain. The phosphate from the 20KLC is removed by myosin light chain phosphatase (MLCP). Since the 20KLC can be phosphorylated and dephosphorylated, it is often called the P-light chain (*4*).

2.2. Thin Filament

The major component of the thin filament is actin. The actin molecule is small with a molecular weight of 42,000 daltons. The globular monomeric form (G-actin) can be polymerized into filament (F-actin). The filaments are 4.2 nm in diameter and their exact length in smooth muscle is not known. In the thin filaments, the monomers appear as spherical particles that form a twisted double stranded bead.

Actin binds to myosin and in low ionic strength solution (e.g., 0.02*M* KCl) it activates the Mg^{2+}-ATPase activity of myosin, though in smooth muscle, the actin-activation requires other factors (see below). In high ionic strength solution (0.5*M* KCl), the actin inhibits the K^+-EDTA-stimulated ATPase activity of myosin.

Actin is a conserved molecule (*7*). Actin isolated from skeletal muscle can activate the ATPase activity of smooth and nonmuscle cell myosin and vice versa (*8, 9*), although minor differences in the kinetics of actomyosin ATPase have been reported.

In addition to actin, the thin filament also contains tropomyosin. Tropomyosin has a molecular weight of 70,000–75,000 daltons (*10*). It shows two subunits (α and β) with molecular weights of 32,000–35,000 daltons and 33,000–35,000 daltons on sodium dodecyl sulfate polyacrylamide gel electrophoresis. Tropomyosin is a long thin molecule that attaches end to end to form a very thin filament on the surface of the actin filament. Each actin strand carries its own tropomyosin filament, which is a coiled coil (double coil) (*11*). It is believed to lie in or at the edge of the groove between the paired strands of the actin filament (*11*). Its specific location in smooth muscle during contraction and relaxation is not known; however, the movements of tropomyosin play an essential role in muscle regulation (*12, 13*).

3. Proteins Involved in the Phosphorylation and Dephosphorylation of Myosin

All types of muscles contain a calcium dependent kinase, myosin light chain kinase (MLCK), which catalyzes the transfer of the terminal phosphate from ATP to one of the 16,000–27,000-dalton mol wt light chains of myosins (*4–6*). In the case of smooth muscle, the 20,000-dalton light chain is phosphorylated and the specific residue involved has been identified as serine-19 (*14*). The studies of Yazawa et al. (*15*) on proteins isolated from skeletal muscle and Dabrowska et al. (*16*) on proteins isolated from chicken gizzard showed that the MLCK is composed of a high molecular weight (80,000 daltons for skeletal muscle MLCK and 105,000 daltons for gizzard MLCK) component and a low molecular weight (17,000 daltons) component. Both subunits were shown to be required for MLCK activity. The smaller subunit was subsequently identified as the calcium-binding protein calmodulin (*16, 17*). The MLCK has a strong affinity (apparent $K_{0.5} = 10^{-9}M$) to Ca^{2+}-calmodulin (*18*). The MLCKs isolated from a variety of sources have been shown to be calmodulin-dependent (*19–25*). The MLCK isolated from different sources had different molecular weights (80–155,000 daltons). These differences in molecular weights of the larger subunits of the MLCK could be due to partial proteolytic degradation (*24, 26, 27*).

MLCK itself can be phosphorylated by a cAMP-dependent protein kinase (*28*) at two sites in the apoenzyme and one site in the holoenzyme (calmodulin–kinase complex) (*29*). Phosphorylation of the apoenzyme reduces its affinity for calmodulin and so the activity of MLCK can also be modulated by phosphorylation.

Dephosphorylation of the myosin and of MLCK is achieved by one or more phosphatases (*30, 31*). Pato and Adelstein (*32*) purified two phosphatases that preferentially dephosphorylated the 20,000 dalton light chain of gizzard myosin. These phosphatases, named phosphatase I and II, differ in their molecular weights and polypeptides on sodium dodecyl sulfate polyacrylamide gels (*32, 33*). Furthermore, phosphatase I does not require Mg^{2+} for activity and it dephosphorylates MLCK at a higher rate than phosphatase II. Both of these enzymes appear to play a role in the phosphorylation and dephosphorylation of myosin. Dephosphorylation of the MLCK increases the activity of the MLCK, hence increasing the level of phosphorylation of the myosin light chain. Dephosphorylation of the myosin light chain by the phosphatases lowers the level of myosin phosphorylation unless MLCK activity keeps up with the dephosphorylation.

4. Effect of Phosphorylation of Myosin Light Chain on the Properties of Myosin

The phosphorylation of the 20,000-dalton light chains of myosin is associated with functional changes, as exhibited by the actin-activated ATPase activity, and structural changes, which leads to alterations in the conformation of the myosin molecule and the ability of the myosin to form synthetic filaments.

4.1. Functional Changes

The K^+-EDTA-stimulated ATPase activity of myosins isolated from smooth muscles of different sources and species is usually over 1 μmol pi/mg per min (37°C) and the Ca^{2+}-activated activity is one half of the K^+-EDTA-stimulated activity (*34–38*). These myosin ATPase activities measured under nonphysiological ionic conditions (0.5*M* KCl) are not altered by phosphorylation (*35, 36*).

Actin inhibits the K^+-EDTA-stimulated activity of myosin. This effect of actin is also not affected by phosphorylation (Chacko, unpublished observation). In contrast, the ATPase activity of myosin measured in low salt (0.02–0.05*M* KCl) in the presence of actin is 10–20-fold higher for phosphorylated myosin (*34–39*). The level of activation of the myosin ATPase activity by actin depends on the degree of myosin light chain phosphorylation (*36, 37*). Both linear (*36, 38*) and nonlinear (*40, 41*) correlations between phosphorylation and actin-activated ATPase activity have been reported. The correlation appears to depend on the manner in which the phosphorylation is achieved. Sellers and Adelstein reported nonlinear correlation between the degree of phosphorylation and ATPase activity during phosphorylation; however, the relationship was linear during dephosphorylation (*42*).

There appears to be some cooperativity between the two heads of the myosin during phosphorylation. However, this depends on the state of myosin during phosphorylation (*41, 43*). Intramolecular cooperativity between the two heads of the myosin has been shown with scallop myosin by Szent-Gyorgyi and his collaborators (*44, 45*) using single-headed and double-headed myosin and soluble heavy meromyosin. Similar experiments on smooth muscle myosin are lacking and further work is needed to elucidate the intramolecular cooperativity of myosin in smooth muscle.

4.2. Structural Changes

Phosphorylation of the 20KLCs that are located in the globular head region of the myosin molecule appears to affect the stability of the filaments formed by the myosin. Onishi et al. (*46*) found that the thick filaments of gizzard myosin are disassembled by ATP. Suzuki et al. (*47*) showed that the phosphorylation of myosin light chain played a role in controlling myosin filament assembly. This observation was confirmed by Scholey et al. (*48*) using thymus myosin. It is difficult to visualize how light chain phosphorylation associated with the myosin heads controls the interactions of the tail region that is believed to be involved in the filament formation. However, recent studies by Onishi and Wakabayashi (*49*) and Trybus et al. (*50*) using gizzard and aortic myosin, respectively, showed that the myosin molecule disassembled by ATP showed intramolecular hairpin "loops" in the tail. These myosin molecules were unphosphorylated. The phosphorylated myosin molecule showed an extended tail. The filaments formed by the phosphorylated myosin are not disassembled by ATP. Hence, the phosphorylation of the myosin light chain appears to change the conformation of the myosin molecule, which determines the stability of the filaments formed. This important discovery made from studies utilizing isolated myosin molecules needs to be investigated with intact smooth muscle. It is important to note that thick filaments can be observed by electron microscopy in the relaxed smooth muscle with a low level of phosphorylation (*51*).

5. Effect of Tropomyosin on the Actin-Activated ATP Hydrolysis

Tropomyosin, a protein located on the thin filament, potentiates the actin-activated ATP hydrolysis by smooth muscle myosin when the myosin is phosphorylated (*35–37, 52*). The maximum effect of tropomyosin is observed at a tropomyosin:actin molar ratio of 1:6 at 37°C (*37*). The full activation of the ATPase activity of the reconstituted actomyosin requires tropomyosin. The addition of tropomyosin to actomyosin reconstituted with unphosphorylated myosin fails to increase the ATPase activity to the level of the actomyosin formed with phosphorylated myosin (*37*). Therefore, tropomyosin appears to exert its effect on actin-activated ATP hydrolysis only after the myosin is "turned on" by phosphorylation.

Tropomyosins isolated both from skeletal and smooth muscles are capable of potentiating the actin-activated ATP hydrolysis (*37*). Recent experiments (*53*) showed a difference in the degree of potentiation when tropomyosin from different sources were used (gizzard > skeletal muscle > pulmonary artery). This difference was observed regardless of the source of actin used.

The potentiation of actin-activated ATP hydrolysis of smooth muscle myosin by tropomyosin appears to be cooperative as observed by Bremel and Weber (*54*) for skeletal muscle myosin. However, the potentiation in the skeletal muscle system was observed under conditions that favor the formation of rigor compex. In fact, under conditions (high ATP) in which tropomyosin potentiates the smooth muscle actomyosin system (*35–39, 52*), the ATPase activity of skeletal muscle actomyosin is either inhibited (*55*), or not affected by tropomyosin (*56*). Though it appears that the tropomyosin potentiates the cofactor activity of actin to activate the ATPase activity of active smooth muscle myosin (i.e., phosphorylated), further experiments are required to understand the mechanism by which this potentiation is accomplished.

6. Direct Effect of Mg^{2+} and Ca^{2+} on Actin-Activation of Phosphorylated Myosin

Ca^{2+} and Mg^{2+} influence the interaction between myosin and actin via phosphorylation as discussed before. In addition, these divalent cations have a direct effect on actin-activation of the phosphorylated myosin. The direct effect of Ca^{2+} on the actin-activation was first shown for guinea-pig vas deferens myosin phosphorylated and purified free of MLCK and MLCP (*35*). A crude actomyosin preparation containing endogenous MLCK and MLCP was incubated in the presence of Ca^{2+} and Mg-ATP to obtain phosphorylated myosin and in the absence of Mg-ATP and Ca^{2+} to prepare unphosphorylated myosin. The level of myosin phosphorylation was controlled by varying the time required for phosphorylation or by lowering the ATP concentration (*37*). The phosphorylated or dephosphorylated myosin was then purified further by gel filtration chromatography to separate the myosin from other proteins (*35*). The kinase activity in the myosin fraction incubated under the assay conditions of the actin-activation was tested using [γ–^{32}P] ATP for unphosphorylated myosin. For phosphorylated myosin, [γ–^{32}P] ATP and exogenous myosin or light chain were used for determining the kinase activity. The amount of phosphorylated light chain was determined by urea gel electrophoresis (*36*). The phosphatase activity of the purified myosin was determined by incubating the phosphorylated myosin under the conditions of the ATPase assay. Exogenous myosin phosphorylated with [γ–^{33}P] was utilized as a substrate for determining the phosphatase activity of the unphosphorylated myosin (*36*).

The phosphorylated myosin free of MLCK and MLCP was reconstituted with pure actin containing tropomyosin and the ATPase activity was measured in the presence or absence of Ca^{2+}. These experiments carried out with vas deferens myosin by Chacko et al. (*35*) gave evidence for the direct enhancement by Ca^{2+} in the actin-activation of smooth muscle myosin. Furthermore, since the unphosphorylated and phosphorylated myosin utilized for this study was free of MLCK and MLCP, the experiments with vas deferens myosin (*35*) provided a direct proof for the requirement for phosphorylation for actin-activation of myosin ATPase (*57*).

The direct effect of Ca^{2+} was also observed by Rees and Frederiksen (*58*) for phosphorylated aortic myosin. However, Hartshorne and his colleagues did not observe the direct calcium effect for gizzard myosin (*59*). Using myosin isolated from swine pulmonary artery Chacko and Rosenfeld (*36*) showed that the phosphorylated myosin had the highest actin-activated ATPase activity when myosin bound 2 mol Ca^{2+}/mol. The calcium binding and ATPase activity were both dependent on the free Mg^{2+} concentration. At 2–3 m*M* free Mg^{2+} concentration and *p*Ca 5 the arterial myosin bound 2 mol of Ca^{2+} and the ATPase activity of the reconstituted actomyosin in *p*Ca 5 was higher than in *p*Ca 8, where myosin did not bind calcium. Both actin-activation and the calcium sensitivity were more pronounced when actin-containing tropomyosin was used. At higher Mg^{2+} concentration, the myosin bound very little Ca^{2+} and the calcium sensitivity also disappeared. The direct effect of Ca^{2+} and Mg^{2+} was shown only when the myosin was phosphorylated, although phosphorylation had no effect on calcium binding by myosin (*60*).

The dual effect of Ca^{2+} on actin activation of myosin ATPase calcium once shown to be present only for myosin isolated from vas deferens (*35*) is now observed for two other mammalian arterial myosins (*36, 58*). In contrast to the observation by Hartshorne and his colleagues (*59, 61*), a recent report by Nag and Seidel has shown that calcium regulates the actin-activated ATPase activity of gizzard myosin even after phosphorylation (*62*). Heaslip and Chacko (*63*) showed that the calcium sensitivity of stably phosphorylated gizzard myosin was independent of the type of actin used for the actin activation; however, tropomyosin was essential for the full activation of the acto-myosin ATPase and the expression of the maximal Ca^{2+} sensitivity. However, as in the case of myosin from swine pulmonary artery (*35*), the calcium sensitivity was observed only at Mg^{2+} concentrations lower than that which was used by Hartshorne and his colleagues (*59, 61*).

7. Evidence for the Actin-Linked System in the Regulation of Actomyosin ATPase

The evidence for the existence of a thin filament regulation comes from the work of Ebashi and his colleagues (*64–71*). The most important feature of the proposed thin filament regulation is the quiescence of myosin–actin–ATP interaction (*65*) in the absence of other activators, similar to the fundamental basis of the phosphorylation theory (*35, 38, 39*). However, in the actin-linked system, regulatory proteins called leiotonins activate the thin filament in quantities only one-hundredth that of actin by weight (*64*). The actin-activation of myosin ATPase and superprecipitation, an in vitro assay of the contractile event, takes place independent of myosin phosphorylation (*70*). In both types of regulation, tropomyosin is essential for the full activation of superprecipitation or actomyosin ATPase (*66, 67*).

The leiotonins were isolated into two components: leiotonin A with a molecular weight of 80,000 daltons and leiotonin C with a molecular weight of 17,000 daltons (*67*). Leiotonin C can be replaced with calmodulin for superprecipitation (*67*). Though leiotonin C was found to be distinct from calmodulin based on the differences in their amino acid composition, peptide maps and activation of phosphodiesterase (*67*), the isolation of leiotonin C from smooth muscle is difficult (Nonomura, personal communication). The apparent difficulty in the isolation and purification of leiotonin C free of calmodulin from smooth muscle may result from the smaller quantity of this protein compared to calmodulin in smooth muscle preparation (Mikawa, personal communication). The possibility that the leiotonin C is a modified form (limited proteolytic degradation, modification during preparative procedures, and so forth) of calmodulin is not ruled out.

Mikawa (*71*) found that glutaraldehyde was able to cross-link or "freeze" the thin filaments of smooth muscle in the active form, capable of activating actomyosin ATPase. A recent report by Ebashi (*57*) confirmed the finding by Chacko et al. (*35*) that the myosin phosphorylated and purified free of kinase and phosphatase interacts vigorously with gizzard and rabbit skeletal muscle actin in the absence of Ca^{2+} and leiotonins. Chacko et al. (*72*) showed that the leiotonin preparation neither enhanced the ATPase activity nor conferred calcium sensitivity on the actomyosin system composed of stably phosphorylated myosin, actin, and tropomyosin.

Evidence for an actin-linked regulatory system for smooth muscle myosin has also been shown by Marston et al. (*73*). However, in this case it was suggested that proteins similar to skeletal muscle troponin I and troponin C were involved (*74*). Factors capable of activating the actomyosin have also been isolated from gizzard by Persechini et al. (*75*) and Cole et al.(*76*).

8. Summary

All the evidences indicate that the actomyosin system in smooth muscle is in a quiescent state and that it is activated by Ca^{2+}. The mode of activation and the regulation of actomyosin ATPase is a subject of major controversy. Evidence from the investigators who favor the phosphorylation theory indicates that the myosin ATPase under physiological condition is activated by actin only when the 20,000-dalton light chain of myosin is phosphorylated. The phosphorylation of the myosin by a Ca^{2+}-dependent kinase has been reported in a number of smooth muscles (*35–39, 77–79*). The kinase involved in phosphorylation has a high affinity to the actin (*80*) and it requires calmodulin (*16*). Inactivation of the actomyosin ATPase is effected by dephosphorylation of myosin by myosin phosphatase, the activity of which dominates in the absence of Ca^{2+}. Hence, Ca^{2+} regulates the actomyosin ATPase via phosphorylation–dephosphorylation of the myosin.

Although phosphorylation is associated with the development of tension (*81–86*), there is no simple quantitative relationship between the extent of phosphorylation and tension developed by the muscle (*82-84*). It is clear that additional factors must play a role in the regulation of contraction. The modulation of actin-activated ATP hydrolysis by phosphorylated myosin caused by divalent cations (*35,36,58,62*) appears to be important in this regard. The calcium dependence of phosphorylation and actin-activation of phosphorylated myosin should be studied in more detail to understand how calcium can play a dual role in the regulation of contraction.

While the role of phosphorylation on the actin-activation of stably phosphorylated myosin is clear, a better understanding of its role in a system where there is cyclic phosphorylation and dephosphorylation requires further work. Recent experiments by Chacko et al. (*72*) indicated that an actomyosin system reconstituted with the thiophosphorylated gizzard myosin (stable phosphorylation), gizzard actin and gizzard tropomyosin showed very high ATPase activity independent of calcium concentration. Addition of leiotonin preparation to this reconstituted actomyosin with stable phosphorylation neither conferred calcium sensitivity nor increased the ATPase activity. Although the leiotonin factor had no effect on the actin activation of stably phosphorylated myosin, its role in the actomyosin system where there is cyclic phosphorylation and dephosphorylation cannot be ruled out.

The role of actin-linked regulation on the actomyosin ATPase is complicated by the presence of kinase and phosphatase in the leiotonin preparation. Further purification of the leiotonins and the demonstration of their effects on the regulation of actomyosin is needed to establish the existence of an actin-linked regulation of actomyosin ATPase in smooth muscle.

Acknowledgment

This work was supported by grants from the National Institute of Health (HL 22264 and IF06-TW00776).

References

1. Hartshorne, D. J., and A. Gorecka. Biochemistry of contractile proteins of smooth muscle. In *Handbook of Physiology—The Cardiovascular System, Vol. II, Vascular Smooth Muscle,* pp 93–120, 1980. Eds.: Bohr, Somlyo, and Sparks. American Physiological Society, Bethesda, 1980, pp 93–120.
2. Lowey, S. In *Fibrous Proteins. Industrial and Medical Aspects.* Ed.: D. A. D. Parry and L. K. Gramen. Academic, New York, 1:1–25, 1979.
3. Kendrick-Jones, J., E. M. Szenti-Kiralyi, and A. G. Szent-Gyorgyi. Regulatory light chains in myosins. *J. Mol. Biol.* 104:747–775, 1976.
4. Perrie, W. T., L. B. Smillie, and S. V. Perry. A phosphorylated light-chain component of myosin from skeletal muscle. *Biochem. J.* 135:151–164, 1973.
5. Frearson, N., B. W. W. Focant, and S. V. Perry. Phosphorylation of a light-chain component of myosin from smooth muscle. *FEBBS Lett.* 63:27–32, 1976.
6. Daniel, J. L., and R. S. Adelstein. Isolation and properties of platelet myosin light chain kinase. *Biochem.* 15:2370, 1976.
7. Elzinga, M., J. H. Collins, W. M. Kuehl, and R. S. Adelstein. Complete amino-acid sequence of actin of rabbit skeletal muscle. *Proc. Nat. Acad. Sci. USA.* 70:2687–2691, 1973.
8. Strezelecka-Golaszecuska, H., E. Prochiniewicz, E. Nowak, and S. Zmoozynski. Chicken gizzard actin: polymerization and stability. *Eur. J. Biochem.* 104:41–52, 1980.
9. Gordon, D. J., E. Eisenberg, and E. D. Korn. Characterization of cytoplasmic actin isolated from *Acanthamoeba castellanii* by a new method. *J. Biol. Chem.* 251:4788–4786, 1976.
10. Stone, D., and L. B. Smillie. The amino acid sequence of rabbit skeletal muscle tropomyosin. *J. Biol. Chem.* 253:1137–1148, 1978.
11. McLachlan, A. D. and M. Stewart. Tropomyosin coiled-coil interactions: evidence for an unstaggered structure. *J. Mol. Biol.* 98:293–298, 1975.
12. Weber, A., and J. M. Murray. Molecular control mechanisms in muscle contraction. *Physiol. Rev.* 53:612–673, 1973.
13. G. N. Philips, Jr., J. P. Fillers, and C. Cohen. Motions of tropomyosin. *Byosphys. J.* 32:485–502, 1980.
14. Jakes, R., F. Northrop, and J. Kendrick-Jones. Calcium binding regions of myosin 'regulatory' light chains. *FEBBS Lett.* 70:229–234. 1976.
15. Yazawa, M., H. Kuwayama, and K. Yagi. Modulator protein as a Ca^{2+}-dependent activation of rabbit skeletal myosin light-chain

kinase. Purification and characterization. *J. Biochem. Tokyo* 84: 1253–1258, 1978.

16. Dabrowska, R., J. M. F. Sherry, D. K. Aromatorio, and D. J. Hartshorne. Modulator protein as a component of the myosin light chain kinase from chicken gizzard. *Biochem.* 17:253–258, 1978.
17. Hartshorne, D. J., R. F. Siemankowski, and M. O. Aksoy. Ca regulation in smooth muscle and phosphorylation: some properties of the myosin light chain kinase. In *Regulatory Mechanisms of Muscle Contraction.* Eds.: Ebashi, Maruyama, Endo. Jap. Sci. So. Press, Tokyo/ Springer-Verlag, Berlin, pp 287–301, 1980.
18. Hathaway, D. R., and R. S. Adelstein. Human platelet myosin light chain kinase requires the calcium binding protein calmodulin for activity. *Proc. Natl. Acad. Sci. USA.* 76:1653–1657, 1979.
19. Adelstein, R. S., M. A. Conti, D. R. Hathaway, and C. B. Klee. Phosphorylation of smooth muscle myosin light chain kinase by the catalytic subunit of adenosine 3′:5′-monophosphate-dependent protein kinase. *J. Biol. Chem.* 253:8347–8350, 1978.
20. Dabrowska, R., and D. J. Hartshorne. A Ca^{2+}- and modulation-dependent myosin light chain kinase from nonmuscle cells. *Biochem. Biophys. Res. Commun.* 85:1352–1359.
21. Nairn, A. C., and S. V. Perry. Calmodulin and myosin light-chain kinase of rabbit fast skeletal muscle. *Biochem. J.* 179:89–97, 1979.
22. Waisman, D. M., T. J. Singh, and J. H. Klang. The modulatory-dependent protein kinase. A multifunctional protein kinase activatable by the Ca^{2+}-dependent modulator protein of the cyclic nucleotide system. *J. Biol. Chem.* 253:3387–3390, 1978.
23. Yerna, M.J., R. Dabrowska, D. J. Hartshorne, and R. D. Goldman. Calcium-sensitive regulation of actin–myosin interactions in baby hamster kidney (BHK-21) cells. *Proc. Natl. Acad. Sci. USA.* 76:184–188, 1979.
24. Bremel, R. D., and M. E. Shaw. Conversion of calcium-sensitive myosin light-chain kinase to a calcium-insensitive form. *FEBBS Lett.* 88:242–246, 1978.
25. Bhalla, R. C., R. V. Sharma, and R. C. Gupta. Isolation of two myosin light-chain kinases from bovine carotid artery and their regulation by phosphorylation mediated by cyclic AMP-dependent protein kinase. *Biochem. J.* 203:583–592, 1982.
26. Walsh, M. P., B. Vallet, F. Autric, and J. G. Demaille. Purification and characterization of bovine cardiac calmodulin-dependent myosin light chain kinase. *J. Biol. Chem.* 254:12136–12144, 1979.
27. Walsh, M. P., S. Hinkins, I. L. Flink, and D. J. Hartshorne. Bovine stomach myosin light-chain kinase: purification, characterization, and comparison with the turkey gizzard enzyme. *Biochem.* 32:6890–6896, 1982.
28. Adelstein, R. S., M. A. Conti, D. R. Hathaway, and C. B. Klee. Phosphorylation of smooth muscle myosin light chain kinase by the catalytic subunit of cAMP-dependent protein kinase. *J. Biol. Chem.* 253:8347–8350, 1978.

29. Conti, M. A., and R. S. Adelstein. Phosphorylation by cyclic adenosine 3′:5′-monophosphate-dependent protein kinase regulates myosin light chain kinase. *Fed. Proc.* 39:1569–1573, 1980.
30. Morgan, M., V. S. Perry, and J. Ottaway. Myosin-light-chain phosphatase. *Biochem. J.* 157:687–697, 1976.
31. Pato, M. D., and R. S. Adelstein. Characterization of myosin phosphatase. I. from smooth muscle. *Biophys. J.* 33:278a, 1981.
32. Pato, M. D., and R. S. Adelstein. Purification and characterization of a multisubunit phosphatase from turkey gizzard smooth muscle. *J. Biol. Chem.* 258:7047–7054, 1983.
33. Pato, M. D., and R. S. Adelstein. Characterization of a Mg^{2+}-dependent phosphatase from turkey gizzard smooth muscle. *J. Biol. Chem.* 258:7055–7058. 1983.
34. Yamaguchi, M., Y. Miyazawa, and T. Sekine. Preparation and properties of smooth muscle myosin from horse esophagus. *Biochem. Biophys. Acta.* 216:411–421, 1970.
35. Chacko, S., M. A. Conti, and R. S. Adelstein. The effect of phosphorylation of smooth muscle myosin on actin activation and Ca^{2+} regulation. *Proc. Natl. Acad. Sci.* 74:129–133, 1977.
36. Chacko, S., and A. Rosenfeld. Regulation of actin-activated ATP hydrolysis by arterial myosin. *Proc. Natl. Acad. Sci. USA* 78:292–296, 1982.
37. Chacko, S. Effect of phosphorylation, Ca^{2+} and tropomyosin on actin-activated ATPase activity of mammalian smooth muscle myosin. *Biochemistry* 20:702–707, 1981.
38. Sobieszek, A., and J. V. Small. Myosin-linked calcium regulation in vertebrate smooth muscle. *J. Mol. Biol.* 102:75–92, 1976.
39. Gorecka, A., M. O. Aksoy, and D. J. Hartshorne. The effect of phosphorylation of gizzard myosin on actin activation. *Biochem. Biophys. Res. Commun.* 71:325–331, 1976.
40. Hartshorne, D. J., and A. J. Persechini. Phosphorylation of myosin as a regulatory component in smooth muscle. *Ann. N.Y. Acad. Sci.* 356:130–141, 1980.
41. Sellers, J. R., P. B. Chock, and R. S. Adelstein. Cooperative interactions between the two heads of smooth muscle myosin. *Biophy. J.* 41:153a, 1983.
42. Sellers, J. R., and R. S. Adelstein. Cooperativity and the reversible phosphorylation of smooth muscle heavy meromyosin. *Biophy. J.* 37:262a, 1982.
43. Persechini, A., and Hartshorne. Ordered phosphorylation of the two 20,000 molecular weight light chains of smooth muscle myosin. *Biochem.* 22:470–476, 1983.
44. Chantler, P. D., J. R. Sellers, and A. G. Szent-Gyorgyi. Cooperativity in scallop myosin. *Biochem.* 20:210–216, 1981.
45. Sellers, J. R., P. D. Chantler, and A. G. Szent-Gyorgyi. Hybrid formation between scallop myofibrils and foreign regulatory light-chains. *J. Mol. Biol.* 144:223–245, 1980.

46. Onishi, H., H. Suzuki, K. Nakamura, K. Takahashi, and S. Watanabe. Adenosine triphosphatase activity and "thick filament" formation of chicken gizzard myosin in low salt media. *J. Biochem.* 83:835–847, 1978.
47. Suzuki, H., H. Onishi, D. Takahashi, and S. Watanabe. Structure of chicken gizzard myosin. *J. Biochem.* 84:1529–1542, 1978.
48. Scholey, J. M., K. A. Taylor, and J. Kendrick-Jones. Regulation of non-muscle myosin assembly by calmodulin-dependent light chain kinase. *Nature* (London) 287:233–235, 1980.
49. Onishi, H., and T. Wakabayashi. Electron microscopic studies of myosin molecules from chicken gizzard muscle I: the formation of the intramolecular loop in the myosin tail. *J. Biochem.* 92:871–879, 1982.
50. Trybus, K. M., T. W. Huiatt, and S. Lowey. A bent monomeric conformation of myosin from smooth muscle. *Proc. Nat'l. Acad. Sci. USA* 79:6151–6155, 1982.
51. Somlyo, A. V. and T. M. Butler. Myosin filaments have non-phosphorylated light-chains in relaxed smooth-muscle. *Nature* 294:567–569, 1981.
52. Sobieszek, A. and J. V. Small. Regulation of the actin myosin interaction in vertebrate smooth muscle: activation via a myosin light-chain kinase and the effect of tropomyosin. *J. Mol. Biol.* 12:559–576, 1977.
53. Nath, N., A. Carlos, and S. C. Seidel. Actin-activated ATPase activity of phosphorylated myosin from pulmonary artery at low Mg^{2+} concentration; effect of actin and tropomyosin from different sources. *Biophys. J.* 41:154a, 1983.
54. Bremel, R. D., J. M. Murray, and A. Weber. Manifestations of cooperative behavior in the regulated actin filament during actin-activated ATP hydrolysis in the presence of calcium. *Cold Spring Harb. Symp. Grant. Biol.* 37:267–275, (Cold Spring Harbor Laboratory, Cold Spring Harber 1972).
55. Eaton, B. L., D. R. Kominz, and E. Eisenberg. Correlation between the inhibition of acto-heavy meromyosin ATPase and the binding of tropomyosin to F-actin: effect of Mg^{2+}, KCl, troponin I and troponin C. *Biochemistry.* 14:2718–2725, 1975.
56. Sobieszek, A., and J. V. Small. Effect of muscle and nonmuscle tropomyosin in reconstituted skeletal muscle actomyosin. *Eur. J. Biochem.* 118:533–539, 1981.
57. Ebashi, S. *Regulation of muscle contraction in cell and muscle motility.* Eds.: R. M. Dowben and J. W. Shay. Plenum Publishing Corporation, pp 79–87, 1983.
58. Rees, D. D., and D. W. Frederiksen. Calcium regulation of porcine aortic myosin. *J. Biol. Chem.* 256:357–364, 1981.
59. Sherry, J. M. F., A. Gorecka, M. O. Aksoy, R. Dabrowska, and D. J. Hartshorne. Roles of calcium and phosphorylation in the regulation of the activity of gizzard myosin. *Biochem.* 17:4417–4418, 1978.
60. Rosenfeld, A., and S. Chacko. Phosphorylation and calcium binding by myosin. *Fed. Proc.* 40:1786, 1981.
61. Hartshorne, D. J., and U. Mrwa. Regulation of smooth muscle actomyosin. *Blood Vessels.* 19:1–18, 1982.
62. Nag, S., and J. C. Seidel. Dependence on Ca^{2+} and tropomyosin of the actin-activated ATPase activity of phosphorylated gizzard myosin

in the presence of low concentrations of Mg^{2+}. *J. Biol. Chem.* 258:6444–6449, 1983.

63. Heaslip, R. J., and S. Chacko. Comparison of the ATPase activities of phosphorylated and thiophosphorylated chicken gizzard myosin. *Biophys. J.* 45:43a, 1984.
64. Ebashi, S., Y. Nonomura, S. Nakamura, H. Nakasone, and K. Kohama. Regulatory mechanism in smooth muscle: actin-linked regulation. *Fed. Proc.* 41:2863–2867, 1982.
65. Ebashi, S. Regulation of muscle contraction. *Proc. R. Soc. Lond.* B207:259–286, 1980.
66. Mikawa, T., T. Toyo-oka, Y. Nonomura, and S. Ebashi. Essential factor of gizzard troponin fraction. *J. Biochem.* 81:273–275, 1977.
67. Mikawa, T., Y. Nonomura, M. Hirata, S. Ebashi, and S. Kakiuchi. Involvement of an acidic protein in regulation of smooth muscle contraction by tropomyosin leiotonin system. *J. Biochem.* 84:1633–1636. 1978.
68. Hirata, M., T. Mikawa, Y. Nonomura, and S. Ebashi. Ca^{2+} regulation in vascular smooth muscle, II. Ca^{2+} binding of aorta leiotonin. *J. Biochem.* 87:369–378, 1980.
69. Nonomura, Y., T. Mikawa, and S. Ebashi. Ca^{2+}-sensitive thin filament from chicken gizzard smooth muscle. *Proc. Japan Acad.* 56(A):178–183, 1980.
70. Mikawa, T., Y. Nonomura, and S. Ebashi. Does phosphorylation of myosin light chain have direct relation to regulation in smooth muscle. *J. Biochem.* 82:1789–1791, 1977.
71. Mikawa, T.: "Freezing" of the calcium-regulated structures of gizzard thin filaments by gluteraldehyde. *J. Biochem.* 85:879–881, 1979.
72. Chacko, S., R. J. Heaslip and S. Ebashi. The effect of leiotonin on stably phosphorylated smooth muscle myosin. *Fed. Proc.* 43:1551.
73. Marston, S. B., R. M. Trevett, and M. Walters. Calcium ion-C regulated thin filaments from vascular smooth muscle. *Biochem. J.* 185:355–365, 1980.
74. Walters, M., and S. B. Marston. Phosphorylation of the calcium ion-regulated thin filaments from vascular smooth muscle. *Biochem. J.* 197:127–139, 1981.
75. Persechini, A. U. Mrwa, and D. J. Hartshorne. Effect of phosphorylation on the actin-activated ATPase activity of myosin. *Biochem. Byophys. Res. Comm.* 98:800–805, 1981.
76. Cole, H. A., J. A. Grand, and S. V. Perry. Non-correlation of phosphorylation of the P-light chain and the actin activation of the ATPase of chicken gizzard myosin. *Biochem. J.* 206:319–328, 1982.
77. Moreland, R. S., and G. D. Ford. The influence of magnesium on calcium-activated, vascular smooth muscle actomyosin ATPase activity. *Arc. Biochem. Biophy.* 208:325–333, 1981.
78. DiSalvo, J., E. Gruenstein, and P. Silver. Ca^{2+} dependent phosphorylation of bovine aortic actomyosin. *Proc. Soc. Exp. Med.* 158: 410–414, 1978.
79. Small, J. V. and A. Sobieszek. Ca-regulation of mammalian smooth muscle actomyosin via a kinase-phosphatase-dependent phosphorylation and dephosphorylation of 20,000-Mr light chain of myosin. *Eur. J. Biochem.* 76:521–530, 1977.

80. Dabrowska, R., S. Hinkins, M. P. Walsh, and D. J. Hartshorne. The binding of smooth muscle myosin light chain kinase to actin. *Biochem. and Biophy. Res. Comm.* 107:1542–1531, 1982.
81. Driska, S. P., Aksoy, M. O., and Murphy, R. A.: Myosin light chain phosphorylation associated with contraction in arterial smooth muscle. *Am. J. Physiol.* 240:C222–C233, 1981.
82. Barron, J. T., M. Barany, and K. Barany. Phosphorylation of the 20,000 dalton light chain of myosin of intact arterial smooth muscle in rest and in contraction. *J. Biol. Chem.* 254:4954–4956, 1979.
83. de Lanerdolle, P., and J. T. Stull. Myosin phosphorylation during contraction and relaxation of tracheal smooth muscle. *J. Biol. Chem* 255:9993–10000, 1980.
84. Hoar, P. E., W. G. L. Kerrick, and P. S. Cassidy. Chicken gizzard: relation between calcium-activated phosphorylation and contraction. *Science* 204:503–506, 1979.
85. Butler, T. M., and M. J. Siegman. Chemical energetics of contraction in mammalian smooth muscle. *Fed. Proc.* 41:204–208, 1982.
86. Ruegg, J. C., and R. J. Paul. Vascular smooth muscle calmodulin and cyclic AMP dependent protein kinase alters calcium sensitivity in porcine carotid skinned fiber. *Circ. Res.* 50:394–399, 1982.

Chapter 8

Studies on Skinned Fiber Preparations

Kaushik D. Meisheri, J. Casper Rüegg, and Richard J. Paul

1. Introduction

Smooth muscle contractility may be discussed in terms of the following basic processes (schematically shown in Fig. 1): (a) excitation of the cell membrane, (b) excitation-contraction (E-C) coupling, which includes the release of Ca^{2+} from internal stores and increased transmembrane Ca^{2+} transport resulting ultimately in an increase in the cytoplasmic free Ca^{2+} concentration, (c) the supply of ATP for contraction and its resynthesis by cell metabolism, and (d) the chemomechanical energy transformation with ATP hydrolysis, catalyzed by the myosin ATPase as the driving force. To these basic processes one might add for consideration the homeostatic mechanisms controlling osmotic pressure, salt concentration, pH, and so forth of the cytoplasm surrounding the contractile structures. Because of the inherently complex nature of the interactions among the above processes, experiments on intact muscle can provide only suggestive information regarding the contractile machinery itself. On the other extreme, studies of the contractile mechanism with isolated contractile proteins lack the structural integrity of the contractile and regulatory systems that could be of critical importance in the intact muscle. The skinned fiber preparation represents a "bridge" between the

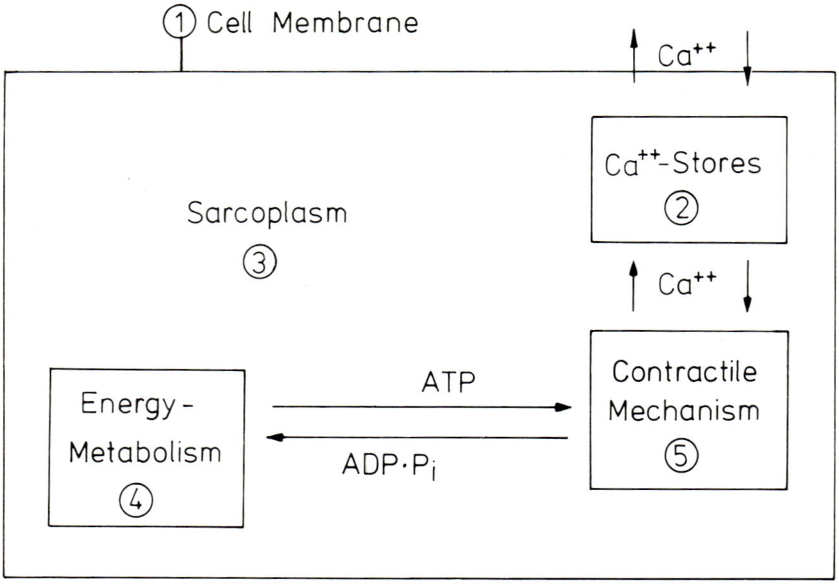

Fig. 1. Schematic presentation of the smooth muscle cell. Various processes participating in the excitation–contraction coupling that may be affected by the skinning procedures are indicated. For details, see text. (1) Excitable cell membrane which can be removed or made highly permeable (to various ions, especially Ca^{2+}, and small molecular weight compounds) by membrane skinning procedures such as glycerol extraction and treatment with detergents like saponin and Triton X-100. (2) Ca^{2+} stores (internal membrane systems of sarcoplasmic reticulum for release and uptake of Ca^{2+}) are removed by Triton X-100, but not by saponin treatment (cf. Section 3). (3) Sarcoplasm: Small cytoplasmic crystallites and soluble proteins are extracted by the skinning procedure. (4) Energy metabolism apparatus (e.g., mitochondria): partly extracted or damaged by the skinning procedure. (5) Contractile machinery: actin and myosin filaments, and regulatory proteins; could be functionally isolated by skinning (but may also be partly extracted or damaged).

above two extremes. In skinned muscle preparations, the physiological behavior (i.e., tension development) of the isolated contractile machinery can be studied under conditions in which the environment (including the energy supply) surrounding the contractile elements can be directly controlled, and in which the functional contributions of the cell membrane (as well as the intracellular membrane structures, when desired) can be excluded. This is the decisive advantage of the skinned fiber preparation over the two experimental systems described for studying the relations between structure and function of the contractile machinery.

Attempts for isolation of the contractile structure and hence the energy converting machinery in muscle date back to 1949, when Szent-Györgyi first described a method of preparing glycerinated skeletal muscle fibers. Similar glycerol extraction procedure has been applied to smooth muscle (*11, 37, 119*). Hasselbach and Ledermair (*46*) showed that the glycerol-extracted uterine smooth muscle could contract with ATP as the only energy source if ATP could be hydrolyzed by the myosin ATPase. If the ATPase was inhibited by salyrgan, the fibers relaxed. Thus the ATP as an energy source for smooth muscle contraction could be established using skinned fiber preparations long before the demonstration of ATP splitting in living smooth muscle (cf. ref. *12*). In addition, the rate of ATP splitting could be directly measured in skinned fibers (*101*). The energy cost required to maintain tension was found to be much less than in the skeletal muscle and corresponded to the high holding economy of living smooth muscle (*see* ref. *92*).

The importance of demonstrating the adequacy of the isolated contractile machinery of smooth muscle as a model for the living muscle was recognized early. This is of particular significance with respect to absolute force, speed of contraction, length–tension relation, and certain features of active state such as the quick release recovery phenomenon. When an isometrically contracted living or glycerinated fiber bundle of smooth muscle is suddenly released by, for example, 10% of the initial length, the tension drops to zero (because of the discharge of elastic elements); the subsequent spontaneous recovery of tension (as the contractile elements contract stretching the elastic elements in series) is as fast in glycerinated fibers as in living muscle. The recovery of tension does not occur in fibers that develop rigor tension after depletion of ATP. In the presence of ATP, the recovered tension is a function of initial muscle length. By plotting force development vs muscle length, a tension–length diagram is obtained with a characteristic force optimum above the resting length. This relation between length and tension is similar in living and glycerinated skinned fibers. By improving the skinning procedure (see below), it has been possible to obtain skinned fibers that develop as much tension, and when unloaded, shorten as fast as living muscle.

The advantages of skinned fiber preparations and the types of hypotheses that may be better suited for testing with the skinned fiber preparations will be briefly introduced here and discussed individually in the later sections. An obvious advantage offered by the skinned fiber preparation is that the Ca^{2+}-dependency of the smooth muscle contractile machinery can be precisely established under known experimental conditions. In smooth muscle, the composition of the cytoplasm and the concentration of free Ca^{2+} ions are not precisely known. In skinned fibers, on the other hand, the composition of the medium surrounding the myofibrils can be controlled and altered at will. Thus the relationship between free Ca^{2+} concentration and tension could be established by using glycerinated smooth muscle fibers. The Ca^{2+} dependence of isometric force, determined using Ca^{2+} ion concentration buffered with the Ca-EGTA chelating system, was found to be similar for smooth and skeletal fibers (*37*, see below).

In fibers skinned with glycerol or by the use of detergents such as Triton X-100, the contractile machinery is functionally isolated from the plasma membrane and also from the internal Ca^{2+} stores. On the other hand, using a mild detergent treatment, such as with saponin, it is possible to retain internal membrane structures that retain the ability to store and release Ca^{2+} (*31*). One can thus obtain a system consisting of the contractile apparatus together with the Ca^{2+} stores that may participate in the E-C coupling. Such models can prove, as will be illustrated later, quite useful in investigating the mechanism(s) of E-C coupling and, in particular, that of intracellular Ca^{2+} release.

It has been found that membrane skinning renders the smooth muscle contractile machinery accessible even to larger protein molecules. Thus it has become possible to directly study the effects of calmodulin, cyclic AMP-dependent protein kinases, and elucidate their role in regulation of smooth muscle contraction and relaxation (*61, 113*). Further examples of this type of experimental approach include a myosin light chain kinase preparation that has been shown to produce contraction (121, 122) and a myosin light chain phosphatase preparation shown to enhance relaxation (*93*) in skinned smooth muscle. Such studies are not possible in intact smooth muscle where the plasma membrane provides a formidable diffusional barrier to the movement of large proteins. Similarly, for testing of current theories regarding the regulation of smooth muscle contractility, the skinned fiber technique has provided the means of establishing a direct relationship between Ca^{2+}-induced tension development and myosin light chain phosphorylation levels (*51*), as well as between Ca^{2+}-

induced tension development and ATPase activity (*42*). Lastly, skinned fibers studies are particularly suitable for testing whether difference among various smooth muscles and alterations in contractility found in disease state (e.g., hypertension) are caused by changes in the contractile machinery itself.

The following recent reviews on skinned fibers are useful in providing a broad range of perspectives on the subjects discussed here (*33, 34, 59, 79, 97*).

2. Methodology

2.1. Skinning Procedures

A summary of various skinning procedures that have been applied to smooth muscles is presented in Table 1; for the sake of brevity and clarity, not all studies have been included. This table also provides for comparison of the Ca^{2+} sensitivity of these skinned smooth muscle preparations. An earlier study (*11*) on glycerol-extracted uterine muscle is available, but no *p*Ca–tension relation was presented. A study on the long term effects of glycerol extraction on dog intestinal smooth muscle is also available (*125*). The saponin skinning procedure, first used in guinea pig taenia coli, has since been successfully employed in guinea pig pulmonary artery (*31*), guinea pig mensenteric artery (*54, 56*) and porcine coronary artery (*55*), in some cases with minior modifications. The method of skinning with Triton X-100 at 4°C has also been applied to porcine coronary artery (*94*) and guinea pig tracheal smooth muscle (*114*). Reports are also available on Triton X-100 skinning at room temperature in rabbit pulmonary artery (*15*), and rabbit ileum (*17*). In a recent report on chicken gizzard (*121*), gentle homogenization followed by 1% Triton X-100 treatment for 20 min at room temperature was used to obtain a skinned preparation. An attempt to skin mammalian smooth muscle by microdissection of the cell membrane was unsuccessful (*33*).

2.2. Criteria for Acceptable Models

The following is a discussion of different criteria that have been examined with various skinning methods in smooth muscle (cf. Table 1) to demonstrate that the surface membrane has become freely permeable to substances such as Ca^{2+}, Mg^{2+}, ATP, and EGTA after the skinning. The discussion also includes the limitations of the studies with skinned smooth muscle in general.

Table 1
Various Skinning Methods Used for Smooth Muscle[a]

Tissue	Skinning procedure	$[Ca^{2+}]$ for 50% of max. tension,[b] μM	Reference
1. Hog carotid artery	50% v/v glycerol extraction	0.3	*37*
2. Hog carotid artery	Freeze glycerination with 50% v/v glycerol	0.25	*86*
3. Guinea pig taenia coli	50–100 μg/mL saponin for 25 min at room temp.	1.0	*31*
		0.5	*100*
4. Rabbit mesenteric artery	500 μg/mL saponin for 6 min at room temp.	0.6	*44*
5. Rabbit taenia coli	1% w/v Triton x-100 for 16 h. at room temp.	1.2	*40*
6. Guinea pig taenia coli	1% w/v Triton x-100 for ½ h at 4°C, then freeze glycerination as in *(3)*	1.5[c]	*113*
Hog carotid artery		0.3[c]	*95*
7. Chicken gizzard	Mechanical skinning	10.0[c]	*17*
8. Rabbit ileum	Staphylococcal α-toxin, 100 μg/ml, 5–10 min	10.0	*14*

[a]This table does not include all studies published on the subject; see text for additional information.
[b]Approximate values derived from the curves.
[c]Values without addition of exogenous calmodulin.

2.2.1. General Criteria—Electron Microscopy, Membrane Potential, and Pharmacological Responses

Only limited data are available with regard to the electron microscopy (EM) of the skinned smooth muscle. Haeusler et al. (*44*) have presented EM data on the saponin skinned rabbit mesenteric artery indicating a folding of the plasma membrane, homogenously compact contractile apparatus, and the absence of any significant number of intact cells. Also, information regarding the effectiveness of saponin in producing holes in the cell membrane is available in nonsmooth muscle cells (*31*). Somlyo et al. (*112*) have demonstrated the presence of a functional SR in saponin skinned vascular smooth muscle using EM and electron probe analysis. Recently Spedding (*115*), using the Triton X-100 skinning method as described by Sparrow et al.(*113*), has provided electromicroscopic evidence for the absence of the plasma membrane in skinned taenia coli smooth muscle.

Loss of the electrical potential difference across the plasma membrane after skinning has been demonstrated in cardiac tissue (*124*). Such information is needed in smooth muscle but could prove difficult because of the small cell size.

The absence, after skinning, of a contractile response to an agonist (which is normally observed in the intact preparation) may be beneficially used for demonstrating the functional removal of the plasma membrane. Histamine, norepinephrine, or depolarization that elicit contraction in living carotid artery have no effect on the "freeze glycerinated" preparation (*85*) or the Triton X-100 treated preparations (Paul and Rüegg, unpublished observations). Endo et al. (*31*) reported that contractions in guinea pig taenia coli induced by carbachol, which acts as the plasma membrane, could no longer be obtained after the saponin treatment. However, this may not be used as an absolute criterion since Haeusler et al. (*44*) using a modified saponin skinning procedure found that norepinephrine still produced a contraction after the skinning of the rabbit mesenteric artery. Caffeine, thought to act directly on the sarcoplasmic reticulum (SR), can be more definitively used to indicate the presence of a functional internal Ca^{2+} store. Using this criterion saponin treatment has been reported to leave the SR intact (*31, 99*). In contrast, caffeine could not induce contraction in Triton skinned taenia coli under the appropriate conditions (low Ca^{2+} buffer with 0.1 m*M* EGTA; cf. section 3.) indicating the removal of SR in addition to the surface membrane (Meisheri and Rüegg, 1983).

2.2.2. Responses To Ca^{2+}

Isometric tension recordings of skinned fibers are usually obtained by first incubating the fiber in a Mg-ATP-containing solution with EGTA to maintain free Ca^{2+} below $10^{-8}M$. When the free Ca^{2+} concentration is increased by the use of appropriate Ca-EGTA buffer (*see* individual studies for details) the fiber develops graded tension. The threshold Ca^{2+} concentration required to produce contraction can be in the range of $3–5 \times 10^{-7}M$ (for example, *see* refs. *86, 100*), and even lower when calmodulin is added (*95, 113*). In one study on mechanically skinned chicken gizzard fibers (*51*), the threshold Ca^{2+} required for activation was between $5 \times 10^{-6}–10^{-5}M$. It is not clear whether this difference can be explained solely on the basis of different stability constants used for the calculation of free Ca^{2+} (*see* below). Since the intact smooth muscle develops, in contrast, negligible tension in these Ca^{2+} ranges, these findings suggest a substantially increased membrane permeability in skinned preparations.

The Ca^{2+}-induced contraction after the skinning has been found to be as high as, or even higher than, the depolarization-induced contraction in the same fiber before skinning (*44, 100*). Glycerinated preparations have been found to produce comparatively less tension (*37, 74*). However, the freeze-glycerinated procedure was recently found to give much better results (*85, 86*). Triton X-100 skinned preparations appear to be similar in terms of force production (*4, 5*) to the freeze-glycerination method. It should be pointed out that the maximum force development by Ca^{2+} activation in the skinned fibers is still comparatively less than the absolute force development observed in intact smooth muscle with the most effective stimulus (*78*). It is not clear whether these differences are real or simply reflect systematic differences in determination of the fiber cross-section area, a parameter used to normalize force.

Saida and Nonomura (*100*) reported that the speed of maximal Ca^{2+}-induced contraction in the saponin-skinned taenia coli was as fast as the high K^{+}-induced activation before skinning. Schneider et al. (*102*) in Triton skinned taenia coli and Peterson (*86*) in freeze-glycerinated hog carotid artery found using "Ca^{2+} jump" technique (whereby a solution of low buffer capacity is switched to one of a very high capacity, allowing the free Ca^{2+} concentration to increase very rapidly), that the time course of Ca^{2+}-activated tension development was identical to the rate of tension development observed in the intact tissue. Tension recovery following a "quick release" during the steady state of a maximial Ca^{2+}-induced contraction is also very rapid and comparable to that found in the intact muscle (*102*). Maximal shortening velocities have also been reported to be similar in skinned and living smooth muscles (*4, 5, 82*). One should be alerted that other factors such as pH, [Mg^{2+}], and temperature may also be important. Thus, depending on a particular smooth muscle preparation and a particular skinning procedure, optimal conditions may have to be established for obtaining acceptable Ca^{2+} responses.

It has been found that while under maximal Ca^{2+} activation, the plateau tension in skinned fibers can be maintained for relatively long periods of time. For example, Triton skinned taenia coli, when activated by $10^{-5}M$ Ca^{2+}, can maintain a stable plateau tension for up to 2 h. In contrast it has been observed that when contraction–relaxation cycles are repeated, force development in subsequent contractions declines particularly at low Ca^{2+} concentrations. Since the addition of calmodulin can partially reverse such a decline, it appears that calmodulin and other as yet unknown factors might be gradually lost from the skinned fiber preparation during the experimental period. Loss of force may also be ascribable to an activation of Ca^{2+}-dependent proteinases, which may be prevented by the addition of inhibitors, such as leupeptin.

2.2.3. ATP Diffusion In Skinned Fibers

Development of rigor tension and relaxation from rigor with ATP removal and readdition respectively can be used to demonstrate a high permeability to ATP in the skinned fiber. This is of particular significance if the time course of the tension changes are similar to that predicted for ATP diffusion into and out of the fiber. Furthermore, force development in the skinned fiber during Ca^{2+} activation should be dependent on the ATP supply from the outside. Endo et al (*31*) reported that in the guinea pig taenia coli, removal of Mg or ATP induced a small rigor tension after saponin skinning that was abolished on reintroduction of Mg or ATP. Hoar et al (*51*) showed in mechanically skinned chicken gizzerd fibers that a rigor state was produced by the removal of ATP in the presence of Ca^{2+} that prevented shortening in the presence of Ca^{2+} or relaxation in the absence of Ca^{2+}. Readdition of ATP reversed the rigor state. They also demonstrated using (γ-^{32}P) ATP that the fibers were permeable to ATP. The diffusion coefficient for ATP in the skinned hog carotid artery has been determined by Peterson (*86*). By using labeled ATP, he showed that the ATP permeation throughout a skinned fiber of 0.4 mm thickness occurred with a half-time ($t_{1/2}$) of 45 s. The time course of tension development following the addition of ATP under the same conditions was similar with $t_{1/2}$ of 69 s (cf. Fig. 5 in ref. *86*). A similar value for the diffusion coefficient of ATP for glycerinated skeletal muscle fibers has been reported by Bowen (*10*). In contrast, the value for the skinned insect flight muscle is smaller (*67*) suggesting that ATP diffuses more readily into the loose structures of skinned smooth muscle than this highly structured striated muscle. Because of the diffusional limitation of the ATP supply, the force per cross-section area in skinned fibers depends on ther ATP concentration (for example, *see* refs. *86* and *100*) and the fiber thickness.

In experiments with the skinned fibers, preparations should not be thicker than about 0.2 mm (and preferably much thinner) to guarantee an adequate ATP supply to the inside of the fiber. For fibers of greater thickness, an ATP-free core in the center of the fiber may develop leading to the generation of a rigor state. The relation between fiber thickness and ATP concentration difference is described by the Meyerhoff-Schultz equation which may be rearranged to give: $\Delta C = AR^2/4D = A'/4D$; where ΔC is the ATP concentration difference between the outside and inside of the fiber (mol/L), R is the radius of the fiber bundle, D is a diffusion coeficient (about 10^{-6} cm^2/s) and A is the ATP splitting rate per second, which may be expressed as mol/cm^3 fiber (A) or as mol/cm fiber length (A'). Hasselbach and Ledermair (*46*) found that for force development in a glycerinated uterine fiber bundle of 0.2 mm thickness, an ATP concentration of 1 m*M* was adequate, whereas in the case of arterial muscle, Peterson (*86*) found it to be 1–2 m*M*. In order to minimize the creation of an ATP concentration gradient, addition to the experimental solutions of an ATP regenerating system, consisting of phosphocreatine and creatine kinase, or alternatively, phosphoenolpyruvate and pyruvate kionase have often been found to be useful (*42, 53, 113*).

2.2.4. Other Considerations.

One of the limitations in making valid comparisons of the Ca^{2+}-sensitivity of the various skinned smooth muscle preparations is the use of different stability constants by various workers for the calculation of various free ion concentrations (*see,* for example, refs. *29, 35, 53,* and *91*). Perhaps, uniformity in this respect can be achieved by the acceptance by various workers in the future of the computation program such as, for example, that described by Fabiato and Fabiato (*36; see also* ref. *117*).

Other methodological considerations include the use of dithioerythritol (DTE, 0.5–2 m*M*) in the skinning and experimental solutions to prevent possible oxidation of –SH groups and the addition of 1 m*M* azide to inhibit bacterial growth during the storage and experimental period. A slightly acidic pH (6.7–6.8) has been chosen in many experiments to avoid extraction of the contractile proteins that are less soluble below pH 7. It may also be noted that most of the skinned fiber studies have been done at room temperature (20–25°C); the use of higher temperatures (35–37°C, as for the intact mammalian tissue) has been found to increase the rate of fiber deterioration. Peterson (*86*), for example, found that long term incubation at 37°C led to a failure of the skinned fiber to relax completely upon subsequent removal of Ca^{2+}. Similarly, at high temperatures, skinned taenia coli were found to become Ca^{2+}-insensitive within 1–2 h, and gradually develop tension even in the absence of Ca^{2+} (Rüegg, unpublished observations).

2.2.5. Summary

Skinned smooth muscle fibers with mechanical properties comparable to those of the living muscle have been developed using a variety of procedures. The enhanced permeability of the skinned preparations, a major advantage in experimental design, can also be a disadvantage in terms of the potential loss of contractile or regulatory proteins. With proper attention to the development of adequate criteria for evaluation of skinned fiber behavior, skinned fibers provide a structured analog of living muscle in which the myofibrillar environment is under experiment control.

3. Intracellular Ca^{2+} Store

One of the primary contributions of the studies on skinned smooth has been elucidation of the presence of intracellular Ca^{2+} store(s) as well as their functional significance. The data obtained from such studies have generally been complementary to the data obtained from structural and pharmacological studies on the intact smooth muscle (*see* Chapters 1 and 4 of this book).

It has been shown by the use of contractility and ^{45}Ca flux measurements that agonists such as carbachol, norepinephrine, and histamine can produce contraction of intact smooth muscle by releasing Ca^{2+} from an intracellular store, and that caffeine releases Ca^{2+} from the same intracellular store. Thus, for example, carbachol and caffeine deplete the same internal Ca^{2+} store in guinea pig taenia coil (*18, 32*) whereas norepinephrine and caffeine share a common internal Ca^{2+} pool in rabbit aorta (*25*). The saponin-skinned smooth muscle has proven valuable in providing complementary evidence for the presence of internal Ca^{2+} store. It has been shown that caffeine retains its ability to produce contraction in skinned smooth muscle from which only the plasma membrane has been functionally removed (*32, 99*). The ability of an agonist such as carbachol to produce contraction under this condition is lost presumably due to the loss of an effective agonist-receptor coupling. However, in one report using a modification of the saponin-skinning procedure, it has been observed that skinned vascular smooth muscle can also produce contraction in response to an agonist, nonrepinephrine (*44*).

The saponin-skinned smooth muscle preparation has proven useful in further characterizing this intracellular Ca^{2+} store. Advantage is taken of the fact that caffeine at high concentrations (10–25 m*M*) can completely deplete the store. In order to observe a contractile effect of caffeine in skinned fibers a solution containing low Ca^{2+} ($<10^{-8}M$) in a weak Ca^{2+} buffer (0.1 m*M* EGTA) has to be used so that the Ca^{2+} released from the store by caffeine, instead of being bound by the buffer system, is made available to the contractile elements. Once the store is emptied by caffeine, it can be reloaded by an exposure to higher Ca^{2+} concentration ($>10^{-7}M$) again in the presence of 0.1 m*M* EGTA. The skinned fiber develops a contraction during this procedure, as only a part of the Ca^{2+} is taken up by the store. The amount of Ca^{2+} taken up is then determined by returning the fiber to a low Ca^{2+} solution ($<10^{-8}M$) and subsequent exposure to caffeine. The magnitude of this contractile response to caffeine (area under the caffeine-induced contraction curve) has been used as an indicator of the Ca^{2+} uptake by the store (*31, 32, 99*). The limitations of this index of Ca^{2+} uptake have been discussed previously (*34, 99*). A very low EGTA buffer system may not effectively maintain a constant Ca^{2+} concentration. Also, the possibility that caffeine, at the high concentration (20 m*M*) used in some studies (*99*), alters the Ca^{2+} sensitivity of the myofilaments has not been excluded although it has been shown that 5 m*M* caffeine did not affect the *p*Ca–tension relationship in skinned coronary smooth muscle (*55*). Using the above index it was found that the loading of Ca^{2+} into the store is dependent on the Ca^{2+} concentration and the time of exposure to Ca^{2+} (*32, 99*). Increasing the $[Ca^{2+}]$ decreases the time of Ca^{2+} exposure required. At $1 \times 10^{-6}M$

Ca^{2+} the store can be fully loaded in 3–5 min. Pretreatment with procaine inhibits caffeine-induced Ca^{2+} release (*32, 55*). In contrast, cyclic AMP and cyclic AMP-dependent protein kinase were shown to enhance caffeine-induced Ca^{2+} release by increasing Ca^{2+} accumulation in the store during Ca^{2+} loading (*54*).

Saida (*99*) has presented evidence in saponin-skinned taenia coli for a Ca^{2+}-induced Ca^{2+} release from this store. When the amount of Ca^{2+} remaining in the storage site was estimated by caffeine-induced contraction at different $[Ca^{2+}]$, it was found to be decreased as the Ca^{2+} concentration increased from 10^{-6} to $10^{-4}M$. This Ca^{2+}-induced Ca^{2+} release was partially inhibited when Mg^{2+} was increased from 0 to 10 m*M*. An indirect evidence for Ca^{2+}-induced Ca^{2+} release is also available in skinned smooth muscle of guinea pig mesenteric artery (*54*).

Another interesting and significant observation is that the replacement of potassium with choline (*55*) or of propionate with chloride (*99*) produced a contraction suggesting "depolarization"-induced Ca^{2+} release in skinned smooth muscle. Saida found the Ca^{2+} release from the store to be dependent on the chloride concentration. This depolarization-induced Ca^{2+} release is from the same store from which caffeine causes Ca^{2+} release.

From the above discussion it can be argued that since Ca^{2+}, cyclic AMP, and ionic composition all can affect the Ca^{2+} release from this internal store, it is likely to be of physiological significance. However, the precise manner in which this internal Ca^{2+} store participates in contraction of intact smooth muscle during excitation by various agonists remains to be established.

There appear to be several parallels between the characteristics of the Ca^{2+} release in smooth muscle and that observed in cardiac and/or skeletal muscle (*32, 34*). Caffeine, Ca^{2+}, cyclic AMP, and depolarization all have been shown to affect internal Ca^{2+} stores in cardiac or skeletal muscle. Based on this analogy, the likely candidate for this internal Ca^{2+} store in smooth muscle is also thought to be the sarcoplasmic reticulum, SR (*32*). The deposition of Ca^{2+} in the SR of saponin-skinned vascular smooth muscle has been recently demonstrated (*112*).

In skinned smooth muscle, the plasma membrane is no longer an effective barrier for maintaining the ionic gradients necessary for the membrane potential. When under this condition, changing ionic composition can still produce a depolarization-induced Ca^{2+} release as described above, then this would effectively exclude the plasma membrane as the source of Ca^{2+} release. Confirmation of the "depolarization"-induced Ca^{2+}

release would be quite useful in this regard. In contrast, Haeusler et al. (*44*), based on their study on saponin skinned rabbit mesenteric artery, favor the internal Ca^{2+} store to be the inner surface of plasma membrane. They presented evidence that NE-induced contraction in skinned arteries could be inhibited by pretreatment with caffeine. However, X-537 A (which itself caused a maximal contraction presumably by its known action of Ca^{2+} release from the SR and/or mitochondria) did not prevent NE contraction, suggesting that NE did not release Ca^{2+} from such structures.

4. Control of Actomyosin Interaction

4.1. Mechanism of Ca^{2+} Activation

A second major area in which skinned fiber studies have played a significant role is the elucidation of the mechanism(s) by which Ca^{2+} controls the interaction between actin and myosin. Contractile activity in smooth muscle, as for its skeletal muscle counterparts, is regulated by the intracellular calcium ion concentration. However, over the past five years, it has become increasingly clear that the mechanism by which Ca^{2+} regulates contractile protein interaction in vertebrate smooth muscles differs fundamentally from that of skeletal muscle. Due to the newness of this hypothesis and the explosive growth of information, the mechanisms discussed here will not have the certainty usually accorded to topics in review articles. However the major controversies are clearly defined and the focus here will be on the use of skinned fibers to differentiate between conflicting hypotheses.

Much of our knowledge of the regulation of smooth muscle contractile activity by Ca^{2+} is based on studies of isolated actomyosin. This is treated in detail in Chapter 7 (Chacko) and in a recent review of Marston (*69*). Conflicting viewpoints can be readily seen in the proceedings of two recent symposia (*111, 120*). There is, however, general agreement as to nature of the overall differences between skeletal and smooth muscle. Purified skeletal muscle myosin is always active, that is, its actin-activated ATPase is high and independent of Ca^{2+}. The well-characterized troponin–tropomyosin system based on the thin (actin-containing) filaments inhibits skeletal actomyosin ATPase. Ca^{2+} binds to troponin and removes this inhibition. Purified smooth muscle actomyosin, on the other hand, is inactive and requires a Ca^{2+}-dependent activator for its ATPase activity. The nature of the Ca^{2+}-dependent activator is currently the focus of considerable controversy. The two major schemes for Ca^{2+} activation of vertebrate smooth muscle actin–myosin interaction are summarized in Fig. 2. Although these hypotheses are currently the focus of considerable attention, it is well to keep in mind other possibilities, including the direct interaction of Ca^{2+} and myosin (*19*), Ca^{2+} regulation of *thin* filaments (*68*), and some other Ca^{2+} dependent mechanisms (*72, 110*).

The leiotonin system proposed by Ebashi and colleagues (*30*) is thought to be located on the thin filament and consists of an 80 kdalton regulatory moiety, leiotonin A, and a 17 kdalton Ca^{2+}-binding moiety, leiotonin C. The mechanism of the activation by leiotonin is unknown. However, as it can be calculated that there would be only five leiotonin molecules per thin filament (*45*), it is thus unlikely that leiotonin regulation is similar to the structural

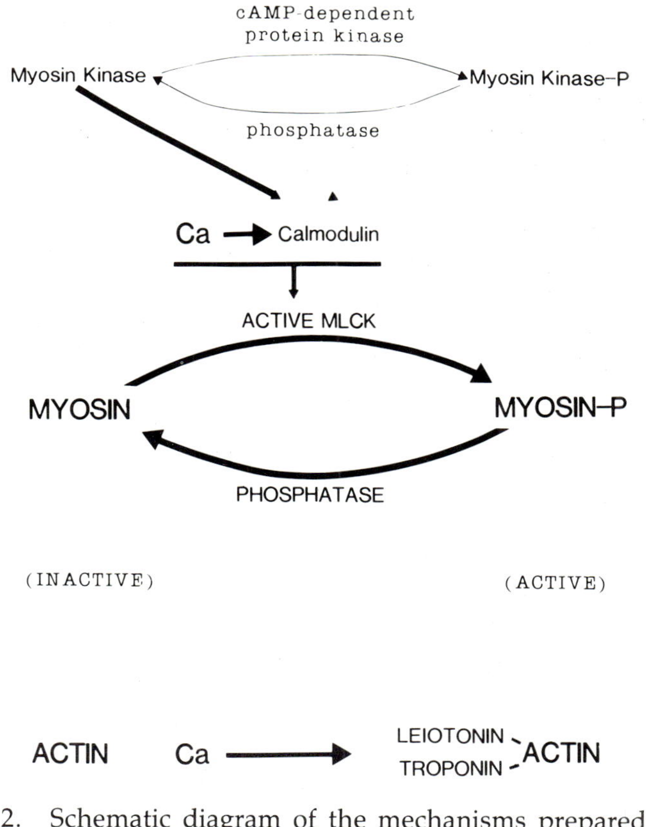

Fig. 2. Schematic diagram of the mechanisms prepared for the Ca^{2+}-activation of actin–myosin interaction in smooth muscle (*see* Sections 4 and 5).

hindrance mechanism proposed for troponin–tropomyosin systems.

The second major hypothesis for the Ca^{2+}-activation of smooth muscle actomyosin involves phosphorylation of a 20 kdalton myosin light chain. In this scheme phosphorylated myosin is active, i.e., shows an actin-activated ATPase or superprecipitation reaction. The Ca^{2+} sensitivity is conferred by a myosin light chain kinase, the activity of which is dependent on the calcium–calmodulin complex concentration.

There appears to be formidable experimental evidence supporting both the leiotonin and myosin phosphorylation theories of Ca^{2+} regulation of smooth muscle actomyosin activity. Although these theories are not necessarily mutually exclusive, proponents of the leiotonin hypothesis, although acknowledging myosin phosphorylation, do not attribute a regulatory role to this phenomenon (*30*).

A further level of complexity as to the nature of Ca^{2+} regulation of actomyosin interaction in the intact structured smooth muscle arises from evidence on living smooth muscle. Studies of the time course of myosin light chain phosphorylation (MLC-P_i) are consistent in that phosphorylation has been shown to precede the development of isometric force (cf. ref. *77*). This is a necessary condition for regulation via phosphorylation. However, during the maintenance phase of isometric force, several groups have reported that the MLC-P_i level decreases (*77*). The extent of the decrease in MLC-P_i is somewhat controversial. Murphy and his colleagues report that during sustained contractures, MLC-P_i levels return to the prestimulated values. This observation plus the fact that these tissues remain sensitive to changes in bathing media Ca^{2+} during sustained contractures led this group to postulate the existence of two distinct Ca^{2+} regulatory systems in smooth muscle (*77*). Observation of a similar decline in the velocity of contraction measured at various times during isometric contraction (*28*), led to the hypothesis that MLC-P_i was primarily involved in regulation of the cross-bridge cycling rate, which is manifested in terms of velocity and economy (isometric force per unit ATP utilization rate). Phosphorylation of skeletal muscle myosin is also reported to be involved in control of cross-bridge cycle rate, however, phosphorylation plays the opposite role in that increasing levels of MLC-P_i are associated with slower velocity and increased economy of tension maintenance (*24, 65*). Butler and Siegman (*13*), on the other hand, do not find any consistent relation between MLC-P_i levels and economy in either living skeletal or smooth muscle preparations. From our perspective, the measurement of MLC-P_i levels in living smooth muscle appears to be a major source of the current controversy as to the role of MLC-P_i in smooth muscle.

Skinned smooth muscle preparations, in addition to retaining three-dimensional structural integrity, have provided a unique model for testing these hypotheses because the components of the regulatory system itself can be altered under experimental control. As alluded to in the Introduction, skinned smooth muscle fiber model systems were used to demonstrate the dependence of isometric force on trace levels of Ca^{2+} nearly 20 years ago. Recent experimental evidence has shown that the ubiquitous Ca^{2+}-binding protein, calmodulin, plays a role in this Ca^{2+}-sensitivity. Addition of calmodulin has been shown to increase the Ca^{2+}-sensitivity of isometric force in skinned chicken gizzard (*17*), taenia coli (*113*), and vascular smooth muscle (*94, 95*) preparations. Furthermore, the shifts in Ca^{2+}-sensitivity are consistent with the hypothesis that the Ca^{2+}–calmodulin complex is the regulatory moiety. In addition, the demonstration that compounds which bind to calmodulin and inhibit calmodulin-dependent enzymes, such as the phenothiazine derivatives (*66*), also inhibit contraction in skinned smooth muscle (*23, 62*), further implicating the Ca^{2+}–calmodulin complex in regulation of smooth muscle contractility. As myosin light chain kinase is known to be activated by Ca^{2+}–calmodulin (*50*), myosin phosphorylation was also implicated in control of isometric force in skinned fibers. This hypothesis received further support in studies using the thiophosphorylated ATP analog, adenosine 5′-*O*(3–thiotriphosphate) ($ATP_{\gamma}S$). On studies of isolated proteins, $ATP_{\gamma}S$ was shown to be an effective substrate for kinases, but a poor substrate for phosphatases (*41*). $ATP_{\gamma}S$ thus provides a tool for effective irreversible thiophosphorylation of myosin (*105*). Incubation of skinned chicken gizzard fibers in the presence of Ca^{2+} and $ATP_{\gamma}S$ in the absence of ATP led to a high level of thiophosphorylation of myosin (*51*). Addition of ATP to the skinned fibers following $ATP_{\gamma}S$ incubation leads to an irreversible, Ca^{2+}-independent contracture. The Ca^{2+}-independence of this contracture has been ascribed to the fact that myosin was irreversibly thiophosphorylated. Similar results have been reported for skinned fibers prepared from rabbit ileum (*14*), taenia coli (*113*), and vascular smooth muscle (*85*). Taken as a whole, these data indicate that a Ca^{2+}–calmodulin-dependent kinase is involved in the regulation of smooth muscle contractility. Recent evidence strongly suggests that this kinase, in fact, is the myosin light chain kinase (MLCK). CTP, which can serve as a substrate for the actomyosin NTPase but not MLCK, does not support contraction in skinned fibers, but does so following thiophosphorylation (*60*). Furthermore, Hartshorne and his collaborators, utilizing a partial trypsin-digestion procedure, have isolated an 80 kdalton fragment of MLCK that retains kinase activity but is independent of Ca^{2+} (*123*). Skinned fiber preparations incubated in the presence of this Ca^{2+}-independent kinase develop isometric force in the absence of Ca^{2+} (*121, 122*). This evidence indicates that myosin phosphorylation is a major pathway for initiation of contractile activity in smooth muscle. A corollary of this mechanism is that enhanced phosphatase activity would lead to smooth muscle relaxation. Rüegg et al. (*93*) demonstrated that incubation of skinned vascular smooth muscle with an enriched phosphatase fraction led to an acceleration of the rate of relaxation.

As is the case for studies of living fibers, measurement of MLC-P_i in skinned fibers is technically difficult. Although MLC-P_i has been shown to increase with isometric force, basal levels and the change in MLC-P_i at maximum force remain controversial. In skinned hog carotid artery, Chatterjee and Murphy (*20*) find a basal MLC-P_i of 0.1 mol P_i/mol myosin and 0.6 mol during a Ca-induced contracture. Gardner and DiSalvo (*39*), on the other hand, report values of 0.3 and 0.8, respectively. In skinned chicken gizzard smooth muscle, Gagelmann et al. (*38*) reported basal levels of MLC-P_i of 0.1 and contracture levels of 0.6, whereas Cassidy et al. (*14*) reported that only 0.2 mol P_i/mol myosin was found under conditions of maximum force development. Proponents of the myosin-phosphorylation hypothesis must reconcile these data with studies on the actin-activated myosin ATPase of chicken gizzard, which indicate that little activity was observed in the range of 0–50% myosin phosphorylation, whereas a further increase to 60–70% caused a dramatic activation (*83*). Differences including not only the methodology for analysis of MLC-P_i, but also the means of fixing the level of MLC-P_i in the fiber and extraction techniques, may underlie these discrepancies. In spite of the uncertainty in the absolute values of MLC-P_i, skinned fiber studies have shown that the level increases with Ca^{2+}-induced force and, furthermore, the increase in MLC-P_i is maintained. The latter finding is different than that reported for MLC-P_i in living smooth muscle (*77*) and suggests that the decline in MLC-P_i with contraction duration may reflect intracellular Ca^{2+} changes.

As a whole, studies on skinned fibers have provided a substantial amount of evidence indicating that myosin phosphorylation is a major pathway for control of smooth muscle contractility. It is anticipated that studies on skinned fibers will play a major role in the identification of other potential Ca^{2+} control mechanisms and elucidation of the role of these systems in the modification of the cross bridge cycle rate (*see* Section 5 below) in addition to the control of the number of activated bridges is discussed in this section.

4.2. Cyclic Nucleotides in Smooth Muscle Relaxation

Although controversial, it is generally held that the relaxation of smooth muscle by β-adrenoceptor agonists and cAMP-phosphodiesterase inhibitors is mediated via an increase in the cytoplasmic cAMP levels (*26, 63*). According to current views, cAMP can produce relaxation of smooth muscle by a reduction in the cytoplasmic free Ca^{2+} concentration or by inhibition of actin–myosin interaction because of phosphorylation of MLCK. Either or both mechanisms may operate during β-stimulation to produce smooth muscle relaxation (for reviews, *see* refs. *2* and *9*).

With regard to the mechanism of lowering Ca^{2+}, β-adrenoceptor stimulation and/or cAMP have been shown in a variety of smooth muscles to reduce cytoplasmic Ca^{2+} activity by affecting various Ca^{2+} flux mechanisms (for references, *see 71*). One of the Ca^{2+} flux mechanisms that may play an important role is intracellular Ca^{2+} sequestration, presumably by the SR (*18, 76*). As discussed before (cf. Section 3), in saponin-skinned smooth muscle, the SR is kept functionally intact and thus can serve as a useful experimental preparation to investigate the role of the SR in cAMP-induced relaxation of smooth muscle. Itoh et al. (*54*), using saponin-skinned rabbit mesenteric artery, demonstrated that caffeine-induced release of Ca^{2+} from the intracellular store was enhanced when the fibers were pretreated with cAMP and cAMP-dependent protein kinase (cAMP-PK) during the Ca^{2+} loading procedure. This indicated that cAMP and cAMP-PK caused an enhanced accumulation of Ca^{2+} in the SR. Furthermore, this effect on the Ca^{2+} accumulation by the SR occurred at the concentration of cAMP plus cAMP-PK that had no direct effect on the contractile elements. The authors concluded that the effect of cAMP on Ca^{2+} accumulation by the SR is physiologically more important than the direct action of cAMP on the contractile elements (see below).

Considerable information concerning the mechanism of the direct effect of cAMP on the contractile elements has been reported. The observation that the catalytic subunit of cAMP-dependent protein kinase (cAMP-PK, csu) can phosphorylate MLCK, and thus reduce its activity, was originally presented by Adelstein et al. (*3*) in an isolated smooth muscle contractile protein system (*see* Fig. 2). This has been extended to show the inhibitory effect of cAMP and cAMP-PK, csu on myosin light chain phosphorylation, actin-activated myosin ATPase, and myosin superprecipitation in various smooth muscles (*8, 21, 75, 108, 109*). The demonstration that cAMP and/or cAMP-PK, csu can have a direct inhibitory effect on the contractile proteins in terms of tension development using skinned smooth muscle preparation has provided a crucial "physiological" test for the above mechanism. The inhibitory effects of cAMP or the kinase on Ca^{2+}-activated tension development were first reported in skinned taenia coli (*75, 96*) and in skinned chicken gizzard (*61*). Further studies from this laboratory (*94, 95, 114*) have extended these observations of cAMP-PK, csu-induced inhibition in other smooth muscles as well (skinned porcine carotid and coronary artery, and skinned guinea pig tracheal smooth muscle). It has also been demonstrated in Triton-skinned taenia coli that the pretreatment with cAMP-PK, csu shifted the pCa–tension relationship curve to the right (*113*). However, in all of the above studies, the inhibitory effect has been observed only at high concentrations (0.1–0.5 m*M*) of cAMP and/or cAMP-PK, csu. In a recent study, we have now demonstrated relaxation of skinned smooth muscle with micromolar concentration ($10^{-6}M$) of cAMP alone (*70*). This indicates that the cascade system (including cAMP-PK) required for manifestation of the cAMP relaxing effect has remained functionally intact in the skinned fiber, and that the cAMP diffusion into the fiber is not a significant problem. There has been one report on the saponin-skinned rabbit mesenteric artery (*54*) in which the pretreatment with cAMP ($3 \times 10^{-6}M$) plus cAMP-PK had no effect on the pCA–tension relation, although $10^{-4}M$ cAMP plus cAMP-PK did reduce the plateau tension induced by the maximal Ca ($10^{-5}M$).

The concept that cAMP can produce relaxation of smooth muscle by affecting two different mechanisms raises a physiologically relevant question about the relative significance of the "direct" relaxing effect of cAMP on the contractile elements in relation to the effect of cAMP on lowering Ca^{2+} in producing smooth muscle relaxation. We have addressed this question by studying in skinned taenia coli the relaxing effect of cAMP at various fixed Ca^{2+} concentrations (*70; see* Fig. 3, left). There is a steep dependence of cAMP-induced relaxation on Ca^{2+} within the physiologically important range of 10^{-7}–$10^{-6}M$ Ca^{2+}. This suggests that the effect of cAMP on lowering of the cytoplasmic free Ca^{2+} concentration may play an important role in determining the extent of the effect of cAMP on the contractile proteins in producing smooth muscle relaxation.

Another area elucidated by studies on skinned fibers has been the mechanism of the direct effect of cAMP. As discussed by Adelstein (*2*), the direct effect of cAMP is achieved by the phosphorylation of MLCK, reducing its activation by the

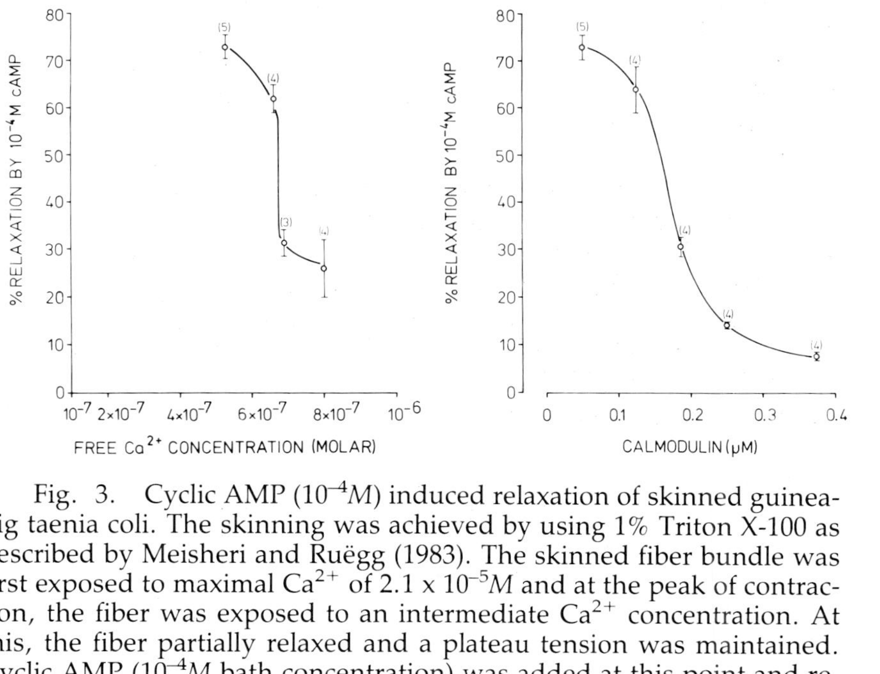

Fig. 3. Cyclic AMP ($10^{-4}M$) induced relaxation of skinned guinea-pig taenia coli. The skinning was achieved by using 1% Triton X-100 as described by Meisheri and Ruëgg (1983). The skinned fiber bundle was first exposed to maximal Ca^{2+} of 2.1 x $10^{-5}M$ and at the peak of contraction, the fiber was exposed to an intermediate Ca^{2+} concentration. At this, the fiber partially relaxed and a plateau tension was maintained. Cyclic AMP ($10^{-4}M$ bath concentration) was added at this point and relaxation was computed as the per cent of plateau tension maintained before the addition of cyclic AMP. Fig. 3, left, shows the effect of increasing Ca^{2+} concentrations at a given calmodulin concentration of 0.05 μM, and Fig. 3, right, shows the effect of increasing calmodulin concentrations at a given Ca^{2+} concentration of 5.3 x $10^{-7}M$. The data is given as mean SEM of three to five observations. Each observation was obtained from a different fiber bundle. All experiments were done at 25°C and pH 6.7. For compositions of the solutions used, *see* Meisheri and Ruëgg (1983).

Ca^{2+}–calmodulin complex, and thus reducing actin–myosin interaction (Fig. 2). Consistent with the above postulate, it has been found that increasing calmodulin prevents the inhibition by cAMP-PK, csu of Ca^{2+}-activated tension development in skinned fibers (*61, 113*). We have observed a steep dependence of cAMP-relaxation on calmodulin concentration in skinned taenia coli at a constant $[Ca^{2+}]$ (Fig. 3, right). Kerrick and Hoar (*61*), in addition, observed that cAMP-PK failed to produce any effect when the skinned fibers were pretreated with Ca^{2+} and ATP$^{\gamma}$S to produce irreversible thiophosphorylation of the myosin light chains. The implication is that the decrease in the MLC phosphorylation is a prerequisite for cAMP-PK induced relaxation. The data on the effects of cAMP or cAMP-PK on MLC phosphorylation in skinned fibers are currently not available and they are important for establishing whether cAMP exerts its observed direct effects in the skinned fibers exclusively via an effect on MLCK.

There has been some evidence in intact smooth muscle that cGMP may also play a role in smooth muscle relaxation, although this remains to be unequivocally established (*27, 64, 103*). A study from this laboratory (*89*) showed that cGMP-dependent protein kinase inhibited Ca^{2+}-activated tension development in the skinned trachael smooth muscle. It was also found that in the skinned taenia coli, cGMP at $10^{-4}M$ has only a slight relaxing effect under the same conditions in which $10^{-4}M$ cAMP produces about 70% relaxation as shown in Fig. 3, left (*70*). More data are required to draw a meaningful conclusion; nonetheless, they indicate that the skinned fiber preparations can be useful for establishing comparative effectiveness of agents thought to have direct effect on the contractile elements.

4.3. Catch Mechanism

Tonic molluscan smooth muscles contract when stimulated with acetylcholine, but they fail to relax rapidly when the stimulation ceases. During such a slow relaxation, these muscles are said to be in "catch." Force is maintained at least ten times more economically than in the preceding contraction (*6, 7*). In contrast to active contraction, the tension is maintained passively in catch since active state is absent or small; there is little, if any, recovery after a quick release (*57*). Serotonin has been found to unlock this catch and induce rapid relaxation (*118*). This effect of serotonin is thought to be mediated via cyclic AMP, which may cause phosphorylation of regulatory proteins or paramyosin (*1*). These experiments on intact molluscan smooth muscle cannot distinguish whether the action of serotonin, via cAMP, is on lowering the intracellular free Ca^{2+} concentration or a direct action on the contractile proteins. Chemically skinned fibers have been used to test these alternative hypotheses. When a skinned molluscan muscle fiber is first contracted by increasing Ca^{2+} to $10^{-6}M$, in an ATP-containing solution, and then the Ca^{2+} concentration is suddenly reduced to $10^{-8}M$ (using an EGTA buffer), the fiber relaxes slowly, exhibiting a catchlike state. In contrast to the contraction in the presence of $10^{-6}M$ Ca^{2+}, there is no recovery of tension after a quick release, and the rate of ATP hydrolysis is as low as in the resting state (*43*). Such a preparation can be relaxed rapidly by the addition of cAMP without any change in the ATP splitting rate (*22, 88*). A similar enhancing effect on relaxation is exerted by the catalytic subunit of cAMP-dependent protein kinase, as well as by an increase in pH to 7.5 that is known to loosen the paramyosin structure (*98*). The low ATP splitting rate suggests that catch tension is maintained without cross-bridge cycling. Moreover, even during the contraction preceding the catch, cross-bridge cycling (and ATP splitting rate) must be extremely slow compared with that in skeletal muscle (*104*), suggesting a high holding economy. In summary, then, the skinned fiber experiments have shown that whereas active contraction is switched on and off by changes in the intracellular free Ca^{2+}, the catch state can be regulated by cAMP in a manner independent of Ca^{2+}.

Recently, evidence has been obtained for the existence of catchlike state or latch state in vertebrate smooth muscle. For example, a tetanically stimulated taenia coli relaxes very slowly after cessation of the stimulation (*107*). The ATP consumption of living muscle during such a slow relaxation was found to be not significantly different from that in the resting muscle. Moreover, active state as determined by quick release techniques was absent. Skinned fibers have been used to determine whether the slow relaxation is caused by a slow removal of Ca^{2+} or some other process, for example, dephosphorylation of myosin. When a Ca^{2+}-induced contraction in Triton-skinned guinea-pig taenia coli is terminated by rapid lowering of Ca^{2+}, a fast relaxation as expected from the Ca-removal hypothesis is not obtained. Instead skinned taenia coli fibers relax slowly; tension is apparently maintained passively as there is little recovery after a quick release (*102*). This catchlike state resembles rigor; however, unlike rigor, it can be rapidly abolished by the addition of inorganic phosphate (*102*).

The ATP splitting rate in skinned taenia coli has been determined during contraction and relaxation with a temporal resolution of 10 s (*42*). The skinned fiber is suspended in a bath containing a NADH-coupled ATP regenerating system that allows both tension and ATP hydrolysis to be measured simultaneously. Any decrease in NADH concentration due to the production of ADP is continuously measured with a microscope spectrofluorometer. Immediately after Ca^{2+} removal, ATPase activity was found to decrease to resting levels, suggesting that the tension maintenance in skinned taenia coli during slow relaxation may be caused by noncycling (i.e., non-ATP splitting) attached cross bridges of the type suggested by Siegman et al. (*106*) and Murphy et al. (*77*). Addition of inorganic phosphate accelerates relaxation without a change in the ATP splitting rate, implying that the noncycling cross bridges become rapidly detached. Such skinned fiber studies may prove useful in elucidating the intracellular mechanisms regulating formation and detachment of the noncycling latch bridges.

5. Effect of Ca^{2+} on Mechanical and Mechanochemical Properties of Smooth Muscle

In the previous section, the opportunity for control of the myofibrillar environment provided by skinned fiber preparations was utilized in studies focusing on the mechanism underlying Ca^{2+} control of actin–myosin interaction, i.e., smooth muscle contractility. Using similar experiments, but with a slightly different focus, the nature of the actin–myosin interaction itself can be probed. From this perspective, the effects of environmental changes on the mechanical properties and ATP hydrolysis of skinned fibers are interpreted at the level of the cross bridge and cross-bridge cycle. Studies of muscle mechanics and energetics of this kind are dominated by skeletal muscle work, largely because of the ease of correlating functional evidence with the intensely studied and well-defined structure of striated muscle. It is only in comparatively recent times that studies on living smooth muscle mechanics and energetics have achieved the sophistication and precision characteristic of skeletal muscle studies. A discussion of the underlying theories and current status is beyond the scope of this chapter; for details, the reader is referred to the recent reviews of Hellstrand and Paul (*49*), Paul (*80*), and Peterson (*84*). Studies of mechanics and energetics on skinned muscle preparations are limited in number, particularly in the case of smooth muscle. Thus the aim of this section is to present examples of the types of questions studied and the potential of skinned fiber techniques, rather than to present definitive answers.

The relation between isometric force and muscle length has been a major piece of evidence underlying the development of the "sliding-filament" theory of muscle contraction. Pfitzer et al. (*87*) reported that the length–tension relation in glycerinated hog carotid artery was comparable to that obtained in the living artery. Though not studied in detail, these results suggest that the integrity of the plasma and SR membrane systems is not of major importance to the transmission of force from cell to cell.

Recently much attention has been focused on the "instantaneous stiffness" of muscle fibers as it is thought to reflect the number of attached cross bridges (*52*). To measure this stiffness, a rapid (~1 ms) change in length (Δl) is imposed on the fiber and the corresponding change in tension (ΔP) recorded; stiffness is defined as the ratio of ΔP to Δl. Stiffness measurements reported by Pfitzer et al. (*87*) again showed little difference between glycerinated and living hog carotid artery. Stiffness was found to be proportional to tension, when tension was altered by varying the initial tissue length. In other words stiffness and force had similar dependence on muscle length. In these experiments, stiffness was determined by plotting the peak tension obtained after quick stretch or release as a function of the length change. The linear slope obtained in the region of 0.5% release to 2% stretch indicates that the immediate stiffness (dP/dL) is Hookean as in the case of living (*52*) and skinned skeletal muscle fibers (*126*). In smooth muscle, the linear slope extrapolated to zero force at 1–2% release is independent of the extent of Ca-activation (*87*). These results are very similar to those obtained in skeletal muscle and although not unequivocal, the straightforward interpretation of these results is that instantaneous stiffness does indeed reflect the number of attached cross bridges in smooth muscle at a particular instant in time. Skinned fiber preparations allowed stiffness as well as force to be measured as a function of Ca^{2+}; this is shown in Fig. 5a of Pfitzer et al. (*87*). Again stiffness was proportional to force, indicating that Ca^{2+} under these conditions controls the number of cross bridges.

In addition to the number of cross bridges, it is possible that Ca^{2+} also affects the rate of the cross-bridge cycle as well. This has been a subject of considerable controversy in the skeletal muscle field (cf. refs. *58* and *90*). In these studies contractile velocity is used as an index of the cross-bridge cycle rates. Arner (*5*), using an isotonic lever system, compared force–velocity characteristics in living and Triton-skinned guinea-pig taenia coli. Under optimal activation conditions no differences in the extrapolated V_{max} or the Hill curvature parameter *a*/*P*o were found. Using similar techniques, Arner (*4*) also reported the Ca^{2+} dependence of V_{max} in rat portal vein. In contrast to the living preparation (*48*) V_{max} in the skinned preparation was dependent on Ca^{2+}. Paul et al. (*82*) also reported that unloaded shortening velocity V_{us}, was dependent on Ca^{2+} in both skinned guinea-pig taenia coli and porcine carotid artery. Moreover, they reported that V_{us} was also dependent on calmodulin, in a manner consistent with the hypothesis that the regulatory moeity was the Ca^{2+}-calmodulin complex. While both isometric force and V_{us} were dependent on Ca^{2+}-calmodulin, the dependencies were not identical in that isometric force could be saturated before V_{us}. This finding provides a mechanism for the decline in shortening velocity with contraction duration under conditions in which isometric force remains constant in living smooth muscle (cf. ref. *77*). In contrast to the living muscle, decreases in shortening velocity with contraction duration are relatively small (*82*). This observation, made under constant Ca^{2+}, supports the hypothesis that the changes in mechanical and metabolic parameters (cf. ref. *82*) observed in living smooth muscle with time reflect the time course of changes in intracellular Ca^{2+} (cf. ref. *73*).

Measurement of rate of ATP splitting provides independent information in terms of both the level of actin–myosin interaction and the nature of the interaction. Although little work has been reported for skinned smooth muscle the ATPase activity of skinned fibers has been shown to be proportional to the level of tension (*85, 101*). Recent information (*42, 47*) has indicated that the relation between ATPase and tension may not be this straightforward. Much more information is required before definitive statements can be made, and judging by the recent activity in this area, such information may be available in the near future.

Skinned fiber preparations provide a valuable tool in analyzing the effects of the myofibrillar environment on the control of actin–myosin interaction and the nature of the interaction itself. Ca^{2+} plays a major role in control of both the number of bridges and the rate of cross-bridge cycling. Studies on skinned smooth muscle fibers have implicated calmodulin, most likely through a Ca^{2+}–calmodulin complex, as being an integral component of these regulatory process.

Although the effects of Ca^{2+} are the primary focus of this chapter and book, it is clear that skinned fibers provide a methodology for assessing the effects of many other factors on the control and nature of the actin–myosin interaction. Although that is beyond the scope of this work, we would like to call your attention to several recent studies of interest in this regard: Sr, in particular, its ability to substitute for Ca^{2+} (*51, 62*); pH (*74*); Mg (*4, 37, 46, 100*); PO_4 (*102*) and VO_4 particular with regard to inhibition of actin–myosin interaction (*85*).

ACKNOWLEDGMENTS

Dr. Meisheri was at the University of Heidelberg as a fellow of the Alexander von Humboldt Stiftung during the course of this study. The work cited in this study was supported by grants from the Deutsche Forschungsgemeinschaft (J. C. Ruëgg) and NIH Grants 23240 and 22619 (R. J. Paul). Dr. Paul is the Established Investigator of the American Heart Foundation.

References

1. Achazi, R. K. Phosphorylation of molluscan paramyosin. *Pflügers Arch.* 379:179–201, 1979.
2. Adelstein, R. S. Calmodulin and the regulation of the actin–myosin interaction in smooth muscle and nonmuscle cells. *Cell* 30:349–350, 1982.
3. Adelstein, R. S., M. A. Conti, D. K. Hathaway, and C. B. Klee. Phosphorylation of smooth muscle myosin light chain kinase by the catalytic subunit of adenosine 3′–5′ monophosphate dependent protein kinase. *J. Biol. Chem.* 253:8347–8350, 1978.
4. Arner, A. Force-velocity relation in chemically skinned rat portal vein: effects of Ca^{2+} and Mg^{2+}. *Pflügers Arch.* (in press), 1983.
5. Arner, A. Mechanical characteristics of chemically skinned guinea-pig taenia coli. *Pflügers Arch.* 395:277–284, 1982.
6. Baguet, F. and J. M. Gillis. The respiration of the anterior byssus retractor muscle of mytilus edulis (ABRM) after a phasic contraction. *J. Physiol. (Lond.)* 188:67–82, 1967.
7. Baguet, F. and J. M. Gillis. Energy cost of tonic contraction in a lamelli-branch catch muscle. *J. Physiol.* (*Lond.*) 198:127–143, 1968.
8. Bhalla, R. C., R. V. Sharma, and R.C. Gupta. Isolation of two myosin light-chain kinases from bovine carotid artery and their regulation by phosphorylation mediated by cyclic AMP-dependent protein kinase. *Biochem J.* 203:583–592, 1982.

9. Bolton, T. B. Mechanisms of action of transmitters and other substances on smooth muscle. *Physiol. Rev.* 59:607–718, 1979.
10. Bowen, W. J. Glycerol treated muscle as working model of contraction and diffusion of ATP through it. In: *Biochemistry of Muscle Contraction.* Ed. J. Gergely. Little, Brown and Company, Boston, 1964. pp 441–447.
11. Briggs, A. H. Characteristics of contraction in glycerinated uterine smooth muscle. *Am. J. Physiol.* 204:739–742, 1963.
12. Butler, T. M. and R.E. Davies. High-energy phosphates in smooth muscle. Bohr, D. F., Somlyo, A. P., and Sparks, H. V., Jr., eds. *Handbook of Physiology, Section 2, The Cardiovascular System, Vol. II, Vascular Smooth Muscle.* Bethesda: Am. Physiol. Soc. pp 237–252, 1980.
13. Butler, T. M. and M. J. Siegman. Chemical energy usage and myosin light chain phosphorylation in mammalian smooth muscle. *Federation Proc.* 42:57–61, 1983.
14. Cassidy, P., P. E. Hoar, and W. G. L. Kerrick. Irreversible thiophosphorylation and activation of tension in functionally skinned rabbit ileum strips by $[^{3S}]$ $ATP_{\beta}S$. *J. Biol. Chem.* 254:11148–11153, 1979.
15. Cassidy, P., P. E. Hoar, and W. G. L. Kerrick. Inhibition of Ca^{2+} -activated tension and myosin light chain phosphorylation in skinned smooth muscle strips by the phenothiazines. *Pflügers Arch.* 387:115–120, 1980.
16. Cassidy, P. and W. G. L. Kerrick. Superprecipitation of gizzard actomyosin, and tension in gizzard muscle skinned fibers in the presence of nucleotides other than ATP. *Biochim. Biophy. Acta.* 705:63–69, 1982.
17. Cassidy, P.S., W. G. L. Kerick, P. E. Hoar, and D. A. Malencik. Exogenous calmodulin increases Ca^{2+} sensitivity of isometric tension activation and myosin phosphorylation in skinned smooth muscles. *Pflügers Arch.* 392:115–120, 1981.
18. Casteels, R. and L. Raeymaekers. The action of acetylcholine and catecholamines on an intracellular calcium store in the smooth muscle cells of the guinea-pig taenia coli. *J. Physiol.* 294:51–68, 1979.
19. Chacko, C. and A. Rosenfeld. Regulation of actin-activated ATP hydrolysis of arterial myosin. *Proc. Natl. Acad. Sci. USA* 79:292–296, 1982.
20. Chatterjee, M. and R. A. Murphy. Ca^{2+}-dependent stress maintenance without myosin phosphorylation in chemically skinned swine carotid media. *Federation Proc.* 42(5): abstract in press, 1983.
21. Conti, M. A. and R. S. Adelstein. The relationship between calmodulin binding and phosphorylation of smooth muscle kinase by the catalytic subunit of 3′:5′ cAMP-dependent protein kinase. *J. Biol. Chem.* 256:3178-3181, 1981.
22. Cornelius, F. Tonic contraction and the control of relaxation in a chemically skinned molluscan smooth muscle. *J. Gen. Physiol.* 79:821–834, 1982.
23. Crosby, N. D., and J. Diamond. Effects of phenothiazines on calcium induced contractions of chemically skinned smooth muscle. *Proc. West. Pharmacol. Soc.* 23:335–338, 1980.
24. Crow, M. T. and M. J. Kushmerick. Phosphorylation of the myosin light chains of mouse fast-twitch muscle is associated with a reduction in the actomyosin turnover rate. *Science* 217:835–837, 1982.

25. Deth, R. and R. Casteels. A study of releasable Ca fractions in smooth muscle cells of the rabbit aorta. *J. Gen. Physiol.* 69:401–416, 1977.
26. Diamond, J. Role of cyclic nucleotides in control of smooth muscle contraction. In: *Advances in Cyclic Nucleotide Research.* Eds.: George W. J. and Ignarro, L. J. Raven Press, N. Y., 1978, pp 327–340.
27. Diamond, J. and R. A. Janis. Increases in cyclic GMP levels may not mediate relaxant effects of sodium nitroprusside, verapamil, and hydralazine in rat vas deferens. *Nature (Lond.)* 271:472–473, 1978.
28. Dillon, P. F., M. O. Askoy, S. P. Driska, and R. A. Murphy. Myosin phosphorylation and the cross-bridge cycle in arterial smooth muscle. *Science* 211:495–497, 1981.
29. Donaldson, S. K. B. and W. G. L. Kerick. Characterization of the effects of Mg^{2+} on Ca^{2+} and Sr^{2+} activated tension generation of skinned skeletal muscle fibers. *J. Gen. Physiol.* 66:427–444, 1975.
30. Ebashi, S., Y. Nonomura, S. Nakamura, H. Nakasone, and K. Kohama. Regulatory mechanism in smooth muscle: actin-linked regulation. *Federation Proc.* 41:2863–2867, 1982.
31. Endo, J., T. Kitazawa, S. Yagi, M. Iino, and Y. Kabuta. Some properties of chemically skinned smooth muscle fibers. In: *Excitation-Contraction Coupling in Smooth Muscle.* Eds.: Casteels, R., Godfriand, T., and Rüegg, J. C. Elsevier-North Holland, Amsterdam, 1977, pp 199–210.
32. Endo, M., T. Kitazawa, and S. Yagi. Different features of responses of the sarcoplasmic reticulum in cardiac and smooth muscles. In: *Muscle Contraction, Its Regulatory Mechanisms.* Eds.: Ebashi et al. Japan Sci. Soc. Press, Tokyo, 1980, pp 447–463.
33. Fabiato, A. Skinned fibers from skeletal, cardiac and smooth muscles: introduction. *Federation Proc.* 41:2223–2224, 1982.
34. Fabiato, A. and F. Fabiato. Calcium release from the sarcoplasmic reticulum. *Circ. Res.* 40:1119–1129, 1977.
35. Fabiato, A. and F. Fabiato. Effects of pH on the myofilaments and sarcoplasmic reticulum of skinned cells from cardiac and skeletal muscles. *J. Physiol. (Lond.)* 276:233–255, 1978.
36. Fabiato, A. and F. Fabiato. Calculator programs for computing the composition of the solutions containins multiple metals and ligands used for experiments in skinned muscle cells. *J. Physiol. (Paris)* 75:463–505, 1979.
37. Filo, R. S., D. F. Bohr, and J. C. Rüegg. Glycerinated skeletal and smooth muscle: calcium and magnesium dependence. *Science* 147:1581–1583, 1965.
38. Gagelmann, M., U. Mrwa, G. Pfitzer, M. Troschka, C. Obst, R. Herrmann, and J. C. Rüegg. Comparison of force and myosin light chain phosphorylation in skinned smooth muscle fibers. *J. Muscle Res. and Cell Motility,* 3:478, 1982.
39. Gardner, J. P. and J. DiSalvo. Temporal relations between isometric force and myosin light chain (MLC) phosphorylation in skinned porcine carotid arteries. *Federation Proc.* 42:736 (Abstract), 1983.
40. Gordon, A. R. Contraction of detergent-treated smooth muscle. *Proc. Natl. Acad. Sci. USA* 75:3527–3530, 1978.

41. Gratecos, D. and E. H. Fischer. Adenosine 5′-O(3-thiotriphosphate) in the control of phosphorylase activity. *Biochem. Biophys. Res. Comm.* 58:960–967, 1974.
42. Guth, K. and J. Junge. Low Ca^{2+} impedes cross-bridge detachment in chemically skinned taenia coli. *Nature* 300:775–776, 1982.
43. Guth, K., M. Gagelmann, and J. C. Rüegg. Skinned smooth muscle: time-course of force and ATPase activity during contraction cycle. *Experientia* 40:174–176, 1984.
44. Haeusler, G., J. G. Richards, and S. Thorens. Noradrenaline contractions in rabbit mesenteric arteries skinned with saponin. *J. Physiol. (Lond.)* 321:537-556, 1981.
45. Hartshorne, D. and U. Mrwa. Regulation of smooth muscle actomyosin. *Blood Vessels.* 19: 1–18, 1982.
46. Hasselbach, W. and O. Ledermair. Der Kontraktionzyklus der isolierten kontraktilen struckturen der Uterusmuskulatur und seine besonderheiten. *Arch. Ges. Physiol.* 267:532–542, 1958.
47. Hellstrand, P. and A. Arner. Quantitative analysis of ATP turnover in relation to Ca^{2+}-activated tension in chemically skinned guinea pig taenia coli. *Biophys. J.* 41:247a (abstract), 1983.
48. Hellstrand, P., B. Johansson, and A. Ringberg. Influence of extracellular calcium on isometric force and velocity of shortening in depolarized venous smooth muscle. *Acta Physiol. Scand.* 84:528–537, 1972.
49. Hellstrand, P. G. and R. J. Paul. Vascular smooth muscle: relations between energy metabolism and mechanics. In: *Vascular Smooth Muscle: Metabolic, Ionic and Contractile Mechanisms.* Eds.: Crass, M. F. III and Barnes, C. D. Academic Press, New York, 1982, pp 1–36.
50. Hidaka, H., M. Naka, and T. Yamaki. Effect of novel specific myosin light chain kinase inhibitors on Ca^{2+}-activated Mg^{2+}ATPase of chicken gizzard actomyosin. *Biochem. Biophys. Res. Commun.* 90:694–699, 1979.
51. Hoar, P. E., W. G. L. Kerrick, and P. S. Cassidy. Chicken gizzard: relation between calcium-activated phosphorylation and contraction. *Science* 204:503–506, 1979.
52. Huxley, A. F. and R. M. Simmons. Proposed mechanism of force generation in striated muscle. *Nature (Lond.)* 233:533–538, 1971.
53. Iino, M. Tension responses of chemically skinned fiber bundles of the guinea-pig taenia coli under varied ionic environments. *J. Physiol. (Lond.).* 320:449–467, 1981.
54. Itoh, T., H. Izumi, and H. Kuriyama. Mechanisms of relaxation induced by activation of β-adrenoceptors in smooth muscle cells of the guinea pig mesenteric artery. *J. Physiol. (Lond.).* 326:475–593, 1982a.
55. Itoh, T., M. Kajiware, K. Kitamura, and H. Kuriyama. Roles of stored calcium on the mechanical response evoked in smooth muscle cells of the porcine coronary artery. *J. Physiol. (Lond.).* 322:107–125, 1982b.
56. Itoh, T., H. Suzuki, and H. Kuriyama. Effects of sodium depletion on contractions evoked in intact and skinned muscles of the guinea-pig mesenteric artery. *Jap. J. Physiol.* 31:831–847, 1981.
57. Jewell, B. R. The nature of phasic and tonic responses of the anterior byssus retractor muscle of Mytilus. *J. Physiol. (Lond.).* 149:154–177, 1959.

58. Julian, F. J. and R. L. Moss. Effects of calcium and ionic strength on shortening velocity and tension development in frog skinned muscle fibres. *J. Physiol. (Lond.)*. 311:179–199, 1981.
59. Kerrick, W. G. L. Myosin light chain kinase in skinned fibers. In: *Calcium and Cell Function, Vol. III.* Academic Press, Inc., N. Y., 1980, pp 279–295.
60. Kerrick, W. G. L., R. L. Bridenbaugh, P. S. Cassidy, and P. Hoar. Ca^{2+} phosphorylation dependent mechanisms regulating contraction in skinned muscle fibers. In: *Smooth Muscle Contraction.* Ed.: Stephens, N. L. Dekker, New York (in press), 1983.
61. Kerrick, W. G. L. and P. E. Hoar. Inhibition of smooth muscle tension by cyclic AMP-dependent protein kinase. *Nature* 292:253–255, 1981.
62. Kerrick, W. G. L., P. E. Hoar, and P. S. Cassidy. $Ca^{2\pm}$activated tension: the role of myosin light chain phosphorylation. *Fed. Proc.* 39:1558–1563, 1980.
63. Kramer, G. L. and J. G. Hardman. Cyclic nucleotides and blood vessel contraction. In: *Handbook of Physiology, Vol. II, Vascular Smooth Muscle.* Eds.: Bohr, D. F., Somylo, A. P., and Sparks, H. V. Jr. Amer. Physiol. Soc., 1980, pp 179–200.
64. Kukovetz, W. R., S. Holzmann, A. Wurm, and G. Poch. Evidence for cyclic GMP-mediated relaxant effects of nitro-compounds in coronary smooth muscle. *Arch. Pharmacol.* 310:129–138, 1979.
65. Kushmerick, M. J. and M. T. Crow. Regulation of energetics and mechanics by myosin light chain phosphorylation in fast-twitch skeletal muscle. *Federation Proc.* 42:14–20, 1983.
66. Levin, R. M. and B. Weiss. Selective binding of antipsychotics and other psychoactive agents to the calcium-dependent activator of cyclic nucleotide phosphodiesterase. *J. Pharmac. Exp. Ther.* 203:454–459, 1979.
67. Mannherz, H. ATP-spaltung and ATP-diffusion in oscillierenden extrahierten muskelfasern. *Pflügers Arch.* 303:230–248, 1968.
68. Marston, S. B., R.M. Treven, M. Walters. Calcium ion-regulated thin filaments from vascular smooth muscle. *Biochem. J.* 185:355–365, 1980.
69. Marston, S. B. The regulation of smooth muscle contractile proteins. *Prog. Biophys. Molec. Biol.* 41:1–1, 1982.
70. Meisheri, K.D., and J. C. Ruegg. Dependence of cyclic-AMP induced relaxation on Ca^{2+} and calmodulin in skinned smooth muscle of guinea pig taenia coli. *Pflügers Arch.* 399:315–320, 1983.
71. Meisheri, K. D. and C. van Breemen. Effects of β-adrenergic stimulation on calcium movements in rabbit aortic smooth muscle: relationship with cyclic AMP. *J. Physiol. (Lond.)* 331:429–441, 1982.
72. Merkel, L., K. D. Meisheri, and G. Pfitzer. The variable relation between myosin light chain phsophorylation and actin-activated ATPase activity in chicken gizzard smooth muscle: modulation by tropomyosin. *Eur. J. Biochem.* 183:429–434, 1984.
73. Morgan, J. P. and K. G. Morgan. Vascular smooth muscle: the first recorded Ca^{2+} transients. *Pflügers Arch.* 395:75–77, 1982.
74. Mrwa, U., I. Achtig, and J. C. Rüegg. Influences of calcium concentration and pH on the tension development and ATPase activity of the arterial actomyosin contractile system. *Blood Vessels* 11:277–286, 1974.

75. Mrwa, U., M. Troschka, and J. C. Ruegg. Cyclic AMP-dependent inhibition of smooth muscle actomyosin. *FEBS Lett.* 107:371–373, 1979.
76. Mueller, E. and C. van Breemen. Role of intracellular Ca^{2+} sequestration in β-adrenergic relaxation of smooth muscle. *Nature (Lond.)* 281:682–683, 1979.
77. Murphy, R. A., M. O. Aksoy, P. F. Dillon, W. T. Gerthoffer, and K. E. Kamm. The role of myosin light chain phosphorylation in regulation of the cross-bridge cycle. *Federation Proc.* 42:51–56, 1983.
78. Murphy, R. A., S. P. Driska, and D. M. Cohen. Variations in actin to myosin ratios and cellular force generation of vertebrate smooth muscles. In: *Excitation–Contraction Coupling in Smooth Muscle.* Eds. Casteels et al. Elsevier-North Holland, Amsterdam, 1977, pp 417–424.
79. Natori, R. Skinned fiber, past and present. In: *Muscle Contraction, Its Regulatory Mechanisms.* Eds.: Ebashi et al. Japan Sci. Soc. Press, Tokyo, 1980, pp 19–29.
80. Paul, R. J. Smooth muscle: mechanochemical energy conversion, relations between metabolism and contractility. In: *Physiology of the Gastrointestinal Tract.* Eds. Johnson, L. R. et al., 1981, pp 269–288.
81. Paul, R. J. Coordination of metabolism and contractility in vascular smooth muscle. *Federation Proc.* 42:62–66, 1983.
82. Paul, R. J., G. Doerman, C. Zeugner, and J. C. Rüegg. Dependence of unloaded shortening velocity (V_{us}) on Ca^{2+}, calmodulin and contraction duration in "chemically skinned" smooth muscle. *Circ. Res.* 53:342–351, 1983.
83. Persechini, A. and D. J. Hartshorne. Cooperative behavior of smooth muscle myosin. *Federation Proc.* 41:2868–2872, 1982.
84. Peterson, J. W. Relation of stiffness, energy metabolism and isometric tension in vascular smooth muscle. In: *Mechanisms of Vasodilation.* Eds.: Vanhoutte, P. M. and Leusen, I. Karger, Basel, 1978, pp 79–88.
85. Peterson, J. W. Vanadate ion inhibits actomyosin interaction in chemically skinned vascular smooth muscle. *Biochem. Biophy. Res. Comm.* 95:1846–1853, 1980.
86. Peterson, J. W. Rate limiting steps in the tension development of freeze glycerinated vascular smooth muscle. *J. Gen. Physiol.* 79:437–452, 1982.
87. Pfitzer, G., J. W. Peterson, and J. C. Rüegg. Length dependence of calcium-activated isometric force and immediate stiffness in living and glycerol extracted vascular smooth muscle. *Pflüegers Arch.* 394:174–181, 1982a.
88. Pfitzer, G. and J. C. Rüegg. Molluscan catch muscle: regulation and mechanics in living and skinned anterior byssus retractor muscle of *Mytilus edulis. J. Comp. Physiol.* 174:137–142, 1982b.
89. Pfitzer, G., J. C. Rüegg, V. Flockerzi, and F. Hofmann. cGMP-dependent protein kinase decreases calcium sensitivity of skinned cardiac fibers. *FEBS Lett.* 149:171–175, 1982c.
90. Podolsky, R. J. and L. E. Teichholz. The relation between calcium and contraction kinetics is skinned muscle fibers. *J. Physiol.* (*Lond.*). 211:19–35, 1970.
91. Portzehl, H., P. C. Caldwell, and J. C. Rüegg. The dependence of contraction and relaxation of muscle fibers from the crab *Maia squinado* on the internal concentration of free calcium ions. *Biochim. Biophy. Acta.* 79:581–591, 1964.

92. Rüegg, J. C. Smooth muscle tone. *Physiol. Rev.* 51:201–248, 1971.
93. Rüegg, J. C., J. diSalvo, and R. J. Paul. Soluble relaxation factor from vascular smooth muscle: a myosin light chain phosphatase. *Biochem. Biophy. Res. Comm.* 106:1126–1133, 1982.
94. Rüegg, J. C., K. Meisheri, G. Pfitzer, and C. Zeugner. Skinned coronary smooth muscle: calmodulin, calcium antagonists, and cAMP influence contractility. *Basic Res. in Cardiology,* in press. 1983b.
95. Rüegg, J. C. and R. J. Paul. Vascular smooth muscle: calmodulin and cyclic AMP dependent protein kinase alter calcium sensitivity in porcine carotid skinned fibers. *Circ. Res.* 50:394–399, 1982.
96. Rüegg, J. C., M. P. Sparrow, and U. Mrwa. Cyclic AMP mediated relaxation of chemically skinned fibers of smooth muscle. *Pflügers Arch.* 390:198–201, 1981.
97. Rüegg, J. C., M. P. Sparrow, U. Mrwa, M. Schneider, and G. Pfitzer. Calcium and calmoduin dependent regulatory mechanisms in chemically skinned smooth muscle. In: *Symposium on Smooth Muscle Contraction.* Ed.: N. Stephens. Marcel Dekker Inc., New York, in press.
98. Rüegg, J. C. and H. H. Weber. Kontraktionszyklus and sperrtonus. In: *Perspectives in Biology.* Eds.: Cori, C. F., Fogilai, V. G., Leloir, L. F. and Ochoa, S. Elseiver, Amsterdam, 1963, pp 301–320.
99. Saida, K. Intracellular Ca release in skinned smooth mucle. *J. Gen. Physiol.* 80:191–202, 1982.
100. Saida, K. and Y. Nonomura. Characteristics of Ca^{2+} and Mg^{2+} induced tension development in chemically skinned smooth muscle fibers. *J. Gen. Physiol.* 72:1–14, 1978.
101. Schädler, M. Proportionale aktivierung von ATPase aktivität und kontraktions-spannung durch calciumionen in isolierten kontraktilen strukturen. *Pflügers Arch. ges. Physiol.* 296:70–90, 1967.
102. Schneider, M., M. Sparrow, and J. C. Rüegg. Inorganic phosphate promotes relaxation in chemically skinned smooth muscle of guinea-pig taenia coli. *Experientia* 37:980–982, 1981.
103. Schultz, K. D., E. Bohme, V. A. W. Kreye, and G. Schultz. Relaxation of hormonally stimulated smooth muscle tissues by the 8-bromo derivative of cyclic GMP. *N.S. Arch. Pharmacol.* 306: 1–9, 1979.
104. Schumacher, T. Zum mechanismus der ökonomischen haltleistung eines glatten muskels (Byssus retractor anterior, *Mytilus edulis*). *Pflügers Arch.* 331:77–89, 1972.
105. Sherry, J. M. F., A. Gorecka, M. O. Aksoy, R. Dabrowska, and D. J. Hartshorne. Roles of calcium and phosphorylation in the regulation of the activity of gizzard myosin. *Biochemistry, N. Y.* 17:4411–4418, 1978.
106. Siegman, M. J., T. M. Butler, S. U. Mooers, and R. E. Davies. Calcium-dependent resistance to stretch and stress relaxation in resting smooth muscles. *Am. J. Physiol.* 231:1501–1508, 1976.
107. Siegman, M. J., T. M. Butler, S.U. Moores, and R. E. Davies. Chemical energetics force development, force maintenance, and relaxation in mammalian smooth muscle. *J. Gen. Physiol.* 76:609–629, 1980.
108. Silver, P. J. and J. DiSalvo. Adenosine 3′:5′-monophosphate mediated inhibition of myosin light chain phosphorylation in bovine aortic actomyosin. *J. Biol. Chem.* 254:9951–9954, 1979.
109. Silver, P. J., M. J. Holroyde, R. J. Solaro, and J. DiSalvo. Ca^{2+}, calmodulin, and cyclic AMP-dependent modulation of actin-myosin interactions in aorta. *Biochim. Biophy. Acta.* 674:65–70, 1981.

110. Sobue, K., K. Morimoto, M. Inui, K. Kanda, and S. Kakiuchi. Control of actin-myosin interaction of gizzard smooth muscle by calmodulin- and caldesmon-linked flip-flop mechanism. *Biomedical Res.* 3:188–196, 1982.

111. Solaro, H. J., J. DiSalvo, and R. J. Paul. Coordination of metabolism and contractility by phosphorylation in cardiac, skeletal, and smooth muscle: introduction. *Federation Proc.* 42:62–66, 1983.

112. Somlyo, A. P., A. V. Somlyc, H. Shuman, and M. Endo. Calcium and monovalent ions in smooth muscle. *Federation Proc.* 41:2883–2890, 1982.

113. Sparrow, M. P., U. Mrwa, F. Hofmann, and J. C. Rüegg. Calmodulin is essential for smooth muscle contraction. *FEBS Lett.* 125:141–145, 1981.

114. Sparrow, M. P., G. Pfitzer, M. Gagelmann, and J. C. Rüegg. Effect of calmodulin, Ca^{2+}, and cAMP protein kinase on skinned trachael smooth muscle. *Am. J. Physiol.* 246:308–314, 1984.

115. Spedding, M. Direct inhibitory effects of some calcium antagonists and trifluoroperazine on the contractile proteins in smooth muscle. *Br. J. Pharmacol.* 79:225–231, 1983.

116. Szent-Györgyi, A. Free energy relations and contraction of actomyosin. *Biol. Bull. (Woods Hole).* 96:140–161, 1949.

117. Tsien, R. Y. and T. J. Rink. Neutral carrier ion-selective microelectrodes for measurement of intracellular free calcium. *Biochim. Biophy. Acta.* 599:623–638, 1980.

118. Twarog, B. M. Responses of a molluscan smooth muscle to acetylcholine and 5-hydroxytryptamine. *J. Cellular Comp. Physiol.* 44:141–164, 1954.

119. Ulbrecht, G., and M. Ulbrecht. Der isolierte arbeitscyclus glatter muskulatur. *Zeitschrift für Naturforschung.* 7:434–443, 1952.

120. van Breeman. C. Calcium switch in vertebrate smooth muscle. *Federation Proc.* 41:2863–2904, 1982.

121. Walsh, M. P., R. Bridenbaugh, D. J. Hartshorne, and W. G. L. Kerrick. Phosphorylation-dependent activated tension in skinned gizzard muscle fibers in the absence of Ca^{2+}. *J. Biol. Chem.* 257: 5987–5990, 1982.

122. Walsh, M. P., R. Bridenbaugh, W. G. L. Kerrick, and D. J. Hartshorne. Gizzard Ca^{2+}independent myosin light chain kinase: evidence in favor of the phosphorylation theory. *Federation Proc.* 42:45–50, 1983.

123. Walsh, M. P., R. Dabrowska, S. Hinkins, and D. J. Hartshorne. Calcium-independent myosin light chain kinase of smooth muscle. Preparation by limited chymotryptic digestion of the calcium ion dependent enzyme, purification and characterization. *Biochemistry* 21:1919–1925, 1982a.

124. Winegrad, S. Studies of cardiac muscle with a high permeability to calcium produced by treatment with ethylenediaminetetraacetic acid. *J. Gen. Physiol.* 58:71–93, 1971.

125. Yabu, H., I. Uchida, and E. Miyazaki. Participation of native tropomyosin in the ATP-contraction of an intestinal glycerinated muscle bundle. *Jap. J. Physiol.* 21:465–473, 1971.

126. Yamamoto, T. and J. W. Herzig. Series elastic properties of skinned muscle fibres in contraction and rigor. *Pflügers Arch.* 373:21–24, 1978.

Chapter 9

Smooth Muscle Subcellular Fractionation

A. M. Kidwai

1. Introduction

For a long time, anatomists recognized the existence of well-defined boundaries to cells and organelles within cells, but the techniques for subcellular fractionation for biochemical studies were not well developed until Hogeboom et al. (*11, 12*) published the liver cell subcellular fractionation. Since then, almost all the animal and plant tissues have been fractionated into subcellular components by homogenization followed by differential or density gradient centrifugation. However, smooth muscle subcellular fractionation posed problems because of high collagen content, resulting in the problems of disruption and fractionation. Various groups were involved in the studies on smooth muscle microsomal and mitochondrial fractions, but it was not until 1971 when the detailed method of isolation and characterization of plasma membrane and other subcellular fractions was published (*14*). Recently, a great deal of interest has been shown in the smooth muscle cell surface calcium uptake and transport (*3, 4, 26*) and hence there is a requirement for pure plasma membranes for such studies.

In the present chapter an attempt will be made to review various homogenization and fractionation techniques suitable for smooth muscle. A few techniques will be dealt with in detail, along with other studies carried out on the subcellular fractions.

2. Homogenization Technique

The prerequisite to subcellular fractionation is the disruption of the cells. The underlying means for cell disruption is probably to stress the membrane (*30, 31, 33*). The disruption of the cells can be achieved by the use of a Potter Elvehjem homogenizer, a Duall tissue grinder, a Dounce tissue grinder, a Sorvall Omni Mixer, a pressure homogenizer, and a Polytron. For homogenization of smooth muscle, the glass–glass homogenizer, the Sorvall Omni Mixer, the glass–Teflon homogenizer (with greater clearance), and the Polytron have been used. The Polytron is the most extensively used apparatus at present (*17*).

The homogenizing medium for subcellular fractionation is usually 0.25*M* sucrose, which provides osmotic protection to such organelles as nuclei, mitochondria, and lysosomes.

Use of the above techniques for disruption of smooth muscle cells results in the breakage of plasma membranes that then form small vesicles. The vesicles from ruptured cells can occur either in the natural orientation, that is, outside-out, or invert themselves into inside-out vesicles. The surface membrane breaks down into various size fragments depending upon the force applied (*5*). A description of the various apparatuses follows.

2.1. Potter Elvehjem Homogenizer

Potter (*29*) described an apparatus made of a Teflon or glass pestle that fitted into a thick-walled glass tube. The clearance (meaning the difference between the internal diameter of the tube and the external diameter of the pestle) can be chosen according to convenience. Tough material (e.g., smooth muscle) can be homogenized in a greater-clearance apparatus drawn by a motor. Soft material can be homogenized in a tight fitting apparatus by hand only. Tissues such as liver and brain can be easily homogenized.

2.2. Duall Tissue Grinder

This is essentially a Potter-Elvehjem homogenizer except that the shape of the homogenizer is changed in such a way as to have a conical and a cylindrical surface in the same apparatus: conical for the initial grinding and cylindrical for the final breaking of the cells.

2.3. Dounce Tissue Grinder

This consists of two pestles, one with large clearance and the other with small clearance, for use in the same tube. This process does not damage nuclei while breaking up the tissue. It is available from Kontes Glass Company, Vineland, NJ, USA.

2.4. Potter-Elvehjem All-Glass Homogenizer

This type has two ground glass surfaces for greater shear. It can be motor-driven using flexible tubing that acts as a friction clutch. It is also available from Kontes.

2.5. Disintegrinder

This consisted of a glass tube and a Teflon pestle with integral stainless-steel cutting blades that permit disintegrating, mincing, and homogenizing in one operation. It is also available from Kontes.

2.6. Sorvall Omni Mixer Homogenizer

This stainless-steel cup and blades driven by a motor up to a speed of 50,000 rpm that produce cutting and breakage of the cells. This is called a liquid shear method.

2.7. Willem's Polytron Apparatus

The kinematic high-frequency and ultrasonic processing technique is used for homogenization of tough material. The instrument was developed in a physicochemical institute in Lucerne, Switzerland and is widely used for smooth-muscle disintegration. It is available from Brinkmann Instruments Co., Westbury, NY.

2.8. Pressure Homogenization

A cell disruption bomb is used for explosive decompression. Nitrogen at high pressure penetrates individual cells, and on the release of pressure the cells disrupt. It is available from Past Instrument Co., Moline, IL, USA.

3. Detailed Technique for Homogenization of Smooth Muscle

Smooth muscle homogenization poses a special problem because the cells are enmeshed in collagen fibers rendering toughness to the tissue. Homogenization of smooth muscle can be achieved by an all-glass Potter-Elvehjem homogenizer, but the resulting homogenate contains silica particles that interface with electron microscopy. A Sorvall Omni Mixer-type homogenizer for vascular tissue homogenization and a Polytron apparatus for uterine, gastric fundus, and mesenteric vein homogenization were used (*14, 23, 24, 37*).

3.1. Rat Myometrium

This method is based on studies by several workers. Estrogen-dominated rats were sacrificed by a blow on the head and cervical dislocation. The myometrium was stripped off, cut into pieces, and homogenized in 0.25*M* sucrose in a Polytron PT 20 for 5 s, three times at 15,000 rpm (*14, 28*).

3.2. Vascular Smooth Muscle

Vascular smooth muscle poses special problems during isolation and homogenization. Various workers have tried to isolate the microsomal fraction from mesenteric vein, arteries and aortas (*37*). In the case of the aorta the separation of smooth muscle is not as difficult as in the case of the mesenteries. Briefly, the method for mesenteries as described by Kwan is as follows (*23*).

3.2.1. Tissue Isolation

The mesenteric vascular bed (including mesenteries, fat, veins, arteries, and lymph nodes) is removed and immersed in ice-cold 0.25*M* sucrose containing 10 m*M* MOPS and 0.05% bovine serum albumin (BSA), pH 7.5. The mesenteric vascular bed is subjected to the following procedure.

3.2.2. Tissue Trimming

Tissue trimming is done on a cold plate with a cushion of several layers of filter paper. Most of the fat can be removed by gentle pressing and scraping with a plastic rod. This step also helps in removing the blood trapped in the blood vessels to the filter paper. After removal of the gross fat and lymph nodes, the mesenteries near the superior mesenteric vein are cut open so that the vein can be carefully pulled away from them, leaving them adherent to the filter paper.

The arteries are then homogenized in a Potter Elvehjem homogenizer in sucrose buffer solution. The homogenization helps to remove the fat as a turbid suspension.

3.2.3. Homogenization of Mesenteries

The arteries isolated as described above are finely minced and homogenized in a Polytron PT 20 at 15,000 rpm for 15 s with 12 up-and-down strokes.

3.3. Vas Deferens Smooth Muscle

Kwan et al. (*25*) have isolated subcellular fractions from rat vas deferens. Each vas deferens was first cleaned to remove loosely attached fat and mesenteric tissues, then cut open longitudinally to expose the lumen surface. The epithelial lining of the lumen was peeled off with a pair of forceps. The smooth-muscle layer was then scraped away from the serosal layer that contained the connective tissues and blood vessels. The isolated smooth muscle layer of vas deferens was rinsed several times with 0.25*M* sucrose containing 10 m*M* imidazole at pH 7.0, finely minced and homogenized using a Polytron PT 20 at a tissue/volume ratio of 1 g/15 mL for 5 s.

3.4. Dog Trachealis

Remove the entire trachea from the laryngeal to the bronchial end, and place it in Krebs-Ringer solution in the presence of 95% O_2 and 5% CO_2 at room temperature for several hours. The muscle is then washed with sucrose-MOPS and homogenized using the Polytron PT 20 for 3 s repeated five times. This process helps complete homogenization of the muscle (*9*).

3.5. Intestinal Smooth Muscle

The method used by Wibo et al. (*38*) for the intestinal smooth muscle of guinea pigs was as follows. Equilibrate ileum segments in Krebs medium (37°C) with 95% O_2/5% CO_2. Separate the longitudinal muscle layer and transfer it into ice-cold 0.25*M* sucrose buffered at pH 7.4 with 5 m*M* Tris HCl. Mince the tissue (1.5–2.0 g) with scissors and homogenize with 15–20 mL of buffered sucrose in an all-glass Potter-Elvehjem-type homogenizer at 2°C by four strokes of the pestle rotating at 1500 rpm.

4. Fractionation Techniques

Once the tissue is properly homogenized into its components (structural or morphological entity, e.g., nuclei, mitochrondria, and so on), the separation of these components is a major step to achieve, in order to study the properties of these components and get an insight into the interaction of various substances to these components and even to the constituents (a biochemical entity, e.g., enzyme, phospholipid, etc.) of the particular components of the cell.

4.1. Differential Centrifugation

The homogenate is subjected to low-speed centrifugation to remove nuclei and cell debris. The supernatant is then centrifuged at a higher speed to sediment cellular components that are lighter, e.g., mitochrondria and lysosomes. The next centrifugation is performed at a much higher speed to sediment smaller components such as microsomes. If the speed is further increased, a postmicrosomal fraction can be sedimented. The remaining supernatant represents the soluble fraction of the cell cytoplasm containing soluble proteins, ions, and other smaller molecular weight substances. This technique has been widely used in almost all kinds of tissues, including smooth muscle. However, this method yields somewhat contaminated fractions.

4.2. Density Gradient Centrifugation

Density gradient can be either continuous or discontinuous. A description of each kind follows.

4.2.1. Discontinuous Gradient

In this technique suitable solutions of sucrose or Ficoll are placed in a centrifuge tube. With the heavy solution in the bottom of the tube, the lighter solutions can be layered on top of each other, thus generating a discontinuous gradient. The material to be fractionated is loaded on the top of this gradient and the tube is centrifuged. By force of gravity, the cellular components move down the gradient, but only to the level of the gradient of equal or higher density than their own individual densities. At the end of the run, the bands are collected and centrifuged to remove the gradient material and suspended in the proper buffer (*17*).

4.2.2. Continuous Gradient

Two solutions, one of low and the other of high density, are taken and mixed in a machine so as to generate a gradient from high density to low density in a centrifuge tube. The desired material is loaded and centrifuged. In this system, various components are distributed all along the gradient as there is no sharp demarcation of density in the tube. A large number of fractions are collected and analyzed for protein, enzyme, and so on.

4.3. Zonal Ultracentrifugation

The density gradient methods mentioned above are capable of handling a limited quantity of material at a time. Anderson (*1*) devised zonal rotors capable of separating large quantities of material. The rotor is a hollow cylinder with a volume of nearly 2 L, divided longitudinally by four vanes into sector-shaped compartments. A gradient is eatablished in the rotor while it is spinning at low speed by pumping in a solution of increasing density. The sample is introduced and the rotor speed increased. The various classes of particles separate, and speed is then reduced and the content recovered by pumping in dense sucrose solution.

4.4. Continuous Sample Flow Density Gradient

Cline (2) has described this technique in which a thin layer of sample continuously moves through the rotor at a rate that allows the particulate of interest to be sedimented from the flowing sample. If the flow rate is fast, it permits only the largest of particles to be trapped, whereas slower flow rates permit both larger and smaller material to be collected. Both types are then banded isopycnically. Thus size versus density separation is made continuously.

4.5. Free-Flow Electrophoresis

Heidrich et al. (*10*) used the free-flow electrophoresis system (available from Brinkman Instruments Inc.) to separate homogenate into microsomes and mitochondria according to their electrophoretic mobility.

4.6. Salt Extraction Method

A widely used technique is extraction of muscle homogenate with salts. Lithium bromide (LiBr) has been used to extract the intracellular contents from skeletal muscle, followed by low speed centrifugation to collect sarcolemma (*19*).

4.7. Toluene–Lithium Bromide Method

Zaidi et al. (*39*) reported a modification of the LiBr extraction in which the homogenized frog skeletal muscle preparation was made permeable to salts and high molecular weight substances by treating the homogenate with toluene. Toluene increased the permeability of the muscle cell fragments, resulting in the rapid extraction of intracellular material. The sarcolemma thus isolated was suitable for enzymatic and binding studies.

5. Isolation of Various Subcellular Components

5.1. Cellular Components

5.1.1. Sarcolemma

Sarcolemma is the outermost layer of muscle cells and is made up of three parts, viz. (a) collagen fibrils, (b) basement membrane, and (c) plasma membrane. The sarcolemma has been isolated from skeletal muscle by various workers (*19, 39*). In smooth muscle the surface membrane isolated so far by density gradient centrifugation probably does not contain basement membrane. It seems the preparation is purely plasma membrane. At what stage the basement membrane is lost has not been studied (*16*).

5.1.2. Plasma Membrane

Plasma membrane from smooth muscle can be isolated in the form of vesicles either by differential or density gradient centrifugation. The microsomal fraction consists of plasma membrane vesicles along with other membrane fragments. Use of density gradient centrifugation introduced by Kidwai et al. (*16*) and subsequently modified by various workers (*28*) enables one to isolate a pure plasma membrane preparation along with other subcellular fractions.

5.1.3. Basement Membrane

The importance of basement membranes in various tissues has recently been realized. It might be involved in a homeostatic repair mechanism by providing a surface upon which platelets adhere and aggregate when this extracellular matrix is exposed to the circulating blood through injury. In addition basement membrane thickens in diabetic glomeruli and is involved in polycystic renal disease and anti-TMB nephritis (*32*). To date no method is reported for isolation of basement membrane from smooth muscle.

5.1.4. Endoplasmic Reticulum

Smooth muscle endoplasmic reticulum can be isolated either by differential centrifugation as a microsomal fraction or by density gradient centrifugation in a more or less pure state as described in an earlier section. One step density gradient is enough to separate plasma membrane and endoplasmic reticulum from smooth muscle homogenate (*14*).

5.1.5. Mitochondria

Smooth muscle mitochondria can be easily isolated by preparing a 10% homogenate in 0.25*M* sucrose followed by differential centrifugation. Centrifuge the homogenate at 1000*g* for 5 min to remove nuclei. The supernatant on centrifugation at 10,000*g* for 15 min sediments mitochondria (*28*).

5.1.6. Nuclei

Tata (*36*) described the isolation of nuclei from uterus. He homogenized uterus smooth muscle in 0.32*M* sucrose–3 m*M* $MgCl_2$, centrifuged at 700*g* for 10 min, suspended the pellet in 0.25*M* sucrose–1 m*M* $MgCl_2$ and loaded on the top of 2.4*M* sucrose–1 m*M* $MgCl_2$ solution in swinging out rotor centrifuge tubes. The preparation was centrifuged at 5000*g* for 45 min. The pellet was suspended in 0.25*M* sucrose containing 1 m*M* $MgCl_2$.

5.1.7. Soluble Fraction

The soluble fraction from smooth muscle can be easily obtained by homogenizing the tissue in 0.25*M* sucrose using a glass–glass homogenizer or Polytron. A 10% homogenate is prepared and centrifuged either differentially to isolate other subcellular fractions or centrifuged at 100,000*g* for 1 h. The supernatant represents the soluble fraction containing soluble proteins, enzymes, ions, nucleotides, and amino acids, along with other small molecular weight substances and probably a postmicrosomal fraction as well.

5.2 Detailed Techniques Commonly in Use for Smooth Muscle Fractionation

The methods described above for the fractionation of various tissue homogenates can be applied to smooth muscle homogenate. A differential centrifugation technique has been used for the isolation of nuclei, mitochondria and microsomes. The first detailed method of isolation of subcellular fractions from rat myometrium published in 1971 used density gradient centrifugation (*14*). This method used a minimum number of centrifugation steps, whereas later modifications were introduced using differential centrifugations followed by a density gradient step (*28, 8*).

5.2.1. Rat Uterine Smooth Muscle Fractionation: Method I

Briefly the method is as follows: Myometrium from the uterine horns of 12 rats was separated from endometrium, cut into small pieces, and homogenized using a Polytron PT 20 for 4 s at maximum speed. The homogenate was filtered through gauze cloth. Unbroken tissue was rehomogenized and filtered, and the final homogenate was made 20% in 0.25*M* sucrose. The homogenate was centrifuged at 105,000*g* for 30 min. The sediment was suspended in a small volume of 0.25*M* sucrose and layered on the top of the continuous or discontinuous gradient. The discontinuous gradient was made by layering 40, 33, and 30% sucrose solutions. It was centrifuged at 100,000*g* for 2 h. At the end of the run, three bands were collected that represented the plasma membrane, endoplasmic reticulum, and mitochondria (*14*).

5.2.2. Method II, Based on the Modification Introduced by Matlib et al.

Nineteen female Wistar rats weighing 160–200 g each were injected with 500 μg of diethylstilbestrol in 0.1 mL of sesame oil for 2 d. The animals were sacrificed on the third day by a blow on the head. The uterine horns were removed and put in Krebs solution of the following composition: KCl 4.6, $MgSO_4$ 1.16, NaH_2PO_4 1.16, $CaCl_2$ 2.5, NaCl 115.5, $NaHCO_3$ 21.9, and glucose 11.1 m*M*. Uterine horns were trimmed of fat and loosely bound connective tissue, and the endometrium was stripped off with most of the circular muscle. This entire process can be carried out on filter paper moistened with Krebs-Ringer solution and maintained at 4°C temperature on a cold plate.

About 3 g myometria is used to incubate at 37°C for 30 min in the presence of 95% O_2 and 5% CO_2. The myometria are removed and transferred to 250 m*M* sucrose solution containing 20 m*M* MOPS at 4°C in a petri dish. They are minced, divided into three portions, and homogenized using a Polytron PT 20 three times for 10 s at 2-s intervals at about 15,000 rpm. The homogenate was centrifuged at 1000*g* for 10 min to remove nuclei, contractile proteins, whole cells, and unbroken tissues. The pellet was discarded and centrifuging continued at 10,000*g* for 15 min. The resulting pellet of crude mitochondria can be suspended in 7 mL of sucrose MOPS solution and centrifuged at 113,000*g* for 30 min in an ultracentrifuge.

The resulting microsomal pellet can be further fractionated on a density gradient either by linear gradient using 20–45% sucrose in a 10-mL centrifuge tube or by discontinuous gradient; the discontinuous gradient can be prepared by layering 4 mL of 50% sucrose (w/w), 2 mL of 35% sucrose, and 3 mL of 30% sucrose. One mL of microsomal fraction was layered on the top of the gradient and centrifuged at 112,000*g* for 2 h in a swing-out rotor. Three distinct and two diffuse bands can be seen at the end of the run and removed by Pasteur pipets. They were diluted with cold distilled water to attain a sucrose concentration of 0.5% and then centrifuged at 100,000*g* for 30 min. The fractions near the top of the gradient were plasma membrane vesicles and as the density increased the endoplasmic reticulum fractions were obtained (*28*).

5.2.3. Isolation of Plasma Membrane from Mesenteric Arteries and Veins

The method is based on the work published by Kwan et al. (*22, 23*). The fractionation of the arterial and veins homogenate is basically the same as that described for rat uterus smooth muscle except that the density gradient consists of 46.5, 38.5, and 26.5% sucrose, the microsomal fraction is suspended in 0.25*M* sucrose, 10m*M* MOPS, and 0.05% bovine serum albumin (pH 7.5) and loaded on the top of the gradient, and the whole is centrifuged for 90 min at 113,000*g*. The plasma membrane marker enzyme 5′-nucleotidase is distributed all along the gradient but maximum activity is present in the lighter fractions.

5.2.4. Isolation of Subcellular Membrane Fractions from Smooth Muscle of Rat Vas Deferens

Rat vas deferens smooth muscle can be homogenized as described in the earlier section and microsomal fractions obtained by the method of Kwan et al. (*25*). The membrane fractions are separated on a discontinuous gradient using 15, 30, 40, and 60% sucrose. The light fractions were rich in plasma membrane marker enzymes.

5.2.5. Isolation of Trachealis Membrane Fractions

Grover et al. (*9*) isolated membranes from dog trachealis using a modified sucrose gradient. They prepared microsomes as described in the case of uterus smooth muscle (*28*). The microsomal fraction was loaded on the gradient made of 20, 30, 40, and 50% sucrose. Four bands from the interphase were collected. Fraction two was enriched in plasma membrane enzymes.

5.2.6. Isolation of Intestinal Smooth Muscle Membrane

The homogenate from the ilium muscle of guinea pigs was centrifuged at 1000*g* for 5 min. The supernatant was centrifuged at 20,000*g* for 30 min followed by 100,000*g* centrifugation for 60 min. The microsomal pellet was suspended in sucrose Tris. The microsomes were subfractionated with isopycnic centrifugation on a gradient of density range from 1.08 to 1.26. The sample was layered on the gradient and centrifuged for 12 h at 100,000*g* and subfractions were collected (*6, 38*).

6. Characterization of Subcellular Components

The subcellular components isolated by various techniques described in the earlier section have to be characterized and examined for possible contamination by other components or constituents that do not belong to the component desired. For characterization the following criteria have been used.

6.1. Morphology

Some constituents can be identified by their morphology either by light or electron microscopy. For example, nuclei can be identified by light microscopy and mitochondria by the electron microscopy. Electron microscopy also helps to visualize vesicles of plasma membrane and of endoplasmic origin, although it is hard to identify the vesicles of different origin. The rough endoplasmic reticulum has the characteristic appearance of ribosomes attached to the vesicles. The vesicles originating from plasma membrane and smooth endoplasmic reticulum are little different. It was observed by some authors that the thickness of the membrane from various sources differs markedly (*15*).

6.2. Enzymatic Markers

Enzymes are very useful markers since they are localized in subcellular components according to their function. For example, mitochondrial enzymes have oxidative and plasma membrane enzymes have transport functions. The well known examples are succinic dehydrogenase and $Na^+ + K^+$-ATPase.

Basement membrane is not known to contain any enzyme. It can only be characterized by its morphology, chemical composition, and solubility in detergents (*20*).

Plasma membrane is known to be very rich in various enzymes, e.g., $Na^+ + K^+$-ATPase, 5′-nucleotidase, adenylcyclase, K^+-activated *p*-nitrophenylphosphatase, phosphodiesterase, and leucyl-β-naphthylamidase.

Smooth and rough endoplasmic reticulum can be characterized by protein biosynthetic activity (*15, 28*).

Succinic dehydrogenase, cytochrome oxidase, and monoamine oxidase are some of the enzymes commonly used as mitochondrial markers.

Nuclei contain DNA, but there is no established enzymatic marker for its membrane. Various enzyme activities have been tested, for example, glucose-6-phosphatase and DPNH-cytochrome *c* reductase, and so forth, but none of the enzymes were exclusively present in a nuclear envelope (*13*).

Many glycolytic enzymes are present in the soluble fraction, e.g., lactic dehydrogenase, glyceraldehyde-3-phosphate dehydrogenase, and phosphofructokinase.

6.3. Chemical Composition

The chemical makeup of a component is sometimes useful for the characterization of various organelles and membranes.

Basement membrane is characterized by its collagenous nature; its proteins contain higher hydroxyproline and have been shown to be different from the collagen of interstitial collagen that forms fibrils, compared to the one in basement membrane that does not form fibrils.

The main noncollagenous glycoprotein characterized from basement membrane is called laminin. Its molecular size is about 850,000 daltons. Another glycoprotein called entactin has been characterized from basement membrane.

Among the proteoglycans, heparan sulfate and hyaluronic acid account for about 1% of the dry weight of the glomerular membrane (*32*).

Plasma membrane has a characteristic cholesterol–phospholipid ratio that is higher than the membrane isolated from other organelles (*14*).

There is no characteristic chemical marker for mitochondria, whereas rough endoplasmic reticulum and nuclei should contain RNA and DNA, respectively.

6.4. Miscellaneous Markers

6.4.1. Protein Biosynthesis

For endoplasmic reticulum protein synthesis can be used as a marker. It was observed by Kidwai et al. (*16, 18*) and Matlib et al. (*28*) that, in rat skeletal and smooth muscle, tritiated leucine incorporation was highest in density gradient fractions that were deficient in plasma membrane markers. This led to the conclusion that amino acid incorporation was the property of endoplasmic reticulum and can be used as a marker for this component of the cell.

6.4.2. Fluorescent Markers

Many workers have used a fluorescent chemical marker, 4-acetamido-4′-isothiocyanate stilbene-2,2′-disulfonic acid (SITS), because it does not cross the surface of the cell and thus the surface membrane (plasma membrane) can be identified by measuring the fluorescence of various subcellular fractions (*27*). Smooth muscle subcellular fractions were identified on this basis (*14*).

Many chemicals that have affinity to the cell surfaces can be used as markers for plasma membrane. Wheat germ agglutinin and oxytocin were used to identify the plasma membrane of rat myometrium by Matlib et al. (*28*).

7. Special Studies

7.1. Membrane Orientation

During homogenization the membrane fragments orient themselves according to the constitution of the medium. The vesicles are formed either inside-out or outside-out (*34*). The orientation can be determined by the markers present on the inside and outside.

The plasma membrane vesicles formed by homogenization of smooth muscle are either right-side-out or inside-out vesicles or a mixture of the two.

Grover et al. (*7*) have tried to evaluate the orientation and intactness of smooth muscle plasma membrane vesicles.

Rat myometrium plasma membrane was prepared by the method of Matlib et al. (*28*). It was assumed that the intact vesicles are osmotically active to sucrose and also impermeable to EGTA and cationized ferritin. For sidedness the assumptions were that ouabain-sensitive K^+-activated PNPPase as well as galactose-oxidizable TCA-precipitable glycolipids of plasma membrane reside on the external cell surface.

7.1.1. Vesicle Intactness

Vesicles that were exposed to sucrose concentrations higher than the concentration trapped inside them shrank when their volume and surface area were considerred. It was concluded that 70% of the vesicles were osmotically active to sucrose. When cationized ferritin was used to label the vesicles, very few vesicles had the large molecules of this protein trapped inside them, supporting the finding that the majority of the vesicles were intact. Further evidence was obtained by Ca^{2+} uptake and EGTA release studies. It was assumed that EGTA will release Ca^{2+} bound on both the surfaces of broken vesicles and only from the outside in the case of intact vesicles. They concluded that only 25% of the vesicles were broken. This finding was also consistent with other studies concluding the presence of 80% intact vesicles.

7.1.2. Vesicle Sidedness

The sidedness was studied assuming that in isotonic media only the external surface of the right-side-out vesicles and all the surface of broken vesicles are available to nonpermeable external cell surface markers, whereas hypotonic shock would also result in exposure of the inside surface of the inside-out vesicles to the reaction media. If this assumption is correct hypotonic shock would result in enhancement of the measurable external cell surface activities. The external cell surface markers were ouabain-sensitive K^+-activated PNPPase and galactose-oxidizable TCA-precipitable label incorporation. Both the markers were enhanced ~55% in their activity.

The authors computed the values as follows—the vesicles were 20% broken, 40% inside out, and 40% right-side out.

The implications of the orientation studies are numerous; vesicles isolated after homogenization and centrifugation can be used for binding and transport studies. Usually, one assumes that all the vesicles are right-side out and start looking for receptors or transport of various substances. If the vesicles are inside out, the properties for the outer surface will be absent, or if there is a mixture of the two orientations, then the activities will be reduced. Grover et al. (*7*) have clearly shown that vesicles can be in different orientation.

7.2. Membrane Solubilization

Detergents have been successfully used in solubilization of membranes. This approach seems logical because the hydrophobic regions of the membrane constituents remain in a hydrophobic microenvironment, whereas the overall molecule becomes fully hydrophilic by the presence of additional ionic or highly hydrated groups and thus becomes water-soluble.

The protein composition of the myometrial sarcolemma was studied by Krall and Korenman (*21*). They solubilized the membrane protein by dissolving the sample in SDS (0.06*M* Tris-HC1, pH 6.8, 3% SDS, 0.7*M* 2-mercaptoethanol, 10% glycerol) and heating for 10 min at 90°C.

7.3. Electrophoresis

For electrophoresis 30–50 μg of solubilized membrane protein was loaded on 8.75% polyacrylamide slab gels with 0.1% SDS at 20 mA for 4 h. After electrophoresis, gels were fixed by soaking in 3.5% sulfosalicylic acid, 11.5% TCA, and 45% methanol, stained with Coomassie brilliant blue, and destained by diffusion. The apparent molecular weight of the membrane protein was estimated by migration rate relative to five standard proteins in the molecular weight range 12,600–212,000 (*21*).

7.4. Labeling Techniques

Studies on the organization of membrane at molecular level was attempted by using labels for specific groups.

A fluorescent diazonium dye, 4-acetamido-4′-isothiocyanostilbene-2,2′-disulfonic acid (SITS), was used in smooth muscle to localize the plasma membrane after the density gradient centrifugation (*14*).

Lactoperoxidase has been used to identify the surface proteins of a cell. This enzyme has the ability to iodinate tryosine and histidine residues of proteins. It is a high molecular weight protein and does not dissociate into subunits. Only the exposed proteins will be iodinated. If the cell membrane is damaged, this enzyme can iodinate the proteins of the inner surface as well (*21*).

7.5. Labeling of Smooth Muscle Cell Surface Proteins

^{125}I iodination of sarcolemmal proteins according to Krall and Korenman (*21*) is described as follows: Myometrial cell surface proteins can be radioiodinated using a NEN ^{125}I iodosulfanilic acid labeling Kit. ^{125}I iodonsulfanilic acid (^{125}I 1000 Ci/mmol) was diazotized just before use by incubation with 0.05*M* sodium nitrite under acidic conditions at 4°C for 15 min. Membranes (1.5 mg) from the density gradient or pieces of myometrium 0.5 × 3.0 mm (0.003 g) were washed and suspended in 0.475 mL of 0.02*M* sodium phosphate buffer (pH 7.4) with 0.15*M* NaCl. One mCi, 0.025 mL of the diazotized ^{125}I iodosulfanilate was added, and the suspension was incubated on ice. The iodination was complete after 20 min.

Radioiodinated muscle pieces were rinsed with unlabeled 0.001*M* NaI in 0.02*M* sodium phosphate (pH 7.4) and 0.15*M* NaCl and combined with the muscle strips from which they were initially cut for subsequent homogenization and fractionation. The ^{125}I iodosulfanilic acid membrane suspension was diluted to 1 mL with 0.02*M* sodium phosphate buffered saline with 0.001*M* NaI and the membranes concentrated by centrifugation. The pellet was washed twice more by centrifugation with buffered unlabeled NaI and finally dissolved in electrophoresis sample buffer for separation of plasma membrane proteins by SDS-gel electrophoresis followed by radioautography of the gels.

8. Conclusion

Smooth muscle subcellular fractionation techniques were not extensively used by many workers. The tissues most widely studied were liver, brain, and red blood cells. The smooth muscle is a tough tissue because the cells are enmeshed in collagen and are difficult to homogenize and subfractionate (*14*). Many techniques are available for homogenization of various tissues, but smooth muscle homogenization was not achieved by most of these techniques. The successful use of the all-glass homogenizer has the drawback of silica particles being present in the homogenate, which interferes in electron microscopy and ionic measurements. Polytron at controlled time and speed has been successfully used for smooth muscle homogenization. Isolation of subcellular fractions can be achieved by conventional differential concentration techniques (*17*).

The problem of isolation of plasma membrane from smooth muscle has recently attracted attention and a great deal of effort is being made to isolate pure plasma membrane from various smooth muscles. On homogenization the cell surface becomes a complex dispersion of fragments, and the plasma membrane fragments may be found in virtually every subcellular fraction in one procedure or another (*5*).

The first detailed method of isolation of smooth muscle plasma membrane was based on the assumption that the homogenate contained plasma membrane fragments of various sizes and, therefore, the centrifugation steps were reduced to bare minimum using a single-step density gradient centrifugation. In subsequent modifications, the procedure was modified and the plasma membrane was isolated from only one subcellular cut that is not truly representative of the plasma membrane as a whole, and certain fragments not lost during isolation of nuclear, mitochondrial, and microsomal fractions were put to density gradient centrifugation (*28*). A recently developed technique for the use of toluene to make the cells permeable to charged ions and large molecules can be used to isolate intact plasma membranes in their original shape. The drawback of the technique is that the cells have to be dispersed first and then treated with toluene and lithium bromide. The technique is successful with frog skeletal muscle, ameba, intestinal epithelial cell, and so forth.

In order to apply this technique to smooth muscle, methods have to be devised to separate smooth muscle cells either by mechanical means or by enzymatic and chemical treatment.

The criteria for purity of various subcellular fractions have been well worked out in liver and other tissues, but in smooth muscle enzymes like glucose-6-phosphatase are not present in sufficient amounts to be used as an endoplasmic reticulum marker. This was achieved by incorporating tritiated leucine as a measure of protein biosynthesis, which is the property of endoplasmic reticulum (*14, 28*).

Membrane orientation is a problem arising out of drastic homogenization of the tissue. Attempts have to be made to orient the membrane vesicles in the desired orientation.

Once the beginning has been made and various groups have started to work on the subcellular fractions of smooth muscle, it is hoped that the techniques will be refined and new techniques will be developed for isolation of subcellular fraction from smooth muscle.

References

1. Anderson, N. G. The zonal ultracentrifuge—a new instrument for fractionating mixtures of particles. *J. Phys. Chem.* 66:1984–1989, 1962.
2. Cline, G. B. Continuous sample flow density gradient centrifugation. In *Progress in Separation and Purification.* John Wiley & Sons Inc., New York, 1971.
3. Daniel, E. E., C. Y. Kwan, M. A. Matlib, D. Crankshaw and A. M. Kidwai. Characterization and Ca^{2+}accumulation by membrane fractions from myometrium and artery. In *Excitation–Contraction Coupling in Smooth Muscle.* Ed. by R. Casteels, p. 181–188, 1977.
4. Daniel, E. E., A. K. Grover and C. Y. Kwan. Role of isolated smooth muscle plasma membranes in active transport of Ca^{2+}. In *Adv. Pharmacol. Therap. II. Vol. 3.* Edited by Yoshida, H., Y. Hagihara and S. Ebashi, Pergamon Press, Oxford p. 67–70, 1982.
5. DePierre, J. W. and M. L. Karnovsky. Plasma membranes of mammalian cells. A Review of Methods for their characterization and isolation. *J. Cell Biol.* 56:275–303 1973.
6. Grover, A. K., C. Y. Kwan, R. E. Garfield, J. McLean, J. E. T. Fox and E. E. Daniel. Fractionation and Ca^{2+} uptake studies on membranes of

rabbit longitudinal and circular intestinal smooth muscles. *Can. J. Physiol. Pharmacol.* 58:1102–1113 1980.

7. Grover, A. K., J. Crankshaw, R. E. Garfield and E. E. Daniel. Smooth muscle membrane vesicle orientation: A study on intactness and sidedness of rat myometrium plasma membrane vesicles. *Can. J. Physiol. Pharmacol.* 58:1202–1211, 1980.
8. Grover, A. K., C. Y. Kwan, J. Crankshaw, D. J. Crankshaw, R. E. Garfield and E. E. Daniel. Characteristics of calcium transport and binding by rat myometrium plasma membrane subfractions. *Am. J. Physiol.* 239:C66–C74, 1980.
9. Grover, A. K., M. S. Kannan and E. E. Daniel. Canine trachealis membrane fractionation and characterization. *Cell Calcium* 1:135–146, 1980.
10. Heidrich, H. G., Stahn, R. and Hanning, K. Surface charge of rat liver mitochondria and their membranes. *J. Cell Biol.* 46:137–150, 1970.
11. Hogeboom, G. H., W. C. Schneider and G. E. Palade. Cyotechemical studies of mammalian tissues. *J. Biol. Chem.* 172:619, 1948.
12. Hogeboom, G. H. Fractionation of cell components of animal tissues. In *Methods in Enzymology, Vol. 1.* Edited by S. P. Colowick and N. O. Kaplan. Academic Press, N. Y., p. 16–19, 1955.
13. Kasper, C. B. Isolation and properties of the nuclear envelope. In *Methods in Enzymology, Vol. 31A.* Edited by S. Fleischer and L. Packer. Academic Press Inc., N. Y., pp 279–292, 1974.
14. Kidwai, A. M., M. A. Radcliffe and E. E. Daniel. Studies on smooth muscle plasma membrane. I. Isolation and characterization of plasma membrane from rat myometrium. *Biochim. Biophys. Acta* 233:538–549, 1971.
15. Kidwai, A. M., M. A. Radcliffe, E. Y. Lee and E. E. Daniel. Isolation and properties of skeletal muscle plasma membrane. *Biochem. Biophys. Acta* 298:593–607, 1973.
16. Kidwai, A. M. Isolation of plasma membrane from smooth, skeletal and heart muscle. In *Methods in Enzymology, Vol. 31A.* Edited by S. Fleischer and L. Packer. Academic Press Inc., N. Y., pp 134–144, 1974.
17. Kidwai, A. M. Homogenization and fractionation techniques in smooth muscle. In *Methods in Pharmacology, Vol. 3.* Edited by E. E. Daniel and D. M. Paton. Plenum Press, N. Y., pp. 543–554, 1975.
18. Kidwai, A. M. Smooth muscle sarcolemmal biochemistry. In *The Biochemistry of Smooth Muscle.* Edited by N. L. Stephens. University Park Press, Baltimore, pp 585–594, 1977.
19. Kono, T. and S. P. Colowick. Isolation of skeletal muscle cell membrane and some of its properties. *Arch. Biochem. Biophys.* 93:520–533, 1961.
20. Krakower, C. A. and Greenspon, S. A. The isolation of basement membranes. In *Biology and Chemistry of Basement Membrane.* Edited by N. A. Kefalides, Academic Press, Inc., New York, 1–16, 1978.
21. Krall, N. F. and S. G. Korenman. Smooth muscle sarcolemma. Purification and properties of plasma membrane from the rat uterus. *Biochim. Biophys. Acta* 556:105–117, 1979.
22. Kwan, C. Y., R. Garfield and E. E. Daniel. An improved procedure for the isolation of plasma membrane from rat mesenteric arteries. *J. Mol. Cellular Cardiology* 11:639–659, 1979.

23. Kwan, C. Y., R. M. K. W. Lee and E. E. Daniel. Isolation of plasma membranes from rat mesenteric veins: a comparison of their physical and biochemical properties with arterial membranes. *Blood Vessels* 18:171–186, 1981.
24. Kwan, C. Y., Y. Sakai, A. K. Grover and R. M. K. W. Lee. Isolation and characterization of plasma membrane fraction from gastric fundus smooth muscle of the rat. *Molec. Physiol.* 2:107–120, 1982.
25. Kwan, C. Y., R. M. K. W. Lee and A. K. Grover. Isolation and characterization of subcellular membrane fractions from the smooth muscle rat vas deferens. *Molec. Physiol.* 3:53–69, 1983.
26. Kwan, C. Y. Mg^{2+}- or Ca^{2+}-activated ATPase activities of plasma membranes isolated from vascular smooth muscle. *Enzyme* 28:317–327, 1982.
27. Marinetti, G. V. and G. M. Gray. A fluorescent chemical marker for the liver cell plasma membrane. *Biochim. Biophys. Acta* 135:580–590, 1967.
28. Matlib, M. A., J. Crankshaw, R. E. Garfield, D. J. Crankshaw, C. Y. Kwan, L. A. Branda and E. E. Daniel. Characterization of membrane fractions and isolation of purified plasma membranes from rat myometrium. *J. Biol. Chem.* 254:1834–1840, 1979.
29. Potter, V. R. In *Methods in Enzymology, Vol. I.* Edited by Colowick, S. P. and Kaplan, N. O. Academic Press, N. Y., p. 10, 1955.
30. Rand, R. P. Mechanical properties of the red cell membrane. II. Viscoelastic breakdown of the membrane. *Biophys. J.* 4:303–316, 1964.
31. Rand, R. P. The structure of a model membrane in relation to the viscoelastic properties of the red cell membrane. *J. Gen. Physiol.* 52:173S–182S, 1968.
32. Risteli, L. and J. Risteli. Basement membrane research. *Medical Biology.* 59:185–189, 1981.
33. Rosenberg, M. D. The relative extensibility of cell surfaces. *J. Cell Biol.* 17:289–297, 1963.
34. Steck, T. L. Preparation of impermeable inside out and right side out vesicles from erythrocyte membranes. In *Methods in Membrane Biology, Vol. 2.* Edited by E. D. Korn. Plenum Publishing Corporation, New York-London, pp 245–282, 1974.
35. Steck, T. L. and D. F. H. Wallach. The isolation of plasma membranes. In *Methods in Cancer Research, Vol. 5.* Edited by H. Busch, Academic Press, N.Y. pp 92–153, 1970.
36. Tata, J. R. Isolation of nuclei from liver and other tissues. In *Methods in Enzymology.* Edited by S. Fleischer and L. Packer, Academic Press Inc., New York, p. 259, 1974.
37. Wolowyk, M. W., A. M. Kidwai and E. E. Daniel. Sodium-potassium stimulated adenosine triphosphatase of vascular smooth muscle. *Can. J. Biochem.* 49:376–384, 1971.
38. Wibo, W., N. Morel and T. Godfraind. Differentiation of Ca^{2+} pumps linked to plasma membrane and endoplasmic reticulum in the microsomal fraction from intestinal smooth muscle. *Biochim. Biophys. Acta.* 649:651–660, 1981.
39. Zaidi, S. I. M., R. N. Pandey, A. M. Kidwai and C. R. Krishna Murti. A rapid method for preparation of sarcolemma from frog skeletal muscle. *J. Biosci.* 3:293–302, 1981.

Chapter 10

Calcium-Handling Studies Using Isolated Smooth Muscle Membranes

A. K. Grover

1. Introduction

Intracellular calcium ion concentration is the major determinant in the contractile state of smooth muscles (*see* Chapters 2–4). A decrease in this concentration to ~0.1 μM is associated with relaxation and an increase in it correlates with contraction of the smooth muscle. How is the intracellular Ca^{2+} concentration regulated? Studies using tissue strips to answer this question have been presented in Chapters 3 and 4. It is generally agreed that the cellular membranes, namely, plasma membranes (PM), endoplasmic reticulum (ER), and mitochondria are responsible for regulating the intracellular Ca^{2+} concentration. Four types of Ca^{2+} interactions with the membranes have been identified: (a) Ca^{2+} extrusion from cytoplasm by PM or into ER by an ATP-dependent Ca^{2+} transport system, (b) transmembrane Ca^{2+} movements by Na^+–Ca^{2+} exchange, (c) passive Ca^{2+} entry from ER or through PM into cytoplasm, and (d) Ca^{2+} binding to and release from various membranes. Even though these processes have been identified, there is considerable dispute concerning the role of the three different types of membranes (PM, ER, and mitochondria) and the relative importance of each of the processes in the excitation–contraction coupling cycle of the smooth muscle. In this chapter a biochemical approach to these problems, namely, studying Ca^{2+}-handling properties of isolated subcellular fractions will be discussed. The advantage of this approach is that one can study the Ca^{2+} interactions of individual organelles without any interference from other organelles or the extracellular space. The disadvantages of this approach include changes in properties of the membranes during the isolation process, unavailability of pure ER fractions, and lack of information on the sidedness of the membrane vesicles (*see* Chapter 9) used in the experiments. Despite these pitfalls, perhaps this is the only approach that can eventually lead to elucidation of the molecular mechanisms involved. This chapter focuses on the current state of knowledge and problems in interpretation of subcellular distribution, possible mechanisms, modulation, and relative importance of these processes as determined from studies using subcellular fractions isolated from smooth muscle.

2. ATP-Dependent Ca^{2+} Transport

2.1. Subcellular Distribution

The subcellular particulate fractions from a number of smooth muscles have now been shown to accumulate Ca from media containing ATP + Mg. The most sensitive and commonly employed method for these studies is mixing the particulate fraction with reaction mixtures containing ^{45}Ca, ATP, and Mg, separating the membranes by filtration and determining the amount of radioactivity retained by the particulate material. Merits and demerits of this method and of the methods using atomic absorption, Ca^{2+}-sensitive dyes, or Ca^{2+} selectrodes have been reviewed recently (*17*).

The accumulation of Ca^{2+} by the particulate material may correspond to: accumulation of Ca^{2+} by inside-out PM vesicles (uptake of Ca^{2+} from cytoplasm to extracellular space), by right-side-out ER vesicles (uptake of Ca^{2+} from cytoplasm into ER), or by intact or right-side-out mitochondrial membranes (uptake of Ca^{2+} from cytoplasm into mitochondria). The major problem in studying the subcellular distribution of the ATP-dependent Ca^{2+} transport is to distinguish between the Ca^{2+} uptake by PM and ER vesicles since the Ca^{2+} uptake by the mitochondrial membranes can be selectively inhibited by azide or ruthenium red (for a review, *see 17*). The ATP-dependent mitochondrial Ca^{2+} uptake (azide- or ruthenium red-sensitive) has been observed in most smooth muscle subcellular fractionation studies (*17*).

By analogy to the skeletal muscle, in the earlier studies the azide-insensitive ATP-dependent Ca^{2+} uptake by the smooth muscle microsomes was attributed to ER (often referred to as sarcoplasmic reticulum or SR by some workers). This Ca^{2+} uptake was stimulated by the precipitating agents oxalate or phosphate as in the skeletal muscle SR (*35, 58, 61*). However, several studies on subfractionation of microsomes prepared from a number of smooth muscles on sucrose density gradients showed that the distribution of azide insensitive ATP-dependent Ca^{2+} uptake in oxalate-free media correlated extremely well with that of the PM markers 5′-nucleotidase, phosphodiesterase I, K^+-activated ouabain-sensitive *p*-nitrophenylphosphatase and binding of wheat germ agglutinin and oxytocin, and not with the putative ER marker NADPH: cytochrome c reductase (for a review, *see 17*). Such a correlation between the azide-insensitive ATP-dependent Ca^{2+} uptake and the PM marker K^+-activated ouabain-sensitive *p*-nitrophenylphosphatase in the subcellular fractions of rat mymetrium is shown in Fig. 1A. The PM membrane fraction isolated from rat myometrium has also been shown to contain inside-out surface (22). Thus, the evidence that the PM fractions isolated from smooth muscle are capable of ATP-dependent Ca^{2+} uptake is convincing.

The evidence for the presence of ATP-dependent Ca^{2+} transport in ER, however, is indirect since it has not been possible to obtain ER fractions of high purity (*see* Chapter 9). Digitonin binding to PM increases its density (for a review *see 17*). When microsomes from guinea pig ileal smooth muscle were treated with digitonin and subjected to sucrose density gradient centrifugation, the distribution of 5′-nucleotidase was shifted to higher densities. There was, however, no concomitant shift in the distribution of the oxalate-stimulated ATP-dependent Ca^{2+} uptake (*66*). Thus the azide-insensitive oxalate-stimulated Ca^{2+} uptake was attributed to membranes other than PM and hence ER. The azide-insensitive ATP-dependent Ca^{2+} uptake in presence of oxalate did not correlate with the PM marker distribution in the subcellular fractions of rat myometrium (Fig. 1B). The disparity in the distribution of the oxalate-stimulated and -unstimulated Ca^{2+} uptake can be explained by a lower percent recovery of the former during the subfractionation (*25*). Consequently, it can be argued that there are two types of Ca^{2+} uptake: oxalate-unstimulated uptake localized in PM, which is more sta-

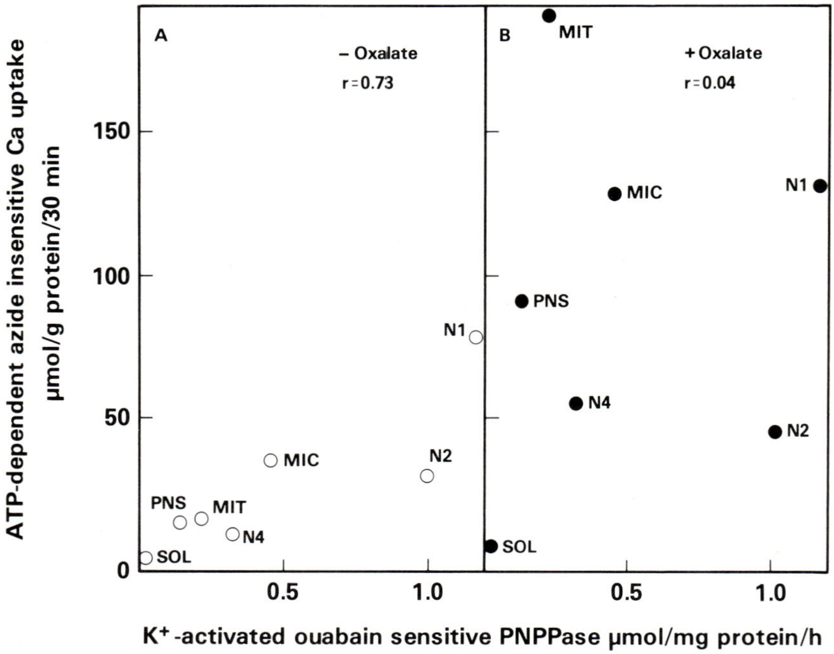

Fig. 1. Correlation of subcellular distribution of azide-insensitive ATP-dependent Ca uptake in rat myometrium in absence (A) and presence of 5 m*M* oxalate based on data in ref. *28*. The fractions PNS (postnuclear supernatant), SOL (soluble), MIT (mitochondrial), and MIC (microsomal) are obtained by differential centrifugation and N1 to N4 by sucrose density gradient centrifugation of MIC as described in ref. *23*.

ble during the isolation process, and the oxalate-stimulated uptake localized in ER, which is less stable. Since this is a very weak argument, it appears that more concrete evidence for the existence of an ATP-dependent Ca^{2+} transport system in ER would require availability of ER fractions of higher purity and establishment of better ER markers. It is emphasized here that the oxalate-stimulated Ca^{2+} uptake cannot be used as an ER marker for the following reasons: (a) the yield of this uptake diminishes during purification for reasons yet unknown (*28*), (b) in skeletal muscle SR a pH-sensitive factor has been shown to be responsible for the oxalate effect and some similar factor(s) may also be involved in smooth muscle as well, and (c) perhaps most important, one cannot use any given property (Ca^{2+} uptake) to study its own distribution.

The smooth muscle mitochondria are capable of azide-sensitive Ca^{2+} uptake (*15, 73*), whereas the azide-insensitive Ca^{2+} uptake in microsomes is predominantly caused by PM, although in presence of oxalate, the Ca^{2+} uptake from ER may be observed as well. In any event, since microsomal membranes would thus contain mitochondria, possibly two types of membranes (PM and ER) contributing to the azide-insensitive Ca^{2+} uptake, and actomyosin (*see* Chapter 9), which is also a Ca^{2+}–Mg–ATPase, it is recommended that instead of microsomes, only relatively pure preparations of PM and ER be used for further studies on the nonmitochondrial Ca^{2+}-uptake.

2.2. Substrate Dependence

The Ca^{2+} uptake by isolated smooth muscle mitochondria can utilize the chemical energy of hydrolysis of ATP or of oxidation of succinate (a respiratory substrate) (*3, 44,* and review *17*). In order to play a key role in relaxation of the smooth muscle, any important Ca-sequestration mechanism should be operative at submicromolar Ca^{2+} concentrations. Batra reported that mitochondria isolated from human myometrium could indeed accumulate Ca^{2+} from media containing as low as 0.01 μM Ca^{2+} (*3*). Malmostrom and Carafoli reported a hyperbolic dependence of the Ca^{2+} uptake by human myometrium on Ca^{2+} concentration with a K_m of 4.4 μM (*44*). In most studies, however, the smooth muscle mitochondrial Ca^{2+} uptake increases with increasing Ca^{2+} concentration and a saturation is not observed up to 100 μM Ca^{2+} (for a review, *see 17*). It is possible that the high K_m for Ca^{2+} in these mitochondrial preparations is caused by the breakdown of the normal function of the organelle from the harsh conditions employed during the isolation process. However, unless this is shown to be true using better preparations, it is concluded that in most smooth muscles, with the exception of human myometrium, the Ca^{2+} uptake process by mitochondria requires Ca^{2+} concentrations that are very high relative to the physiological concentrations of intracellular Ca^{2+}.

The azide-insensitive Ca^{2+} uptake in the microsomal fraction cannot utilize succinate as a substrate and has an obligatory requirement for ATP or another phosphate donor (for a review *see 17*). The Ca^{2+} uptake in 10 min by rat aortic microsomes in presence of oxalate using various nucleotide phosphates was in the order ATP $>>$ ADP $>$ GTP $>$ UTP $>>$ ITP, but the values for the uptake in absence of oxalate were ATP $>$ GTP $>$ ADP $>$ UTP $>$

control > ITP (*65*). In a PM fraction isolated from rabbit mesenteric artery, UTP, ITP, and CTP were poor substitutes for ATP (*63*). Similarly in a PM fraction isolated from rat myometrium, GTP, ITP, acetylphosphate, and *p*-nitrophenylphosphate could not substitute for ATP (*38*). Thus the Ca^{2+} pump in the smooth muscle PM preferentially uses ATP and has a much narrower substrate specificity for the phosphate donor than the Ca^{2+} pump of the skeletal muscle SR (*35, 58*). Grover et. al. showed that the Ca^{2+} pump in rat myometrium PM vesicles had an obligatory requirement for magnesium, but it was unclear whether this represented a requirement for Mg^{2+} or $MgATP^{2-}$ (*26*). In rabbit mesenteric artery PM, a small amount of ATP-dependent Ca^{2+} uptake was also observed without any added magnesium but the addition of magnesium enhanced the amount of ATP-dependent Ca^{2+} uptake tremendously (*63*). Thus it is apparent that the Ca^{2+} pump in the smooth muscle PM requires ATP and magnesium.

The Ca^{2+} concentration dependence of the ATP-dependent Ca^{2+} uptake has been studied recently in our laboratory in several smooth muscle PM preparations using Ca^{2+}–EGTA buffers. The apparent initial velocity of the ATP-dependent azide-insensitive Ca^{2+} uptake was studied at different concentrations of Ca^{2+}, and it was found that in the smooth muscles studies, half maximum uptake occurred at Ca^{2+} concentration of 0.3–0.8 μM. The Ca^{2+}-concentration dependence isotherms were hyperbolic for the PM preparations from rat fundus, vas deferens (unpublished), and dog corpus but showed positive cooperativity towards Ca^{2+} in those from rat myometrium and dog antrum smooth muscle (*26, 27, 33, 56*). Figure 2 schematically shows the Ca^{2+}-dependence of the apparent initial velocity of the Ca^{2+} pump in the rat myometrium PM. Thus it is clear that the Ca^{2+} pump in the PM can operate in the submicromolar Ca^{2+} concentration range.

Ca^{2+} concentration dependence of the azide-insensitive Ca^{2+} uptake has been studied using microsomal or PM preparations from a number of smooth muscles in various laboratories, and the reported values of K_m for Ca^{2+} vary. The possible reasons for the variation include: (a) the variety of stability constants for $CaEGTA^{2-}$ and the p*K* values of EGTA used in the computation of the Ca^{2+} concentrations (*26*; review *17*); also frequently the corrections for the effects of temperature and ionic strength on the stability constants are neglected, (b) slight differences in reaction conditions like differences in pH and Mg^{2+} concentra-

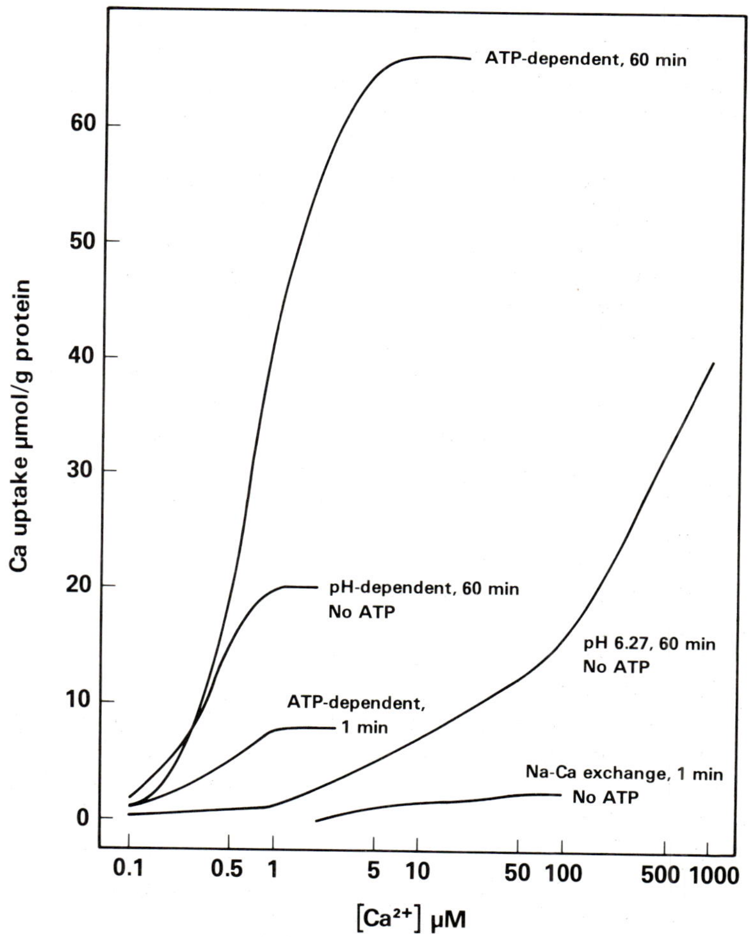

Fig. 2. Ca^{2+} concentration-dependence of various Ca^{2+}-uptake processes in rat myometrium PM fraction N1.

tion, (c) reaction time since usually the Ca^{2+} uptake at a fixed time is determined and it is not specified whether the uptake values are initial rates or uptake capacities. (d) whether oxalate is present in the reaction mixtures or not (*67*), (e) variations in preparative procedures, and (f) tissue variations.

2.3. Evidence for Active Transport

Four independent lines of evidence demonstrate that the ATP-dependent Ca^{2+} uptake by the PM or microsomes isolated from smooth muscle is an active transport process. The first depends

Table 1
Effect of Ionophores on Ca^{2+} Uptake by PM[a]

	A 23187	Monensin
ATP-dependent	↓	↔
Passive	V_o ↑ Equilibrium value ↔	↔
Na^+–Ca^{2+} exchange	↓	↓

[a]The symbols are: ↓ inhibition, ↑ activation, and ↔ none or marginal effects.

upon the use of Ca^{2+} ionophores. In a number of studies it has been shown that Ca^{2+} ionophores A 23187, X-537A and ionomycin inhibit the ATP-dependent Ca^{2+} accumulation (Table 1) by the smooth muscle membrane vesicles (*14, 23, 38, 56, 67, 68*; review *17*). Similarly, once these membrane vesicles have been loaded with Ca^{2+}, addition of the Ca^{2+} ionophores results in loss of the accumulated Ca^{2+}. These data are consistent with the hypothesis that the ATP-dependent Ca^{2+} uptake leads to higher intravesicular Ca^{2+} concentrations than those present in the reaction media, i.e., generation of Ca^{2+} concentration gradients that are expected if the Ca^{2+}-movement were to be caused by active transport. The second line of evidence is that dilution of the vesicles loaded with Ca^{2+} into hypotonic media results in release of a large proportion of intravesicular Ca^{2+} (*23*). The third line of evidence is that passive loading of vesicles to the same uptake levels (see passive binding) requires Ca^{2+} concentration in m*M* range as would be obtained during the ATP-dependent Ca^{2+} uptake from media containing 1 μM Ca^{2+} (*26, 27, 33*). Thus, for dog antrum PM preparation, it has been calculated that transmembraneous gradients as large as 3000 may be created during the ATP-dependent Ca^{2-} uptake (*33*). Even this figure may be an underestimate since it is based on the assumption that all the vesicles would accumulate Ca^{2+}, which may not be true, since mixed orientation of membranes is found in the smooth muscle membrane preparation (Chapter 10), and only the inside-out vesicles would show ATP-dependent Ca^{2+} uptake. The fourth line of evidence is the stimulation of the Ca^{2+} uptake in microsomes and to a smaller degree in the isolated PM preparation by the Ca^{2+}-precipitating agents, oxalate and phosphate. The rationale for this hypothesis has been discussed in detail in recent reviews (*17, 35, 58*). However, as discussed in an earlier section, the oxalate effect is highly variable, and it diminishes as the purity of the membranes increases. In any event, there is overwhelming evidence that the Ca^{2+} uptake by the PM or microsomal fractions isolated from smooth muscles is an active transport process that results in generation of 1000-fold Ca^{2+} concentration gradients across the vesicle membranes. In view of the possible partial loss of activation owing to the isolation process and the underestimation inherent in these calculations, these gradients are probably comparable to those required for maintenance of relaxation of the smooth muscle cell.

2.4. Mechanism

The following scheme depicts the mechanism of action of the Ca^{2+} pump in the skeletal muscle SR vesicles.

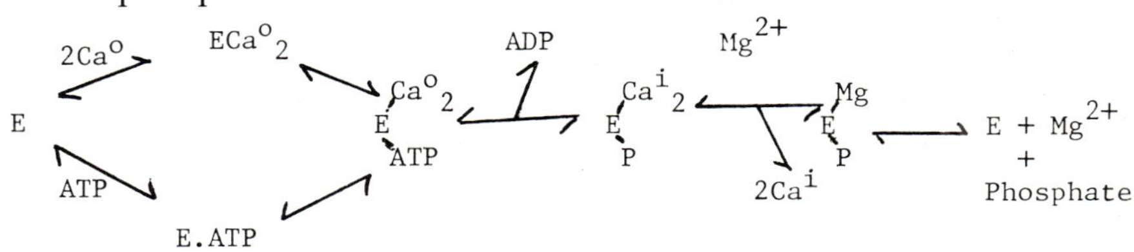

Here E is the Ca^{2+} pump protein, Ca^o and Ca^i refer to Ca^{2+} on the outside and inside of the vesicle respectively. The model can be summarized as follows: (a) Ca^{2+} pump is caused by a Ca–Mg–ATPase; transport of Ca^{2+} across the vesicle membranes represents extra splitting of 1 mol ATP to ADP plus phosphate/2 mol Ca^{2+} transported; (b) the transport involves the formation of a phosphoenzyme intermediate (EP), and (c) the Ca^{2+} pump can operate in the forward direction as shown, or in the reverse direction in which it can utilize the energy of dissipation of a Ca^{2+} concentration gradient to synthesize EP, which in turn can be used for ATP synthesis or the ATP:ADP phosphotransferase reaction. The evidence for these steps has been summarized in recent reviews (*35, 58*).

PM and microsomes isolated from smooth muscles contain a very active Mg–ATPase (*38*, review *17*), which makes it difficult to measure Ca–Mg–ATPase. The problem is compounded by the presence of high activities of ADPase (*43*) and 5′-nucleotidase (review *17*) as well. Thus rat myometrium PM vesicles can hydrolyze ATP to yield nearly 3 mol inorganic phosphate (P_i)/mol ATP in 2 min when ATP concentration is low and in longer periods of time when ATP concentrations are high and small amounts of the membranes are employed (unpublished). Since most studies on Ca–Mg–ATPase in smooth muscle membranes have employed assays in which the P_i liberated during the reaction is measured, it makes the interpretation of literature on this subject extremely difficult. The problems are best illustrated with the example of a recent study using bovine aortic microsomes (*20*). The first problem is that since the microsomes are often contaminated with actomyosin, which is a Ca–Mg–ATPase (review *24*), it is not possible to attribute the measured Ca–Mg–ATPase activity to the Ca^{2+} pump unequivocally. Second, in this study the P_i liberated was up to 1.8 mol/mol ATP and since hydrolysis of ATP to ADP can account for only 1 mol P_i/mol ATP, further hydrolysis products would be formed because of ADPase and 5′-nucleotidase activity. The third, in a plot of P_i liberated vs time at any given time, the P_i liberated in the presence of added EGTA and no added Ca^{2+} was not significantly different from that in presence of the added Ca^{2+}. Yet these workers have succeeded in conducting elaborate kinetic studies on Ca–Mg–ATPase, suggesting that under certain conditions the stoichiometry was as low as 0.03 Ca transported/mol ATP hydrolyzed. Similar Ca–Mg–ATPase measurement using the P_i liberation method has also been reported in several other studies (*38*, review *17*). Calmodulin has been reported to activate the Ca–Mg–ATPase activity of canine trachealis microsomes (*37*). Even though it is known that calmodulin stimulates the Ca–Mg–ATPase activity of smooth muscle actomyosin (*36*), the contribution from the actomyosin Ca–Mg–ATPase was not assessed in the study on the trachealis microsomes. The most reliable assays of Ca–Mg–ATPase, even though carried out on microsomes, and hence problematic because of the multiplicity of the various types of Ca–Mg–ATPase activities, are those of Wuytack and Casteels (*67*). These workers used a coupled enzyme ATP-regenerating system for the assays, and included azide in the reaction media. The ATP-regenerating system instantly converts the ADP formed during the reaction back to ATP, thus ensuring that only ATPase activities are measured and there is no contribution to this from ADPase and 5′-nucleotidase activities. Presence of azide partially inhibits the basal Mg–ATPase (*43*) and thus decreases the masking effect of this enzyme. Using this system, Ca^{2+} uptake as well as Ca–Mg–ATPase in the porcine coronary artery, were examined. The two processes occurred with K_m values of 0.34 ± 0.03 and 1.17 ± 0.15 μM for Ca^{2+} and Hill coefficients of 1.69 ± 0.09 and 1.23 ± 0.17, respectively. Thus, whereas the ATP-dependent Ca^{2+} uptake was nearly saturated at a Ca^{2+} concentration of 1 μM, the Ca–Mg–ATPase activity had less than half the maximum value. The stoichiometry for Ca^{2+}-transport/

ATP hydrolyzed was reported to be 0.75. Thus, there are discrepancies between the Ca^{2+} pump and the Ca–Mg–ATPase activities even in the best available data. As far as the stoichiometry is concerned the values in the literature range from 0.03 to 0.75 Ca^{2+}/ATP (*20, 67*). Thus a low concentration of the pump sites relative to the Mg–ATPase activity and multiplicity of the Ca–Mg–ATPase present, and perhaps of the Ca^{2+} pump (if the PM and the ER Ca^{2+} pump were to differ), make it difficult to conduct these studies.

Raeymaekers et al. (*52*) have tried to solve this problem by preparing a "Ca^{2+} pump-enriched fraction" from pig antral smooth muscle microsomes. They loaded the microsomal membranes with Ca^{2+} in presence of ATP, Mg, and oxalate and centrifuged them in 53% sucrose (density = 1.253 g/mL) to obtain a pellet consisting of membrane vesicles loaded with Ca-oxalate. Similarly, they also obtained Ca-phosphate loaded membrane preparation. The latter was capable of further ATP-dependent Ca^{2+} uptake (*53*). It could also carry out the ATP:ADP exchange and the ATP:P_i exchange reactions that are known to be partial reactions of the skeletal muscle SR Ca–Mg–ATPase. These membranes could also be phosphorylated using γ ^{32}P-ATP, and the phosphorylation product was hydroxylamine-sensitive, but only slightly sensitive to low pH. On the basis of these studies, the following reaction scheme was proposed:

$$\mathrm{Ca_{out} + E + ATP \rightleftharpoons Ca\text{-}E\text{-}ATP \overset{ADP}{\rightleftharpoons} Ca\text{-}E \sim P \rightleftharpoons Ca_{in} + P_i}$$

This scheme is analogous to the one proposed for the skeletal muscle SR Ca^{2+} pump. The stoichiometry of the Ca^{2+} pump, however, is unknown. It is interesting that Raeymaeker et al. assumed that the membrane vesicles used in this study were from ER. However, any marker data to support this contention was not presented (*52*). Since, accoridng to Wibo et al. (*66*), only the ER fractions were to show the oxalate-stimulated uptake, this assumption may be correct, but is as yet unproven. This would also mean that whether the Ca^{2+} pump from the smooth muscle PM can carry out these partial reactions or not remains to be tested.

Another approach to obtaining membranes enriched in the Ca–Mg–ATPase has been that of Wuytack et al. (*69*) who washed the pig antrum smooth muscle microsomes with KCl to remove the actomyosin contamination, partially solubilized the microsomes by detergent (deoxycholate) treatment and centrifugations, and reconstituted the resulting material by removing the detergent by dialysis. The resulting preparation was enriched in Ca–Mg–ATPase that could also be activated with calmodulin but not enriched in NaK–ATPase, Mg–ATPase, or the ATP-dependent Ca^{2+} uptake. It is unclear whether the detergent treatment partially inactivated the NaK–ATPase and the Mg–ATPase or led to removal of these activities in other fractions since the overall yields of these activities were not given. Similarly the reasons for lack of enrichment of the ATP-dependent Ca^{2+} uptake is also not known. The reason may have been poor integrity of the reconstituted membranes or a possible uncoupling of the Ca–Mg–ATPase from the Ca^{2+} transport activity. In another instance Wuytack et al. (*70*) solubilized the microsomes with Triton X-100 and purified the resulting material by affinity chromatography on calmodulin–sepharose. A fraction enriched in Ca–Mg–ATPase was obtained. It could be phosphorylated with ^{32}P-γ-ATP on a protein subunit of 140,000 daltons. This protein subunit has now been reconstituted into phospholipid vesicles and exhibits Ca–Mg–ATPase as well as ATP-dependent calmodulin-activated A 23187-sensitive Ca-uptake (Wuytack et al., personal communication). Antibodies were raised to the Ca–Mg–ATPase from pig antrum as well as for those from pig erythrocytes and skeletal muscle. The antibodies to the Ca–Mg–ATPase from erythrocyte or pig antrum cross reacted with this protein subunit, but those for the Ca–Mg–ATPase from the skeletal muscle did not (*71*). These studies are a major breakthrough in this field.

Thus it appears that some progress is being made in this area, but a substantial amount of work is needed to answer the following questions: Are there two types of Ca^{2+}-pump activities, one in PM and the other in ER? If yes, how do these differ and when is each pump operative in the intact smooth muscle? What are the partial reactions of the Ca^{2+}-pump proteins? What is the stoichiometry? Does the movement of the cation Ca^{2+} result in a net change movement, i.e., is the Ca^{2+} pump electrogenic or is the resulting membrane electrical potential dissipated because of the countermovement of H^+ or other cations or because of countermovement of anions?

2.5. Modulation

There are two major reasons for studying modulation of the Ca^{2+}-pump activity of isolated smooth muscle membranes. The first is to determine whether loss of some modulators during the membrane isolation process results in changed properties of the pump, and the second is to study how transient changes in concentrations of some regulating agents during the excitation–contraction cycle could affect functioning of the Ca^{2+}-transport system.

Batra et al. (*4*) reported that diethylstilbesterol, ethinyl estradiol, and estradiol-17-β inhibited the Ca^{2+} uptake by human myometrium mitochondria, whereas progesterone had no effect on the mitochondrial Ca^{2+}-uptake activity.

There have been a number of studies in which effects of several agents on the ATP-dependent Ca^{2+} uptake by microsomal or PM fractions isolated from smooth muscles was examined. A summary of the literature on this subject is given in Table 2. The agents examined include: cyclic nucleotide phosphates, protein kinase, calmodulin and its antagonists, peptides, and a number of other pharmacological agents including catecholamine, prostaglandins, and Ca^{2+} antagonists. There are however several difficulties in interpretation of these studies. First, most of these studies have been conducted on microsomes, and hence suffer from the problems discussed in earlier sections. Second, in most studies the effect of these agents at fixed Ca, Mg, ATP concentrations, and time is examined. Thus, it is unclear whether the effects are on the maximal initial velocity of the Ca^{2+} pump, affinities for the substrates, Ca^{2+}-uptake capacities of the membranes, or the rate of loss from the vesicles of the Ca^{2+} stored by the Ca^{2+} pump. The third difficulty, perhaps originating from the first two, is that the results reported from various laboratories are sometimes conflicting. For example, one group reports no effect of protein kinase on the Ca^{2+} uptake by rat aortic microsomes, whereas another group reports an enhancement (*5, 7*). Thus more extensive studies using better membrane preparations and well-defined conditions are needed.

In addition to the studies shown in Table 2, there are several others in which effects of Na^+, K^+, Cd^{2+}, Mn^{2+}, La^{3+} and other ions on the ATP-dependent Ca^{2+} uptake have been examined. The literature on this subject has been reviewed recently (*17*).

3. Na^+–Ca^{2+} Exchange

Na^+–Ca^{2+} exchange has been studied in detail in squid axon (review *6*), and in PM vesicles isolated from canine cardiac muscle (*9, 50, 51, 55*). More recently, it has also been reported to occur in PM isolated from skeletal muscle (*21*). The characteristics of the

Table 2
Modulation of ATP-Dependent Ca-Uptake in Smooth Muscle Microsomes as PM[a]

Modulator	Effect	Tissues and remarks	References
cAMP	↔	Pig coronary artery, rabbit and canine aorta, bovine trachealis, rat myometrium	*1, 2, 19, 39, 48, 57, 62, 71*
cAMP	↑	Rat aorta, myometrium	*5, 47, 49, 65*
cGMP	↔	Pig coronary artery, rabbit aorta	*62, 72*
Protein kinase	↔	Rat aorta	*5*
Protein kinase	↑	Rat aorta, only when oxalate present, ouabain inhibited this effect	*7*
Protein kinase	↑	Rabbit aorta	*18, 62*
cAMP and protein kinase	↑	Rabbit and rat aorta	*5, 19, 62*
cGMP and protein kinase	↑	Rabbit aorta	*62*
Calmodulin	↑	Pig coronary artery, canine trachealis, in EGTA-treated microsomes shifted Ca^{2+}-dependence curve to left	*37, 45, 68*
Chlorpromazine	↓	Rat myometrium	*16*
TMB-6	↓	Rat aorta	*13*
Oxytocin	↔	Rat uterus	*2*
Oxytocin	↓	Bovine uterus	*10*
Angiotensin	↔	Pig coronary artery, rat myometrium	*54, 72*
Bradykinin	↓	Pig coronary artery	*72*
Papaverine	↑	Rat myometrium, rabbit taenia coli	*39, 49, 60*
Aspaminol	↓	Rat myometrium, rabbit taenia coli	*39, 59*
Nitroprusside	↔	Canine aorta	*42*
Epinephrine	↑	Rat uterus	*2*
Ergotamine	↔	Rat uterus	*2*
Caffeine	↔	Rat uterus	*2*
Canbachol	↔	Trachealis	*56*
PGE_2	↔	Rat aorta	*64*
$PGE_{2\alpha}$	↓	Bovine uterus	*10*
$PGF_{2\alpha}$	↔	Rat aorta, myometrium	*54, 65*
PGF_2	↑	Bovine uterus	*10*
Progesterone	↑	Human myometrium	*12*
D600	↔	Canine aorta	*42*

(*continued*)

Table 2 (*continued*)

Modulator	Effect	Tissues and remarks	References
D600	↑	Rat myometrium, rabbit taenia coli	*16, 59*
Verapamil	↓	Rat myometrium	*16*
YC-93	↑	Rat myometrium	*49*
Phosphatidyl serine	↓	Canine aorta only in presence of oxalate	*40*

[a]Note: The symbols are: activation ↑, inhibition ↓, and none or marginal effects ↔.

Na^+–Ca^{2+} exchange in the cardiac PM vesicles are summarized in Table 3. The exchange occurs via an antiporter with stoichiometry of >3 Na^+/Ca^{2+}, and is thus electrogenic. Its kinetic properties are also influenced by artificially generated electrical potentials across the PM vesicles. Thus it differs from the mitochondrial Na^+–Ca^{2+} exchange that has a stoichiometry of 2 Na^+/Ca^{2+} and is thus electrically neutral.

PM vesicles isolated from rat myometrium can also carry out Na^+–Ca^{2+} exchange (*25, 34*). The vesicles loaded with Ca^{2+} in the presence of ATP when diluted in media containing Na^+ lost the stored Ca^{2+} more rapidly than those diluted in media without Na^+. Thus Na^+ supported release of Ca^{2+} occurred from the inside-out vesicles loaded with Ca^{2+}. Conversely when Na^+ loaded vesicles were diluted in media containing $^{45}Ca^{2+}$ without

Table 3
Properties of Na–Ca Exchange Dependent Ca-Influx by PM Vesicles from Myometrial and Cardiac Muscles[a]

	Rat myometrium	Cardiac muscle
pH, optimum	≥6.67	>"
V_o, μmol/g/min	2–4	200–300
Maximal value, μmol/g	~2	80
Uphill movement	Yes	Yes
Na^+ specificity	Li^+ poor substrate	Li^+ does not substitute
K_m Ca^{2+}, μ*M*	7.0	1.5 to 20
Stoichiometry	Unknown	3.12 Na^+/Ca^{2+}
Electrogeneity	Unknown	Internal positive potential
Valinomycin	No effect	Stimulation
H^+ uncouplers	No effect	Stimulation

[a]Note: The data for this table are from ref. *34*.

any ATP, these showed higher Ca^{2+} uptake than the control membranes not loaded with Na^+. Thus a higher concentration of Na^+ on one side of the PM vesicle membrane caused the movement of Ca^{2+} to that side either in the Ca^{2+}-efflux or in the Ca^{2+}-influx experiment. Sucrose, Rb^+, Cs^+, and K^+ could not substitute for Na^+, and Li^+ substituted poorly for these effects (*25*). The Na^+-dependent Ca^{2+} uptake could be blocked by monensin and was thus shown to depend on Na^+-concentration gradients. A 23187 caused rapid initial Ca^{2+} uptake and blocked any effects of the Na^+ gradients on the uptake. In subsequent experiments using vesicles loaded with Na^+ as well as Ca^{2+}, influx experiments showed that the Na^+–Ca^{2+} exchange could occur against an electrochemical gradient of Ca^{2+} (*34*). Thus it is clear that the Na^+-dependent Ca^{2+} movement in the rat myometrium PM vesicles could be explained on the basis of an Na^+–Ca^{2+} exchange mechanism similar to that described in the PM vesicles of cardiac muscle. However, the magnitude of the Na^+–Ca^{2+} exchange in these vesicles was only ~2 μmol/g. Properties of Na^+–Ca^{2+} exchange in rat myometrium PM vesicles in relation to other PM-Ca^{2+} interactions in this tissue are compared in Fig. 2. Table 3 provides a comparison of the Na^+–Ca^{2+} exchange in rat myometrium and in cardiac muscle PM vesicles. In myometrium, the Ca^{2+} concentration required for 50% maximal activity was ~7 μ*M*; and Na^+ concentration required for the 50% maximal activity was 10–25 m*M*. Further studies also revealed that the Na^+–Ca^{2+} exchange was a characteristic unique to the PM vesicles in this tissue (*25*). However, low magnitude of the uptake poses a problem in stoichiometry measurement and in determination of its electrogenicity. Na^+–Ca^{2+} exchange similar to that reported in rat myometrium PM vesicles also occurs in rat mesenteric artery PM vesicles (unpublished) and in guinea pig ileal smooth muscle membranes (*46*).

In order to determine the possible role of Na^+–Ca^{2+} exchange in the excitation–contraction coupling cycle, it is important to review the intracellular and extracellular concentrations of Na^+–Ca^{2+} exchange. Intracellular concentration of Ca^{2+} in a resting smooth muscle cell is ~0.1 μ*M* (review *17*). In barnacle muscle single fibers, measurements with arsenazo III showed that appreciable tension was not generated until intracellular Ca^{2+} concentration reached 2 μ*M*, and that a further threefold increase in this concentration accompanied a 20-fold change in tension (*18*). Assuming that a similar situation were to occur in smooth muscle as well, the K_m of the Na^+–Ca^{2+} exchange for Ca^{2+}

would allow it to remove Ca^{2+} from the cell only during the peak contraction. When the smooth muscle is relaxed, i.e., intracellular Ca^{2+} concentration is ~0.1 μ*M*, Na^+–Ca^{2+} exchange would not remove the intracellular Ca^{2+}. Equilibrium considerations of Na^+–Ca^{2+} exchange have been discussed recently (*64*). Assuming an extracellular Ca^{2+} concentration of 1.5 m*M* and a stoichiometry of 3 Na^+/Ca^{2+}, a membrane potential of −50 mV and an $Na^+_{external}$ to $Na^+_{internal}$ ratio of 15 at 37°C, Na^+–Ca^{2+} exchange would be in equilibrium when intracellular Ca^{2+} concentration is ~0.07 μ*M*. However, when the membrane potential drops to −30 mV and Na^+ concentration gradient is reduced from 15 to 10, intracellular Ca^{2+} concentration should be 0.44 μ*M* for Na^+–Ca^{2+} exchange to be at equilibrium. Since in the electromechanical coupling partial depolarization of the membrane and relaxation of Na^+ gradients occur, it is possible that Na^+–Ca^{2+} exchange can contribute to Ca^{2+} entry during such a contraction. This would also be kinetically feasible since extracellular Ca^{2+} concentration (1.5 m*M*) would be saturating for Na^+–Ca^{2+} exchange (K_m for Ca^{2+} ~7 μ*M*), and intracellular Na^+ concentration would also be ~20 m*M*. Similar considerations in cardiac muscle support the concept of Na^+–Ca^{2+} exchange supplying Ca^{2+} to intracellular targets during contraction (*8*). However, to arrive at these conclusions firmly, further work is required on the modulation of Na^+–Ca^{2+} exchange by various agents and by electric potentials, as well as on the stoichiometry of the Na^+–Ca^{2+} exchange in smooth muscle.

4. Passive Ca^{2+} Binding

Ca^{2+} binding has been studied using membranes isolated from smooth muscle with three major aims: (a) as a control for the ATP-dependent Ca^{2+} uptake, (b) Ca^{2+} concentration dependence of passive binding to estimate the Ca^{2+} gradients formed during the ATP-dependent Ca^{2+} uptake and (c) high affinity pH-dependent Ca^{2+} binding to understand its role in the excitation–contraction cycle.

Ca^{2+} concentration dependence of passive binding to dog aortic microsomes was analyzed by Kutsky et al. (*41*) by Scatchard plots and a deviation from hyperbolic kinetics was observed. Sakai et al. (*56*) also observed a similar deviation using dog corpus PM, but reported that the pure binding isotherms, obtained by subtracting the amount of Ca^{2+} trapped in the vesicles from the total passive Ca^{2+} uptake, were hyperbolic. In both these studies Ca^{2+} concentration used was several μM or mM.

PM vesicles isolated from a number of smooth muscles bind Ca^{2+} passively with high affinity at pH ≥ 7, but not at pH ≤ 6 (*24, 31*). Distribution of the binding in subcellular fractions correlates only with that of the PM markers. In rat myometrium PM vesicles this binding occurs with a half maximum at pH 6.8 when Ca^{2+} concentration in the uptake medium is 1 μM. Ca^{2+} dependence of the low affinity uptake (binding plus trapping) at pH 6.27 is shown in Fig. 2. There is very little uptake at Ca^{2+} concentrations $\leq$ 1 μM. This figure also shows the Ca^{2+} concentration dependence of the high affinity pH-dependent passive Ca^{2+} uptake (passive Ca^{2+} uptake near equilibrium at pH 7.07 minus similar uptake at pH 6.27) by the rat myometrium PM vesicles. The pH-dependent Ca^{2+} binding occurs with a K_d for Ca^{2+} ~0.3 μM, and with a Hill coefficient <2, thus being nearly saturated at a Ca^{2+} concentration of 1 μM. Experiments using A 23187 showed that (a) the pH-dependent Ca^{2+} binding occurs predominantly at the intravesicular sites, (b) the rate-limiting step in the uptake is accessibility to the binding sites, and (c) the rate-limiting step in the release of the bound ^{45}Ca or its exchange with ^{40}Ca is the membrane permeability to Ca^{2+} (*31*). Thus this binding is characterized by its pH dependence, high affinity for Ca^{2+}, large cooperativity, occurrence predominantly on the inside surface of the vesicles, and very rapid on–off rates. Similar binding has also been observed in smooth muscle PM preparations from rat fundus (*27*), vas deferens and mesenteric artery (unpublished), dog antrum (*33*), and microsomes from rabbit aortic smooth muscle (*29*). The magnitude of this binding, however, varies from 2–4 μmol/g protein in dog antrum to 20–30 μmol/g in rat myometrium. It is not possible to attribute the discrepancy in the high affinity binding in different tissues totally to differences in PM content of the membrane preparations because the PM content of the two preparations has been estimated to be 60–65% and ~99%, respectively (*23, 32*). The major unanswered questions concerning the pH-dependent Ca binding are the following: Which surface of the PM is responsible for the Ca^{2+} binding? Such a binding could be of paramount importance in regulation of cell function if it were to occur on the inside surface, which is exposed to low Ca^{2+} concentrations. What biochemical groups are responsible for this binding? Is this binding unique to smooth muscle or also present in other tissues? What is its physiological function? It has been suggested that this high affinity Ca^{2+} binding may form a large intracellular Ca^{2+} pool, and that the intracellular Ca^{2+} pool utilized for contraction in response to agonists like norepinephrine is a small component of this larger pool (*29*).

5. Ca^{2+} Release

Several studies have been conducted on the effects of various agents on the release of Ca^{2+} from previously loaded smooth muscle membrane vesicles. A brief summary of these studies is presented in Table 4. Typically, in these studies microsomal membranes are first loaded with $^{45}Ca^{2+}$ in the presence or absence of ATP, then diluted into media with or without EGTA along with the modulators and subsequently the amount of $^{45}Ca^{2+}$ retained by the membranes [$y(t)$] is determined at different time intervals. The difference between the Ca^{2+} uptake and $y(t)$ is the Ca^{2+} released or lost in that time period. In such experiments prostaglandins and oxytocin at high doses, bradykinin, and PCMB enhanced the Ca^{2+} release, aspaminol and papaverine reduced it, and TMB-6 had no effect (Table 4). Similar studies on the effects of various ions have also been conducted (review *17*). However, these studies yield no information on the intravesicular source or the mode of transmembraneous flux of the Ca^{2+} release. To un-

Table 4
Modulation of Ca^{2+} Release from Smooth Muscle Microsomes or PM

Modulator	Effect[a]	Tissue(s) and remarks	References
Prostaglandins	↑	PGE_2, $PGF_{1\beta}$, and PGF_2 in bovine uterus, but doses used very high, nonvascular smooth muscle	*11, 14*
Oxytocin	↑	Bovine uterus, dose used very high	*11*
Bradykinin	↑	Coronary artery	*71*
TMB-6	↔	Rat aorta marginal	*13*
PCMB	↑	Nonvascular smooth muscle	*14*
Aspaminol	↓	Rabbit taenia coli	*59*
Papaverine	↓	Rabbit taenia coli	*59*

[a]Note: The symbols are: ↑ activation, ↓ inhibition and ↔ none or marginal effects.

derstand this release in such experiments, let us consider the various Ca^{2+} pools involved and their modes of release. The Ca^{2+} uptake in presence of ATP, for instance, results in an intravesicular accumulation of Ca^{2+} that consists of free Ca^{2+} inside the vesicles and Ca^{2+} bound to the inner surface of the vesicles, the binding being in rapid equilibrium with, and thus dependent upon, intravesicular free Ca^{2+}. The extravesicular Ca^{2+} binding depends on Ca^{2+} concentration in the uptake medium and may also be regulated by intravesicular bound Ca^{2+}. Thus a loss of Ca^{2+} from such vesicles could correspond to loss of extravesicular bound Ca^{2+} or loss of intravesicular Ca^{2+}. Furthermore, modulation of loss of the intravesicular Ca^{2+} could result from changes in passive leak through the membrane vesicles or from changes in the conformation of the voltage-dependent Ca^{2+} channels. Thus, even though in theory, Ca^{2+} release studies could be used to obtain information on changes in membrane permeability, voltage-dependent Ca^{2+} channels, or in binding characteristics of the membranes, experimental work in this area is minimal perhaps owing to the unavailability of large amounts of well-characterized pure membrane preparations. Only recently a method has been suggested to analyze the efflux data using $y(t)$ and $y(\infty)$ values to get a better estimate of membrane permeability of the smooth muscle PM vesicles to Ca^{2+} (*30*). There have also been recent studies on binding of the Ca^{2+} antagonist, nitrendipine (Chapter 7), to membranes isolated from smooth muscles. However, the effect of this antagonist on Ca^{2+} fluxes in these membranes has not been reported.

6. Summary and Synthesis

Four types of membrane–Ca^{2+} interactions have been studied using various membrane preparations from smooth muscle: ATP-dependent Ca^{2+} transport, Na^+–Ca^{2+} exchange, passive Ca^{2+} binding, and Ca^{2+} release. The ATP-dependent Ca^{2+} transport system is located in mitochondria, PM, and possibly also in ER, and Na^+–Ca^{2+} exchange and a high affinity pH-dependent Ca^{2+} binding are localized in PM. To assess the possible relative importance of the various membrane Ca^{2+} interactions purely on the basis of the available studies on isolated membrane fractions, two questions should be asked. First, how fast can these processes occur, and second, what are the Ca^{2+} affinities of these interactions? Figure 2 answers these questions in part. Assuming that the critical Ca^{2+} concentration for the contraction–relaxation transition lies between 0.1 and 1 μM, it is apparent that the ATP-dependent Ca^{2+} transport by PM, the pH-dependent Ca^{2+} binding and the entry of Ca^{2+} into the cell, and in human, also the mitochondrial Ca^{2+} transport system are of prime importance. The role of Na^{+}–Ca^{2+} exchange in the contraction–relaxation cycle may be only supportive. However, in most areas a substantial amount of work remains to be conducted to understand mechanism and modulation of the various Ca^{2+} handling processes.

Acknowledgments

The author acknowledges his colleagues for their comments and suggestions. He is a Senior Research Fellow of the Ontario Heart Foundation.

References

1. Allen, J. C. Ca-binding properties of canine aortic microsomes: lack of effect of cAMP. *Blood Vessels* 14:91–104, 1977.
2. Batra, S. C. and E. E. Daniel. Effect of multivalent cations and drugs on Ca-uptake by the rat myometrial microsomes. *Comp. Biochem. Physiol. 38A*:285–300, 1971.
3. Batra, S. The role of mitochondrial Ca uptake in contraction and relaxation of the human myometrium. *Biochim. Biophys. Acta* 305:428–432, 1973.
4. Batra, S. The effect of some estrogens and progesterone on Ca-uptake and Ca release by myometrial mitochondria. *Biochem. Pharmacol.* 22:803–809, 1973.
5. Bhalla, R. C., R. C. Webb, D. Singh and T. Brock. Role of cAMP in rat aortic microsomal phosphorylation and Ca uptake. *Am. J. Physiol.* 244:H508–H514, 1978.
6. Blaustein, M. P. The interrelationship between Na and Ca fluxes across cell membranes. *Rev. Physiol. Biochem. Pharmacol.* 70:34–82,1974.
7. Brockbank, K. J. and P. J. England. A rapid method for the preparation of sarcolemma vesicles from rat aorta, and the stimulation of Ca-uptake into the vesicles by cAMP-dependent protein kinase. *FEBS Letters* 122:67–71, 1980.
8. Burt, J. M. Electrical and contractile consequences of Na^{+}/Ca^{2+}gradient reduction in cultured heart cells. *J. Mol. Cell Cardiol.* 14:99–110, 1982.
9. Caroni, P., L. Reinlib and E. Carafoli. Charge movements during the Na^{+}–Ca^{2+} exchange in heart sarcolemmal vesicles. *Proc. Nat. Acad. Sci.* 77:6354–6358, 1980.

10. Carsten, M. E. Prostaglandins and oxytocins: their effects on uterine smooth muscle. *Prostaglandins* 5:33–40, 1974.
11. Carsten, M. E. and J. D. Miller. Effects of prostaglandins and oxytocin on Ca release from a uterine microsomal fraction. *J. Biol. Chem.* 252:1576–1581, 1977.
12. Carsten, M. E. Ca accumulation by human uterine microsomal preparations. Effects of prostesterone and oxytocin. *Am. J. Obstet. Gynecol.* 133:598–601, 1979.
13. Chaturvedi, A. K., E. J. Landon, and B. V. Ramasastry. Influence of 6(N,N-diethylamine)–hexyl-3,4,5-trimethoxybenzoate on the responsiveness of aortae to norepinephrine and Ca movements in microsomes from spontaneously hypertensive and normotensive rats. *Pharmacol.* 17:315–322,1978.
14. Chaturvedi, A. K., S. Fox, B. V. Ramasastry, and E. J. Landon. Uptake and release of Ca by microsomes of non-vascular smooth muscle. *Biochem. Biophys. Res. Comm.* 88:1132–1139, 1979.
15. Clyman, R. I., V. C. Manganiello, and M. Vaughan. Oxygen: influence on cyclic nucleotides and Ca-sequestration in the human umbilical artery. *Chest* 715:247S–249S, 1977.
16. Crankshaw, D. J., R. A. Janis,and E. E. Daniel. The effects of Ca-antagonists on Ca-accumulation by subcellular fractions of rat myometrium. *Can. J. Physiol. Pharmacol.* 55:1028–1032, 1977.
17. Daniel, E. E., A. K. Grover, and C. Y. Kwan. Calcium. In *Handbook of Smooth Muscle,* N. L. Stephens (Ed.), in press.
18. Dubyak, G. R. and A. Scarpa. Sarcoplasmic Ca^{2+} transients during the contractile cycle of single barnacle muscle fibers: measurements with arsenazo III-injected fibers. *J. Muscle Res. and Cell Motility* 3:87–112, 1982.
19. Fitzpatrick, D. F. and A. Szentivanyi. Stimulation of Ca-uptake into aortic microsomes by a cAMP-dependent protein kinase. *Naunyn–Schmiedeberg's Arch. Pharmacol.* 298:255–257, 1977.
20. Ford, G. D. and M. L. Hess. Influence of ATP on SR function of vascular smooth muscle. *Am. J. Physiology* 242:C242–249, 1982.
21. Gilbert, J. R. and G. Meissner. Na–Ca exchange in skeletal muscle sarcolemmal vesicles. *J. Memb. Biol.* 69:77–84, 1982.
22. Grover, A. K., J. Crankshaw, R. E. Garfield and E. E. Daniel. Smooth muscle membrane vesicle orientation. *Can. J. Physiol. Pharmacol.* 58:1202–1211, 1980.
23. Grover, A. K., C. Y. Kwan, J. Crankshaw, D. J. Crankshaw, R. E. Garfield and E. E. Daniel. Characteristics of Ca transport and binding by rat myometrium plasma membrane subfractions. *Am. J. Physiol.* 239: c66–c74, 1980.
24. Grover, A. K. Smooth muscle contractile protein. *J. Sci. Ind. Res.* 40:798–803, 1981.
25. Grover, A. K., C. Y. Kwan, and E. E. Daniel. Na–Ca exchange in rat myometrium membrane vesicles highly enriched in plasma membranes. *Am. J. Physiol.* 240:c175–c182, 1981.
26. Grover, A. K., C. Y. Kwan, and E. E. Daniel. Ca^{2+}concentration dependence of Ca-uptake by rat myometrium plasma membrane enriched fraction. *Am. J. Physiol.* 242:c278–c282, 1982.

27. Grover, A. K., C. Y. Kwan, and E. E. Daniel. Transport of binding of Ca by a plasma membrane enriched fraction from rat gastric fundus smooth muscle. *Mol. Physiol.* 2:301–313, 1982.
28. Grover, A. K. and C. Y. Kwan. Oxalate stimulation of ATP-dependent Ca uptake is diminished during smooth muscle subcellular fractionation. *Life Sciences* 32:2655–2660, 1983.
29. Grover, A. K., J. Crankshaw, C. R. Triggle, and E. E. Daniel. Nature of norepinephrine-sensitive Ca-pool in rabbit aortic smooth muscle: effect of pH. *Life Sciences* 32:1553–1558, 1983.
30. Grover, A. K. Analysis of data on efflux of radioactive ions from isolated membrane vesicles. *Am. J. Physiol.*, (submitted).
31. Grover, A. K., C. Y. Kwan, and E. E. Daniel. High affinity pH-dependent passive Ca binding by rat myometrium plasma membrane enriched fraction. *Am. J. Physiol.* 244:C61–C67, 1983.
32. Grover, A. K., P. Oakes, C. Y. Kwan, and R. E. Garfield. Studies on canine gastric antrum smooth muscle. I. Preparation and characterization of plasma membrane enriched fraction. *Can. J. Physiol. Pharm.*, (submitted).
33. Grover, A. K. and S. N. Sipos. Studies on canine gastric antrum smooth muscle. II. Calcium transport and binding by a plasma membrane enriched fraction. *Can. J. Physiol. Pharm.*, (submitted).
34. Grover, A. K., C. Y. Kwan, P. K. Rangachari, and E. E. Daniel. Na–Ca exchange in a rat myometrium plasma membrane enriched fraction: further evidence and properties. *Am. J. Physiol.*, 244: C158–C165, 1983.
35. Hasselbach, W. The reversibility of the SR Ca pump. *Biochim. Biophys. Acta* 515:23–53, 1978.
36. Hidaka, H., T. Yamaki, M. Naka, T. Tamka, H. Hayashi, and R. Kobayashi. Calcium-regulated modulator protein interacting agents inhibit smooth muscle calcium stimulated protein kinase and ATPase. *Mol. Pharmacol.* 17:66–72, 1980.
37. Hogaboom, G. K. and J. S. Fedan. Calmodulin stimulation of Ca uptake and (Ca^{2+}–Mg^{2+}) ATPase activities in microsomal from canine tracheal smooth muscle. *Biochim. Biophys. Res. Comm.* 99:737–744, 1981.
38. Janis, R. A., D. J. Crankshaw, and E. E. Daniel. Control of intracellular Ca^{2+} activity in rat myometrium. *Am. J. Physiol.* 232:C50–C58, 1977.
39. Koike, K. and I. Takanayagi. Possible mechanisms of stimulatory action of papaverine on Ca-uptake by rat uterine microsomal fraction. *Jap. J. Pharmacol.* 31:757–762, 1981.
40. Kutsky, P. and F. R. Goodman. Calcium incorporation by canine aortic smooth muscle microsomes. *Arch. Pharmacol.* 231:4–20, 1978.
41. Kutsky, P., G. Weiss, and H. Karaki. Delineation of high and low affinity ^{45}Ca incorporation of ^{45}Ca-efflux in canine aortic smooth muscle microsomes. *Gen. Pharmac.* 11:475–481, 1980.
42. Kutsky, P. and G. B. Weiss. Specific alterations in binding and uptake of high and low affinity Ca^{2+} in canine aortic microsomes. *Blood Vessels* 19:53–64, 1982.
43. Kwan, C. Y. and T. Ramlal. On the inhibition of smooth muscle membrane ATPase by Na azide. *Biochem. Int.* 4:439–449, 1982.

44. Malmstrom, K. and E. Carafoli. The interaction of Ca^{2+} with mitochondria from human myometrium. *Arch. Biochem. Biophys.* 182:657–666, 1977.
45. Morel, N., M. Wibo, and T. Godfraind. A calmodulin stimulated Ca-pump in rat aorta plasma membranes. *Biochim. Biophys. Acta* 644:812–822, 1981.
46. Morel, N. and T. Godfraind. Na–Ca exchange in heart and smooth muscle microsomes. *Arch. Int. Pharmacodyn.* 258:319–321, 1982.
47. Nishikori, K., T. Takenaka, and H. Maeno. Stimulation of microsomal Ca uptake and protein phosphorylation by cAMP in rat uterus. *Mol. Pharmacol.* 13:671–678, 1977.
48. Nishikori, K. and H. Maeno. Close relationship between cAMP dependent endogenous phosphorylation of a specific protein and stimulation of Ca-uptake in rat uterus microsomes. *J. Biol. Chem.* 254:6099–6106, 1979.
49. Nishikori, K., T. Takenaka, and H. Maeno. A possible mechanism for relaxation of rat uterine smooth muscle by nicardipine hydrochloride (YC-93), a new potent vasodilator. *Jap. J. Pharmacol.* 31:701–709, 1981.
50. Philipson, K. D. and A. Y. Nishimoto. Na^{+}–Ca^{2+} exchange is affected by membrane potential in cardiac SL vesicles. *J. Biol. Chem.* 255:6880–6882, 1980.
51. Pitts, B. J. R. Stoichiometry of Na–Ca exchange in cardiac sarcolemmal vesicles. *J. Biol. Chem.* 254:6232–6235, 1979.
52. Raeymaekers, L., B. Agostini, and W. Hasselbach. Electron cytochemistry of oxalate stimulated Ca uptake in microsomes from smooth muscle of the pig stomach. *Histochem.* 65:121–129, 1980.
53. Raeymaekers, L. and W. Hasselbach. Ca-uptake, Ca ATPase activity, phosphoprotein formation and phosphate turnover in a microsomal fraction of smooth muscle. *Eur. J. Biochem.* 116:373–378, 1981.
54. Rangachari, P. K., M. G. Pernollet, and M. Worcel. Ca uptake by myometrial membranes. *Eur. J. Pharmacol.* 40:291–294, 1976.
55. Reeves, J. P. and J. L. Sutko. Na–Ca exchange generates a current in cardiac membrane vesicles. *Science* 208:1461–1464, 1980.
56. Sakai, Y., A. K. Grover, J. T. Fox, and E. E. Daniel. Uptake and release of Ca by canine gastric corpus smooth muscle plasma membrane enriched fraction. *Can. J. Physiol. Pharmacol.*, (submitted).
57. Sands, H. and J. Mascali. Effects of cAMP and of protein kinase on the Ca uptake by various tracheal smooth muscle organelles. *Arch. int. Pharmacodyn.* 236:180–191, 1978.
58. Tada, M., T. Yamamoto, and Y. Tonomura. Molecular Mechanism of active Ca transport by sarcoplasmic reticulum. *Physiol. Rev.* 58:1–79, 1978.
59. Takanayagi, I., T. Hisayama, and S. Suzuki. Effects of nonspecific smooth muscle relaxants and Ca-blocker on Ca-release and Ca-binding in microsomal fractions from rabbit taenia coli. *Jap. J. Pharmacol.* 30:641–646, 1980.
60. Takanayagi, I., K. Koike, and T. Hisayama. A dual action of papaverine on Ca uptake by microsomal fraction isolated from rat uterus. *Eur. J. Pharm.* 69:367–370, 1981.

61. Tate, C. A., A. Chu, J. McMillin-Wood, W. B. Van Winkle, and M. L. Entman. Evidence for a Ca-sensitive factor which alters the alkaline pH sensitivity of sarcoplasmic reticulum Ca transport. *J. Biol. Chem.* 256:2934–2939, 1981.
62. Thorens, S. and G. Haeusler. Effects of cAMP on Ca uptake and phosphorylation in membrane fractions of vascular smooth muscle. *Biochim. Biophys. Acta* 512:415–428, 1978.
63. Thorens, S. Ca-ATPase and Ca uptake without requirement for Mg in membrane fractions of vascular smooth muscle. *FEBS Letters* 98:177–180, 1979.
64. Van Breeman, C., P. Aaronson, and R. Loutzenhizer. Na–Ca interactions in mammalian smooth muscle. *Pharm. Rev.* 30:167–208, 1979.
65. Webb, R. C. and R. C. Bhalla. Ca sequestration by subcellular fractions isolated from vascular smooth muscle. *J. Mol. Cell Cardiol.* 8:145–157, 1976.
66. Wibo, M., N. Morel, and T. Godfraind. Differentiation of Ca-pumps linked to plasma membrane and endoplasmic reticulum in the microsomal fraction from intestinal smooth muscle. *Biochim. Biophys. Acta* 649:651–660, 1981.
67. Wuytack, F. and R. Casteels. Demonstration of a (Ca^{2+}–Mg^{2+})–ATPase activity probably related to Ca-transport in the microsomal fraction of porcine coronary artery smooth muscle. *Biochim. Biophys. Acta* 595:257–263, 1980.
68. Wuytack, F., G. De Shutter, and R. Casteels. The effect of calmodulin on the active Ca-ion transport and (Ca^{2+}–Mg^{2+})-dependent ATPase in microsomal fraction of smooth muscle compared with that of erythrocytes and cardiac muscle. *Biochem. J.* 190:827–831, 1980.
69. Wuytack, F., G. De Schutter, and R. Casteels. Partial purification of Ca-Mg-dependent ATPase from pig smooth muscle and reconstitution of ATP-dependent Ca-transport system. *Biochem. J.* 198:265–271, 1981.
70. Wuytack, F., G. De Schutter, and R. Casteels. Purification of Ca-Mg ATPase from smooth muscle by calmodulin affinity chromatography. *FEBS Letters* 129:297–300, 1981.
71. Wuytack, F., G. De Schutter, J. Verbist, and R. Casteels. Antibodies to the calmodulin-binding $Ca^{2\pm}$transport ATPase from smooth muscle. *FEBS Letters* 154:191–195, 1983.
72. Zelck, V., L. Konya, and E. Albrecht. ATP-dependent Ca-uptake by microsomal fraction of pig coronary artery and its dependence on bradykinin and angiotensin. *Acta Biol. Med. Germ.* 32:K1–K5, 1974.
73. Zelck, V. and V. W. E. Karnstedt. ATP-dependent Ca-binding by mitochondrial and microsomal fractions isolated from different smooth muscle. In *Excitation-Contraction Coupling.* R. Casteels (Ed.), Elsevier/North Holland Biomed. Press, pp. 171–180, 1977.

Chapter 11

Receptor Binding Studies on Smooth Muscle Subcellular Fractions

D. J. Crankshaw

1. Introduction

Major advances have been made in the field of radiochemistry during the past twenty years that have allowed the development of a large number of receptor ligands radiolabeled to high specific activity. Indeed growth is so rapid at the present time that it is difficult enough to keep abreast of the radiochemical catalogs, let alone the literature generated by the use of this ever-increasing number of compounds.

Radiolabeled agonists and antagonists have become powerful weapons in the struggle to understand a number of problems related to smooth muscle function; however, like all such commodities they carry a great potential for abuse as well as for use.

Thus, despite the fact that the major problems encountered in the design, practice, and interpretation of receptor binding studies have been adequately discussed in a large number of articles (*17, 38, 39, 49, 57, 75, 78*), many investigators tend to repeat avoidable mistakes, some of which find their way into the literature. The reader wishing to embark on receptor binding studies would be well advised to consult these excellent resources before setting sail. Some rather basic requirements that must be met before "ligand binding" can be considered synonymous with "receptor binding" have become widely recognized; among these are that binding be a saturable phenomenon (since there is only a finite number of receptors), that binding show the stereospecificity demonstrated by receptor interactions, and that binding show the same selectivity among analogs as the receptor-mediated response, i.e., agents inactive at the receptor in question ought not to bind. Where appropriate, there should be agreement between kinetic properties (such as reversibility, time course, and affinity) as deduced from binding experiments and from functional experiments. The purpose of this chapter will not be to reiterate these points, but to address the question: what can receptor binding experiments contribute to an understanding of smooth muscle function?

Receptor binding experiments have been performed on whole tissues, on subcellular fractions ranging in purity from the crudest homogenate to the highly purified membrane preparation, and on dispersed or cultured, intact smooth muscle cells. It is the purpose of this chapter to deal with the middle category that accounts for the bulk of the literature in this area.

2. Significance of Receptor Binding Studies to the Understanding of Smooth Muscle Function

Interpretation of the classical muscle-bath experiments for analyses of dose–response curves and of drug antagonism are in many instances confounded by problems of drug distribution and biotransformation within the test tissues (*46, 75*) as well as by the fact that some agents may have multiple sites of action within a given tissue. Even if these variables can be accounted for, these classical experiments provide no information on the relationship between receptor occupancy and biological response. Radioligand binding experiments can provide the necessary information concerning receptor occupancy that can be used in conjunction with the results from the other types of experiments to establish this relationship. Several important processes can be investigated using the receptor binding technique; the major ones are listed in Table 1.

It is almost impossible to study the effects of thermodynamic and kinetic manipulations on drug/receptor interactions when the index of the binding process is a contractile response. However, when the binding process can be measured directly without reli-

Table 1
Processes That Can Be Investigated Using Receptor Binding Studies as Tools in Smooth Muscle

1.	The binding mechanism between drug and receptor
2.	Receptor–response coupling
3.	Receptor isolation
4.	Receptor distribution
5.	The mechanism of physiological or pathological change in sensitivity

ance on downstream events, such investigations become possible.

One of the major questions in smooth muscle research that is addressed frequently in this book, concerns the mechanisms by which receptor occupancy produces a smooth muscle response. The immediate postreceptor events that lead to smooth muscle excitation or inhibition are in many instances still poorly understood. One approach to solving this problem suggests that the molecular processes involved in immediate postreceptor events could be more easily identified and quantitated in a cell-free system where other interfering processes have been eliminated. Such has proved to be the case for a number of nonsmooth muscle systems, e.g., the effect of acetylcholine on the electric organ of the electric eel (*43–45*) and the effect of neurohypophyseal hormones on the beef renal medulla (*11, 12, 59*). Claims for similar demonstrations in smooth muscle (*15*) have not been substantiated and have been criticized for several reasons (*26*). Such experiments are fraught with difficulties, and several possibilities could account for the inability to demonstrate a hypothetical response in a cell-free preparation. One such possibility is that appropriate receptors have lost their discriminatory properties or have been rendered inaccessible by the isolation process. By performing receptor binding experiments on the same preparations that are used to study immediate postreceptor events, such possibilities can be eliminated. Furthermore, the signal-to-noise ratio for any immediate postreceptor event can be increased by choosing to study a fraction that has been shown to be enriched in the receptor of interest.

Ultimately, if one wishes to isolate the receptor and to study its behavior in isolation, its purification can be followed by its ability to bind a radioligand appropriately.

In tissues where multiple receptor subtypes occur, careful binding experiments using appropriate selective ligands combined with optimal analysis techniques may be able to resolve the relative proportions of receptor subtypes in a way that classical methods cannot.

A number of physiological, pathological, and developmental processes that occur in smooth muscle are, theoretically, amenable to more detailed study by the use of radioligand binding techniques. For instance, the variation in sensitivity of the uterus to a number of excitatory and inhibitory substances with gestation or steroid hormone conditions has recently been investigated by a number of groups (see below) with an aim to understand the role of changes in receptor numbers in regulating these processes. Receptor binding studies in pathological situations that lead to desensitization or to supersensitivity of smooth muscles (e.g., some forms of hypertension) may give clues to the underlying mechanisms of the sensitivity changes.

3. Limitations of Receptor Binding Studies

Apart from the problem alluded to in the introduction to this chapter, i.e., that receptor binding studies must be performed properly in order to produce meaningful results, there are a number of limitations to the usefulness of such studies.

Since the turn of the century and the work of J. N. Langley (*50*) and Paul Ehrlich (*28*) on widely divergent biological systems, the experimental pharmacologist has been aware of the existence of the entity that we now refer to as a receptor. Careful quantitative study has yielded a wealth of information concerning the receptor—note particularly the amount of information available concerning the "analgesic" receptor before the discovery of the endogenous ligand for this receptor, let alone the availability of labeled probes (*6*)—and many pharmacologists express the view that costly radioisotope techniques are not needed to demonstrate the presence of receptors. This view elegantly expresses the major drawback to receptor binding studies—their misuse.

Among those of us who are too clumsy to work with microelectrodes, too pedestrian to develop such techniques as that employing skinned smooth muscle fibers, or too indolent to master the electron microscope, there are few who are unable to acquire the simple skills of homogenization, centrifugation, and throwing in some ligand to see what happens. The consequence of this is a plethora of papers that have little meaning. Apart from those few studies that are concerned specifically with the mechanisms by which ligands bind to their receptors, certainly an important aspect of smooth muscle pharmacology (see below), receptor binding experiments are of little value when performed for their own sake. Thus determination of the relationship between receptor occupancy and response requires not only the receptor binding study, but also the classical muscle-bath experiments to complement it. In such a case, however, the meaning of the final result will be dependent upon restrictions on the interpretation of classical studies mentioned above (*46, 75*). Similarly, experiments concerned with physiological and pathological changes in tissue response require more than simple binding data before adequate explanations of underlying mechanisms can be arrived at. In 1972 Rudinger et al. (*65*) succinctly described a number of phases in the action of a drug while dealing specifically with the peptide hormone, oxytocin. These phases are seen to include permeation of the drug into the tissue, inactivation of the drug by the tissue, initiation of a stimulus by the drug receptor complex, and development of a response as a result of the stimulus; all these are in addition to the binding of the drug to its receptors. It will be interesting to see what exciting adaptive mechanisms such as changes in the efficiency of receptor stimulus coupling are discovered when the current penchant for ascribing changes in tissue sensitivity to those awkward terms "up-regulation" and "down-regulation" subsides and the discrepancies between binding data and changes in responsiveness are analyzed more thoroughly. The terms up- and down-regulation have been coined to denote mechanisms by which the total number of receptors in a cell or organ are either increased or decreased, respectively.

4. Studies of Specific Receptor Systems in Smooth Muscle

Although structures said to be composed of smooth muscle are united in having some similar ultrastructural, biochemical, and biophysical properties, there is tremendous diversity of function within this class. As a consequence of this diversity of function, there is also a diversity of agents for which receptors are thought to exist in smooth muscle, and it is not possible to discuss every one of these adequately. Rather, a few examples have been chosen either because they have been more extensively studied or because they are considered to be of greater physiological and pharmacological interest. Similarly all studies performed on a given receptor system have not been included; two biases have been responsible for the selection discussed. The first bias represents the author's particular interest, while the second reflects his belief that readers will not wish to be directed to the large number of poorly conceived and carried out studies that he has read while preparing this chapter. As far as possible only studies that have attempted to establish the usefulness of the ligand in smooth muscle, or to investigate some of the processes listed in Table 1, or that illustrate some pertinent point have been included. It will be noted that there are relatively few reports that combine receptor binding studies with more classical pharmacological experiments in such a way as to maximize the power of the receptor binding technique.

4.1. The Muscarinic Cholinoceptor

Questions concerning the mechanisms by which both the sympathetic and parasympathetic nervous systems regulate smooth muscle function have played a central role in smooth muscle physiology and pharmacology since the very beginning of these disciplines. It is, therefore, not surprising that one of the earliest successful receptor binding reports on a smooth muscle should be concerned with the labeling of the muscarinic cholinoceptor. Yamamura and Snyder (*82*) described the use of the potent muscarinic antagonist, [^{3}H]quinuclidinyl benzilate (QNB), as a probe for the muscarinic cholinoceptor in guinea pig ileal longitudinal muscle. Binding of QNB to homogenates of this tissue was found to be a saturable and reversible process that could be inhibited by a number of muscarinic agonists and antagonists, but not by nicotinic or noncholinergic drugs. The apparent dissociation constant (K_d) was found to be 0.5 n*M* by saturation analysis and 0.03 n*M* by kinetic analysis, which represents a significant disagreement between the two approaches. The ability of muscarinic drugs to influence QNB binding was compared with their effects on smooth muscle contraction. Good agreement was found even though effects on binding were expressed as the concentration producing 50% inhibition of binding (ID_{50}) rather than as apparent inhibition constant (K_i) (*17*) and pharmacological data were obtained from the literature, in some cases from tissues other than the guinea pig ileal longitudinal muscle. In its conception the work of Aronstam et al. (*5*) represented a major advance in the way receptor binding studies were used and still ranks as an outstanding contribution today. The ability of stereoisomers of muscarinic agonists and antagonists to inhibit the binding of QNB to a microsomal fraction from guinea ileal longitudinal muscle (expressed as pK_i) were compared with their ability to stimulate or inhibit contraction of the muscle (expressed as pD_2 and pA_2). Values of pA_2 and pD_2 were either obtained from the author's previous work on the same muscle preparation, or generated specifically for this report. The binding of QNB to rat brain was also investigated. A total of 20 agents active at the muscarinic cholinoceptor were studied and care was taken to protect agents susceptible to breakdown by cholinesterase by treatment of membranes with diisopropyl fluorophosphate that by itself did not influence QNB binding. Unlike most receptor binding studies in which K_i is calculated from inhibition curves obtained using a fixed concentration of labeled ligand and various concentrations of competitor, K_i in these experiments was calculated from double reciprocal plots obtained at various concentrations of QNB in the presence and absence of a single concentration of competitor.

By this technique the K_d for QNB binding to the muscle membranes was found to be 0.2 nM. For antagonists and weak agonists there was a close correlation between pK_i and pA_2 or pD_2. For strong agonists however, the pD_2 was consistently higher than the pK_i and this was taken to indicate a large spare receptor population for these compounds. In fact it was calculated that the strong agonists occupy only 0.5–5.3% of the putative muscarinic receptors when present at their pD_2 concentrations. Although similar stereospecificities were found for binding and biological activity of isomers of atropine and acetyl-beta-methylcholine, two *cis*-isomers of C_2-substituted 1,3,-dioxolanes showed increased biological activity without change in binding affinity. The possibility that these compounds might produce their biological effects at sites structurally different from the QNB binding site was raised. Furthermore, it was pointed out that all conclusions were drawn on the assumption that all binding sites demonstrated the same affinity for all ligands. However, in brain, muscarinic cholinoceptors appeared to exist in at least two populations that differed in their affinity for agonist, but not antagonists (*9*).

With respect to the possible heterogeneity of muscarinic cholinoceptor sites in smooth muscle, Beld et al. (*7*) examined the inhibition of [^{3}H]dexetimide binding to bovine trachael smooth muscle membranes by a series of methylfurthrethornium derivatives having activities ranging from antagonist to partial agonist to full agonist. They found antagonists to produce "normal" sigmoid inhibition curves whereas agonists produced inhibition curves that were much flatter. When the data were analyzed using computerized nonlinear least squares methods (*32*), they were found to be consistent with the hypothesis that there are two noninterconverting muscarinic binding sites in trachael smooth muscle for which agonists have different but antagonists have the same affinities. The affinity ratio for the two sites was found to decrease from 61 for a potent agonist to 1 for antagonists in parallel with decreasing intrinsic activity. Unfortunately values for intrinsic activity were extrapolated from data obtained on other smooth muscle preparations, which may be dangerous if different preparations have different amounts of receptor reserve. Beld et al., however, argue that the whole concept of receptor reserve is based on the assumption of a simple bimolecular reaction between drug and receptor. They claim that the binding results can be explained by a trimolecular reaction involving drug, receptor, and either a regulatory protein or an effector molecule (*40*).

The binding processes of two muscarinic antagonists, QNB and *N*-methyl-4-piperidinylbenzilate (4-NMPB), have been investigated on crude microsomal membranes from rat small intestine (*41*). In each case, the first-order rate constant for association was dependent upon the concentration of agonist in a hyperbolic manner, suggesting that the interaction involves two equilibria. The first, rapid binding equilibrium, is followed by a slower isomerization of the receptor antagonist complex such that

$$R + A \underset{K_{-1}}{\overset{K_1}{\rightleftharpoons}} AR \underset{K_{-2}}{\overset{K_2}{\rightleftharpoons}} AR^* \qquad (1)$$

where R is the receptor, A is the antagonist, and AR and AR* are two different forms of the antagonist receptor complex. The K_d values for the two antagonists calculated from kinetic data on the basis of this model were in good agreement with the values obtained from equilibrium binding studies. The authors pointed out that the isomerization is a monomolecular reaction and so cannot be shifted completely by addition of excess antagonist. As a result, a proportion of the receptor that is present as the fast-dissociating complex will always escape detection in filtration assays, and will be independent of antagonist concentration. Thus measurements of receptor number using these ligands and the filtration system will be in error.

Further investigations using 4-NMPB and rat intestinal membranes were directed at the kinetics of association and dissociation (*42*). Addition of neither agonist nor antagonist was found to influence K_{-2} in Eq. (1). The antagonist atropine caused a large increase in K_{-1}/K_1 whereas agonists had no effect. The result indicated that competition between antagonists could be explained on the basis of competition for the free receptor while competition between antagonist and agonist could not fit this simple model. Although the authors were unable to propose a plausible model to account for all the observations regarding agonist–antagonist competition in smooth muscle, they felt that their results indicated that the classical model of competition between agonists and antagonists for the free receptor is not applicable to the muscarinic receptor from smooth muscle.

Competition studies between carbamylcholine and QNB on membranes from the guinea pig ileal longitudinal muscle gave results consistent with the hypothesis that the muscarinic receptor in smooth muscle exists in two agonist affinity states (*4*). Treatment of membranes with *N*-ethylmaleimide (NEM) had no effect on antagonist binding, but increased the apparent affinity for carbamylcholine in a manner suggestive of the conversion of sites from the low to the high affinity states. When longitudinal muscle strips were exposed to NEM, the dose–response curves to acetylcholine were shifted to the right, whereas barium curves were unaffected. The suggestion was made (*4*) that NEM may uncouple the receptor macromolecules from the structures involved in receptor response coupling, which in turn could possibly lead to an increase in binding energy and an apparent increase in affinity (*14*). These experiments could not detect the existence of putative superhigh affinity receptors (*10*) in the smooth muscle because of the adverse ratio of receptor concentration to dissociation constant for QNB and the large decrease in free QNB concentration resulting from tissue binding (*17*).

Binding of QNB to homogenates of another smooth muscle, the lower esophageal sphincter of the cat, has been characterized (*62*) and shown to be rapid, reversible, saturable, linear with protein concentration used, and stereospecific. Seventeen muscarinic compounds were tested for their ability to inhibit QNB binding along with a number of agents thought to act at other receptors. Only a small number of compounds were compared for their activity in biological and binding assays and once again ID_{50} was used instead of K_i; nevertheless biological data were obtained on the same tissue as the binding data. A discrepancy between the K_d for QNB obtained by saturation analysis (1 nM) and by kinetic analysis (0.08 nM) was again noted. The novel aspect of this study was the use of the ligand binding technique to map the distribution of receptors within a tissue. Thus QNB binding using a single concentration of 1.5 n*M* was measured at eight different regions along the esophagus. Distribution of binding was found to correspond to the regional variation in response to muscarinic agonists and to parallel the relative amounts of smooth muscle present.

This observation raises the important question of how to express and interpret data concerning regional distribution of receptors. In the case just quoted (*62*), data were expressed per milligram of homogenate protein and should not be extrapolated to indicate any more than that. If there are changes in cell types or in the distribution of cellular to extracellular proteins, or even in the amount of protein per cell, the results with respect to the number of receptors per cell of interest will be confused.

A study that raises a similar problem concerns the loss of cholingeric sensitivity of the expansor secondariorum muscle of the chick during development (*8*). In control animals sensitivity to muscarinic stimulation was shown to be lost in parallel with QNB binding (measured at a single concentration). Destruction of the adrenergic nerve supply prevented the loss of biological and binding activities in young chicks and restored these activities in older chicks. Although binding data were expressed per milligram of protein, the authors were concerned that developmental changes in the muscle might make the different populations not strictly comparable on this basis.

There is a number of reports of attempts to investigate the role of receptor regulation by neuronal activity (*67*) or by homologous hormone (*53*) using the muscarinic cholinoceptor in smooth muscle as a model, and a number of the more rigorous of these will be discussed briefly.

Two interesting reports from Higuchi and coworkers (*34, 35*) have demonstrated that a reduction in the total number of QNB binding sites occurs when guinea pig vas deferens is cultured with muscarinic agonists. This reduction can be blocked by simultaneous incubation with a muscarinic antagonist. The rate at which an agonist reduces the number of binding sites seems to correlate with its efficacy in promoting contraction. Neither norepinephrine nor high potassium ion concentration could produce the changes in QNB binding and response to muscarinic agonists. Drugs considered to be antiendocytotic (e.g., monodansylcadaverine) or antimicrotubular (e.g., vinblastine), but not those considered to be antimicrofilament (e.g., cytocholasin B) or lysosomal enzyme inhibitors (e.g., protease inhibitors), reduced the loss of QNB binding sites on long term exposure to muscarinic agonists. Interestingly, when chloroquine or vinblastine were used to inhibit the loss of QNB binding sites, effects were found on the responses to muscarinic agonists before changes in binding could be detected, even though responses to other contractile agents were unimpaired. The suggestion that the receptor first becomes uncoupled from the stimulus-generating mechanism was made.

Several workers have examined the effect of acute or chronic administration of acetylcholinesterase inhibitors on the number of putative muscarinic receptors in smooth muscle using ligand binding to isolated membranes as a tool. Among these Dawson and Jarrot (*27*) found no effect of 8 d of exposure to carbamate or organophosphate on the number of QNB binding sites in guinea pig ileum; other reports (*19, 20*) claim a reduction in both the number of binding sites and in the K_d for QNB in intestinal muscle from other species, which can be prevented by administration of methylatropine concurrently with the cholinesterase inhibitor.

In work of relevence to the experimental pharmacologist and of some pertinence to the question of receptor reserve, Siegel and Triggle (*66*) examined the effect of the irreversible muscarinic antagonist benzilylcholine mustard (BCM) on both the contractile response of guinea pig ileal longitudinal muscle to muscarinic stimulation and the number of QNB binding sites present in this tissue. Increasing concentrations of BCM produced shifts and subsequent depression of dose response curves without coincident decreases in the maximum binding capacities. Several explanations can be offered for these results, including the problems of agonist and antagonist diffusion into the center of muscle strips used for contractile experiments and the possibility that BCM interacts with a population of receptors not labeled by QNB. Whatever the correct explanation is, the authors point out that the practice of using irreversible antagonists to measure the true dissociation constant for agonists in a system with substantial receptor reserve is made suspect by these results.

4.2. The Alpha-Adrenoceptor

Studies on the binding of ligands to the alpha-adrenoceptor in smooth muscles are somewhat complicated by the fact that work was well underway before the results of classical experiments suggesting that subtypes of alpha-adrenoceptors exist were fully appreciated. Chronologically, therefore, the work falls into two categories, although willingness to embrace the concept of receptor subtypes has not occurred in all minds simultaneously.

The first rigorous examination of the usefulness of a radioligand for labeling alpha-adrenoceptors in smooth muscle came from the group of Lefkowitz (*80*) that has been a major contributor to the present understanding of adrenoceptor mechanisms in a number of systems. The agent, [^{3}H]dihydroergocryptine (DHE), a potent alpha-adrenoceptor antagonist, was studied for its ability to bind to a crude microsomal fraction from rabbit uterus. Binding was rapid, reversible, and saturable with the K_d measured at equilibrium being 8–10 n*M*, whereas that measured kinetically was 3 n*M*. Binding appeared to occur in a noncooperative manner to a single class of sites. A large number of compounds active at adrenoceptors and other sites was studied for their ability to compete with DHE. The K_i was calculated for seventeen phenylethylamine compounds and in most cases was compared with the potency relative to (−)epinephrine in producing contraction. Five out of eleven comparisons used data from tissues other than rabbit uterus. There was reasonable agreement between a compound's effectiveness at the alpha-adrenoceptor and its ability to inhibit DHE binding, although dopamine, histamine, and serotonin all showed some affinity for the sites. The binding of DHE was also shown to be stereospecific. The authors considered the uterine binding site for DHE to have characteristics indistinguishable from the alpha-adrenoceptor in this tissue. Apart from differences in K_d, essentially similar results have been reported for the use of this ligand in dog aorta (*76*) and guinea-pig vas deferens (*37*). In the latter tissue, serotonin and histamine were not found to compete significantly with DHE and there was a considerable discrepancy between K_i and the "physiological K_d" calculated from the dose ratio at a single concentration of antagonist for two imidazoline compounds.

In the brain (*71*) and in platelets (*58*) DHE certainly seems to bind to other than the alpha-adrenoceptor and this may also be the case for some smooth muscles. It will be incumbent upon investigators claiming to use this agent to count alpha-adrenoceptors in smooth muscles to demonstrate that they are indeed counting nothing else. DHE has been used as a tool to examine the distribution of adrenoceptors in the rabbit urinary bladder (*52*)—this will be discussed in the next section.

Attempts to resolve alpha-adrenoceptors in smooth muscle into the alpha$_1$ and alpha$_2$ subtypes were first reported on rabbit uterine membranes (*51*). Binding of the putative alpha$_1$ and alpha$_2$ selective antagonists, [^{3}H]prazosin and [^{3}H]yohimbine, were studied in parallel with the standard nonselective DHE. Two approaches were used to examine the subtype selectivity of twelve pharmacologically active agents. In the first case complex competition curves between the agent under test and DHE were analyzed using nonlinear least squares methods (*32*) assuming a two-site model. In the second case, the K_i was calculated from a uniphasic competition curve obtained against either of the labeled selective antagonists. There was good correlation between the two techniques, but the use of selective radioligands was considered to offer several advantages, the most striking being that a competitor must show a 100-fold selectivity between two sites before the respective affinity can be accurately determined using the DHE approach. In contrast much smaller degrees of selectivity could be reliably measured using the selective ligands without the need for complex computer analysis. Saturation curves with the different ligands revealed 19 fmol of prazosin sites/mg protein with a K_d of 0.5 n*M*, 72 fmol of yohimbine sites with a K_d of 11 n*M*, and 167 fmol of DHE sites with a K_d of 3 n*M*. The number of sites determined by selective agents does not add up to the number determined by DHE. Several points that have relevance to any kind of binding study were stressed, namely, that the sensitivity of the methods is dependent upon the proportion of receptor subtypes present, the number of data points obtained, and the level of nonspecific binding; to this should also be added density of binding sites and the specific radioactivity of the ligand.

The phenomenon of "heterologous receptor regulation" (a term used to describe the regulation of one type of receptor by an agent or agents that operate via a different type of receptor) in the myometrium by estrogen and progesterone has produced a considerable number of papers. The alpha-adrenoceptor has been the topic of a number of these. The primary demonstration (*79*) was that progesterone treatment significantly reduced (threefold) the number of DHE binding sites, but not the K_d for this ligand in estrogen pretreated rabbits. Data were expressed per milligram of protein. Essentially similar results were obtained in another study where results were expressed per mg of DNA (*63*). The response of uterine strips from estrogen-treated rabbits to norepinephrine was a phentolamine-sensitive contraction, whereas the response of this tissue from subsequently progesterone-treated animals was a propranolol-sensitive inhibition (*63*). Similar techniques were used to show an increase in the number of DHE binding sites in immature rabbit myometrium after 24 h in organ culture (*18*). Cultured myometrium was more sensitive than noncultured myometrium to norepinephrine and to methoxamine, but not to acetylcholine, as demonstrated by changes in dose–response relationships. The presence of 1 μM norepinephrine in the culture medium prevented the increase in DHE binding sites whereas the administration of hexamethonium to intact rabbits (which would be expected to reduce sympathetic activity and thus present less norepinephrine to the muscle cells in vivo) resulted in a twofold increase in the number of binding sites when compared to controls. The mechanism by which alpha-adrenoceptor numbers seemed to be repressed in vivo could not be explained.

In one of the better conceived and carried out radioligand binding studies on smooth muscle, Roberts et al. (*64*) demonstrated that the estrogen-induced increase in DHE binding in rabbit myometrium was accompanied by an increase in sensitivity of uterine muscle strips to norepinephrine and methoxamine, but not to acetylcholine. Contributions of changes in activity of amine uptake mechanisms, the activity of catechol-*O*-methyltransferase or of the quantity of contractile proteins were ruled out as playing a significant part in the sensitivity change. Estrogen resulted in an increase in DHE binding sites whether results were expressed on the basis of protein, DNA, 5′-nucleotidase or fluoride-stimulated adenylate cyclase activity.

It is somewhat ironic in view of the many mediocre reports of receptor binding experiments in smooth muscle that the results of such a well-conducted study should be contradicted by those of another. However, Hoffman et al. (*36*) found no increase in the sensitivity of strips of rabbit uterus to norepinephrine or phenylephrine in the presence of beta-adrenoceptor and neuronal-uptake blockade. The estrogen-induced increase in binding sites for DHE was ascribed to a selective increase in the yohimbine-displaceable component (putative alpha$_2$-adrenoceptors). Alpha$_2$-adrenoceptors were not considered to contribute to the catecholamine-induced contraction of rabbit uterine strips since Schild plots of both prazosin and yohimbine against either phenylephrine or norepinephrine were linear and produced pA_2 values to be expected if competition were taking place at alpha$_1$-adrenoceptors. In the worst case there was a fivefold discrepancy between the K_i for prazosin at the prazosin-sensitive DHE binding site and the pA_2 for prazosin and a tenfold discrepancy between the K_i for yohimbine at the prazosin-sensitive DHE binding site and the pA_2 for yohimbine. Alpha$_2$-adrenoceptors in the rabbit's uterus were considered to be neuronal since treatment of intact animals with 6-hydroxydopamine for 2 d resulted in a 50% decrease in the number of yohimbine-sensitive DHE sites measurable on day 4 with little change in the number of prazosin-sensitive sites. The marked discrepancy between these two studies (*36, 64*) could not be attributed to differences in the age of rabbits used (*36*), and no convincing explanations have been offered.

In pregnant rabbit myometrium the number of DHE binding sites was found to decline (2.4-fold) during the second week of gestation, but to progressively return to prepregnancy levels at term (*77*). The response of the myometrium of these rabbits to alpha-adrenoceptor stimulation was not presented.

4.3. The Beta-Adrenoceptor

The consequent advantages of the well-established link between beta-adrenoceptor activation and the stimulation of adenylate cyclase for the study of the beta-adrenoceptor have not been exploited in smooth muscle as they have in other types of tissue [*see* ref. (*56*)] using receptor binding technique. Well-conceived and conducted binding studies on the beta-adrenoceptor in smooth muscle are few. Although a number of studies has been conducted on lung, which undoubtedly contains smooth muscle, the heterogeneity of cell types makes conclusions with regard to the smooth muscle component impossible and these will not be considered here.

The appropriateness of a radioligand for quantitating beta-adrenoceptors has not been established as rigorously as for the previous two classes of receptors. A study that comes close to doing this examined the characteristics of the beta-adrenoceptor antagonist [^{3}H]dihydroalprenolol (DHA) binding to a microsomal fraction from human myometrium (*33*). Binding was rapid, reversible, stereospecific and of high affinity (K_d being approximately 0.5 n*M* by both equilibrium and kinetic methods). Three beta-adrenoceptor agonists competed with DHA in a manner that suggested that binding was to a beta$_2$-adrenoceptor. Competition between DHA and the putative beta$_2$-selective agonist Zinterol was analyzed by nonlinear least squares curve-fitting techniques (*32*) and suggested an 87% beta$_2$, 13% beta$_1$ distribution of DHA binding sites. No functional data were presented. Similar observations of rapid, reversible, and stereospecific binding were obtained for the beta-adrenoceptor antagonist [^{125}I]iodohydr oxybenzylpindolol (IHYP) in rat mesenteric artery membranes (*81*). The K_d for this agent was 0.12 n*M* by equilibrium and 0.08 n*M* by kinetic analysis. There was good agreement between the K_i and the concentration producing 50% activation of adenylate cyclase for three agonists. Although antagonists were examined in binding assays, they were not studied in cyclase assays. Binding of IHYP had to be performed in the presence of 0.1 m*M* phentolamine in order to show stereospecificity.

Unlike DHE binding sites, the number of binding sites for IHYP was not found to alter in the rabbit's myometrium as a result of estrogen or progesterone treatment (*63*). Nor were binding sites for DHA found to increase as a result of organ culture. The predominance of beta-adrenoceptor-mediated responses in myometria from progesterone-treated rabbits could not be explained on the basis of changes in receptor numbers.

There are reports of decreased numbers of beta-adrenoceptors in blood vessels from hypertensive animals without functional studies to indicate the meaning of such observations (*54, 81*).

The alpha-adrenoceptor ligand DHE and the beta-adrenoceptor ligand DHA have been used to quantitate the distribution of adrenoceptors in rabbit urinary bladder (*52*). There were more alpha-receptors in the base (78 fmol/mg) than the body (18 fmol/mg) and more beta-receptors in the body (96 fmol/mg) than the base (42 fmol/mg). The distribution of receptors suggested by the ligand-binding experiments correlated well with the response of isolated strips of the two regions to methoxamine, isoproterenol, and epinephrine. It was concluded that the difference in adrenoceptor-mediated responses of the two regions resulted from the predominance of alpha-receptors in the base and beta-receptors in the body. The rather neat correlation observed here could be fortuitous since firstly, it is unknown how many of DHE binding sites represent pre- and how many represent postsynaptic receptors. Secondly, there is no evidence to indicate that the efficiency of coupling between the two types of receptor and the stimulus-generating mechanism is the same. Consequently markedly different numbers DHE and DHA binding sites may be found in a situation where alpha- and beta-adrenoceptor-mediated responses are approximately equivalent. In fact the myometrium of the pregnant ewe has six times as many DHE binding sites than DHA binding sites (*21*), although an inhibitory beta-adrenoceptor-mediated response predominates (*22*).

4.4. Peptide Receptors

The ever-increasing array of peptides that appear to influence smooth muscle function (*see* Chapter 16) has posed a number of questions for receptor binding studies to tackle. Unfortunately peptides present a number of problems that, although not unique to this class of compounds, have proved less amenable to solution than with other compounds.

Firstly, proteases and peptidases are ubiquitous in tissues and subcellular preparations. An early observation with many peptides is that the ligand is destroyed during binding assay incubations, thus making equilibrium studies difficult. A common approach at this stage is to set about to find the appropriate peptidase inhibitor to include in the assay medium to prevent degradation. Such experiments are then saddled with the attendant problem of interpreting the role of the inhibitor in the observations made.

Peptides are difficult and costly to tritiate and purify as the sparsity of such compounds in the radiochemical catalogs will attest. However, peptides containing tyrosine or phenylalanine residues can be easily iodinated and in many cases iodinated compounds may be preferable to tritiated ones since despite the reduction in isotopic half-life the specific radioactivity that can be attained is much greater. Unfortunately, the introduction of iodine sometimes eliminates (*29*) and often reduces biological activity. The latter situation is not always insoluble, but care must be taken when using an iodinated peptide to represent a native peptide even if the difference in biological activities is very small since the two forms cannot be regarded as a single species. If a peptide cannot be tritiated and it contains no tyrosine or phenylalanine, other strategies for labeling must be adopted (*13*).

Partly as a result of the problems outlined, few adequate receptor binding studies have been performed on subcellular fractions from smooth muscle. In order to illustrate how receptor binding studies can be used to examine some of the questions in Table 1, one peptide that has been intensely studied—oxytocin—will be discussed.

Early work (*70*) demonstrated the existence of binding sites for [^{3}H]oxytocin on particulate fractions from the uterus of the rat and the sow. The K_d was approximately 2 nM by saturation analysis. Binding was potentiated by the divalent cations, magnesium, zinc, cobalt, and nickel, but was uninfluenced by calcium and could be completely eliminated by treatment of membranes with ethylenediaminetetraacetic acid. A number of oxytocin analogs competed for binding to porcine membranes with the order of potency expected from their uterotonic activities in the rat.

Four antagonist analogs of oxytocin were shown to possess K_i values for the inhibition of [^{3}H]oxytocin binding to rat uterine particles that were in very close agreement with the K_d values calculated from published pA_2 values (*69*), thus further substantiating the hypothesis that the [^{3}H]oxytocin binding sites represent oxytocin receptors. The claim that the relative potencies of the antagonists depended upon their affinities for the receptor site rather than their intrinsic activities (*69*) seems rather obvious since antagonists have no intrinsic activity, and the claim (*69*) that the close agreement between K_i and K_d demonstrates a lack of receptor reserve in this tissue is in error since pA_2 gives a measurement of the true K_d of the antagonist receptor complex, which is independent of receptor numbers or reserve.

When the myometrium of estrogen-treated rats was subjected to fractionation and careful analysis (*55*), [^{3}H]oxytocin binding was shown to be concentrated in a plasma membrane-enriched fraction (*23*), thus suggesting that the majority of receptors for this hormone are located on that structure. These observations probably surprise no one, but they do illustrate how careful attention to the balance sheet during fractionation studies can permit the isolation of a fraction that is very much enriched in binding sites and produces a very high signal-to-noise ratio. All too often schema for isolation of binding fractions seem to be taken from the most recent cookbook rather than have any logic behind them. Similarly, these results indicate how the binding technique can be used to prepare a subcellular fraction suitable for studies of the consequence of drug–receptor interactions rather than to use totally uncharacterized fractions from which rather equivocal results have little meaning (*15*).

Oxytocin binding sites have also been characterized in the myometrium of the rabbit (*61*) and the pregnant ewe (*25*). Studies in the rabbit added a comparison of binding and uterotonic activities of a number of oxytocin analogs in the same species. There was good correlation between "relative uterotonic activity" and "relative binding activity." The oxytocin concentration producing half-maximal stimulation was similar to the K_d obtained from binding experiments, suggesting no receptor reserve. In the ewe (*25*), binding of [^{3}H]oxytocin was shown to be reversible, and an agreement between the K_d by equilibrium (1.5 n*M*) and kinetic analyses (3.1 n*M*) was demonstrated. Binding was shown to occur to a single class of noninteracting sites. The EC_{50} value of 0.16 n*M* for oxytocin in this tissue suggested the existence of considerable receptor reserve.

The effects of estrogen and progesterone on oxytocin receptors in the uterus (*60, 68*) have been most thoroughly investigated in the rabbit (*60*). Estradiol treatment of young rabbits was shown to render isolated uteri from these animals sensitive to both oxytocin and the muscarinic agonist acetyl-beta-methylcholine. Subsequent treatment with progesterone abolished the contractile response to oxytocin while leaving the muscarinic response unaltered. An analysis of saturation binding curves demonstrated that changes in the number of [^{3}H]oxytocin binding sites with estrogen and progesterone treatment paralleled the changes in response, the binding affinity being unaffected. The effects of progresterone in reducing the number of binding sites were mimicked by the administration of a sublethal dose of actinomycin D. Progesterone and actinomycin D effects were not additive. A role for receptor regulation in the action of estrogen and progesterone on myometrial excitability was proposed. In light of the possible effects of estrogen and progesterone on the relative amount of membrane protein to total protein or to the number of cells, it would be interesting to know the effects of progesterone treatment on the apparent number of muscarinic cholinoceptors, where the sensitivity does not appear to change.

A question of some physiological significance concerns whether oxytocin receptor number or affinity are regulated during gestation, at parturition, or postpartum, and several publications have examined this topic. In the rat the concentration of oxytocin receptors has been shown to increase shortly before term, to remain elevated during parturition, and to fall steeply shortly postpartum; manipulations that produce premature labor produce premature changes in receptor numbers (*1–3*).

Estimates of oxytocin receptor number were obtained using single-point, subsaturating concentrations (0.6 n*M*, K_d = 2 n*M*) of [^{3}H]oxytocin. Despite the existence of some evidence suggesting that the sensitivity of the rat's myometrium to oxytocin might increase just before the onset of labor (*30*), no evidence was presented that the observed increase in binding was accompanied by a change in excitability in these animals. If this seems mere academic criticism of a rather obvious relationship, the results of Goren et al. (*31*) should be considered.

Although there was an 85% reduction in the number of oxytocin binding sites in the uterus from homozygous diabetes insipidus rats of the Brattleboro strain compared to heterozygous controls, there were no detectable changes in the dose–response curves of isolated uterine horns to oxytocin. Thus control of uterine sensitivity to the hormone appeared to be independent of the receptor number. In adopycytes from similar animals the homozygous group did not respond to oxytocin with increased glucose oxidation despite the fact that the numbers and affinity of binding sites were similar to controls. All binding data were obtained from complete saturation analyses (*31*).

4.5. Receptors for Prostanoids

Despite the potent and diverse effects of the prostaglandins and prostaglandinlike compounds on a number of smooth muscles, binding information with this class of compounds is very sparse and restricted almost exclusively to the uterus. Kimball et al. (*48*) described the binding of [^{3}H]prostaglandin E_1 to a postnuclear supernatant of human myometrium. Binding was saturable with a K_d by saturation analysis of 2 n*M*. Fifteen prostaglandin analogs were examined for their ability to compete with [^{3}H]PGE_1 for binding and the relative affinities of most of these compounds were compared with their relative potencies in either the pregnant or nonpregnant rhesus monkey uterus in vivo. There was some agreement between binding and uterotonic potency, but also some disagreement that was ascribed to the pharmacokinetic behavior of the compounds. Subfractionation studies of human myometrium (*24*) demonstrated that [^{3}H]PGE_1 binding was concentrated mainly in the plasma membrane-enriched fraction. This latter fraction appeared to have three classes of binding sites for PGE_1 with apparent K_d values of 1.6, 2.8, and 110 n*M*, respectively. The predominantly plasma membrane location of prostaglandin sites was confirmed for [^{3}H]$PGF_{2\alpha}$ in the uterus of the laying hen (*72*), but not for PGE_2 in the bovine uterus (*16*). Binding of [^{3}H]$PGF_{2\alpha}$ to membranes from the uterus of the laying hen was shown to be partially reversible at 37°C. Only one class of sites was detected with a K_d of 400 n*M*, the EC_{50} for contraction being 20 n*M*. PGE_1 was markedly less potent than $PGF_{2\alpha}$ in both binding and uterotonic assays (*73*).

An exciting development has been the tentative identification of receptors for PGI_2 using the stable analog [^{3}H]Ciloprost in bovine coronary arteries (*74*). Binding was found to be stable at 0°C, rapid, reversible, and to occur at two sites with respective K_d

values of 21 and 750 n*M*. Relative binding potencies (Ciloprost = 1) of PGE_1, E_2, and D_2 were 0.14, 0.01, and 0.02, respectively. $PGF_2\alpha$ was ineffective in competing with [^{3}H]Ciloprost. $PGF_{2\alpha}$ could not be tested since it did not retain activity under the incubation conditions used; therefore the [^{3}H]Ciloprost could only be tentatively proposed to label sites at which PGI_2 is the endogenous ligand.

5. Future Directions for Receptor Binding Studies in Smooth Muscle Subcellular Fractions

Despite the skeptical view of the misuse of receptor binding studies expressed in the early sections of this chapter, the foregoing section clearly illustrates that much can be learned by the appropriate use and interpretation of such studies. Just as clearly many uncertainties remain and much needs to be done.

In a general sense it must be recognized that the usefulness of a ligand needs to be thoroughly established in each system before it can be put to use; nevertheless, the days ought to be over when the demonstration of saturable, reversible, stereospecific binding of an old ligand in a new smooth muscle represents a valuable contribution. Rather, those interested in binding studies in smooth muscle should concentrate on either developing ligands where none already exists or on using existing ligands to answer some of the questions in Table 1 (a lifetime's work for a number of industrious investigators!). The trend towards the simultaneous presentation of binding data with functional or biochemical studies of the consequence of drug–receptor interaction should be continued. Only in this way can the activation (or inhibition) processes in smooth muscle be fully appreciated and the current myths that exist about both types of study be exposed. Greater attention should be paid to differences and similarities between binding and biological activities and explanations that consider all aspects of drug action on smooth muscles tested. Similarly, in physiological or pathological situations in which receptor regulation is suspected to play a role, both binding and functional data should be carefully analyzed so that the role played by changes in receptors and in other processes can be precisely delineated. The results of the study by Goren et al. (*31*) should serve as a clear warning to those who think otherwise.

More specifically, for the muscarinic cholinoceptor it would be useful to the experimental pharmacologist to have more data of the type presented by Aronstam et al. (*5*) on the muscarinic cholinoceptor in other tissues in order to appreciate the possible uniformity (or lack of uniformity) of the large receptor reserve in such tissues and to confirm the structural requirement of the receptor. The intriguing binding mechanism and the relation between agonists and antagonists, and the two sites or forms of receptor already described requires full investigation in a single tissue. As was stated earlier a study of the effects of estrogen and progesterone on muscarinic cholinoceptor in the uterus would be useful in helping with the interpretation of results using other ligands since progesterone appears not to affect the muscarinic response (*60*).

The most controversial question concerning the alpha-adrenoceptor is centered not directly on binding studies but on the effect of estrogen treatment on the response of the myometrium to catecholamines. In pathological situations, hypertension might be a good example, there is need for adequate quantitation of adrenoceptors combined with carefully carried out functional studies aimed at elucidating the roles of subtypes and the influence of changes in number or affinity on smooth muscle reactivity.

Similar directions to those suggested above are needed for beta-adrenoceptors, and for peptide and prostanoid receptors. Of all systems, binding studies on prostanoid receptors are the most poorly developed. The recently proposed classification of prostanoid receptors arrived at by careful and innovative use of classical techniques (*47*) has not been confirmed by binding studies. Such techniques could be used to determine the tissue distribution of prostanoid receptors and perhaps to design more selective agonists and antagonists. For the really adventurous and innovative, the leukotriene receptors are waiting.

References

1. Alexandrova, M., and M. S. Soloff. Oxytocin receptors and parturition. I. Control of oxytocin receptor concentration in the rat myometrium at term. *Endocrinology* 106:730–735, 1980.
2. Alexandrova, M., and M. S. Soloff. Oxytocin receptors and parturition. II. Concentrations of receptors for oxytocin and estrogen in the gravid and nongravid uterus at term. *Endocrinology* 106:736–738, 1980.
3. Alexandrova, M., and M. S. Soloff. Oxytocin receptors and parturition. III. Increases in estrogen receptor and oxytocin receptor concen-

trations in the rat myometrium during prostaglandin $F_{2\alpha}$-induced abortion. *Endocrinology* 106:739–743, 1980.

4. Aronstam, R. S., and G. O. Carrier. Influence of n-ethylmaleimide on cholinoceptors and resonses in longitudinal muscles from guinea-pig ileum. *Br. J. Pharmac.* 77:89–95, 1982.
5. Aronstam, R. S., D. J. Triggle, and M. E. Eldefrawi. Structural and stereochemical requirements for muscarinic receptor binding. *Mol. Pharmacol.* 15:227–234, 1979.
6. Beckett, A. H., and A. F. Casy. Analgesics and their antagonists biochemical aspects and structure-activity relationships. *Prog. Med. Chem.* 4:171–213, 1965.
7. Beld, A. J., E. J. Klok, and J. F. Rodrigues de Miranda. Inhibition of [^{3}H]dexetimide binding by a homologous series of methylfurthrethonium analogues at the peripheral muscarinic receptor. *Biochem. Biophys. Res. Commun.* 97:430–436, 1980.
8. Bennett, T., T. Y. Lot, and P. G. Strange. The effects of noradrenergic denervation on muscarinic receptors of smooth muscle. *Br. J. Pharmac.* 76:177–183, 1982.
9. Birdsall, N. J. M., and E. C. Hulme. Biochemical studies on musarinic acetylcholine receptors. *J. Neurochem.* 27:7–16, 1976.
10. Birdsall, N. J. M., E. C. Hulme, and A. S. V. Burgen. The character of the muscarinic receptors in different regions of the rat brain. *Proc. Roy. Soc. Ser. B* 207:1–12, 1980.
11. Birnbaumer, L., P. -C. Yang. Studies on receptor mediated activation of adenylyl cyclases. I. Preparation and description of general properties of an adenylyl cyclase system in beef renal medullary membranes sensitive to neurohypophyseal hormones. *J. Biol. Chem.* 249: 7848–7856, 1974.
12. Birnbaumer, L., T. Nakahara, and P. C. Yang. Studies on receptor-mediated activation of adenylyl cyclases. II. Nucleotide and nucleoside regulation of the activities of the beef renal medullary adenylyl cyclase and their stimulation by neurohypophyseal hormones. *J. Biol. Chem.* 249:7857–7866, 1974.
13. Bolton, A. E., and W. M. Hunter. The labelling of proteins to high specific radioactivities by conjugation to a [^{125}I]-containing acylating agent. Application to the radioimmunoassay. *Biochem. J.* 133:529–539, 1973.
14. Burgen, A. S. V. Confirmational changes and drug action. *Fed. Proc.* 40:2723–2728, 1981.
15. Carsten, M. E., and J. D. Miller. Effects of prostaglandins and oxytocin on calcium release from a uterine microsomal fraction. *J. Biol. Chem.* 252:1576–1580, 1977.
16. Carsten, M. E., and J. D. Miller. Prostaglandin E_2 receptors in the myometrium: distribution in subcellular fractions. *Arch. Biochem. Biophys.* 212:700–704, 1980.
17. Chang, K.-J., S. Jacobs, and P. Cautrecasas. Quantitative aspects of hormone-receptor interactions of high affinity. Effect of receptor concentrations and measurement of dissociation constants of labelled and unlabelled hormones. *Biochem. Biophys. Acta* 406:294–303, 1975.

18. Cornett, L. E., A. Goldfien, and J. M. Roberts. Tonic *in vivo* inhibition of rabbit myometrial adrenergic receptors. *Nature* 292:623–624, 1981.
19. Costa, L. G., B. W. Schwab, H. Hand, and S. D. Murphy. Decreased muscarinic binding sites in small intestine from mice treated with neostigmine. *Life Sci.* 29:1675–1682, 1981.
20. Costa, L. G., B. W. Schwab, and S. D. Murphy. Differential alterations of cholinergic muscarinic receptors during chronic and acute tolerance to organophosphorous insecticides. *Biochem. Pharmacol.* 31:3407–3413, 1982.
21. Crankshaw, D. J. The isolated plasma membrane of the myometrium from the pregnant ewe is a multi-receptor preparation. *Proc. Can. Fed. Biol. Soc.* 23:317, 1980.
22. Crankshaw, D. J. Beta-adrenoreceptors in the myometrium of the pregnant ewe. *The Physiologist* 23:134, 1980.
23. Crankshaw, D. J., L. A. Branda, M. A. Matlib, and E. E. Daniel. Localization of the oxytocin receptor in the plasma membrane of rat myometrium. *Eur. J. Biochem.* 86:481–486, 1978.
24. Crankshaw, D. J., J. Crankshaw, L. A. Branda, and E. E. Daniel. Receptors for E type prostaglandins in the plasma membrane of nonpregnant human myometrium. *Arch. Biochem. Biophys.* 198:77–79, 1979.
25. Crankshaw, D. J., E. Romaniuk, and L. A. Branda. Identification and characterization of receptors for oxytocin in the myometrium of the pregnant ewe. *Gynecol. obstet. Invest.* 1:202–213, 1982.
26. Daniel, E. E., D. J. Crankshaw, and C.-Y. Kwan. Intracellular sources of calcium for activation of smooth muscle. In *Trends in Autonomic Pharmacology*. Ed.: S. Kalsner, Urban & Schwarzenberg, Baltimore, Maryland, 1979, pp 443–484.
27. Dawson, R. M., and B. Jarrot. Response of muscarinic cholinoceptors of guinea pig brain and ileum to chronic administration of carbamate or organophosphate cholinesterase inhibitors. *Biochem. Pharmacol.* 30:2365–2368, 1981.
28. Ehrlich, P. Chemotherapeutics: scientific principles, methods and results. *Lancet* 2:445–451, 1913.
29. Flouret, G., S. Terada, F. Yang, S. H. Nakagawa, T. Nakahara, and O. Hechter. Iodinated neurohypophyseal hormones as potential ligands for receptor binding and intermediates in synthesis of tritiated hormones. *Biochem.* 16:2119–2124, 1977.
30. Fuchs, A.-R., and V. F. Poblete, Jr. Oxytocin and uterine function in pregnant and parturitient rats. *Biol. Reprod.* 2:387–400, 1970.
31. Goren, H. J., R. M. Geonzon, M. D. Hollenberg, K. Lederis, and D. O. Morgan. Oxytocin action: Lack of correlation between receptor number and tissue responsiveness. *J. Supramolec. Struct.* 14:129–138, 1980.
32. Hancock, A. A., A. L. DeLean, and R. J. Lefkowitz. Quantitative resolution of beta-adrenergic subtypes by selective ligand binding: application of a computerized model fitting technique. *Mol. Pharmacol.* 16:1–9, 1979.

33. Hayashida, D. N., R. Leung, A. Goldfien, and J. M. Roberts. Human myometrial adrenergic receptors: Identification of beta-adrenergic receptor by [^{3}H]dihydroalprenolol binding. *Am. J. Obstet. Gynecol.* 142:389–393, 1982.

34. Higuchi, H., K. Takeyasu, S. Uchida, and M. Yoshida. Receptor-activated and energy-dependent decrease of muscarinic cholinergic receptors in guinea-pig vas deferens. *Eur. J. Pharmacol.* 75:305–311, 1981.

35. Higuchi, H., K. Takeyasu, S. Uchida, and M. Yoshida. Mechanism of agonist-induced degradation of muscarinic cholinergic receptor in cultured vas deferens of guinea pig. *Eur. J. Pharmacol.* 79:67–77, 1982.

36. Hoffman, B. B., T. N. Lavin, R. J. Lefkowitz, and R. R. Ruffolo, Jr. Alpha adrenergic receptor subtypes in rabbit uterus: mediation of myometrial contraction and regulation by estrogens. *J. Pharmacol. Exp. Ther.* 219:290–295, 1981.

37. Holck, M. I., B. H. Marks, and C. A. Wilberding. Characterization of alpha-adrenergic receptors in guinea-pig vas deferens by [^{3}H]dihydroergocryptine binding. *Mol. Pharmacol.* 16:77–90, 1979.

38. Hollenberg, M. D., and P. Cautrecasas. Distinction of receptor from nonreceptor interactions in binding studies. In *The Receptors. A Comprehensive Treatise, Vol. 1. General Principles and Procedures.* Ed.: R. D. O'Brien. New York, Plenum Press, 1979, pp 193–214.

39. Hollenberg, M. D., and E. Nexo. Receptor binding assays. In *Membrane Receptors: Methods for Purification and Characterization.* Ed. S. Jacobs and P. Cuatrecasas. London, Chapman and Hall, 1981. pp. 2–31.

40. Jacobs, S., and P. Cautrecasas. The mobile receptor hypothesis and "cooperativity" of hormone binding application to insulin. *Biochem. Biophys. Acta.* 433:482–495, 1976.

41. Järv, J., B. Hedlund, and T. Bartfai. Isomerization of the muscarinic receptor-antagonist complex. *J. Biol. Chem.* 254:5595–5598, 1979.

42. Järv, J., B. Hedlund, and T. Bartfai. Kinetic studies on muscarinic antagonist-agonist competition. *J. Biol. Chem.* 255:2649–2651, 1980.

43. Kasai, M. and J.-P. Changeux. *In vitro* excitation of purified membrane fragments by cholinergic agonists. 1. Pharmacological properties of the excitable membrane fragments. *J. Membrane Biol.* 6:1–23, 1971a.

44. Kasai, M. and J.-P. Changeux. *In vitro* excitation of purified membrane fragments by cholinergic agonists. 2. The permeability change caused by cholinergic agonists. *J. Membrane Biol.* 6:24–57, 1971b.

45. Kasai, M. and J.-P. Changeux. *In vitro* excitation of purified membrane fragments by cholinergic agonists. 3. Comparison of the dose-response curves to decamethonium with the corresponding binding curves of decamethonium to cholinergic receptor. *J. Membrane Biol.* 6:58–80, 1971c.

46. Kenakin, T. P. Organ selectivity of drugs: alternatives to receptor selectivity. *TIPS* 3:153–156, 1982.

47. Kennedy, I., R. A. Coleman, P. P. A. Humphrey, G. P. Levy, and P. Lumley. Studies on the characterization of prostanoid receptors: a proposed classification. *Prostaglandins* 24:667–689, 1982.

48. Kimball, F. A., K. T. Kirton, C. H. Spilman, and L. J. Wyngarden. Prostaglandin E_1 specific binding in human myometrium. *Biol. Reprod.* 13:482–489, 1975.
49. Klotz, I. M. Ligand-receptor interactions: what we can and cannot learn from binding measurements. *TIPS* 4:253–255, 1983.
50. Langley, J. N. On the contraction of muscle, chiefly in relation to the presence of "receptive" substances. Part IV. The effect of curare and of some other substances on the nicotine response of the sartorius and gastrocnemius muscles of the frog. *J. Physiol.* 39:235–295, 1909.
51. Lavin, T. N., B. B. Hoffman, and R. J. Lefkowitz. Determination of subtype selectivity of alpha-adrenergic antagonists. *Mol. Pharmacol.* 20:28–34, 1981.
52. Levin, R. M. and A. J. Wein. Distribution and function of adrenergic receptors in the urinary bladder of the rabbit. *Mol. Pharmacol.* 16:441–448, 1979.
53. Lesniak, M. A. and J. Roth. Regulation of receptor concentration by homologous hormone. Effect of human growth hormone on its receptor in IM-9 lymphocytes. *J. Biol. Chem.* 251:3720–3729, 1976.
54. Limas, C. J., and C. Limas. Decreased number of beta-adrenergic receptors in hypertensive vessels. *Biochim. Biophys. Acta* 582:533–536, 1979.
55. Matlib, M. A., J. Crankshaw, R. E. Garfield, D. J. Crankshaw, C-Y. Kwan, L. A. Branda, and E. E. Daniel. Characterization of membrane fractions and isolation of purified plasma membranes from rat myometrium. *J. Biol. Chem.* 254:1834–1840, 1979.
56. Minneman, K. P., R. N. Pittman, and P. B. King. Beta-adrenergic receptor subtypes: properties, distribution, and regulation. *Ann. Rev. Neurosci.* 4:419–461, 1981.
57. Molinoff, P. B., B. B. Wolfe, and G. A. Weiland. Quantitative analysis of drug-receptor interactions: II. Determination of the properties of receptor subtypes. *Life Sciences* 29:427–443, 1981.
58. Motulsky, H. J., and P. A. Insel. [^{3}H]Dihydroergocryptine binding to alpha-adrenergic receptors of human platelets. A reassessment using the selective radioligands [^{3}H]prazosin, [^{3}H]yohimbine, and [^{3}H]rauwolscine. *Biochem. Pharmacol.* 31:2591–2597, 1982.
59. Nakahara, T. and L. Birnbaumer. Studies on receptor-mediated activation of adenylyl cyclases. V. Transient kinetics of the activation of beef renal medullary adenylyl cyclase by neurohypophyseal hormones: estimation of apparent rate constants of the receptor-hormone interaction. *J. Biol. Chem.* 249:7886–7891, 1974.
60. Nissenson, R., G. Flouret, and O. Hechter. Opposing effects of estradiol and progesterone on oxytocin receptors in rabbit uterus. *Proc. Natl. Acad. Sci.* 75:2044–2048, 1978.
61. Nissenson, R. A., G. Flouret, and O. Hechter. Oxytocin receptors coupled to uterine contractions in estrogen-dominated rabbits. *Biochim. Biophys. Acta* 628:209–219, 1980.
62. Rimele, T. J., W. A. Rogers, and T. S. Gaginella. Characterization of muscarinic cholinergic receptors in the lower esophageal sphincter of the cat: binding of [^{3}H]quinuclidinyl benzilate. *Gastroenterology* 77:1225–1234, 1979.

63. Roberts, J. M., P. A. Insel, R. D. Goldfien, and A. Goldfien. Alpha adrenoreceptors but not beta adrenoreceptors increase in rabbit uterus with oestrogen. *Nature* 270:624–625, 1977.
64. Roberts, J. M., P. A. Insel, and A. Goldfien. Regulation of myometrial adrenoreceptors and adrenergic response by sex steroids. *Mol. Pharmacol.* 20:52–58, 1981.
65. Rudinger, J., V. Pliška, and I. Krejčí. Oxytocin analogs in the analysis of some phases of hormone action. *Rec. Prog. Horm. Res.* 28:131–166, 1972.
66. Siegel, H., and D. J. Triggle. Benzilylcholine mustard and spare receptors in guinea pig ileum. *Life Sci.* 30:1645–1652, 1982.
67. Siman, R. G., and W. L. Klein. Cholinergic activity regulates muscarinic receptors in central nervous system cultures. *Proc. Natl. Acad. Sci.* 76:4141–4145. 1979.
68. Soloff, M. S. Uterine receptors for oxytocin: effects of estrogen. *Biochem. Biophys. Res. Commun.* 65:205–212, 1975.
69. Soloff, M. S. Uterine receptors for oxytocin: correlation between antagonist potency and receptor binding. *Br. J. Pharmac.* 57:381–386, 1976.
70. Soloff, M. S. and T. L. Swartz. Characterization of a proposed oxytocin receptor in the uterus of the rat and sow. *J. Biol. Chem.* 249:1376–1381, 1974.
71. Tittler, M., P. Weinreich, and P. Seeman. New detection of brain dopamine receptors with [^{3}H]dihydroergocryptine. *Proc. Natl. Acad. Sci.* 74:3750–3753.
72. Toth, M., G. Asboth, and F. Hertelendy. Selective binding of prostaglandin $F_{2\alpha}$ to membrane fractions enriched in 5′-nucleotidase, Ca^{2+}-ATPase, and Mg^{2+}-(Na^++K^+) ATPase. *Endocrinology* 109:106–112, 1981.
73. Toth, M., D. M. Olson, and F. Hertelendy. Binding of prostaglandin $F_{2\alpha}$ to membranes of shell gland muscle of laying hens: correlations with contractile activity. *Biol. Reprod.* 20:390–398, 1979.
74. Town, M.-H., E. Schillinger, A. Speckenbach, and G. Prior. Identification and characterisation of a prostacyclin-like receptor in bovine coronary arteries using a specific and stable prostacyclin analogue, ciloprost, as radioactive ligand. *Prostaglandins* 24:61–73, 1982.
75. Triggle, D. J., and J. F. Moran. The identification and isolation of adrenergic receptors. In *Trends in Autonomic Pharmacology*. Ed.: S. Kalsner. Urban & Schwarzenberg, Baltimore, Maryland, 1979, pp 387–419.
76. Tsai, B. S., and R. J. Lefkowitz. [^{3}H]Dihydroergocryptine binding to alpha adrenergic receptors in canine aortic membranes. *J. Pharmacol. Exp. Ther.* 204:606–614, 1978.
77. Vallières, J., J. Fortier, and L. Bukowiecki. Alpha- and beta-adrenergic receptors in isolated rabbit uterine membranes: study of receptor alterations by pregnancy. *Biol. Reprod.* 19:318–325, 1978.
78. Weiland, G. A., and P. B. Molinoff. Quantitative analysis of drug-receptor interactions: I. Determination of kinetic and equilibrium properties. *Life Sciences* 29:313–330, 1981.

79. Williams, L. T. and R. J. Lefkowitz. Regulation of rabbit myometrial alpha adrenergic receptors by estrogen and progesterone. *J. Clin. Invest.* 60:815–818, 1978.

80. Williams, L. T., D. Mullikin, and R. J. Lefkowitz. Identification of alpha-adrenergic receptors in uterine smooth muscle membranes by [^{3}H]dihydroergocryptine binding. *J. Biol. Chem.* 251:6915–6923, 1976.

81. Woodcock, E. A., C. A. Olsson, and C. I. Johnston. Reduced vascular beta-adrenergic receptors in deoxycorticosterone-salt hypertensive rats. *Biochem. Pharmac.* 29:1465–1468, 1980.

82. Yamamura, H. I. and S. H. Snyder. Muscarinic cholinergic receptor binding in the longitudinal muscle of the guinea pig ileum with [^{3}H]quinuclidinyl benzilate. *Mol. Pharmacol.* 10:861–867, 1974.

Chapter 12

Calcium-Handling Defects and Smooth Muscle Pathophysiology

C. Y. Kwan

1. Introduction

It is generally agreed that in spite of the functional diversity in the control of contractile activity of smooth muscle at structural, myogenic, neurogenic, and hormonal levels, the final common pathway of these control functions is the regulation of cytoplasmic calcium ion (Ca^{2+}) concentration in the smooth muscle cell. The source of Ca^{2+} available to initiate excitation–contraction coupling in smooth muscles may be extracellular as well as intracellular depending upon the source of smooth muscle, cellular environment, and the nature of stimulus. Thus, the increase of cytoplasmic concentration of Ca^{2+} for contraction may be achieved by the influx of Ca^{2+} via Ca^{2+} channels that open upon depolarization (electromechanical coupling via voltage-dependent Ca^{2+} channels) or receptor occupancy (pharmacomechanical coupling via receptor-operated Ca^{2+} channels) or alternatively by the release of intracellularly sequestered Ca^{2+}. For the initiation and maintenance of the state of relaxation after each contraction, inhibition of Ca^{2+} influx or release would not be adequate in lowering cytoplasmic Ca^{2+} contraction to $10^{-7}M$ or less because of the residual inward leak of Ca^{2+} across the cell membrane. A mechanism for extrusion or sequestration of the cytoplasmic Ca^{2+} must therefore be operative to lower the Ca^{2+}

concentration. Plasma membrane, endoplasmic reticulum, and mitochondria have been proposed as the major Ca^{2+} sequestering sites based upon direct evidence derived from ultrastructural studies (*see* Chapter 1) and studies using isolated subcellular membranes (*see* Chapters 10 and 11). Clearly, intracellular Ca^{2+} ions are compartmentalized in all smooth muscles, but the relative importance of the roles of various Ca^{2+} compartments during excitation–contraction–relaxation cycles in smooth muscle has been and still is a debating issue (*23, 63*).

Since Ca^{2+} plays a vital role in the contractile function of all types of muscles including smooth muscles from tissues of different structural and functional origin, a dysfunction of its regulation under pathological conditions may conceivably lead to an altered contractile state of the smooth muscles. For example, the control of contractile function of vascular smooth muscle is a major determinant of the regulation of vascular tone that obviously is important in the systemic and pulmonary hemodynamic and in the pathophysiology of cardiovascular complications associated with hypertension, coronary artery spasm, shock, and heart failure. The recent introduction of some Ca^{2+} antagonists as powerful vasodilator agents and their clinical use as an effective antihypertensive drug presumably acting by selectively interfering with the Ca^{2+}-handling processes in vascular smooth muscle (*see* Chapter 7), may underscore the role of Ca^{2+} in pathophysiology of hypertensive diseases. Not only the normal but also the pathological changes in the mechanisms of Ca^{2+}-handling have been analyzed more fully in vascular smooth muscle than in any others. This is therefore an appropriate subject for review. Brief overviews on altered Ca^{2+} metabolism in vascular smooth muscle in arterial hypertension have recently been reported (*8, 20, 21, 22, 50, 56*). In this chapter, evidence of abnormal handling of Ca^{2+} by vascular and nonvascular smooth muscle in arterial hypertension is presented, evaluated, and discussed in more detail in connection with the pathogenesis of hypertensive disease.

2. Alterations of Vascular Smooth Muscles in Hypertension

2.1. Structural and Functional Changes

Based upon earlier physiological studies, Folkow and his coworkers (for review, *see* ref. 33) have proposed that a major portion of the increased vascular resistance that could not be abolished by pharmacological relaxation of vascular smooth muscle in chronic hypertension, was caused by an increased wall thickness and decreased lumen diameter. This hypothesis has received considerable attention and such vascular structural changes were supported by direct morphometric analysis that indicated vascular medial hypertrophy or hyperplasia (*71, 79*).

Evidence supporting functional changes in vascular smooth muscle has been derived from the findings of (a) changes in membrane potentials (*43*); (b) changes in ion fluxes across the vascular muscle cell membranes (*50, 101*); (c) changes in Ca^{2+} binding and transport using isolated subcellular membrane fragments or vesicles (*22, 56*), and (d) changes in the activity of cytoplasmic or membrane associated enzymes that are believed to be linked to ion transport or receptor activation (*6, 9, 10, 39*), and (e) changes in receptors for vasoactive ligands (*121*).

Structural changes of arterial muscle have been considered, in general, an adaptive response to elevated blood pressure (for more detailed discussion, *see 21, 33, 48, 85, 120*) whereas some of the functional changes may be intimately associated with the primary events leading to hypertension(to be discussed in Section 2.4). However, recent interesting findings of structural and functional changes in veins (*2, 33–36, 42, 86, 88, 99*) from hypertensive animals suggest that some of these changes are independent of vascular stress induced by elevated blood pressure since venous pressure has been shown to remain low and normal in hypertensive animals (*36*).

2.2. Contraction and Relaxation

Increased myogenic activity including the basal active tension, spontaneous activity, and/or responsiveness to various vasoactive stimuli, has consistently been observed in arteries from hypertensive animals and was attributed to an increased permeability of cell membrane to Ca^{2+} (*30, 82, 100*) since this increased myogenic activity is dependent upon the extracellular Ca^{2+} concentration. Mulvany and his associates (*78, 80*) have observed an increased Ca^{2+} sensitivity in response to norepinephrine (NE) but not to KCl of mesenteric resistance vessels from spontaneously hypertensive rats (SHR) compared to Wistar-Kyoto normotensive rats (WKY). This increased Ca^{2+} sensitivity was proposed to be associated with increased Ca^{2+} permeability induced by NE, whereas the KCl-induced Ca^{2+} permeability appears to be normal. The fact that the increased Ca^{2+} sensitivity induced by NE persists in vessels from SHR at 4 wk of age before the blood pressure becomes elevated, after antihypertensive treatment with

Felodipine or chemical sympathectomy, supports the view that this increased Ca^{2+} sensitivity is an intrinsic defect of vascular smooth muscle independent of vascular structural changes, elevation of blood pressure, and neurogenic influences (*80*). However, mesenteric arteries from renal hypertensive rats did not show the increased Ca^{2+} sensitivity compared to the control, suggesting that this alteration may be genetically determined (*80*). Holloway and Bohr (*45, 46*) have shown that response of vascular smooth muscle to KCl was depressed by elevated Ca^{2+} concentrations from 4 to 18 m*M*. The elevated Ca^{2+} concentration also caused KCl-contracted vessels to relax and this Ca^{2+}-induced relaxation was altered in deoxycorticosterone (DOC) induced hypertensive pigs (*116*). These findings were interpreted to indicate that plasma membranes of vascular smooth muscle require higher concentration of external Ca^{2+} concentration for its membrane-stabilizing action. A contractile response upon exposure to La^{3+} was observed in vascular strips from SHR but not from WKY (*45 98*). The ability of La^{3+} to induce contraction has been observed in nonvascular smooth muscle from SHR as well (*20*). It has been proposed that La^{3+} may act by entering the smooth muscle cells and causing release of intracellular Ca^{2+} stores since La^{3+} ions presumably compete for the Ca^{2+} binding sites owing to their similar size and greater change than Ca^{2+} ions (*119*). This interpretation, however, is in contrast to the previous assumption underlying the use of La^{3+} as a tool to study Ca fluxes in smooth muscle (*112*), that La^{3+} acts by competing for extracellular and surface membrane bound Ca^{2+}, prevents Ca^{2+} entry, and does not enter into smooth muscle. However, these assumptions have previously been questioned (for review, *see* ref. *22*).

Decreased rate of relaxation has been observed in arteries from hypertensive animals (*18, 19, 47*). Since the rate of relaxation after the agonist is eliminated is determined by the rate of *net* removal of Ca^{2+} from cytoplasm and by the rebound of elastic elements stretched by contraction, an increased Ca^{2+} permeability, a decreased Ca^{2+} binding, a decreased rate of Ca^{2+} extrusion or intracellular Ca^{2+} sequestration, or structural changes of the intercellular matrix elements and smooth muscle cells could be involved. The relaxation studies, however, yielded different findings depending on the agonist used for contraction, the experimental conditions under which vascular muscle relaxation is obtained, the nature of the vasculature, and the duration as well as the type of hypertension studied. For example, in KCl-contracted preparations, SHR aorta showed a slower relaxation compared to WKY aorta in normal Krebs solution during the first 14 min, but a faster relaxation was noted after 8 min in Ca^{2+} free solution (*18*). The reduced rate of relaxation is more prominently found in KCl-contracted than in NE-contracted aorta preparations from SHR (*18*). Swamy and Triggle (*102*) reported that relaxation of maximum NE-induced contraction was faster and relaxation of KCl-induced contraction was generally slower with carotid vascular strips from SHR compared to those from WKY. Shibata et al. (*98*) reported that removal of Ca^{2+} resulted in faster rate of decay of NE- or KCl-induced contraction in SHR than in WKY. The aorta strips from SHR and WKY showed similar responses to nitroglycerin and papaverine (*98*), but they differ in sensitivity to diazoxide (*47*) and D600 (*88*). Recently, Cheng and Shibata (*17*) have shown that the relaxation response to isoproterenol, acetylcholine, Mg^{2+}, Mn^{2+}, Co^{2+}, and La^{3+} was less in thoracic aorta, but remained unaltered in abdominal aorta of young and adult SHR compared to the controls. Cohen et al. (*18*) reported that nitroprusside at low (0.1–1.6 μg), but not high (6.2–25 μg) produced a greater degree of vasodilation in blood-perfused renal vasculature from rats with DOC/salt-induced hypertension compared to the controls.

2.3. Calcium Content and Fluxes

Determination of cellular Ca distribution in vascular smooth muscle is complicated by the fact that the cellula Ca content represents only a small fraction of the total tissue Ca and that Ca remained bound or as free ions in intracellular stores, cytoplasm, and the extracellular matrix (*22*). These problems also imposed technical difficulties in Ca fluxes studied using vascular muscle strips.

Reports on Ca content in vascular smooth muscle of hypertensive animals have been variable. Tobian (*106*) reported an increase of total tissue Ca content determined by chemical method in mesenteric arteries of renovascular hypertensive rats compared to the normotensive control. Hinke (*44*) found that the total Ca content of femoral arteries was elevated in DOC/salt-induced hypertensive rats after normal Krebs perfusion, but not after zero-Ca Krebs perfusion. Furuta (*32*) also reported that Ca content increased with development of DOC/salt-induced hypertension in rats. Overbeck and Conrad (*83*) did not observe significant alteration in intracellular Ca content in carotid arteries of renal hypertensive dogs when the Ca content was measured in several subcellular fractions by isotope (^{45}Ca) method. Using atomic absorption spectroscopy method, Jones (*49*) observed that aorta of SHR contained less total Ca than that of Sprague-Dawley normotensive control, but it was not different from that of Wistar control. Bhalla et al. (*10*) reported that the total tissue Ca content in aorta of SHR and WKY was not different at the age of 4–5 wk, but more Ca content in aorta of SHR compared to that of WHY was progressively observed with age.

$^{45}Ca^{2+}$ fluxes in smooth muscles have been extensively studied (*see* Chapter 4), but comparative studies in vascular tissues from hypertensive and normotensive animals were relatively few and inconsistent findings were reported. In rat aortae of SHR in comparison to those of WKY, the total Ca^{2+} uptake was reported to be increased (*97*) or decreased (*123*); the $La^{3\pm}$resistant Ca^{2+} influx was decreased (*97*), unchanged (*123*), or increased (*10*). The efflux of ^{45}Ca was found to be multiexponential as a function of time. The initial efflux was reported to be decreased (*10*), unchanged (*97*), or increased (*123*). Similar study was also reported for aortic strips from hypertensive rabbits (*107*). Because of the complexity of the distribution of Ca in intact vascular strips, interpretation of these results in terms of transmembrane fluxes or in relation to a particular Ca-compartment, has not yielded fruitful information.

2.4. Calcium Accumulation by Isolated Membranes

A relatively more direct approach used cellular organelles that contain the isolated putative sequestration sites to study their Ca^{2+} binding and transport. The organelles can often be isolated as closed membrane vesicles. If so, the transport and binding properties of each cellular membrane component, when properly isolated and adequately enriched, can be defined. Also the experimental conditions can be better controlled, and easily manipulated. This approach also allows the study of mechanisms of Ca^{2+} regulation at a molecular level (*see* Chapters 10 and 11 for more detailed discussion). However, isolation of membrane fraction may suffer from the possibility of functional damage and purification of membranes may preclude interactions between various organelles and membranes. Also, membrane vesicles may vary in their orientation and in their ability to maintain ionic gradient (*see* Section 4.3.). Despite these limitations, much useful information can be obtained. Reduction in ATP-dependent Ca^{2+}-accumulation by microsomal fractions from large arteries of SHR and rats with DOC/salt-induced hypertension compared to appropriate controls have been consistently reported and given considerable attention [*5, 10, 12, 77, 81,* but *see* ref. *(97)*]. However, there are several points that must be considered before interpreting the results obtained from such studies; these are discussed below.

2.4.1. Plasma Membrane vs Endoplasmic Reticulum

Using a fraction enriched in plasma membrane from aortic microsomes on a sucrose density gradient, Wei et al. (*117*) showed that the reduced ATP-dependent Ca^{2+} accumulation by microsomal fraction isolated from SHR aortae can actually be accounted for by the plasma membrane vesicles in spite of contamination by endoplasmic reticulum. Several altered activities of membrane bound enzymes such as alkaline phosphodiesterase I, basal Mg^{2+}–ATPase, 5′-nucleotidase and alkaline phosphatase were all associated primarily with plasma membrane (*53–56, 62, 64–65*). Furthermore, no one has successfully purified an endoplasmic reticulum fraction or related alteration of Ca^{2+} accumulation to the endoplasmic reticulum content of various fractions from vascular smooth muscle. Most of the studies used heterogenous microsomal fractions containing plasma membrane fragments to account for the changes of Ca^{2+} accumulation in the vascular microsomal fraction isolated from hypertensive animals. Evidence obtained from relatively purified plasma membrane fraction isolated from small arteries supports such a conclusion (*56*).

2.4.2. Small vs Large Arteries

Most small arteries have different morphological, electrical and mechanical characteristics compared to large arteries (*100*). If small arteries represent a better model for resistance vessels, studies of Ca^{2+} accumulation by subcellular membrane vesicles isolated from small arteries from hypertensive animals may yield more revelant information. Some early studies suggested that there might be differences in Ca^{2+} transport changes in membranes from large and small arteries (*118*). However, it has recently been shown that the Ca^{2+} accumulation in the presence of ATP by the plasma membrane vesicles isolated from small mesenteric arteries of SHR was significantly reduced (*60, 61*) as in similar fractions from aorta (*117*). Earlier results were apparently the result of contamination by membrane fragments from nonsmooth muscle cells (*61, 117*). A similar decrease in ATP-dependent Ca^{2+} accumulation was obtained in plasma membranes from mesenteric arteries of rats with DOC/salt-induced hypertension (*60, 61*). Thus it appears that Ca^{2+} accumulation by plasma membrane vesicles is similarly altered in both large and small arteries in these two types of hypertension.

2.4.3. Ca^{2+} Binding vs Transport

The experimental evidence that the ATP-supported Ca^{2+} accumulation by isolated smooth muscle membrane vesicles is an active transport process has been described in detail in Chapter 11. One of the methods to differentiate the binding vs the active transport process in ATP-supported Ca^{2+} accumulation by smooth muscle membrane vesicles was the study of the stimulation of Ca^{2+} accumulation by oxalate that acts presumably by precipitating the intravesicular Ca^{2+} when its concentration is built up via the active transport process to exceed the Ca-oxalate solubility product. However, it should be noted that the mode of action of oxalate in smooth muscle membrane vesicles has not been critically evaluated.

Earlier studies using aorta from rats with spontaneous hypertension and rats with DOC/salt-induced hypertension, showed that the oxalate stimulated, ATP-supported Ca^{2+} accumulation by the aortic microsomes from hypertensive animals was significantly decreased (*5, 77, 114*) compared to control microsomes. In these studies, it was assumed that the changes in "oxalate effect" on the Ca accumulation is equivalent to the changes of Ca^{2+} sequestration by smooth muscle endoplasmic reticulum by analogy to that in skeletal muscle. The validity of its use as a membrane marker, in particular for smooth muscle endoplasmic reticulum, has been questioned (*63, 66*) since it does not appear to be confined solely to endoplasmic reticulum and has never been established unequivocally to be a marker for endoplasmic reticulum.

Ca^{2+} ionophores X537A and A 23187 that relax the Ca^{2+} gradient built up across the membrane vesicles during the active transport process, partially inhibited the Ca^{2+} accumulation in the presence of ATP by plasma membrane fractions isolated from either normotensive or hypertensive animals (*61, 62*). Since the ionophore-resistant Ca^{2+} accumulation (defined as Ca^{2+} binding) was similar in both normotensive and hypertensive groups, it was concluded that it was the transport component (defined as ionophore-sensitive, ATP-dependent Ca^{2+} accumulation) that is altered in hypertension.

Changes in Ca^{2+} binding as well as transport may occur in hypertension. A decreased Ca^{2+} binding has been reported in arterial plasma membrane fraction from SHR at 1 m*M* Ca^{2+} concentration (*118*). Recently, a pH-dependent high affinity Ca^{2+} binding by isolated rat myometrium plasma membrane as well as rabbit aortic microsomes has been reported (*38*). Such a high affinity Ca^{2+} binding site has been reported to be reduced in plasma membranes of various tissues isolated from SHR (*26*). This will be discussed further in Section 4.1.

2.4.4. Cause vs Effect of Hypertension

The significance of any vascular abnormalities in hypertension depends upon whether it is a primary event or secondary to elevation of blood pressure. The decrease of ATP-dependent Ca^{2+} transport by plasma membranes of arteries in spontaneous hypertension preceded the elevation of blood pressure in young SHR (*60*). In DOC/salt-induced hypertension, the steroid may be a necessary stimulus maintaining the hypertension in unilaterally nephrectomized rats, given 1% salt in drinking water. If so, withdrawal of DOC treatment after the induction of hypertension should substantially lower the blood pressure. Kwan et al. (*61*) found that during the first 2 wk after DOC withdrawal, 40% of the population of the rats showed decreased blood pressure (from 160–180 to 120–140 mm Hg) and the other 60% remained hypertensive (160–210 mm Hg). Comparison of the biochemical parameters in these two groups of rats showed that diminished ATP-dependent Ca^{2+} transport by the plasma membrane fraction, but not the arterial wet weight or the total plasma membrane protein content was reversed in those rats whose blood pressure fell as a result of DOC withdrawal. These results suggest that decreased Ca^{2+} transport may be associated with factors causing the hypertension that had been initiated and maintained in some animals by DOC.

If alterations of Ca^{2+} handling by vascular smooth muscle were secondary to the elevated arterial pressure in hypertension and reversible, then that chronic effective antihypertensive treatment of hypertensive animals should reverse the abnormalities following normalization of blood pressure. However, it has recently been reported that plasma membrane fraction isolated from mesenteric arteries of SHR after chronic treatment with hydralazine persistently showed decreased ATP-supported Ca^{2+} transport compared to that from WKY under similar treatment (*58*). Zsoter and Wolchinsky (*122*) also failed to demonstrate any evidence of normalization of Ca^{2+} flux in SHR effectively treated with hydralazine. Certainly, if hypertension were prevented throughout the life span of an SHR animal, these changes should be prevented, but this experiment has not yet been done. This is consistent with the contention that most antihypertensive drug treatments only control, but do not cure, hypertension, i.e., they do not correct the underlying causative mechanism.

In SHR, venous pressure does not seem to be affected in hypertension (*36*). However, altered contractility of venous muscle strips isolated from SHR in relation to its dependence on external Ca^{2+} concentration (*88, 89*) and increased ion fluxes across the venous smooth muscle cell membrane (*50*) have been reported. Kwan and Daniel (*57*) reported that plasma membrane fraction isolated from mesenteric veins of SHR showed greater retention of Ca^{2+} in the absence of ATP than that of WKY. They suggested that this increased binding of Ca^{2+} by venous plasma membrane fraction from SHR may reflect increased binding sites for activator Ca^{2+} (possibly on the inner surface); however, detailed study has been hampered by the substantially low yield of membranes from venous muscle. In nonvascular smooth muscles, which are never exposed to increased wall stress, altered contractility and Ca^{2+} transport associated with spontaneous hypertension have also been observed. This is discussed in Section 3. The collective evidence from these studies indicates that abnormalities of plasma membrane fractions isolated from vascular and nonvascular smooth muscle of hypertensive animal in handling Ca^{2+} (binding and/or transport) can hardly be explained solely as secondary changes adaptive to elevated blood pressure and may play a role in initiation as well as maintenance hypertension.

2.4.5. Other Modes of Hypertension

Kwan et al. (*62*) have extended their vascular muscle membrane studies to two types of Goldblatt renovascular hypertension in rats: one-kidney, one-clip and two-kidney, one-clip types. Although these two types of renovascular hypertension in rats are known to be different in their dependence on renin–angiotensin system, the decreased ATP-supported Ca^{2+} transport across the isolated vascular plasma membrane fraction, similarly observed in genetic and DOC/salt-induced hypertension, also occurred in these two types of renovascular hypertension. The reduction of Ca^{2+} accumulation by plasma membrane vesicles in the presence of ATP paralleled the blood pressure level after captopril treatment in these rats. This suggests that changes of the ATP-dependent Ca^{2+} accumulation were associated with the blood pressure control system that was independent from the system(s) affected by captopril, most likely the renin–angiotensin system.

2.4.6. Changes in Phosphorylation–Dephosphorylation

Changes in cAMP metabolism and cAMP-dependent phosphorylation in vascular smooth muscle may be associated with hypertension (*3, 4, 9, 10, 70*). However, reports on the cAMP level, adenylate cyclase activity and cAMP phosphodiesterase activity in vascular smooth muscle from hypertensive animals have been highly inconsistent. Bhalla et al. (*9, 10, 40*) reported reduced cAMP-dependent phosphorylation, cAMP-dependent protein kinase activity, and cAMP-dependent stimulation of ATP-supported Ca^{2+} accumulation using aortic microsomal fraction isolated from SHR compared to that from WKY. The decreased phosphorylation may be further intensified by an increase in phosphoprotein phosphatase activity (*10*). Thus, it was suggested that the phosphorylation-dephosphorylation process may be the regulatory mechanism that was altered leading to reduced Ca^{2+} transport by microsomal membrane vesicles (*9–11*). Increased activity of alkaline phosphatase, which has been reported to possess phosphoprotein dephosphorylase activities as well (*76*), has consistently been observed in membrane fractions of aorta (*109*), tail artery (*110*), mesenteric arteries (*60–62*), and veins (*57*) and other cardiovascular tissues (*54*) of rats with different modes of hypertension. The finding of Kreye and Schlicker (*69*) that alkaline phosphatase obtained commercially reduced the ATP-dependent Ca^{2+} accumulation by plasma membrane enriched fraction from vascular smooth muscle, raised the possiblity that reduced Ca accumulation may be attributed to the phosphoprotein dephosphorylase activity of the alkaline phosphatase. Biochemical properties of alkaline phosphate from vascular smooth muscle (*54*) have not been sufficiently characterized at this time to unravel their physiological function and the role of their alteration in vascular smooth muscle in hypertension.

2.4.7. Other Considerations

The reduced ATP-dependent Ca^{2+} transport by plasma membrane vesicles from vascular smooth muscle has consistently been demonstrated in studies of large as well as small arteries and in different modes of hypertension. This may be caused by one or several of the following mechanisms: (a) increased proportion of right-side-out oriented vesicles; (b) decreased efficiency of the Ca^{2+} pump; and (c) increased permeability of vesicle membranes.

The problems in determining and controlling the orientation of isolated smooth muscle membrane vesicles has not been sufficiently analyzed. Some preliminary studies have been reported to indicate the presence of approximately equal proportions of inside-out and right-side-out oriented vesicles in the isolated plasma membrane fraction from nonvascular smooth muscle (*see* Chapter 11). But these methodologies for estimating inside-out vesicles have not been applied to vascular smooth muscle in hypertension.

The molecular identity of the Ca^{2+}-pump in smooth muscle is obscure. The existence of a (Ca^{2+}-Mg^{2+})-ATPase has been mainly inferred, by analogy to that well characterized in sarcoplasmic reticulum of skeletal muscle as the Ca^{2+} pump in smooth muscle microsomes. However, because of the high basal level of the ATP hydrolytic activity that can be activated by Mg^{2+} and Ca^{2+} to a comparable extent in most smooth muscles (*53, 67*), reports on identifying the (Ca^{2+}-Mg^{2+})- ATPase activity in isolated vascular smooth muscle membranes were not conclusive (*63*). The Ca^{2+}-stimulated extra splitting of ATP in the presence of Mg^{2+} usually represents a very small fraction of the apparent basal Mg^{2+}-ATPase activity (10%). Although reduced basal ATPase activity in the presence of Mg^{2+} or Ca^{2+} has been observed in the extract of aorta or mesenteric arteries of SHR (*53*), there was no definite evidence of association between the Ca^{2+} pump and these ATPase activities. So far then, the decreased Ca^{2+} transport observed in hypertensive artery membranes cannot be assigned to a defective Ca^{2+} pump.

Increased ion permeability of cell membrane of vascular smooth muscle from hypertensive animals has been implicated in a number of studies (for reviews, *10, 50*), but has not been directly demonstrated using isolated plasma membrane vesicles.

3. Alterations of Nonvascular Smooth Muscles in Genetic Hypertension

3.1. Contractility

Caulfield et al. (*16*) in a preliminary finding reported that isolated vas deferens from 3–4 months-old genetically hypertensive rat developed less maximum tension than that from the normotensive Wistar rats in response to applied NE. There was no significant difference between their sensitivities to NE. Such a decrease in maximum response to NE of vas deferens from hypertensive rats was present even after 8 wk of treatment with antihypertensive drugs. These findings clearly indicate that genetically hypertensive rats exhibit altered smooth muscle contractility that is not confined to vascular smooth muscle and cannot be a secondary adaptation to the elevation of blood pressure. Corbett et al. (*20*) have extended such contractility studies by comparing the characteristics of the response of vasa deferentia from SHR and three different strains of normotensive control rats to several agonists, including NE, K^+, Ca^{2+}, and Ba^{2+}. They have observed decreased maximum response but unchanged sensitivity to NE, unchanged maximum response but increased sensitivity to K^+ in SHR tissue compared to the controls. Alterations were also observed in the Ca^{2+} dependence of the maximum responses to $10^{-4}M$ NE or 80 m*M* KCl. The rate of loss of response to KCl, but not to NE, upon replacement of Tyrode's solution by Ca^{2+}-free Tyrode's solution, was increased in SHR tissue. These findings together with the finding of increased threshold sensitivity to Ba^{2+} and greater responses to La^{3+} in Tyrode's and Ca^{2+}-free Tris-Tyrode's solution in vas deferens from SHR compared to that from control groups strongly suggest that the fundamental Ca^{2+}-handling process in vas deferens is altered in SHR.

Altman et al. (*1*) studied the contractile properties of gastric fundus muscle isolated from genetically hypertensive and normotensive male and female rats and found altered contractile function of gastric fundus from SHR. The alterations include increased contractions to Ca^{2+} when depolarized, increased relaxation of the above contractions by diazoxide, and decreased sensitivity to Ba^{2+}-induced contraction (only in male rats). Some of these results were consistent with changes observed in vascular smooth muscle. The studies on vas deferens and gastric fundus muscle strips from SHR support the contention that changes of contractile function observed in vascular smooth muscle cannot be atrributed solely to structural or other changes secondary to increased wall stress.

Sakai et al. (*96*) have recently shown that the prostatic portion of the vasa deferentia from SHR elicited an enhanced maximum tension induced by $10^{-4}M$ NE compared to those from normotensive Wistar rats but not to those from Kyoto-Wistar normotensive rats. Furthermore, the maximum tension induced by NE of vasa deferentia was not altered in DOC/salt hypertension. Thus, it is possible that the alterations observed in nonvascular smooth muscle may be genetically determined and may not necessarily be related to hypertension. However, Greenberg (*34*) has recently studied the responses of the vasculature and the intestinal segments to vasoactive agents during the development of 2-kidney, 1-clip Goldblatt hypertension in dogs. They showed that not only the mesenteric vascular responses (small arteries and veins) to serotonin, angiotensin II, and 9,11$_{\alpha}$-epoxymethanoprostaglandin H_2 were enhanced, intestinal smooth muscle responses to these agents were also enhanced, even before the onset of elevation of blood pressure.

3.2. Calcium Handling by Isolated Membranes

Some of the modifications of the contractile response of gastric fundus and vas deferens muscle strips from SHR to various stimulants were similar to those previously described in vascular smooth muscle in relation to the role of Ca^{2+} in smooth muscle contractile function (*20*). This raised the possibility that handling of Ca^{2+} by subcellular membranes in these nonvascular smooth muscles may be modified as well in SHR. Recently, Kwan et al. (*65*) have confirmed, using isolated microsomal fraction from gastric fundus and vas deferens of SHR and WKY, that the ATP-supported, azide-insensitive Ca^{2+} accumulation by microsomal fractions from SHR was less than that from WKY for these nonvascular smooth muscles. Although the heterogenous microsomal fractions were used, the difference in Ca^{2+} accumulation between SHR and WKY groups was not the result of differential contamination by subcellular membranes based upon the distribution of membrane marker enzyme activities in all isolated subcellular fractions. The alteration of Ca^{2+} accumulation by microsomal fractions from SHR is most likely to be a property of plasma membrane, because changes in enzyme activities, when occurring, were those associated with the plasma membranes (*65*). In addition, the subcellular membranes from the smooth muscle of rat gastric fundus have recently been isolated and purified to the plasma membranes (*68*). It was shown that the azide-insensitive, ATP-dependent Ca^{2+} accumulation by these fractions paralleled the distribution of several plasma membrane associated enzymes.

Thus, qualitatively speaking, the contractile as well as subcellular membrane studies of vascular and nonvascular smooth muscles in genetically hypertensive rats have been reasonably consistent with the notion that Ca^{2+} regulation may be altered at the cell membrane level. A recent study (*96*) of the contractile properties of vas deferens (prostatic portion) and Ca^{2+} accumulation by the isolated microsomal membranes from rats with deoxycorticosterone-salt induced hypertension (*52, 59*) indicated no significant changes in the sensitivities and maximum responses to NE and KCl, rate of relaxation upon removal of the agonist, field stimulated contraction, and ATP-supported Ca^{2+} accumulation by isolated microsomes. The differences between the findings in spontaneous hypertension and deoxycorticosterone-salt induced hypertension suggest that the abnormalities in contractility and Ca^{2+} handling of nonvascular smooth muscle may be genetically determined in SHR, but changes in DOC/salt induced hypertension may be directly related to the hypertensive process and confined to arterial muscle membranes.

4. Ionic Dysfunction as a Generalized Cell Membrane Alteration in Genetic Hypertension

The foregoing discussion has indicated clearly that altered Ca^{2+}-handling is not confined to arterial smooth muscle in genetic hypertension since altered contractile function together with defective Ca^{2+} metabolism was observed not only in venous smooth muscle (*57*), but also in nonvascular smooth muscle (*65*), as well as cardiac muscle (*72*). Most recently, increasing evidence has accumulated to indicate that defect in Ca^{2+}-handling in hypertension can be generalized to other noncardiovascular, nonsmooth muscle tissues (see below). Furthermore, the nature of the ionic dysfunction of the cell membrane is not confined to Ca^{2+} metabolism, but to Na^{2+} as well.

4.1. Dysfunction of Calcium Metabolism

In 1977, Postnov et al. (*94*) reported that erythrocytes of patients with essential hypertension had a reduced Ca store. The decreases of Ca^{2+} binding by the red blood cell membrane was also observed in SHR but not in 2-kidney, 1-clip renovascular hypertensive rats (*92*). This suggests that the defect in Ca^{2+} binding by erythrocyte membrane may be genetically determined in spontaneous hypertension in rats and essential hypertension in humans. Devynek et al. (*25*) have recently studied the mechanisms of altered Ca^{2+}-handling in erythrocyte membranes of SHR. They have observed slightly accelerated passive influx of Ca^{2+} into ATP-depleted erythrocytes and reduced Ca^{2+} binding to a high affinity site at the inner surface of the erythrocyte membranes. Although the ATP supported Ca^{2+} transport by inside-out membrane vesicles is lower in SHR compared to that in WKY, there was no significant difference in the Ca^{2+},Mg^{2+}–ATPase activity in the erythrocyte membranes from SHR and WKY. This is consistent with the hypothesis that the decreased Ca^{2+} accumulation by erythrocyte membranes is caused by increased permeability of cell membrane to Ca^{2+} and other monovalent cations as a result of decrease in membrane-bound calcium (*14, 26, 46, 104, 118*).

Postnov and Orlov (*93*) studied the kinetics of $^{45}Ca^{2+}$ efflux from adipocytes of SHR and observed three distinct pools of exchangeable Ca^{2+}, of which Ca^{2+} content in the intracellular pools of fat cells from SHR was increased compared to those from the normal controls. They further observed that isolated adipocyte plasma membranes, but not endoplasmic reticulum or mitochondria, had lower Ca^{2+} binding capacity in the absence of ATP. ATP-supported Ca^{2+} accumulation was not changed in plasma membrane, was decreased in endoplasmic reticulum, and was increased in mitochondria (*91*). Although these observations are difficult to interpret, the reduced Ca^{2+} binding to vascular smooth muscle cells, red blood cells, and fat cells from SHR might be a common denominator in the dysfunction of Ca^{2+} metabolism in spontaneous hypertension. Recent findings by Devynck et al. (*26*) of reduced Ca^{2+} binding to plasma membrane enriched fractions of cardiac muscle, liver, and brain tissues provided additional support of this contention. Since these findings were obtained from young SHR (3 wk) before the elevation of blood pressure, the reduced Ca^{2+} binding of cell membrane is therefore implicated in the pathogenesis of hypertension and may be of genetic origin.

4.2. Dysfunction of Sodium Metabolism

Dietary salt intake has been considered a major environmental factor in addition to genetic factors in the etiology and pathogenesis of hypertensive diseases. However, there have been considerable variations in the nature and interpretation of the dysfunction of Na^+-handling by vascular smooth muscle (*see* review articles (*21, 50, 101*). In general, most evidence suggests that permeability of vascular muscle cell membrane to Na is increased in genetic hypertension (*50, 101*). But increased as well as decreased Na^+ pump activity in arterial muscle of SHR and DOC/salt-induced hypertensive rats have been reported (*28, 40*). Clearly, it is difficult to explain vascular alterations in hypertension both as a result of decreased and increased Na pump activities. The increased Na^+ pump activity has been attributed to an adaptive response induced by an increase of permeability to Na^+ (*40*) whereas the decreased Na^+ pump activity, usually found in volume-expansion hypertension (*40, 84, 87*), has been attributed to the inhibition by an unidentified humoral substance of the Na^+,K^+–ATPase (*41*). The inconsistent findings may be partly a result of the methodologies used to study the Na^+ pump activity. Measurements of rubidium uptake (*84*), resting membrane potential (*43*) and K^+-induced relaxation of vascular muscle strips (*115*) and cellular Na^+ concentration using selective ion electrodes (*27*) have been employed to study the Na^+ pump activity and changes allow alternative interpretations (*21*). Direct measurement of the Na^+,K^+–ATPase activity using isolated vascular subcellular membrane has not been fruitful so far because of difficulties associated with the low level of this activity in the presence of high level of plasma membrane basal ATPase and ADPase activities (*67*) and low sensitivity of the pump ATPase to inhibition by ouabain in rat tissues. Nevertheless, an alteration in Na pump has also been reported in cardiac muscle and brain from SHR (*7*). An alteration of Na^+fluxes has also been observed in leucocytes and lymphocytes in hypertensive patients (*95, 105*). Although there is much disagreement between findings of different workers, a large body of evidence suggests that Na^+handling (Na pump activity, intracellular Na^+ concentration, and permeability to Na^+) is altered in the erythrocytes of SHR and patients with essential hypertension (*23, 29, 94*). It has been suggested that altered Na^+ transport in erythrocytes may be useful as a "marker" for diagnosis of essential hypertension because alteration was found only in essential hypertension or SHR, not in patients with renovascular hypertension (*31*) or rats with experimental hypertension (*24*). This, however, has recently been challenged by the studies of Swarts et al. (*103*), who found no significant alterations in the ouabain-sensitive and furosemide-sensitive, ouabain-insensitive Na^+ and K^+ fluxes as well as the specific activity of Na^+,K^+–ATPase in erythrocytes of patients with essential hypertension compared to normotensive subjects. Wambach et al. (*113*) on the other hand, have found that ouabain-sensitive Na^+,K^+–ATPase activity was increased not only in the erythrocyte ghosts of patients with essential hypertension, but also in those with secondary hypertension and therefore suggested that Na^+,K^+–ATPase in erythrocyte ghosts is not a marker for essential hypertension.

4.3. Calcium, Sodium, and Hypertension

Blaustein (*12*) has postulated an interesting hypothesis to correlate the changes of Na^+ metabolism in vascular smooth muscle and the elevation of blood pressure. He proposed that Ca^{2+} concentration gradient across the plasma membrane is tightly linked to the Na^+ concentration gradient via a Na^+–Ca^{2+} exchange carrier. A reduction of the Na^+ gradient as a result of a rise in cytoplasmic Na^+ caused by increased membrane leaks to Na^+ will therefore reduce the extrusion of Ca^{2+} via the Na^+–Ca^{2+} exchange carrier leading to a rise in cytoplasmic Ca^{2+} concentration that is in turn translated into elevated vascular tone followed by increased vascular resistance and hence hypertension. However, the role of Na^+–Ca^{2+} exchange in the regulation of cytoplasmic Ca^{2+} concentration in smooth muscle, at this time, is not substantiated. Evidence against the hypothesis that Ca homeostasis is obligatorily linked to the Na^+–Ca^{2+} exchange system has been amply reviewed (*15, 111*). Plasma membrane vesicles isolated from vascular (*23*) as well as nonvascular smooth muscle (*37*) were shown to be capable of Na^+–Ca^{2+} exchange, but its low efficacy compared to that of ATP-supported Ca accumulation by smooth muscle plasma membrane raised doubt about its physiological importance in the Ca^{2+} extrusion process (*see* Chapters 3 and 11 for more detailed discussion).

Some evidence of changes in intracellular Ca^{2+} and contractility as the Na^+-gradients change is consistent with the operation of a Na^+–Ca^{2+} exchange system. Nilsson and Mulvany (*73*) have shown the prolonged incubation (30 min) of mesenteric arterial ring segments from SHR with 1 m*M* ouabain reduced the response to maximal NE stimulation and eliminated the increased sensitivity to Ca^{2+} of NE-induced contraction. Although their data do not exclude the possibility of a depressed Na^+–Ca^{2+} exchange process, they attributed their results to an increased permeability of NE-dependent Ca^{2+} channel.

5. Conclusion and Comments

As most studies indicated, altered Ca^{2+} handling could account for some of the altered vascular function in hypertension. However, because of the heterogeneity of the Ca^{2+}-handling process that includes the Ca^{2+} movements (binding, release, extrusion, sequestration, and exchange), the Ca^{2+} channels (voltage-dependent, receptor-dependent, and their subtypes?), and the Ca^{2+} sites (extracellular, superficially bound inner surface of cell membranes, endoplasmic reticulum, mitochondria, and others?), it is not surprising that the Ca^{2+} metabolism may be differently modified in hypertension in view of the structural and functional diversity of smooth muscles. Since the blood pressure control system is a function of variables representing several dynamic, integrated, and interacting regulatory mechanisms in equilibrium, hypertension has been considered a multifactorial disease. Dysfunction in one or more of these regulatory mechanisms may initiate the process leading to hypertension with concomitant adaptive changes of other regulatory mechanisms to reset the equilibrium. Defects in Ca^{2+} regulation observed in vascular smooth muscle provided a plausible and logical explanation of some of the abnormalities in myogenic control of smooth muscle tone associated with hypertension. Similarly, defects in Ca^{2+} handling by central and/or peripheral nerve systems may have disturbed the transmitter release in relation to the alterations in neurogenic control of vasoconstriction (*13*). If the Ca^{2+} dysfunction is genetically determined, the alterations in Ca^{2+} handling may also occur in some tissues or cells not related to blood pressure regulation. Such a general dysfunction in Ca^{2+} handling may be a plausible working hypothesis of the etiology of genetic hypertension. This hypothesis has recently received strong support (*see* Section 4.2.). A general disturbance of calcium metabolism in SHR has been reflected as decreased serum ionized Ca^{2+} and increased urinary Ca^{2+} excretion (*51*). There is a good correlation between increased daily Ca^{2+} intake, increased serum Ca^{2+}, and lowering of blood pressure in SHR and WKY (*75*). Patients with essential hypertension ingest less calcium than the normotensive subjects (*74*). These findings were independent of Na^+ or K^+ intake. McCarron et al. (*74*) suggested that inadequate Ca intake may be a previously unrecognized factor in the development of hypertension. It remains to be tested whether long term dietary Ca supplementation will reverse the generalized membrane defect in Ca^{2+} handling in genetic hypertension.

Acknowledgments

The author appreciates the support by Ontario Heart Foundation and helpful discussion with and suggestions from his colleagues during the preparation of this chapter.

References

1. Altman, J., F. Daponte, and M. Worcel. Evidence for a smooth muscle abnormality in Okamoto spontaneous hypertension. *Br. J. Pharmacol.* 59:621–625, 1977.

2. Altman, A. and G. Simon. Abnormal venous wall composition in spontaneously hypertensive rats. *Proc. Soc. Exp. Biol. Med.* 167:97–102, 1980.
3. Amer, M. S. Cyclic nucleotides in disease: on the biochemical etiology of hypertension. *Life Sci.* 17:1021–1038, 1976.
4. Andersson, R., K. Nilsson, J. Wikberg, S. Johansson, E. Mohme-Lundhollm, and L. Lundholm. Cyclic nucleotides and the contraction of smooth muscle. *Adv. Cyc. Nucl. Res.* 5:491–518, 1975.
5. Aoki, K., K. Yamashita, and K. Hotta. Calcium uptake by subcellular membrane from vascular smooth muscle of spontaneously hypertensive rats. *Jap. J. Pharmacol.* 26:624–627, 1976.
6. Aoki, K., N. Yamashita, K. Tazumi, and K. Hotta. ATPase activity and Ca^{2+} binding activity of subcellular membrane of arterial smooth muscle in spontaneously hypertensive rats. *Jap. Heart J.* 15:180–181, 1974.
7. Ayachi, C. C. Ho and Aks, Ho. Increased brain Na^+,$K^{\pm}$ATPase activity in the spontaneously hypertensive rats. *Fed. Proc.* 38:700, 1979.
8. Berthelot, A., A. Gairard, M. Goyault, and F. Pernot. Relation between calcium and cardiovascular reactivity in mineralocorticoid-induced hypertension in the rat. *Br. J. Pharmacol.* 70:301–306, 1980.
9. Bhalla, R. C., R. V. Sharma, and S. Ramanathan. Possible role of phosphorylation–dephosphorylation in the regulation of calcium metabolism in cardiovascular tissues of SHR. *Hypertension* 2:207–214, 1980.
10. Bhalla, R. C., R. C. Webb, D. Singh, T. Ashley, and T. Brock. Calcuim fluxes, calcium binding and adenosine cyclic 3′, 5′-monophosphate-dependent protein kinase activity in the aorta of spontaneously hypertensive and Kyoto-Wistar-normotensive rats. *Mol. Pharmacol.* 14:468–477, 1978.
11. Bhalla, R. C., R. Webb, D. Singh, and T. Brock. Role of cAMP in rat aortic microsomal phosphorylation and calcium uptake. *Am. J. Physiol.* 234:H508–H514, 1978.
12. Blaustein, M. P. Sodium ions, calcium ions, blood pressure regulation and hypertension: a reassessment and a hypothesis. *Am. J. Physiol.* 232:C165–C173, 1977.
13. Bohr, D. F. What makes the pressure go up? A hypothesis. *Hypertension* 3(Suppl.II) 160–165, 1981.
14. Caroni, P. and E. Carafoli. Modulation by calcium of the potassium permeability of dog heart sarcolemmal vesicles. *Proc. Natl. Acad. Sci.* (USA) 79:5763–5767, 1982.
15. Casteels, R. and G. Droogmans. Does Na-Ca exchange play a role in the regulation of contraction in vascular smooth muscle? In *Vascular Neuroeffector Mechanisms.* J. A. Bevan, T. Godfraind, R. A. Maxwell and P. M. Vanhoutte, Eds. Raven Press, New York. pp 237–239, 1980.
16. Caulfield, M. P., G. Paterson, and A. R. Wayyes. Altered response to noradrenaline of vasa deferentia from genetically hypertensive rats. *J. Physiol.* 266:18p–19p, 1977.
17. Cheng, J. B. and S. Shibata. Vascular relaxation in the spontaneously hypertensive rat. *J. Cardiovasc. Pharmacol.* 3:1126–1140, 1981.

18. Cohen, D. M., R. C. Webb, and D. F. Bohr. Nitroprusside-induced vascular relaxation in DOCA-hypertensive rats. *Hypertension* 4:13–19, 1982.
19. Cohen, M. L. and B. A. Berkowitz. Decreased vascular relaxation in hypertension. *J. Pharmacol. Exp. Ther.* 196:396–, 1976.
20. Corbett, D. A., M. T. Gottberg, V. C. Swamy, C. R. Triggle, and D. T. Triggle. Reactivity of vasa deferentia from spontaneously hypertensive and normotensive Wistar rats. *Can. J. Physiol. Pharmacol.* 58:656–665, 1980.
21. Daniel, E. E. Role of altered vascular smooth muscle function in hypertension. In *Vasodilatation*. P. M. Vanhoutte and I. Leusen, eds. Raven Press, New York. pp 381–390.
22. Daniel, E. E., and C. Y. Kwan. Control of contraction of vascular muscle—relation to hypertension. *Trends Pharmacol. Sci.* 2:270–223, (1981).
23. Daniel, E. E., A. K. Grover, and C. Y. Kwan. Isolation and properties of plasma membrane from smooth muscle. *Fed. Proc.* 41:2898–2904, 1982.
24. De Mendonca, M., M. L. Grichois, R. P. Garay, J. Sassard, D. Ben-Ishay and P. Meyer. Abnormal net Na^+ and K^+ fluxes in erythrocytes of three varieties of genetically hypertensive rats. *Proc. Natl. Acad. Sci. (USA)* 77:4283–4286, 1980.
25. Devynck, M-A, M. G. Pernollet, A-M. Nunez, and P. Meyer. Analysis of calcium handling in erythrocyte membranes of genetically hypertensive rats. *Hypertension* 3:397–403, 1981.
26. Devynck, M. A., M. G. Pernollet, A. M. Nunez, and P. Meyer. Calcium binding alteration in plasma membrane from various tissues of spontaneously hypertensive rat. *Clin. Exp. Hypertension* 3:797–808, 1981.
27. Folkow, B. Physiological aspects of primary hypertension. *Physiol. Rev.* 62:347–504, 1982.
28. Friedman, S. M. and M. Nakashiwa. Evidence for enhanced Na transport in hypertension induced by DOCA in the rat. *Can. J. Physiol. Pharmacol.* 56:1029–1035, 1978.
29. Friedman, S. M., M. Nakashima, and R. A. Mcindoe. Glass electrode measurement of net Na^+ and K^+ fluxes in erythrocytes of the spontaneously hypertensive rat. *Can. J. Physiol. Pharmacol.* 55:1302–1310, 1977.
30. Fitzpatrick, D. F. and A. Szentivanyi. The relationship between increased myogenic tone and hyperreponsiveness in vascular smooth muscle of spontaneously hypertensive rats. *Clin. Exp. Hypertension* 2:1023–1037, 1980.
31. Garay, R. P., P. Hannaert, G. Dagher, C. Nazaret, I. Maridonneau, and P. Meyer. Abnormal erythrocyte Na^+, K^+ cotransport system. A proposed genetic marker of essential hypertension. *Clin. Exp. Hypertension* 3: 851–860, 1981.
32. Furuta, Y. Studies on sodium and calcium contents of cardiovascular tissue in experimental hypertension. *Jap. Circ. J.* 41:19–28, 1977.
33. Greenberg, S. Venous function in human and experimental hypertension. *Methods Find. Exp. Clin. Pharmacol.* 3:233–245, 1981.

34. Greenberg, S. Vascular responses of the perfused intestine to vasoactive agents during the development of two-kidney, one-clip Goldblatt hypertension in dogs. *Circ. Res.* 48:895–906, 1981.
35. Greenberg, S., E. C. Palmer, and W. M. Wilborn. Pressure-independent hypertrophy of veins and pulmonary arteries of spontaneously hypertensive rats. *Clin. Sci. Mol. Med. (Suppl.)* 55:31–36, 1978.
36. Greenberg, S. and W. Wilborn. Functional and structural changes in veins in spontaneous hypertension. *Arch. Int. Pharmacodyn. Ther.* 258:208–233, 1982.
37. Grover, A. K., C. Y. Kwan, and E. E. Daniel. Na–Ca interactions in smooth muscle: evidence for Na–Ca exchange by rat myometrium plasma membrane vesicles. *Am. J. Physiol.* 240:C175–C182, 1981.
38. Grover, A. K., C. Y. Kwan, and E. E. Daniel. High affinity pH dependent passive Ca binding by myometrial plasma membrane vesicles. *Am. J. Physiol.* 244:C61–C67, 1983.
39. Gupta, R. C., R. C. Bhalla, and R. V. Sharma. Altered distribution and properties of c-AMP-dependent protein kinase enzymes in spontaneously hypertensive rat aorta. *Biochem. Pharmacol.* 31:1837–1841, 1982.
40. Haddy, F., M. Pamnani, and D. Clough. The sodium–potassium pump in volume expended hypertension. *Clin. Exp. Hypertension* 1:295–336, 1978.
41. Haddy, F. J. and H. W. Overback. Minireview: the role of humoral agents in volume expanded hypertension. *Life Sci*: 19:935, 948, 1976.
42. Harris, A., V. C. Swamy, D. J. Triggle, and D. H. Waters. Responses to noradrenaline of portal vein strips from normotensive and spontaneously hypertensive rats. *J. Anton. Pharmac.* 1:61–66, 1980.
43. Hermsmeyer, K., P. W. Abel, and A. J. Trapani. Norepinephrine sensitivity and membrane potentials of caudel arterial muscle in DOCA-salt DAHL and SHR hypertension in the rat. *Hypertension* 4(Suppl.II) 49–51, 1982.
44. Hinke, J. A. M. Effect of Ca^{2+} upon contractility of small arteries from DOCA-hypertensive rats. *Circ. Res.* 18–19(Suppl. I) 123, 1966.
45. Holloway, E. T. and D. F. Bohr. Reactivity of vascular smooth muscle in hypertensive rats. *Circ. Res.* 33:678–685, 1973.
46. Holloway, E. T., M. D. Sitrin, and D. F. Bohr. Calcium dependence of vascular smooth muscle from normotensive and hypertensive rats. In *Hypertension.* J. Genest and E. Koiw, eds. Springer Verlag, Berlin, pp 400–408, 1972.
47. Janis, R. A. and D. J. Triggle. Effect of diazoxide on aortic reactivity to calcium in spontaneously hypertensive rats. *Can. J. Physiol. Pharmacol.* 51:621–626, 1973.
48. Johansson, B. Vascular smooth muscle reactivity. *Ann. Rev. Physiol. 43:359–370, 1981.*
49. Jones, A. W. Altered transport in large and small arteries from SHR and the influence of Ca. *Circ. Res.* 34–35 (Suppl. I):117–122, 1974.
50. Jones, A. W. Ionic dysfunction and hypertension. *Adv. Microcirc.* 11:134–159, 1982.
51. Kesteloot, H. and J. Geboers. Calcium and blood pressure. *The Lancet* (No. 8276) 1:813–815, 1982.

52. Kwan, C. Y. Development of a deoxycorticosterone-salt induced hypertension in rats without unilateral nephrectomy. *IRCS Med. Sci.* 10:610, 1982.
53. Kwan, C. Y. Mg^{2+} or Ca^{2+}activated ATPase activities of plasma membranes isolated from vascular smooth muscle. *Enzyme* 28:317–327, 1982.
54. Kwan, C. Y. Elevated alkaline phosphatase activity in cardiovascular tissues of spontaneously hypertensive rat. *IRCS Med. Sci.* 11:690–691, 1983.
55. Kwan, C. Y. Characteristics of plasmalemma alkaline phosphatase of rat mesenteric artery. *Blood Vessels* 20:109–121, 1983.
56. Kwan, C. Y. and E. E. Daniel. Calcium transport by plasma membrane vesicles isolated from vascular smooth muscle of normal and hypertensive rats. In *Vasodilatation.* P. M. Vanhoutte and I. S. Leusen, eds. Raven Press, New York, pp 405–415, 1981.
57. Kwan, C. Y. and E. E. Daniel. Biochemical abnormalities of venous plasma membrane fraction isolated from spontaneously hypertensive rats. *Eur. J. Pharmacol.* 75:321–324, 1981.
58. Kwan, C. Y. and E. E. Daniel. Arterial muscle membrane abnormalities of hydralazine-treated spontaneously hypertensive rats. *Eur. J. Pharmacol.* 82:187–190, 1982.
59. Kwan, C. Y. and Grover, A. K. Membrane abnormalities occur in vascular smooth muscle but not in non-vascular smooth muscle from rats with deoxycorticosterone-salt induced hypertension. *J. Hypertension* 1:257–265, 1983.
60. Kwan, C. Y., L. Belbeck, and E. E. Daniel. Abnormal biochemisty of vascular smooth muscle plasma membrane as an important factor in the initiation and maintenance of hypertension in rats. *Blood Vessels* 16:259–268, 1979b.
61. Kwan, C. Y., L. Belbeck, and E. E. Daniel. Abnormal biochemistry of vascular smooth muscle plasma membrane isolated from hypertensive rats. *Mol. Pharmacol.* 17:137–140, 1980.
62. Kwan, C. Y., L. Belbeck, and E. E. Daniel. Characteristics of arterial plasma membrane in renovascular hypertension in rats. *Blood Vessels* 17:131–140, 1980.
63. Kwan, C. Y., G. D. Ford, and M. L. Hess. ATP, sarcoplasmic reticulum and vascular smooth muscle. *Am. J. Physiol.* 244:C115–C118, 1983.
64. Kwan, C. Y., R. E. Garfield, and E. E. Daniel. An improved procedure for the isolation of plasma membrane from rat mesenteric arteries. *J. Mol. Cell. Cardiol.* 11:639–659, 1979a.
65. Kwan, C. Y., A. K. Grover, and Y. Sakai. Abnormal biochemistry of subcellular membranes isolated from non-vascular smooth muscles of spontaneously hypertensive rats. *Blood Vessels* 19:273–283, 1982.
66. Kwan, C. Y., A. K. Grover, C. R. Triggle, and E. E. Daniel. On the oxalate stimulation of ATP-dependent calcium accumulation by smooth muscle subcellular membranes. *Biochem. Int.* 6:713–722, 1983.
67. Kwan, C. Y. and T. Ramlal. On the inhibition of smooth muscle membrane ATPase by sodium azide. *Biochem. Int. 4:439–449, 1982.*

68. Kwan, C. Y., Y. Sakai, A. K. Grover and R. M. K. W. Lee. Isolation and characterization of plasma membrane fraction from gastric fundus smooth muscle of the rat. *Mol. Physiol.* 2:107–120, 1982.
69. Kreye, V. A. W. and E. Schlicker. Effects of vasodilator drugs, alkaline phosphatase and cyclic AMP-dependent protein kinase on the ^{45}Ca-uptake of sarcolemmal microsomes from human umbilical arteries. *Br. J. Pharm.* 70:537–544, 1980.
70. Kukowitz, W. R., G. Poch, and S. Holzmann. Cyclic nucleotides and relaxation of vascular smooth muscles. In *Vasodilatation.* P. M. Vanhoutte and I. Leusen, eds. Raven Press, New York, pp 339–353, 1981.
71. Lee, R. M. K. W., J. B. Forrest, R. E. Garfield, and E. E. Daniel. Ultrastructural changes in mesenteric arteries from spontaneously hypertensive rats. *Blood Vessels* 20:72–91 1983.
72. Limas, C. J. and J. N. Cohn. Defective calcium transport by cardiac sarcoplasmic reticulum in spontaneously hypertensive rats. *Circ. Res.* 40(5) (Suppl. I):I62–I69, 1979.
73. Nilsson, H. and M. J. Mulvany. Prolonged exposure to ouabain eliminates the greater norepinephrine-dependent calcium sensitivity of resistance vessels in spontaneously hypertensive rats. *Hypertension* 3:691–697, 1981.
74. McCarron, D. A., C. D. Morris, and C. Cole. Dietary calcium in human hypertension. Science 217:267–269, 1982.
75. McCarron, D. A., N. N. Yung, B. A. Ugoretz, and S. Krutzik. Disturbances of calcium metabolism in the spontaneously hypertensive rat. *Hypertension* 3 (Suppl. I):I162–I167, 1981.
76. Mellgren, R. L., G. R. Slaughter, and J. A. Thomas. Dephosphorylation of phosphoproteins by *E. coli* alkaline phosphatase. *JBC* 252(17)6082–6089, 1977.
77. Moore, L., L. Hurwitz, G. R. Davenport, and E. J. Landon. Energy-dependent calcium uptake activity of microsomes from the aorta of normal and hypertensive rats. *Biochim. Biophys. Acta* 413:432–443, 1975.
78. Mulvany, M. J. and N. Nyborg. An increased calcium sensitivity of mesenteric resistance vesicles in young and adult spontaneously hypertensive rats. *Br. J. Pharmacol.* 71:583–596, 1980.
79. Mulvany, M. J., P. K. Hansen, and C. Aalkjaer. Direct evidence that the greater contractility of resistance vessels in spontaneously hypertensive rats is associated with a narrower lumen, a thicker media and a greater number of smooth muscle cell layers. *Circ. Res.* 43:854–864, 1978.
80. Mulvany, M. J., N. Korsgaard, and N. Nyborg. Evidence that the increased calcium sensitivity of resistance vessels in spontaneously hypertensive rats is an intrinsic defect of their vascular smooth muscles. *Clin. Exp. Hypertension* 3:749–761, 1981.
81. Mushlin, P. S., B. V. R. Sastry, R. C. Boerth, M. J. Surber, and E. J. Landon. Dithiothreital-induced alterations of blood pressure. Vascular reactivity and aortic microsomal calcium uptake in spontaneously hypertensive rats. *J. Pharmacol. Exp. Ther.* 207:331–339, 1978.

82. Noon, J. P., P. J. Rice, and R. J. Baldessrini. Calcium leakage as a cause of high resting tension in vascular smooth muscle from spontaneously hypertensive rats. *Proc. Natl. Acad. Sci. (USA)* 75:1605–1607, 1978.
83. Overbeck, H. W. and L. L. Conrad. Intracellular distribution of calcium-45 in arteries of normal and hypertensive dogs. *Proc. Soc. Exp. Biol. Med.* 127:565–570, 1968.
84. Overbeck, H. W., M. B. Pamnani, T. Akera, T. M. Brody, and F. J. Haddy. Depressed function of a ouabain-sensitive sodium-potassium pump in blood vessels from renal hypertensive dogs. *Circ. Res.* 38(Suppl. II) 48–52, 1976.
85. Owens, G. K. and S. M. Schwartz. Alterations in vascular smooth muscle mass in the spontaneously hypertensive rat: role of cellular hypertrophy, hyperploidy and hyperplasia. *Circ. Res.* 51:280–289, 1982.
86. Pamnani, M. and H. Overbeck. Abnormal ion and water composition of veins and normotensive arteries in coarctation hypertension in rats. *Circ. Res.* 38:375–738, 1978.
87. Pamnani, M. B., D. L. Clough, and F. J. Haddy. Altered activity of the sodium-potassium pump in arteries of rats with steroid hypertension. *Clin. Sci. Mol. Med.* 55:415–435, 1978.
88. Pang, C. C. Y. and M. C. Sutter. Differential effects of D600 on contractile response of aorta and portal vein from spontaneously hypertensive rats. *Blood Vessels* 18:120–127, 1981.
89. Pegram, B. L. and B. Ljung. Neuroeffector function of isolated portal vein from spontaneously hypertensive and Wistar-Kyoto rats: Dependence on external calcium concentration. *Blood Vessels* 18:89–99, 1981.
90. Postnov. Y. V. and S. N. Orlov. Features of intracellular calcium distribution in the adipose tissue of spontaneously hypertensive rats (SHR). *Experientia* 35: 1481, 1979.
91. Postnov, Y. V. and S. N. Orlov. Evidence of altered calcium accumulation and calcium binding by membranes of adipocytes in spontaneously hypertensive rat. *Pfluger's Arch.* 385:85–89, 1980.
92. Postnov, Y. V., S. N. Orlov, and N. I. Pokudin. Decrease of calcium binding by the red blood cell membrane in spontaneously hypertensive rats and in essential hypertension. *Pfluger's Arch.* 379:191–195, 1979.
93. Postnov, Y. V., S. N. Orlov and N. I. Pokudin. Alteration of the intracellular calcium pool of adipose tissue in spontaneously hypertensive rats—no effect of peripheral immunosympathectomy. *Pfluger's Arch.* 390:256–259, 1981.
94. Postnov, Y. V., S. N. Orlov, A. S. Shevchenko, and A. M. Adler. Altered sodium permeability, calcium binding and Na,K-ATPase activity in red blood cell membrane in essential hypertension. *Pfluger's Arch.* 371:263–269, 1977.
95. Poston, L., R. B. Jones, P. J. Richardson, and P. J. Hilton. The effect of antihypertensive therapy on abnormal leucocyte sodium transport in essential hypertension. *Clin. Exp. Hypertension* 3:693–702, 1980.
96. Sakai, Y., C. Y. Kwan, A. K. Grover, and T. Ramlal. Altered contractile and calcium handling properties of vas deferens in spontaneously

hypertensive rat (SHR), but not in rats with deoxycorticosterone-salt induced hypertension (DHR). *Fed. Proc. (Absts.)* 42:894, 1983.

97. Shibata, S., M. Kuchii, and T. Taniguchi. Calcium flux and binding in the aortic smooth muscle from the spontaneously hypertensive rat. *Blood Vessels* 12:279–289, 1975.

98. Shibata, S., K. Kurahachi, and M. Kuchii. A possible etiology of contactility impairment of vascular smooth muscle from spontaneously hypertensive rats. *J. Pharmacol. Exp. Ther.* 185:406–417, 1973.

99. Simon, G. Altered venous function in hypertensive rats. *Circ. Res.* 38:412–418, 1976.

100. Somlyo, A. P. and A. V. Somlyo. Vascular smooth muscle I. Normal structure, pathology, biochemistry and biophysics. *Pharmacol. Rev.* 20:197–272, 1968.

101. Swales, J. D., Ion transport in hypertension: review. *Biosci. Report* 2:969–990, 1982.

102. Swamy, V. C. and D. J. Triggle. The responses of carotid vascular strips from spontaneously hypertensive and normotensive rats. *Can. J. Physiol. Pharmacol.* 58:53–59, 1980.

103. Swarts, H. G. P., S. L. Bonting, J. J. H. H. M. DePont, M. A. H. S. Stekhoen, T. A. Thien, and A. V. Laar. Cation fluxes and Na^+–K^+ activated ATPase activity in erythrocytes of patients with essential hypertension. *Hypertension* 3:641–649, 1981.

104. Romero, P. J. Role of membrane-bound Ca in ghost permeability to Na^+ and K^+. *J. Mem. Biol.* 29:329–343, 1976.

105. Thomas, R. D., R. P. S. Edmondson, P. I. Hilton, and N. F. Jones. Abnormal sodium transport in leucocytes from patients with essential hypertension and the effect of treatment. *Clin. Sci.* 48:1695–1709, 1975.

106. Tobian, L. and G. Chesley. Calcium content of arteriolar walls in normotensive and hypertensive rats. *Proc. Soc. Expt. Biol. Med.* 121:340–343, 1966.

107. Tomera, J. F. and C. Harakal. Calcium fractions in aortic segments from hypertensive rabbits. *J. Cardiovasc. Pharmacol.* 4:592–600, 1982.

108. Triggle, C. R. and D. J. Triggle. An analysis of the actions of cations of the lanthanide series on the mechanical responses of guinea pig ileal longitudinal muscle. *J. Physiol. (London)* 254:39–54, 1976.

109. Twietmeyer, T. A., R. C. Bhalla, and J. A. Maynard. Acid and alkaline phosphatase activities and calcium transport in aortic smooth muscle from DOCA hypertensive rats. *J. Mol. Cell. Cardiol.* 10:1131–1140, 1978.

110. Twietmeyer, T. M., J. A. Maynard, and R. C. Bhalla. Localization of acid and alkaline phosphatase activities in the blood vessels of normotensive and hypertensive rats. *Angiology* 30:317–326, 1979.

111. Van Breemen, C., P. Aaronson, and R. Loutzenhiser. Na-Ca interactions in mammalian smooth muscle. *Pharmacol. Rev.* 30:167–205, 1979.

112. Van Breemen, C., O. Hwang, and B. Siegel. The lanthanum method. In *Excitation–Contraction Coupling in Smooth Muscle.* R. Casteels, T. Godfraind, J. C. Ruegg, eds. North-Holland, Amsterdam. pp 243–252, 1977.

113. Wamback, G. and Helber, A. Na–K–ATPase in erythrocyte ghosts is not a marker for primary hypertension. *Clin. Exp. Hypertension* 3:663–674, 1981.

114. Webb, R. C. and R. C. Bhalla. Altered calcium sequestration by subcellular fractions of vascular smooth muscle from spontaneously hypertensive rats. *J. Mol. Cell. Cardiol.* 8:651–660, 1976.
115. Webb, R. C. and D. F. Bohr. Potassium relaxation of vascular smooth muscle from spontaneously hypertensive rats. *Blood Vessels* 16:71–79, 1979.
116. Webb, R. C. and D. F. Bohr. Regulation of vascular tone, molecular mechanisms. *Prog. Cardiovasc. Dis.* 24:213–242, 1981.
117. Wei, J. W., R. A. Janis, and E. E. Daniel. Ca accumulation and enzymatic activities of subcellular fractions from aortae and ventricles of genetically hypertensive rats. *Circ. Res.* 39:133–140, 1976.
118. Wei, J. W., R. A. Janis, and E. E. Daniel. Studies on subcellular fractions from mesenteric arteries of spontaneously hypertensive rats: alterations in both calcium uptake and enzyme activities. *Blood Vessels* 13:293–308, 1976.
119. Weiss, G. B. Cellular pharmacology of lanthanum. *Ann. Rev. Pharmacol.* 14:343–345, 1974.
120. Winquist, R. J., R. C. Webb, and D. F. Bohr. Vascular smooth muscle in hypertension. *Fed. Proc.* 41:2387–2393, 1982.
121. Woodcock, E. A. and C. I. Johnston. Cardiovascular adrenergic receptors in experimental hypertension in the rat. *Circ. Res. 46:*(Suppl. I) I–45–146, 1980.
122. Zsoter, T. T. and C. Wolchinsky. Effect of antihypertensive on calcium kinetics in the aorta of spontaneously hypertensive rats. *Arch. Int. Pharmacodyn.* 234:287–293, 1978.
123. Zsoter, T. T., C. Wolchinsky, N. F. Henein, and L. C. Ho. Calcium kinetics in the aorta of spontaneously hypertensive rats. *Cardiovasc. Res.* 11:353–357, 1977.

Chapter 13

Adrenergic Interactions in Smooth Muscle Contractility

C. R. Triggle

1. Introduction

The objective of this chapter is to present a review of how adrenergic mechanisms affect smooth muscle. In particular the direct and indirect effects of norepinephrine and epinephrine on smooth muscle will be discussed with special reference to the nomenclature and distribution of adrenoceptors and the role and sources of Ca^{2+} used to mediate the excitatory and inhibitory events initiated by activation of these receptors.

2. Classification of Receptor Types: Development of Current Concepts

A general summary of this section is presented in Table 1, while the following represents a brief elaboration of the development of our knowledge of the tissue and cellular distribution of the adrenoceptor associated with smooth muscle.

2.1. α- and β-Adrenoceptors

The basis for our current classification of the sites at which norepinephrine and epinephrine act was the observation pre-

Table 1
Classification of Adrenoceptors: Development of Current Concepts

Receptor type	Location	Proposed by	Date	Key references
α, β	Postsynaptic smooth, cardiac muscle	Alhquist	1948	*1, 2*
β_1	Postsynaptic cardiac muscle			
β_2	Postsynaptic smooth muscle	Levy	1967	*55*
α_2	Presynaptic	Farnebo and Malmfors	1971	*35*
α_2	Postsynaptic	Docherty, McDonald, and McGrath	1979	*83*
γ	"Synaptic"	Hirst and and Neild	1981	*43,44*

sented by Ahlquist in 1948 (*1*), and recently reviewed (*2*), that the order of potency of a series of catecholamines differed from tissue to tissue. In fact, two orders of potency were noted and reflected by excitation in the uterus, blood vessels, and the nictitating membrane associated with mediation by the α-receptor, and inhibition in the uterus, blood vessels, and bronchioles, and excitation in cardiac muscle associated with activation of the β-receptor. The concept of a "dual adrenoceptor" hypothesis was thus introduced. Support for this concept was later provided by the availability of relatively selective adrenoceptor antagonists such as phenoxybenzamine and phentolamine for the α-adrenoceptor, and dichloroisoproterenol and propranolol for the β-receptor.

2.2. β-Adrenoceptor Subtypes

In 1967 Lands et al. (*55*) proposed that there existed two subtypes of β-adrenoceptors, namely β_1 and β_2, with the β_1-adrenoceptor being primarily associated with cardiac muscle and β_2 in smooth muscle and other systems. Lands' hypothesis was supportive of earlier publications; however, a number of anomalies to the original β_1; β^2 hypothesis have been noted with the result that currently it is believed that the relative proportion of the two beta adrenoceptors in a tissue may vary considerably from animal to animal as well as tissue to tissue. For more details the reader is referred to a recent review (*28*).

2.3. Presynaptic Adrenoceptors

During the last decade considerable attention has been focused upon the manner by which nerve endings respond to substances in the extracellular fluid with an appropriate change in action potential-evoked neurotransmitter release and this literature has been extensively reviewed (*56, 85*), In brief, it has been noted that noradrenergic terminal axons respond to α-adrenoceptor agonists with a depression of transmitter release and to α-adrenoceptor antagonists with facilitation of release. Conversely β-adrenergic agonists facilitate transmitter release and β-adrenoceptor antagonists depress release.

An important development in the study of the presynaptic α-adrenoceptor has been the demonstration that this receptor type differs from the classical postsynaptic α-adrenoceptor. The presynaptic α-adrenoceptor, designated α_2, differs in its structural requirements compared to the postsynaptic adrenoceptor, designated the α_1-adrenoceptor.

For the α_1-adrenoceptor agonists: norepinephrine > naphazoline > tramazoline > phenylephrine > methylnoradrenaline > clonidine; and for α_1-antagonists: prazosin > phentolamine > yohimbine > rauwolscine. For the α_2-adrenoceptor agonists: tramazoline > α-methylnorepinephrine > clonidine > naphazoline > norepinephrine >> phenylephrine and for α_2-antagonists: rauwolscine > yohimbine > phentolamine > prazosin.

2.4. Postjunctional α_2-Adrenoceptors

Until recently the anatomical distinction between the postsynaptic location of the α_1-adrenoceptor and the presynaptic α_2-adrenoceptor seemed to be acceptable. In a number of systems norepinephrine-mediated responses are resistant to prazosin, indicating that norepinephrine is not acting through an α_1-adrenoceptor. On the basis of the results obtained from the use of selective α_2-agonists, reported to have no affinity for α_1-adrenoceptors (*90*), such as M-7 (*34*), the azepine derivatives B-HT 920 and B-HT 933 (Fig. 1) (*53,83,91*) and UK-14,304 (*23*), there is evidence for the "postjunctional" location of not only the α_1-adrenoceptor but also an adrenoceptor similar to the α_2 type. Two recent critical reviews (*63,64*) provide an extensive coverage of this subject. It should be stressed that much of the evidence in support of the existence of the postjunctional α_2-adrenoceptor is based upon in vivo studies using the pithed rat preparation whereas the evidence from in vitro studies, with some exceptions

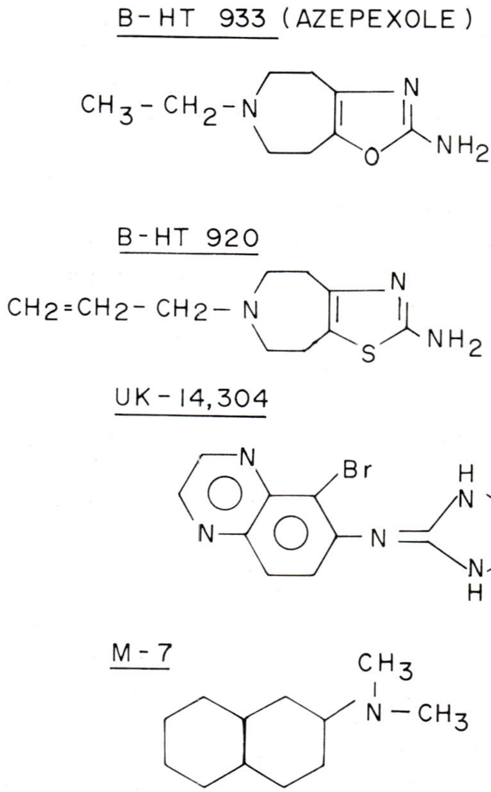

Fig. 1. Selective agonists for the α_2-adrenoceptor.

(*29*), is less convincing. Thus, prazosin-resistant responses to norepinephrine are apparent in human palmar arteries (*49*), but as pointed out by McGrath (*62*), these adrenoceptors do not clearly conform to the α_2-type. Since highly selective α_2-antagonists are not available, questionable conclusions may be drawn when results from the use of yohimbine, piperoxane, or rauwolscine are relied upon.

Thus there is evidence that suggests more than one type of postjunctional adrenoceptor, but it may be premature to label the non-α_1, as an α_2-adrenoceptor. Suitably designed studies that utilize both functional and radioligand binding techniques, as discussed in Chapter 16, may help to resolve the question of the existence of both postjunctional α_1- and α_2-adrenoceptors.

2.5. Intrasynaptic α- and γ-Adrenoceptors

A currently very important question is the nature of the adrenoceptor activated by the norepinephrine released by electrical stimulation of the adrenergic fiber. Langer et al. (*59*) have described the postjunctional synaptic (intrasynaptic) adrenoceptor in the canine vasculature as being of the α_1-type on the basis of their studies with the autoperfused hind limb preparation. The synaptic or "intrasynaptic" receptor is defined as that population of receptors preferentially activated by neuronally released transmitter. It has been reported that reserpine (*52,93*) and 6-hydroxydopamine (*93*) treatment will increase the detection, by both functional and radiologand binding studies, of α_2-adrenoceptors. In addition, denervation of rat vasa deferentia (*41*) causes a decrease in the number of binding sites for the α_1-agonist 3H-WB4101. These results again may be interpreted as indicating a neuronal influence on the postjunctional α_1-adrenoceptor, perhaps reflecting its "intrasynaptic" location. However, it has been reported (*32*) that in the rabbit portal vein, but not the aorta, a proportion of the α-adrenoceptors mediating nerve-evoked contractions are of the α_2-type. Such results indicate that one cannot expect a universally similar distribution of postjunctional α_1- and α_2-adrenoceptors.

Hirst and Neild (*43, 44*) have studied the responses of arterioles taken from the submucosa of the small intestine of young guinea pigs to ionophoretically applied norepinephrine. Two distinct types of responses could be detected: one response consisted of a localized constriction not associated with a change in arteriolar membrane potential, which was abolished by phentolamine. The second type of response was a depolarization similar to the excitatory junction potentials (ejp), and this response was phentolamine-resistant. Similar results have been reported by Holman and Surprenant (*45*), and in addition, nifedipine-sensitive nerve-evoked action potentials are present in the guinea-pig vas deferens, but not in the rat-tail artery preparation, although contractions of the tail artery were decreased (*80*). Hirst and Neild (*43*) propose that the postjunctional adrenoceptors mediating the depolarization should be designated γ and that these γ-receptors represent the intrasynaptic adrenoceptor, whereas the α-adrenoceptors are extrasynaptically located. Studies by Cheung (*27*), Hotta (*46*), and McGrath (*62*) also suggest that the "intrasynaptic" adrenoceptor may be a speized receptor or that the neurotransmitter responsible for the ejps is not norepinephrine.

3. The Adrenergic Neuron

The adrenergic fibers innervating smooth muscle are non-myelinated postganglionic sympathetic fibres that, in general, form a ground plexus within the smooth muscle. Typically, the axons exhibit periodic swellings, 1–2 μm in diameter, called varicosities, which contain large numbers of norepinephrine-containing vesicles. Detailed descriptions of the innervation of smooth muscle have been provided by Bennett (*8*) and Burnstock and Costa (*22*).

In general two patterns of innervation of smooth muscle are observed. The first type consists of small axon bundles such as that to the mammalian gastrointestinal tract and the uterus. The second type of innervation, typical in the vas deferens and seminal vesicles, consists of single varicose nerve fibers, naked of Schwann cells, passing between the smooth muscle cells and forming close-contact with the muscle cells. Intermediate forms of innervation also occur in which both small axon bundles and single bare axons innervate the smooth muscle, and typically this is present in the guinea-pig vas deferens, the sphincter and dilator of the rabbit iris, the mouse urinary bladder, and the cat nictitating membane. In vascular smooth muscle, a multiaxonal plexus is usually associated with the tunica media at the adventitiomedial border and does not usually penetrate the muscle coat of the blood vessel. A major exception to this generalization is the observation that axons appear to ramify to the endothelium of the rabbit facial vein (*69*). The neuroeffector separation varies from the close contacts seen in the rat vas deferens, 10 nm, to the wider cleft separations of 500 nm or more seen in large blood vessels. Wider neuroeffector cleft separations are associated with vascular smooth muscle and the mammalian gastrointestinal tract; indeed, the closest apposition between adrenergic nerves and smooth muscles in blood vessels is between 50 and 80 nm (*22*). Further details of the innervation of vascular smooth muscle are provided by several recent reviews (*9, 66, 85*).

The mechanisms regulating the synthesis of norepinephrine have been well documented (*10*) and are diagrammatically depicted in Fig. 2. The conversion to norepinephrine occurs in the storage vesicles that themselves are formed in the cell body and

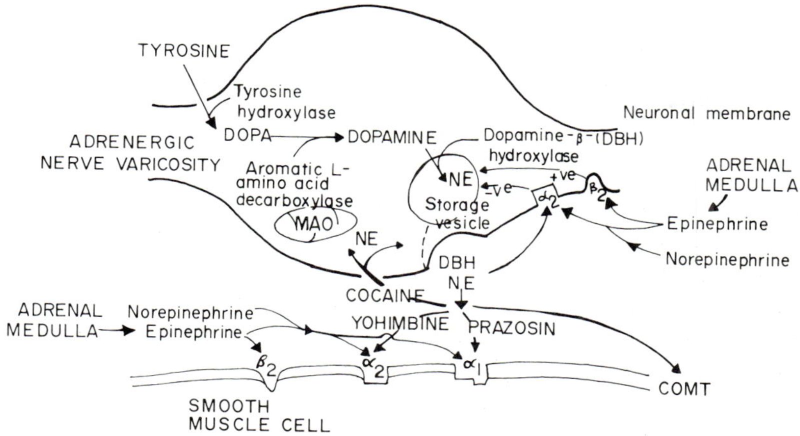

Fig. 2. The adrenergic neuron depicting a synthetic pathway for norepinephrine and distribution of adrenoceptors.

carried to the periphery by axoplasmic transport. An increase in the neuroplasmic concentration of norepinephrine exerts a negative feedback on the activity of tyrosine hydroxylase, whereas increased stimulation of the fiber or depletion of norepinephrine increases tyrosine hydroxylase activity.

It is generally accepted that norepinephrine release occurs by exocytosis, although it is not clear whether partial release or total release of the norepinephrine content of a vesicle occurs (*33,36,47,79,84*). The arrival of the action potential and the resultant changes in membrane permeability to Ca^{2+} resulting in an influx of Ca^{2+} through a Ca^{2+} channel triggers changes in membrane structure and/or a contractile process leading to the exocytotic release (*33*). Some modifications in this general concept of exocytosis may have to be considered if the evidence that newly synthesized acetylcholine is preferentially released (*47*) extends to the mechanisms regulating norepinephrine release.

As indicated in Fig. 2, several processes contribute to the removal of norepinephrine from the junctional cleft.

The norepinephrine in the region of the adrenergic varicosities can be taken up by the neuronal cell membrane uptake, by a Na^{+}-dependent, cocaine-and ouabain-sensitive neuronal pump (*48,67*). The effects of inhibition of the neuronal uptake mechanism may, depending upon the density of innervation, cause an augmentation and prolongation of the contractile response, a retardation of relaxation processes, and reduction of appearance of DOPEG (3,4-dihydroxyphenylglycol).

Nonvesicular norepinephrine, such as that taken up into the neuron by uptake$_1$, is subjected to the enzymic action of type A monoamine oxidase(MAO-A) with the formation of DOPEG. Extraneuronal MAO-B transforms norepinephrine into DOMA (3,4-dihydroxymonadelic acid), or, to a lesser extent, DOPEG. MAO also transforms normetanephrine into MOPEG or VMA (3-methoxy-4-hydroxyphenylglycol and 3-methoxymandelic acid). Catechol-*O*-methyltransferase (COMT) present extraneuronally, for instance, in the smooth muscle cells, transforms extraneuronally accumulated norepinephrine into normetanephrine, which can be further subjected to enzymic degradation by MAO. Norepinephrine reaching the smooth muscle cells is subjected to extraneuronal binding and subsequently, to varying extents, can be taken up into the smooth muscle cell by an active carrier process (uptake$_2$), which can be inhibited by a variety of agents including phenoxybenzamine, normetanephrine, and certain steroids (*38,72*). Extraneuronally accumulated norepinephrine is then degraded by COMT and subsequently by MAO-B.

Associated with the noradrenergic neuron are presynaptic adrenoceptors that can regulate the release, and hence availability, of norepinephrine to the postjunctional adrenoceptors. The initial indication that such mechanisms were functional was noted in 1957 by Brown and Gillespie (*13*), who reported that the α-adrenoceptor antagonist phenoxybenzamine enhanced the overflow of norepinephrine elicited by nerve stimulation in the perfused cat spleen. It was not until 1971, however, that the explanation for these results was explicitly proposed by Farnebo and Malmfors (*35*) in a feedback mechanism whereby a transmitter can control its own release or biosynthesis. It was later proposed (*24*) that the presynaptic receptors involved specifically in the regulation of the release of the same transmitter substance should be referred to as auto receptors. The mechanisms involved are debated in the following section of this chapter.

4. Indirect Actions of Adrenoceptor Agonists and Antagonists

As depicted in Fig. 3, there are a number of presynaptic receptor types for which there is evidence of involvement in the regulation of norepinephrine release. In this chapter, only those of the α- and β-type are considered, but the involvement of other mecha-

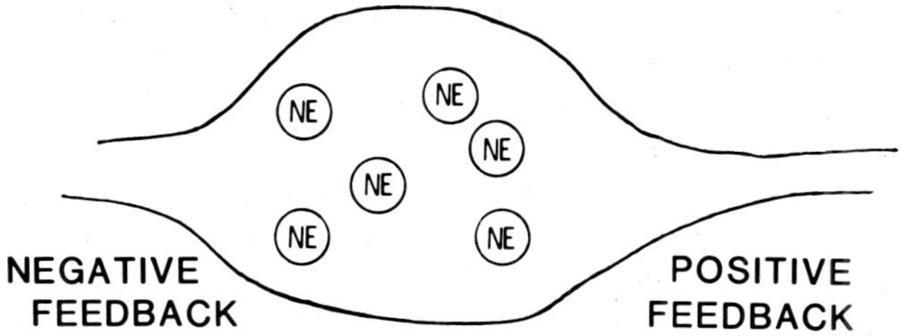

Negative feedback:

Norepinephrine (α_2)
Dopamine (dopamine receptors?)
Histamine (H_2)
5-Hydroxytryptamine
Adenosine
Acetylcholine (muscarinic)
Prostaglandins E_1, E_2
K^+
H^+
Osmolarity
Enkephalin

Positive feedback:

Epinephrine (β_2)
Angiotensin II
Prostaglandins B_1, B_2*

* There is a high degree of organ and species heterogeneity and other eicosanoids undoubtedly are also involved.

Fig. 3. Various excitatory (+) and inhibitory (−) influences on adrenergic neurotransmission.

nisms has recently been reviewed (*56,85*) and in addition presynaptic receptors are associated with other neurone types (*24,75*).

4.1. Role of α-Adrenoceptors in a Negative Feedback Mechanism

Presynaptic α-adrenoceptors which mediate an inhibitory effect on neurotransmitter release have been described in a variety of smooth muscle tissues. That catecholamines have a presynaptic inhibitory effect on cholinergic transmission in the myenteric plexus has been described and recently reviewed, and in addition, presynaptic α-adrenoceptors located at the noradrenergic nerve terminals mediating the autoinhibition of norepinephrine release have been found at almost all noradrenergic axons at which they have been postulated to exist. At both cholinergic and noradrenergic nerve endings, the relative potencies of agonists at the presynaptic α-adrenoceptors indicate that the presynaptic α-receptor is of the α_2-subtype (*94*).

It has been suggested that inhibition of the neuronal membrane Na^+-K^+-ATPase, as a result of an influx of Ca^{2+} into the neuron, would result in an enhancement of release (*68*), and hence an inhibition of release may result from a depression of Ca^{2+} entry or an enhanced rate of Ca^{2+} efflux indirectly mediated by the activation of the neuronal Na^+ pump by α-adrenoceptor agonists. Such a phenomenon has been demonstrated in the adrenal medulla (*40*). Support for the hypothesis that an inhibition of Ca^{2+} entry may be involved in the presynaptic receptor-mediated inhibition of transmitter release has been provided by Alberts (*3*) and Langer (*57*). Indirect support for the concept that the autoregulation of norepinephrine release results from a reduction of Ca^{2+} entry into the varicosity is provided by the evidence that the Ca^{2+} independent release of norepinephrine induced by tyramine is not modified by α-adrenoceptor agonists or antagonists (*56*).

The possible role of an increase in neuronal levels of cGMP or a decrease in cAMP in the link between presynaptic α_2 activation and the inhibition of norepinephrine release has also been discussed (*39,96*). Alternatively, the activation of presynaptic inhibitory α-adrenoceptors may prevent the spread of depolarization through the network of varicosities as a result of local hyperpolarization triggered by an increase in K^+ conductance and a resultant localized hyperpolarization and conduction block.

4.2. Role of β-Receptors in a Positive Feedback Mechanism

Evidence that the facilitation of the stimulation evoked release of norepinephrine may be mediated through presynaptic β-adrenoceptors was first proposed by Langer et al. (*58*), and recently reviewed (*56*). Based on the use of relatively selective agonists it would appear that the presynaptic facilitatory β-adrenoceptors are of the β_2-type (*77*). Whether the facilitation of norepinephrine release by β_2-agonists results from an increase in Ca^{2+} influx into the varicosity has not been established (*78*).

The physiological functions for the presynaptic α- and β-receptors are thought to reflect a component of a number of regulatory mechanisms that contribute to the "fine-tuning" of adrenergic tone. The system activated by presynaptic β_- receptors is said to be activated by low concentrations of presumably circulating epinephrine, whereas the activation of the presynaptic

α_2-receptor requires a higher concentration of norepinephrine or epinephrine in the cleft.

As is depicted in Fig. 3, there are a variety of other excitatory and inhibitory substances that affect norepinephrine release via presynaptic actions. Of course, it does not necessarily follow that all such mechanisms are active at any given adrenergic varicosity.

This general hypothesis of the physiological roles for an autoinhibitory and a facilitatory process at adrenergic varicosities is not universally accepted. In particular, the work of Kalsner (*51*), which has been hotly debated (*50,71*), has raised a number of serious reservations concerning the acceptability of the role of feedback modulation of adrenergic transmission.

5. Direct Actions of Adrenergic Drugs

5.1. Excitatory α-Adrenoceptors

Excitatory α-adrenoceptors are widely associated with smooth muscle; indeed the excitatory actions of catecholamines are mediated almost exclusively through α-adrenoceptor activation. A possible major exception may be the preferential activation by sympathetic neuronal activation of excitatory γ-receptors (*45*). Excitatory receptors are associated with vascular smooth muscle, myometrium, vas deferens, and gastrointestinal, as well as other visceral, smooth muscle beds. In the gastrointestinal tract, excitatory receptors are associated with the esophagus, stomach and the ileum, particularly the terminal portion (*see* ref. *18*).

The relationship between membrane potential and the tension development initiated by α-adrenoceptor activation has been reviewed (*11, 15, 18, 54*). It is apparent that, in general, action potential generation is not an absolute requirement for linking receptor activation and smooth muscle contraction (*11*). It has been reported that certain smooth muscles do not normally produce action potentials (*25*) and, in addition, some pharmacological procedures can cause the initiation of action potentials without resultant tension development (*6*).

In addition, in the absence of extracellular Ca^{2+}, action potentials can still be initiated, but tension development will not occur (*5*). Considerable differences in the ionic basis of α-mediated depolarization have also been reported. In the guinea-pig vas deferens α-adrenoceptor activation mainly increases the Na^+ conductance (*60*), whereas in the myometrium (*81*), Cl^- conductance is primarily increased; in the rat portal vein, α-activation leads to an increase in Ca^{2+}, Cl^-, and Na^+ conductances (*74*). In the rabbit pulmonary artery (*42*), norepinephrine initiated depolarization has been associated with a decrease in K^+ conductance.

The relationship between the excitatory junction potential (ejp), the muscle action potential, and muscle contraction is also an area of considerable interest with tissue differences again being apparent. It has been demonstrated that the ejps can summate (*4*) and that an action potential is produced when the threshold is exceeded (*42*); however, in both the guinea-pig vas deferens and vascular smooth muscle, α_1 and α_2-antagonists, although reducing the contractile event, do not reduce the ejps (*27, 32, 43, 44*). This phenomenon, as discussed earlier (Section 2.5), has lead to the concept of the γ-receptor that is preferentially activated by neurally released norepinephrine or a "cotransmitter" (*43*). In the guinea-pig vas deferens, but not in the rat-tail artery, the action potential is depressed by nifedipine; in both instances, the contractile process is nifedipine sensitive (*37*). The postsynaptic events in the guinea pig vas deferens are intriguing because the contractile event to nerve stimulation consists of an adrenergic and a nonadrenergic component (*62*) and nifedipine selectively blocks the nonadrenergic component (*37*).

In the pithed rat preparation, vasoconstriction induced by the stimulation of postsynaptic α_2-adrenoceptors with B-HT 920 was markedly inhibited by verapamil, D600, and nifedipine as well as Mn^{2+}, Ni^{2+}, and Co^{2+} (*89*). Conversely, vasoconstriction produced by α_1-selective agonists was only minimally affected by Ca^{2+} antagonists. These results have lead to the hypothesis that, at least in vascular smooth muscle, contraction initiated by stimulation of the postsynaptic α_2 adrenoceptor is associated with an influx of extracellular Ca^{2+}, whereas a transmembrane influx of Ca^{2+} ions is probably not required in order to link α_1-adrenoceptor activation with vasoconstriction. Exceptions may exist in certain canine vascular beds (*86*) although Vanhoutte and Rimele (*87*) have also recently concluded that the varying degrees of dependency of α-adrenoceptor activation on Ca^{2+} entry depends upon the functional characteristics of the smooth muscle preparation rather than the pharmacological characteristics of their α-adrenoceptors. This conclusion is supported by the recent study by Cauvin et al. (*26*), wherein it is reported that in the superior mesenteric artery, and the mesenteric arcade of the rabbit, there is a decrease in the contribution of intracellular

Ca^{2+} release to norepinephrine-mediated activation, whereas that of Ca^{2+} influx increases, as one moves from conduit to resistance vessels. Furthermore, van Breemen and coworkers (*88*) have demonstrated, at least in the rabbit aorta, that the sensitivity of norepinephrine-induced ^{45}Ca influx and contraction to the inhibitory action of nisoldipine and diltiazem decreases dramatically as the norepinephrine concentration is increased. Since many investigators routinely investigate the effects of varying concentrations of the antagonist on the maximal response to the agonist, it is apparent that ambiguous results may be presented with respect to the true sensitivity of the norepinephrine response to the Ca^{2+} channel antagonist (*see* Chapter 4).

The association between the action of α-adrenoceptor agonists and the turnover of cellular phosphatidylinositol has been extensively investigated and reviewed (*70*). It has been proposed by Michell (*65*) that phosphatidylinositol breakdown, the PI response, represents an early event through which receptor activation leads to the cellular mechanisms, and for a number of tissues the evidence seems to support this hypothesis. Whether this hypothesis also applies to smooth muscle is not yet clear, although, in the rabbit aorta (*92*), α_1-adrenoceptors are involved in phosphatidylinositol turnover and Ca^{2+} mobilization whereas α_2-adrenoceptors are apparently linked in an inhibitory fashion to adenylate cyclase. In parallel with the general association between α_1 activation and the PI response, it is also apparent that there is a general association between α_2 activation and the inhibition of adenylate cyclase with the resultant fall in cellular levels of cAMP (*64*). The molecular mechanism by which α_2-adrenoceptor agonists inhibit adenylate cyclase has not been clearly defined, although the dependence of inhibition on the presence of GTP (*92*) could be inferred as support for the role of a guanine nucleotide-binding regulatory component in the mediation between α_2 activation and adenylate cyclase inhibition. The relationship between the α_2-adrenoceptor and adenylate cyclase and the greater ability of β-adrenoceptor activation to overcome α_2 rather than α_1-mediated events are of interest because of the association between adrenoceptor activation and the stimulation of adenylate cyclase. Alternatively, this relationship may reflect the similar anatomical distribution of α_2- and β-adrenoceptors. McGrath (*63*) has indicated that the behavior of the postsynaptic α-adrenoceptor, as an α_1- or an α_2-type, depends upon a number of factors and has described a dramatic switch in the pressor effect of epinephrine from α_1 to α_2 mediation as acidosis is induced, Conceivably, the pH-mediated change may result from a diminished Ca^{2+} influx and a reduced activity of adenylate cyclase.

5.2. Inhibitory α-Adrenoceptors

α-Adrenoceptors mediating inhibitory actions on smooth muscle are classically associated with the gastrointestinal tract (*18*). Exceptions, however, are noted in the rat colon, human taenia coli and rabbit taenia coli where α-mediated events lead to contraction (*7*). The relaxation associated with the activation of inhibitory α-adrenoceptors is, unlike that following activation of inhibitory β-receptors, rapid in onset and recovery, short in duration and readily antagonized by local anesthetics (*12, 19*).

The mechanical inhibition of the guinea-pig taenia coli preparation by epinephrine has been associated with hyperpolarization and abolition of the spontaneous spike discharge (*14*). It has been suggested that the epinephrine-induced hyperpolarization could be explained by stimulation of an electrogenic Na^{2+} pump (*16*), although this has been questioned in favor of increases in K^+ and Cl^- conductances as indicated by the double sucrose-gap technique (*20*).

It has been demonstrated that α-adrenoceptor-mediated inhibition of smooth muscle activity depends upon the presence of Ca^{2+}, since the inhibitory effect decreases as extracellular Ca^{2+} is reduced (*21*), Conceivably, the supportive role for Ca^{2+} in the inhibitory effect of α-adrenoceptor activation reflects a modulatory role for Ca^{2+} in regulating the K^+ conductance of the plasma membrane: the binding of Ca^{2+} to a site on the gating mechanisms for the K^+ channel leading to an opening of the channel (*17*). In addition, Ca^{2+} may be involved in the binding of the agonist to the α-adrenoceptor (*82*).

5.3. Inhibitory β-Adrenoceptors

Activation of β-adrenoceptors in smooth muscle leads to a reduction in the force of the contractile response and, in myogenically active preparations, a reduction in spontaneous action potential generation. For reasons stated in Section 2.2, no attempt is made to distinguish β_1- from β_2-mediated events.

In taenia coli β-adrenoceptor activation leads to a suppression of the pacemaker potential (*20*) without significant hyperpolarization, although in the rat myometrium (*30*), hyperpolarization occurs but may not be a prerequisite for relaxation. In the myometrium the hyperpolarization is related to an increase in K^+ conductance (*61*).

The role of a Na^{2+}, a Ca^{2+} pump, or a Na^+–Ca^{2+} exchange process in contributing to the hyperpolarization has been considered, and the temptation is to relate the hyperpolarization to metabolic changes resulting from activation of adenylate cyclase (*see* ref. 18). In addition to pump or exchange activation, β-adrenoceptor activation, could also lead to a Ca^{2+}-dependent increase in K^+ conductance in an analogous manner as has been suggested for the inhibitory α-adrenoceptor (*18*).

It is clear that β-adrenoceptor leads to an elevation of cAMP, but whether such an elevation is a prerequisite for tissue relaxation is still a critical question (*73*).

5.4. Summary

Some of the key points concerning the properties and cellular events mediated by activation of the various types of adrenoceptors discussed in this chapter are summarized in Table 2.

6. Conclusions and Future Directions for Research

The mechanisms by which adrenergic drugs can influence smooth muscle contractility are complex since, in addition to species and tissue differences, there is a current trend, perhaps premature, to subtype the adrenoceptors involved in both the direct and indirect effects of sympathetic nerve stimulation and adrenoceptor agonists on smooth muscle activity. The preferential activation of an "intrasynaptic" γ-adrenoceptor, if indeed it is activated by norepinephrine, by sympathetic stimulation, whereas exogenously applied norepinephrine activates α_1- and α_2- and other (?) adrenoceptors, is an exciting area for further clarification, as is the basis for the membrane events initiated by such receptor activation, and the apparent selective inhibition of postsynaptic α_2-adrenoceptor-mediated events by Ca^{2+} antagonists. Such findings would also need to be reconciled with the role of the PI response and the involvement of cAMP, and GMP in the cellular mediation of adrenergic events. The observation that the relative importance of postsynaptic α_1- versus α_2-mediated events can be modified by the acid–base status as well as circulating glucocorticoids, may explain some of the apparent discrepancies between in vivo and in vitro studies of

Table 2
Adrenoceptor Subtypes: Distribution and Pharmacological and Physiological Properties

Receptor	Distribution	Selective Agonist	Selective Antagonist	Membrane permeability conductance changes	Dependence on extracellular Ca^{2+}, Ca_e	Comments
α_1 Excitatory	Postsynaptic Intrasynaptic	Phenylephrine Methoxamine	Prazosin	Depolarization (↑ Cl; ↑ Na; ↑ Ca; ↓ K, but tissue-dependent)	Utilization of bound Ca^{2+} (*89*, but *see* refs. *26*, *86–88*)	Subtypes? Depends on frequency of stimulation? Sensitivity ↑ as pH ↑ (*63*)
α_1 Inhibitory	Postsynaptic	Phenylephrine Methoxamine	Prazosin	Hyperpolarization (↑ K conductance)	Inhibition ↓ as Ca_e ↓ (*21*)	
α_2	Presynaptic	Clonidine BHT-920	Yohimbine Rauwolscine	Hyperpolarization (↑ K conductance)	Inhibition of Ca^{2+} entry (*3*)	Distinct from post-α_2?
α_2	Postsynaptic Extrasynaptic	Clonidine BHT 920	Yohimbine Rauwolscine		Utilization of Ca_e (but *see* refs. *26*, *86–88*)	Demonstrated in vivo pithed rat; less clear in vitro. Sensitivity ↑ as pH ↓ (*63*)

γ	Postsynaptic (selectively stimulated by neuronal activation; *see* refs. *27, 43–46, 66, 80*)	?	?	Action potentials not affected by nifedipine in rat-tail artery; but abolished in guinea-pig vas deferens (*37*)		Not activated by NE, but by cotransmitter? (*43*)
β_1	Postsynaptic (primarily cardiac muscle, *55*)	H110/38	Practolol			
β_2	Postsynaptic (primarily smooth, muscle, *55*) Presynaptic (*see* ref. *77*)	Salbutamol	Butoxamine	Inhibitory action in myometrium related to ↑ conductance (*61*)	Ca-dependent ↑ in K conductance (*17*)	Mixed receptor (β_1/β_2) population in tissues

adrenoceptor mechanisms in smooth muscle. In addition, the possible, again disputed, physiological role of presynaptic α- and β-adrenoceptor mechanisms in modulating neuronal, sympathetic and otherwise, control of smooth muscle in vivo may also complicate the ease by which conclusions from in vitro studies can be applied to functions in the whole organism.

Ultimately, the same techniques that are being applied to answer the above questions should also provide further insights into how, or whether, basic changes in the adrenergic control of smooth muscle occur in various physiological settings such as fetal development, aging, and stress, and whether alterations in adrenergic control of smooth muscle can initiate or contribute to the maintenance of such pathophysiological states as asthma, hypertension and the cardiovascular complications of diabetes mellitus.

References

1. Ahlquist, R. P. Study of adrenotropic receptors. *Am. J. Physiol.* 153:586–600, 1948.
2. Ahlquist, R. P. Adrenoceptors: from figment to fact. In *Trends in Autonomic Pharmacology*. Ed. S. Kalsner. Volume 2, Urban and Schwarzenberg, Baltimore-Munich, 1982, pp 3-13.
3. Alberts, P. Mechanisms of facilitation and muscarinic or α-adrenergic inhibition of acetylcholine and noradrenaline secretion from peripheral nerves. *Acta. Physiol. Scand. Suppl.* 506, 1982.
4. Ambache, N., and M. A. Zar. Evidence against adrenergic motor transmission in the guinea-pig vas deferens. *J. Physiol.* 216:359–389, 1971.
5. Axelsson, J. Dissociation of electrical and mechanical activity in smooth muscle. *J. Physiol. London* 158:381–398, 1961.
6. Axelsson, J. Mechanical properties of smooth muscle and the relationship between mechanical and electrical activity. In *Smooth Muscle.* Eds. G. Bulbring, A. F. Brading, A.W. Jones, and T. Tomita. Arnold, London, 1970, pp 289–315.
7. Belisle, S. and D. J. Gagnon. Stimulating action of catecholamines on isolated preparations of the rat colon and human and rabbit taenia coli. *Br. J. Pharmac.* 41:361–366, 1971.
8. Bennett, M. R. *Autonomic Neuromuscular Transmission.* Cambridge University Press, 1972.
9. Bevan, J. A., R. D. Bevan, and S. P. Duckles, Adrenergic regulation of vascular smooth muscle. In: *Handbook of Physiology. The Cardiovascular System. Vascular Smooth Muscle.* Eds. D. F. Bohr, A. P. Somlyo, and H. V. Sparks, Jr. Bethesda, Maryland, Am. Physiol. Soc., 1980, Section 2, Vol. II, Chap. 18, pp 515–566.

10. Blaschko, H. Catecholamine biosynthesis. *Br. Med. Bull.* 29:105–109, 1973.
11. Bolton, T. B. Mechanisms of action of transmitters and other substances on smooth muscle. *Physiol, Revs.* 59:606–718, 1979.
12. Brody, T.M., and J. Diamond. Blockade of the biochemical correlates of contraction and relaxation in uterine and intestinal smooth muscle. *Ann. N.Y. Acad. Sci.* 139:772–780, 1967.
13. Brown, G. L., and Gillespie, J. S. The output of sympathetic transmitter from the spleen of the cat. *J. Physiol.* 138:81–102, 1957.
14. Bulbring, E. Changes in configuration of spontaneously discharged spike potentials from smooth muscle of the guinea–pig's taenia coli. The effect of electronic currents and of adrenaline, acetylcholine and histamine. *J. Physiol.* 135:412–425, 1957.
15. Bulbring, E., A. F. Brading, A. W. Jones, and T. Tomita. *Smooth Muscle.* E. Arnold Ltd., London, 1970.
16. Bulbring, E. and H. Kuriyama. Effect of changes in the external sodium and calcium concentrations on spontaneous electrical activity in smooth muscle of guinea–pig taenia coli. *J. Physiol.* 166:29–58, 1963.
17. Bulbring, E. and H. Kuriyama. The action of catecholamines on guinea–pig taenia coli. *Phil. Trans. R. Soc. B. 265:115–121, 1973.*
18. Bulbring, E., H. Ohashi, and T. Tomita. Adrenergic mechanisms. In *Smooth Muscle: An Assessment of Current Knowledge.* Eds. E. Bulbring, A. F. Brading, A. W. Jones, and T. Tomita. E. Arnold, London, 1981, pp 219–248.
19. Bulbring, E. and T. Tomita. Increase of membrane conductance by adrenaline in the smooth muscle of guinea–pig taenia coli. *Proc. R. Soc. B.* 172:89–102, 1969.
20. Bulbring, E., and T. Tomita. Suppression of spontaneous spike generation by catecholamines in the smooth muscle of the guinea–pig taenia coli. *Proc. R. Soc. B.* 172:103–119, 1969.
21. Bulbring, E., and T. Tomita. Effects of calcium, barium and manganese on the action of adrenaline in the smooth muscle of the guinea–pig taenia coli. *Proc. R. Soc. B.* 172:121–136, 1969.
22. Burnstock, G., and M. Costa. *Adrenergic neurons.* John Wiley and Sons, New York, 1975.
23. Cambridge, D. UK-14,304, a potent and selective agonist for the characterization of α–adrenoceptor subtypes. *Eur. J. Pharmacol.* 72:413–415, 1981.
24. Carlsson, A. Dopaminergic autoreceptors. In *Chemical Tools in Catecholamine Research.* Eds. O. Almyren, A. Carlsson, J. Engel. North-Holland/Elsevier, Amsterdam and New York, 1975, pp 219–25.
25. Casteels, R., K. Kitamura, H. Kuriyama, and H. Suzuki. The membrane properties of the smooth muscle cells of the rabbit main pulmonary artery. *J. Physiol. London,* 271:41–61, 1977.
26. Cauvin, C., K. Saida and C. Van Breemen. Effects of Ca antagonists on Ca fluxes in resistance vessels. *J. Cardiovasc. Pharmacol.* 4:S287 –S290. 1982.
27. Cheung, D. W. Two components in the cellular response of rat tail arteries to nerve stimulation. *J. Physiol.* 328:461–468, 1982.

28. Daly, M. J. and G. P. Levy. The subclassification of β–adrenoceptors: evidence in support of the dual β–adrenoceptor hypothesis. In *Trends in Autonomic Pharmacology,* Volume 1. Ed. S. Kalsner. Urban and Schwarzenberg, Baltimore–Munich, 1979, pp 347–385.
29. De Mey, J., and P. M. Vanhoutte. Uneven distribution of postjunctional α_1-like adrenoceptors in canine arterial and venous smooth muscle. *Circ. Res.* 48:875–884, 1981.
30. Diamond, J., and J. M. Marshall. Smooth muscle relaxants: dissociation between resting membrane potential and resting tension in rat myometrium. *J. Pharmac. Exp. Ther.* 168:13–20, 1969.
31. Docherty, J. R., and J. C. McGrath. An examination of factors influencing adrenergic transmission in the pithed rat, with special reference to noradrenaline uptake mechanisms and postjunctional α–adrenoceptors. *Naunyn–Schmideberg's Arch. Pharmacol.* 313:101–111, 1980.
32. Docherty, J. R. and K. Starke. Postsynaptic α–adrenoceptor subtypes mediating nerve–evoked contractions in rabbit blood vessels in vitro. *Br.J. Pharmacol, 74:803P, 1981.*
33. Drapeau, P. and M. P. Blaustein. Calcium and neurotransmitter release: what we know and don't know. In *Trends in Autonomic Pharmacology,* Ed. S. Kalsner. Urban and Schwarzenberg, Baltimore-Munich, 1982, Volume 2, pp 117–130.
34. Drew, G. M. Postsynaptic α_2–adrenoceptors mediate pressor responses to 2*N*,*N*–dimethylamino-5,6-dihydroxy-1,2,3,4-tetrahydronaphthalene (M-7), *Eur. J. Pharmacol. 65:85–87, 1980.*
35. Farnebo, L. 0. and T. Malmfors. ^{3}H-noradrenaline release and mechanical response in the field stimulated mouse vas deferens. *Acta Physiol. Scand. Suppl.* 371:1–18, 1971.
36. Folkow, B., J. Haggendal, and B. Lisander. Extent of release and elimination of noradrenaline at peripheral adrenergic nerve terminals. *Acta Physiol. Scand. Suppl.* 307:1–38, 1967.
37. French, A. M. and N. C. Scott. A comparison of the effects of nifedipine and verapamil on rat vas deferens. *Br. J. Pharmac.* 73:321–323, 1981.
38. Gillespie, J. S. Extraneuronal uptake of catecholamines in smooth muscle and connective tissue. In *The Mechanism of Neuronal and Extraneuronal Transport of Catecholamines.* Ed. David M. Paton. New York, 1976, pp 325–354.
39. Greengard, P. Cyclic nucleotides, phosphorylated proteins and the nervous system. *Federation Proc.* 38:2208–2217, 1979.
40. Gutman, Y., P. Boonyaviroj, and L. Eckstein. Mechanism of PGE and alpha-andrenergic effects on release of catecholamines. In *Presynaptic Receptors.* Eds. S. Z. Langer, K. Starke, and M. L. Dubocovich. Pergamon Press, Oxford, 1979, pp 341–345.
41. Hata, F., K. Takeyasu, Y. Morikawa, R. -T. Lai, H. Ishida, and H. Yoshida. Role of α–adrenergic receptors in denervation supersensitivity of rats vas deferens. *Jap. J. Pharmacol.* 31:383–390, 1981.
42. Hirst, G. D. S. Neuromuscular transmission in arterioles of the guinea–pig submucosa. *J. Physiol (London)* 273:263–275, 1977.

43. Hirst, G. D. S. and T. O. Neild. Evidence for two populations of excitatory receptors for noradrenaline on arteriolar smooth muscle. *Nature, Lond.* 283:767–768, 1980.
44. Hirst, G. D. S., and T. O. Neild. Localization of specialized noradrenaline receptors at neuromuscular junctions on arterioles of the guinea–pig. *J. Physiol.* 313:343–350, 1981.
45. Holman, M. E., and A. M. Surprenant. An electrophysiological analysis of the effects of noradrenaline and α–receptor antagonists on neuromuscular transmission in mammalian muscular arteries. *Br. J. Pharmac. Chemother.* 71:651–661, 1980.
46. Hotta, Y. Some properties of the junctional and extrajunctional receptors in the vas deferens of the guinea-pig. *Agents and Actions* 1:13–21, 1969.
47. Israel, M., Y. Dunant, and R. Manaranche. The present status of the vesicular hypothesis. *Prog. Neurobiol.* 13:237–275, 1979.
48. Iversen, L. L. *The Uptake and Storage of Noradrenaline in Sympathetic Nerves*. Cambridge University Press, 1967.
49. Jauernig, R. A., R. F. W. Moulds, and J. Shaw. The action of prazosin in human vascular preparations. *Arch. Int. Pharmacodyn. Ther.* 231:81–89, 1978.
50. Kalsner, S. The presynaptic receptor controversy. *Trends in Pharmacological Sciences* 3: 11–16, 1982.
51. Kalsner, S. Feedback regulation of transmitter release through adrenergic presynaptic receptors: time for a reassessment. In *Trends in Autonomic Pharmacology*. Urban and Schwarzenberg, Baltimore–Munich 1982, Vol. 2, pp 385–425.
52. Kobinger, W., and L. Pichler. α_1 and $_2$ adrenoceptor subtypes: selectivity of various agonists and relative distribution of receptors as determined in rats. *Eur. J. Pharmacol.* 72:313–321, 1981.
53. Kobinger, W. and L. Pichler. α_2 adrenoceptor agonistic effect of B–HT 920 in isolated perfused hindquarters of rats. *Eur. J. Pharmacol.* 76:101–105, 1981.
54. Kuriyama, H., Y. Ito, H. Suzuki, K. Kitamura, and T. Itoh. Factors modifying contraction–relaxation cycle in vascular smooth muscle. *Am. J. Physiol.* 243:H641–H662, 1982.
55. Lands, A. M., A. Arnold, J. P. McAuliff, F. P. Luduena, and T. G. Brown. Differentiation of receptor systems activated by sympathominetic amines. *Nature (London)* 214:597–598, 1967.
56. Langer, S. Z. Presynaptic regulation of the release of catecholamines. *Pharmacol. Reviews* 32:337–362, 1981.
57. Langer, S. Z., M. L. Dubocovich, and S. M. Celuch. Prejunctional regulatory mechanisms for noradrenaline release, elicited by nerve stimulation. In *Chemical Tools in Catecholamine Research*. Eds. O. Almgren, A, Carlsson, and J. Engel. North Holland Publishing Company, Amsterdam, 1975, Vol. II, pp 183–196.
58. Langer, S. Z., M. A. Enero, E. Alder-Graschinsky, M. L. Dubocovich, and S. M. Celuch. Presynaptic regulatory mechanisms for noradrenaline release by nerve stimulation. In *Proceedings of the Symposium on Central Action of Blood Pressure*. Eds. D. S. Davies and J. L. Reid. Pitman Medical, London, 1975, pp 135–151.

59. Langer S. Z., N. B. Shepperson, and R. Massingham. Preferential noradrenergic innervation of alpha–adrenergic receptors in vascular smooth muscle. *Hypertension* 3 (Supp. I):112–118, 1981.
60. Magaribuchi, T., Y. Ito, and H. Kuriyama. Effects of catecholamines on the guinea–pig vas deferens in various ionic environments. *Jap. J. Physiol.* 21:691–708, 1971.
61. Marshall, J. M. Modulation of smooth muscle activity by catecholamines. *Federation Proc.* 36:2450–2455, 1977.
62. McGrath, J. C. Adrenergic and 'non-adrenergic' components in the contractile response of the vas deferens to a single indirect stimulus. *J. Physiol.* 283:23–39, 1978.
63. McGrath, J. C. Evidence for more than one type of postjunctional α–adrenoceptor. *Biochem. Pharmacol.* 31:467–484, 1982.
64. McGrath, J. C. The variety of vascular α–adrenoceptors. *Trends in Pharmacol. Sci.* 4:14–18, 1983.
65. Michell, R. H. Inositol phospholipids and cell surface receptor function. *Biochim. Biophys. Acta.* 415:81–147, 1975.
66. Neild, T. O. and E. Zelcer. Noradrenergic neuromuscular transmission with special reference to arterial smooth muscle. *Prog. Neurobiol.* 19:141–158, 1982.
67. Paton, D. M. Characteristics of uptake of noradrenaline by adrenergic neurons. In *The Mechanism of Neuronal and Extraneuronal Transport of Catecholamines.* Ed. David M. Paton. Raven Press, 1976, pp 49–66.
68. Paton, W. D. M., E. S. Vizi, and M. A. Zar. The mechanism of acetylcholine release from guinea–pig intestine and longitudinal muscle strips. *J. Physiol.* 215:819–848, 1971.
69. Pergman, B. L., R. D. Bevan, and J. A. Bevan. Facial vein of the rabbit: neurogenic vasodilation mediated by α-adrenergic receptors. *Circ. Res.* 39:854–860, 1976.
70. Putney, J. W. Jr. Phosphatidylinositol metabolism and α-adrenoceptor mechanisms, In *Adrenoceptors and Catecholamine Action, Part B.* Ed. G. Kunos. John Wiley and Sons, 1983, pp 51–64.
71. Rand, M. J., M. W. McCulloch, and D. F. Story. Feedback modulation of noradrenergic transmission. *Trends Pharmacol. Sci.* 3:8–11, 1982.
72. Salt, P. J. Inhibition of noradrenaline uptake$_2$, in the isolated rat heart by steroids, clonidine, and methoxylated phenylethylamines. *Eur. J. Pharmacol.* 20:324–340, 1972.
73. Schield, C. R., T. W. Honeyman, and E. S. Fay. Mechanisms of α-adrenergic relaxation of smooth muscle. *Nature,* 277:32–36, 1979.
74. Shuba, M. F., A. V. Gurkovskaya, N. J. Klevetz, N. G. Kochemasova, and V. V. Taranenko. Mechanism of the excitatory and inhibitory actions of catecholamines on the membrane of smooth muscle cells. In *Physiology of Smooth Muscle.* Eds. E. Bulbring and M. F. Shuba. Raven Press, New York, 1976, pp 347–355.
75. Starke, K. Presynaptic receptors. *Ann. Rev. Pharmacol. Toxicol.* 21:7–20, 1981.
77. Stjarne, L. and J. Brundin. β_2-adrenoceptors facilitating noradrenaline secretion from human vasoconstrictor nerves. *Acta. Physiol. Scand.* 97:88–93, 1976.

78. Stjarne, L. Facilitation and receptor-mediated regulation of noradrenaline secretion by control of recruitment of varicosities as well as by control of electro-secretory coupling. *Neuroscience* 3:1147–1155, 1978.
79. Su, C., and J. A, Bevan. Adrenergic transmitter release and distribution in blood vessels. In *Physiology and Pharmacology of Vascular Neuroeffector Systems.* Eds. J. A. Bevan, R. F. Furchgott, R. A. Maxwell, and A. P. Somlyo. Karger, Basel, 1971, pp 13–21. (Proc. Symp. Physiol. Pharmacol. Vascular Neuroeffector Systems, Interlaken, 1969).
80. Surprenant, A., T. 0. Neild, and M. E. Holman. Effects of nifedipine on nerve–evoked action potentials and consequent contractions in rat tail artery. *Pflugers Arch. Eur. J. Physiol.* 396:342–349, 1983.
81. Szurszewski, J. H, and E. Bulbring. The stimulant action of acetylcholine and catecholamines on the uterus. *Phil. Trans. R. Soc. B.* 265:149–156, 1973.
82. Takagi, K., I. Takayanagi, and C. S. Liao. The effect of calcium and magnesium ions on drug receptor interaction. *Eur. J. Pharmacol.* 19:330–342, 1972.
83. Timmermans, P. B. M. W. M., and P. A. Van Zwieten. The postsynaptic α_2-adrenoceptor. *J. Auton. Pharmac.* 1:171–183, 1981.
84. Trifaro, J. M. and L. Cubeddux. Exocytosis as a mechanism of noradrenergi transmitter release. In *Trends in Autonomic Pharmacology.* Ed. S. Kalsner. Urban and Schwarzenberg, Baltimore–Munich, 1979, pp 195–249.
85. Vanhoutte, P. M., T. J. Verbueren, and R. C. Webb. Local modulation of adrenergic neuroeffector interaction in the blood vessel wall. *Physiol. Reviews* 61:151–247, 1981.
86. Vanhoutte, P. M. Heterogeneity of postjunctional vascular α-adrenoceptors and handling of calcium. *J. Cardiovac. Pharmacol.* 4:591–596, 1982.
87. Vanhoutte, P. M. and T. J. Rimele. Calcium and α-adrenoceptors in activation of vascular smooth muscle. *J. Cardiovac. Pharmacol.* 4:5280–5286, 1982.
88. Van Breemen, C., O. Hwang, and C. Cauvin. Ca antagonist inhibition of norepinephrine stimulated Ca influx in vascular smooth muscle. In *Calcium Modulators.* Eds. T. Godfraind, A. Albertini, and R. Paoletti. Elsevier Biomedical Amsterdam, pp 185–198.
89. Van Meel, J. C. A., A. Dejonge, H. 0. Kalkman, B. Wilffert, P. B. M. W. M. Timmermans, and P. A. Van Zwieten. Organic and inorganic calcium antagonists reduce vasoconstriction in vivo mediated by postsynaptic α_2-adrenoceptors. *Naunyn-Schmiedeberg's Arch. Pharmacol.* 316:288–293, 1981.
90. Van Meel, J. C. A., A. Dejonge, P. B. M. W. M. Timmermans, and P. A. Van Zwieten. Selectivity of some α-adrenoceptor agonists for peripheral and central α_1- and α_2-adrenoceptors in the normotensive rat. *J. Pharmac. Exp. Ther.* 219:760–767, 1981.
91. Van Rossum, J. M. Different types of sympathomimetic α-receptors. *J. Pharm. Pharmac.* 17: 202–216, 1965.

92. Villalubos-Mulina, R., M. Uc, E. Hong, and J. A. Garcia-Sainz. Correlation between physphotidylinositol labelling and contraction in rabbit aorta: effect of alpha-1 andrenergic activation. *J. Pharmacol. Exp. Ther.* 222:258–261, 1982.
93. Watanabe, Y., R. T. Lai, H. Maeda, and H. Yoshida. Reserpine and sympathetic denervation cause an increase of postsynaptic α_2-adrenoceptors. *Eur. J. Pharmacol.* 80:105–108, 1982.
94. Wikberg, J. E. S. The pharmacological classification of adrenergic α_1 and α_2 receptors and their mechanisms of action. *Acta Physiol. Scand. Suppl.* 468:1–84, 1979.

Chapter 14

Cholinergic Interactions and Vascular Smooth Muscle Tone

Tony J-F. Lee

1. Introduction

Acetylcholine (ACh) has been shown to consistently contract almost all mammalian nonvascular smooth muscles examined (*15, 21, 24, 60*). In contrast, the action of ACh on vascular smooth muscle tone has been found to be inconsistent. ACh, intravascularly administered, has been shown to induce vasodilation of most vascular beds (*22*), whereas it constricts several isolated blood vessels (*53, 56, 73, 75, 121*). However, the theory has prevailed that ACh is a vasodilator substance or transmitter and that the direct effect of ACh on vascular smooth muscle cells is relaxation (*22, 126*). This concept has recently been challenged by the findings about the obligatory role of endothelial cells in ACh-induced vasodilation (*54, 85, 89, 91*).

Several excellent reviews on the action of ACh on the nonvascular smooth muscle can be found elsewhere (*15, 21, 24*). This article will concentrate on the recent progress in cholinergic mechanisms in controlling vascular smooth muscle tone. Possible explanations for the discrepancy in findings between ACh actions on vascular and nonvascular smooth muscles will be discussed.

2. Acetylcholine-Induced Vasodilation Depends on Endothelial Cells

Since its discovery, ACh has been shown to induce dilation of various vascular beds in vivo. It has been known that to demonstrate a vasodilator response in vitro, the isolated blood vessel preparation should have active muscle tone, either contracted by vasoactive substance, e.g., norepinephrine, or of intrinsic origin (*73, 89, 95, 96*). Nevertheless, in the presence of active muscle tone, some isolated blood vessels still constrict upon application of ACh (*73, 121*). The dependence of endothelial cells in ACh-induced vasodilation was first described in the rabbit aortic strip by Furchgott and Zawadzki (*54*). These authors demonstrated that the rabbit aortic strips with intact endothelium relaxed upon application of ACh up to $10^{-6}M$. In contrast, the aortic strips without intact endothelial cells constricted exclusively upon application of ACh at any concentration. Since then, similar findings have been obtained in cerebral and several other peripheral vascular beds of several species (*26, 27, 34, 85, 87, 89, 91*). (*See* Fig. 1.) Thus, in the presence of active muscle tone and intact endothelium, isolated blood vessels (except some veins) that previously have been claimed to constrict, dilate exclusively upon application of ACh at low concentrations (*26, 27, 34, 54, 85, 89*). Furchgott et al. (*55*) further report that the ACh-induced relaxation can be transferred from a rabbit aortic strip with intact endothelium to other strips where intima have been disrupted. We have also observed that the dilator, but not the constrictor responses of the isolated arteries with intact endothelial cells become tachyphylaxic to the repeated application of ACh after the preparation is given a maximal concentration of ACh to induce maximum vasodilation (*89*). These results suggest that the endothelial cells may release a chemical factor or substance to mediate relaxation of vascular smooth muscle.

3. Direct Effect of Acetylcholine on Vascular Smooth Muscle

Since ACh-induced vasodilation of most vascular beds is dependent exclusively on endothelial cells, it is necessary to disrupt or remove the endothelial cells before one can examine the direct ef-

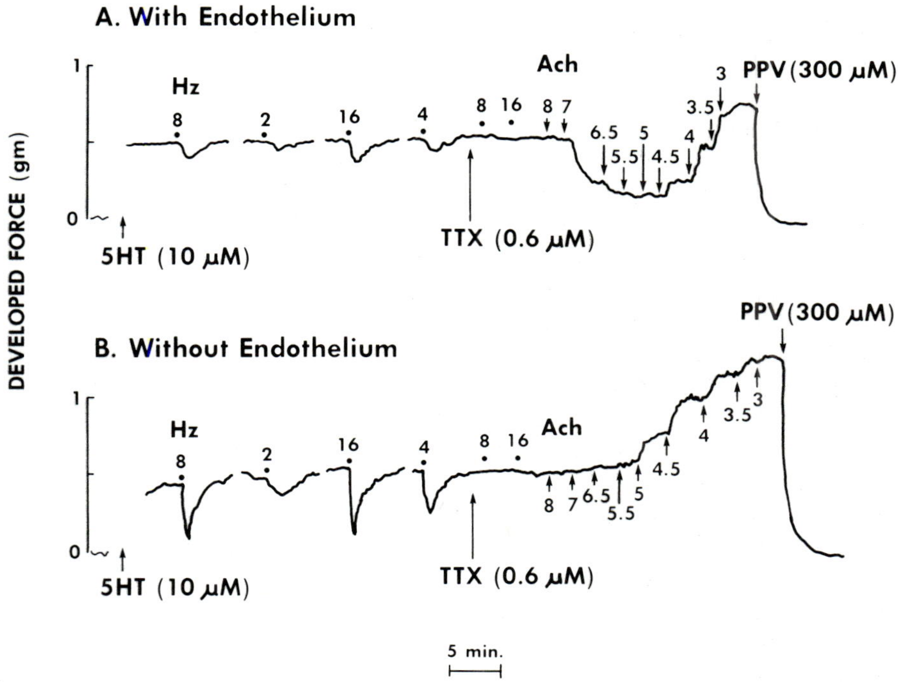

Fig. 1. Response of a segment of cat middle cerebral artery from a control (A) and a segment of the artery in which the endothelium was rubbed (B) to acetylcholine (ACh) and transmural nerve stimulation (TNS). Trains of 200 biphasic square wave pulses of 0.3 ms duration at different frequencies (2, 4, 8, and 16 Hz) were delivered at supramaximal voltage. The active muscle tone was produced by serotonin (5-HT, 10 μ*M*). TNS induced a frequency-dependent relaxation in both control and endothelium rubbed arteries. The TNS-induced dilator response was abolished by tetrodotoxin (TTX, 0.6 μ*M*). ACh induced relaxation in control artery. In contrast, ACh induced constriction exclusively in the artery upon which the endothelium has been rubbed. At the end of each experiment, papaverine (PPV) was given to induce maximum relaxation. Number with arrow indicates the negative log molar concentration of ACh in the bath (*85*). (By permission of Elsevier Biomedical Press BV, Amsterdam.)

fect of ACh on smooth muscle cells. The blood vessels without endothelial cells, as already mentioned, only constricted upon application of ACh. The constriction persisted even after the artery was denervated by cold storage (*89*), suggesting that the constriction did not result from activation of vasoconstrictor nerves. Furthermore, the isolated cultured aortic smooth muscle cells from rat have been shown to contract upon application of ACh (*114*). These results indicate that the direct ACh-stimulation of smooth muscle cells of rat and rabbit aortic strips and several other blood vessels are constriction rather than dilation. Both ACh-induced vasodilation and constriction appear to be mediated by muscarinic receptors since both responses were blocked by atropine (*26, 27, 54, 85, 89, 91*).

4. Calcium and Endothelium-Mediated Relaxation

It is known that muscarinic receptors in most tissues appear to be coupled to membrane calcium-channels such as in exocrine cells (*35, 112*) and in smooth muscle (*14*). Zawadzki et al. (*141*) have shown that the calcium ionophore A 23187 can also elicit endothelium-dependent relaxations of rabbit aortic rings; it is even more potent than ACh. A 23187 is a well-documented ionophore with specificity for divalent cations, especially calcium (*113*). Zawadzki et al. (*141*) also suggested that A 23187 may activate the same pathway as ACh in endothelial cells, with resulting release of the same dilator substance. Furthermore, calcium-channel blockers or removal of calcium from the incubation buffer have been shown to inhibit methacholine and A 23187-induced relaxation of rabbit aortic strip (*120*). These results suggest that calcium is probably an important regulator or stimulus for the production and/or release of the endothelial dilator substance(s).

The nature of the dilator substance(s), and the exact mechanism that involves calcium in regulating or stimulating the production and/or release of this substance is not yet known. Furchgott et al. (*55*) proposed that muscarinic receptor activation leads to mobilization of calcium in the region of phospholipase A_2, activating this enzyme to hydrolyze phosphatides with the liberation of arachidonic acid and lysophosphatides. Oxidation of the liberated arachidonic acid by lypoxygenase leads to the formation of an active intermediate product(s), possibly a hydroperoxide or free radical that diffuses from the endothelial cells to the smooth muscle cells of the medial layer. The intermediate product(s) is considered the dilator substance(s) (*55*). This mechanism is consistent with the calcium dependence observed in the relaxation response. Phospholipase, especially phospholipase A_2, appears to be activated in a calcium-dependent manner by calcium ionophore A 23187 in many tissues (*67, 76, 112*).

5. Calcium and Vascular Smooth Muscle Contraction

It is generally accepted that calcium ions mobilized to activate the contractile apparatus in vascular smooth muscle may originate from two different sources. The biphasic nature of the contractile responses of various vascular smooth muscles, arteries, and veins to catecholamines has been well documented (*8, 19, 122*). The different phases of contractile responses have been thought to be caused by different mechanisms of activation by a different source of calcium. The initial phasic contraction has been attributed to the release of intracellular calcium stores, and the following tonic phase of influx of calcium from the extracellular fluid (*9, 12, 59, 80, 129*). The dependence of contraction of blood vessels from different regions upon extra- or intracellular calcium is variable. For example, calcium channel blockers have been shown to have differential effects in antagonizing NE-induced vasoconstriction in different regions such as peripheral vs cerebral arteries (*9*). The role of calcium in ACh-induced contraction of vascular smooth muscle has not been as adequately described as that in NE-induced vasoconstriction. However, the ACh-induced contraction of vascular smooth muscle may also depend on both intra- and extracellular calcium as is found in the nonvascular smooth muscle (*15, 17, 18, 25, 30*). Although the role of endothelial cells in ACh-induced response was not taken into consideration by these authors, Watkins and Davidson (*136, 137*) reported that ACh at maximal concentration ($10^{-5}M$) activated the rabbit aortic strip in a manner similar to that induced by NE. They demonstrated that the contractile response of the isolated rabbit aortic strip upon application of ACh was biphasic. In the absence of extracellular calcium, the phasic component did not appear to the affected but the tonic component was greatly decreased. These results suggest that the ACh-induced biphasic response in the rabbit aortic strip is probably caused by activation of intra- and extracellular calcium pools.

6. Acetylcholine and Adrenergic Neuron Interaction

6.1. Peripheral Blood Vessels

There are two types of cholinergic receptors located on adrenergic nerve terminals innervating a variety of smooth muscles including the cat, rabbit, and guinea-pig heart, rat mesenteric artery, guinea-pig mesenteric artery, central ear artery of the rabbit, and dog saphenous vein (*130, 138*). Activation of the nicotinic receptors appears to enhance NE release, whereas activation of the muscarinic receptors appears to decrease the release of NE to nerve stimulation and other procedures requiring calcium to induce release (nicotine, KCl, and so forth). Since inhibition of NE release via activation of the muscarinic receptors is seen at lower concentrations of ACh than by nicotinic receptors, the muscarinic receptors may be more important physiologically. It is possible that prejunctional muscarinic receptors on adrenergic nerve terminals have an important role in modulating NE release from adrenergic neurons, expecially in those regions where adrenergic and cholinergic nerve terminals are in close apposition to each other such as in vas deferens, lung, and several vascular beds (*22, 130, 138*). Thus, under a physiological condition, a decrease in NE release caused by presynaptic muscarinic receptor stimulation will result in a decreased smooth muscle tone and therefore vasodilation. However, in vascular preparations, there must be careful interpretation of data since ACh-induced relaxation can result from activation of muscarinic receptors in the endothelial cells in addition to the adrenergic nerve terminals. In order to demonstrate that a decrease in muscle tone (or vasodilation) is caused by activation of presynaptic muscarinic receptors, the isolated blood vessels without endothelial cells have been used. Figure 2 indicates that in the presence of active muscle tone induced by NE, the rabbit ear artery with intact endothelium relaxed upon application of ACh. The ACh-induced relaxation was blocked by atropine. The rabbit ear artery without endothelial cells also constrict in response to exogenous NE and upon transmural nerve stimulation (TNS). ACh relaxed the TNS-induced constriction, but further elevated the NE-induced tone. It has been well documented that the TNS-induced constriction of the rabbit ear artery can be blocked by guanethidine and phentolamine (*33, 94*), suggesting that constriction is primarily caused by NE release from the adrenergic nerves. The ACh-induced relaxation of the ear arteries with active muscle tone induced by TNS is also blocked by atropine. This result demonstrates that ACh most likely inhibits adrenergic nerve activity to decrease NE transmitter release by stimulating presynaptic muscarinic receptors.

6.2. Cerebral Blood Vessels

Results from ultrastructural studies also demonstrate that cerebral blood vessels of several species receive adrenergic and non-

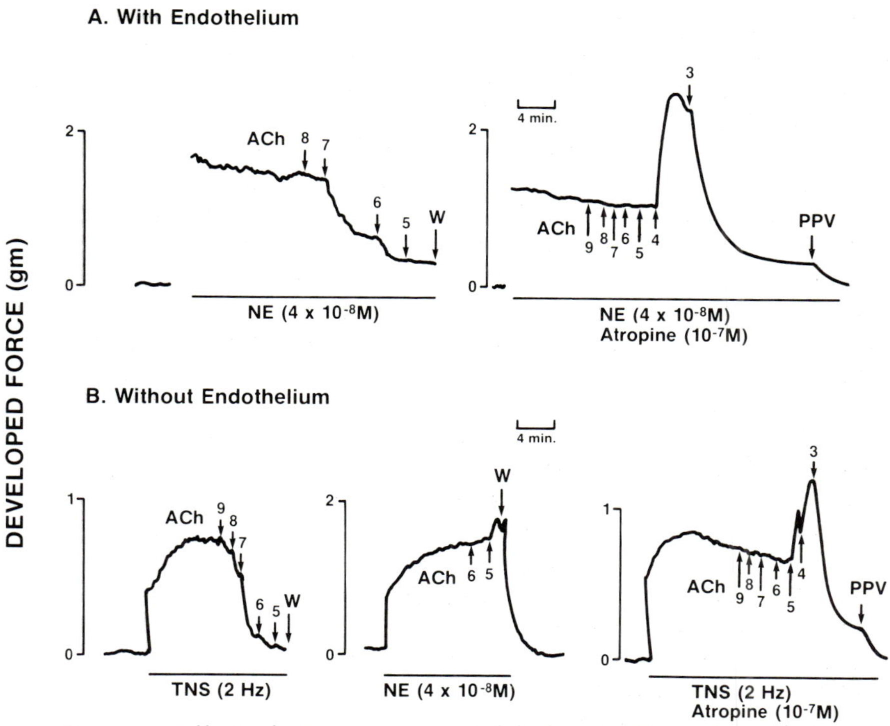

Fig. 2. Effect of atropine on acetylcholine (ACh)-induced relaxation of rabbit central ear arteries with (A) and without (B) endothelial cells in the presence of active muscle tone induced by norepinephrine (NE) and transmural nerve stimulation (TNS) (*89*). The endothelial cells were removed by rubbing the endothelium with a stainless-steel rod as described previously (*89*). In the presence of active muscle tone induced by NE, the artery with intact endothelial cells (A) relaxed upon application of ACh, which was blocked by atropine. In the presence of active muscle tone induced by TNS (2 Hz, 0.3 ms duration), the artery without endothelial cells (B) relaxed upon application of ACh. However, in this preparation, ACh further constricted the arteries with active muscle tone induced by NE. The ACh-induced relaxation was blocked by atropine. All experiments were carried out in the presence of yohimbine ($3 \times 10^{-8}M$). Wash (W), papaverine (PPV), $3 \times 10^{-4}M$. Number with arrow indicates the negative log molar concentration of ACh in the bath.

adrenergic nerves (*72, 86, 88, 94, 108*), and the nerve terminals of both types are frequently seen opposing each other in close range. Although direct evidence for ACh released from the nonadrenergic nerves is lacking, the possibility that ACh may modulate NE release from the adrenergic nerves has been examined by several investigators.

Edvinsson et al. (*41*) reported that nicotine decreased [^{3}H]NE release in the cat pial arteries, an effect that was blocked by a high concentration of hexamethonium. A similar result was demonstrated in rabbit cerebral blood vessels (*2*). Thus inhibition of NE release in these preparations was thought to be mediated by nicotinic rather than muscarinic receptors. This is different from that found in most peripheral tissues (*130, 138*). Using a similar technique, Duckles and Kennedy (*39*) also reported that ACh decreased TNS-induced endogenous NE release from the cat middle cerebral artery and rabbit basilar artery. In both species, the effect of ACh was not blocked by hexamethonium, but was blocked by atropine. These authors did observe that in cat middle cerebral arteries, atropine ($10^{-7}M$) added after the addition of ACh could not completely antagonize the effects of ACh even when the dose of atropine was increased to $5 \times 10^{-7}M$. However, when the order of drug addition was altered and atropine ($10^{-7}M$) was added before ACh, the ability of ACh to decrease NE release was blocked. These data suggest that ACh acts through muscarinic rather than nicotinic receptors on the presynaptic adrenergic nerve terminals. Although measuring the release of endogenous NE may more truly reflect the movement of endogenously synthesized transmitter, the difference between findings by Edvinsson et al. (*41*) and Duckles and Kennedy (*39*) remains to be solved.

7. Calcium and Muscarinic Presynaptic Inhibition of Transmitter Release

The effect of ACh inhibition of NE release from the presynaptic nerve terminals is seen after stimulation by potassium buffer, nicotinic agonists, and nerve stimulation, but is not seen after tyramine. All three forms of stimulation (nerve stimulation, high potassium, and nicotinic agonists) have been shown to require extracellular calcium for their releasing effect, in contrast to tyramine and solution depleted of sodium ion, which do not (*52, 105, 138*). Furthermore, in guinea-pig vas deferens, lowering extracellular calcium potentiates, whereas raising extracellular calcium masks, the inhibitory effect of muscarinic agonists in NE release induced by nerve stimulation (*36, 69, 138*). These results suggest that calcium may play a critical role in the muscarinic inhibitory process. On the contrary, Vanhoutte et al. (*130*) demonstrated in the isolated dog saphenous veins that altering the calcium concentration of the solution did not affect the degree of muscarinic inhibitory response during nerve stimulation. These authors (*131*) also demonstrated that calcium ionophore A 23187 caused release of [^{3}H]NE and contractions that, surprisingly, were augmented by ACh. There is no apparent explanation for this increase in [^{3}H]NE release and contraction. These authors suggested that the prejunctional effect of ACh was not a result of direct interference with the availability of calcium for the electrosecretory process in adrenergic nerve terminals (*131*). Furthermore, ACh reduces the overflow of [^{3}H]NE caused by partial depolarization, but not that evoked by full depolarization of the nerve endings with high K^+ solution (*132, 133, 135*). These and other authors speculate that activation of the muscarinic receptors on the neuronal cell membrane of the adrenergic varicosities causes hyperpolarization that prevents or delays the occurrence of action potentials (*64, 119*). However, Gothert (*62*) reported that methacholine inhibits transmitter release evoked by introduction of calcium to rabbit hearts fully depolarized by perfusion with high-potassium Tyrode's solution. At the present time, the exact mechanism by which ACh inhibits the release of NE from the adrenergic nerve is not resolved.

8. Muscarinic Receptors on Adrenergic Nerve, Smooth Muscle, and Endothelial Cells

There is no question that muscarinic receptors are located on presynaptic adrenergic nerve terminals, postsynaptic smooth muscle cells, and endothelial cells (*54, 85, 89, 130, 138*). An important question is: Are they similar types of muscarinic receptors? In an attempt to pharmacologically characterize the receptor responsible for the inhibition of NE release upon nerve stimulation of the rabbit ear artery, Steisland et al. (*122*) compared the potencies of several muscarinic agonists and measured the dissociation constant value of atropine activity as an antagonist. It was observed that the average relative potencies of three different agonists (ACh, methacholine, and carbachol) in inhibiting the electrically induced release of NE are very close for these three agonists in producing contraction of the rabbit stomach muscle strip in the presence of cholinesterase inhibitors. The calculated mean dissociation constant for atropine of the prejunctional receptor complex was $8.0 \times 10^{-10}M$. A similar dissociation constant or the prejunctional atropine–receptor complex of $4.5 \times 10^{-10}M$, was obtained for the superfused guinea pig vas deferens (*97*). These calculated values are very similar to those found in the PA_2 values or dissociation constant for atropine in other smooth muscle-mediated contractions (*100*).

As has already been mentioned, stimulation of muscarinic receptors on the smooth muscle causes vasoconstriction and that on the endothelial cells causes vasodilation. Calculation from the data reported by Lee (*89*) indicates that the PA_2 values for atropine in blocking both ACh-induced relaxation and constriction of cerebral arteries are between 9.0 and 9.6. These values approximate those calculated from the data reported by Furchgott et al. (*55*) and Chand and Altura (*26*), suggesting that the cholinergic receptors are probably similar on the endothelial and smooth muscle cells in these cerebral and peripheral blood vessels. Thus, limited data to date suggest that the muscarinic receptors on presynaptic adrenergic nerve that mediate inhibition of NE release, on smooth muscle cells that mediate contraction, and on endothelial cells that mediate relaxation probably are not different. However, examination of more vascular beds using some of the newer cholinergic agonists and antagonists is needed in order to reveal if subtypes of muscarinic receptors exist.

Several investigators have demonstrated the presence of muscarinic cholinergic receptors in cerebral blood vessels using biochemical binding techniques (*45, 65, 110*). These radioligand binding studies provide direct evidence for the presence of muscarinic receptors in cerebral blood vessels. According to Estrada and Krause (*45*), there is only one binding site for [^{3}H]QNB in pial vessels, and the affinity of [^{3}H]QNB for the pial vessel receptors is similar to that found for muscarinic receptors in several other nonvascular preparations (*115, 140*). There are, however, two binding sites with different affinities for the cholinergic agonists in pial vessels. This finding of high and low binding sites for agonists has been a general characteristic found in other nonvascular preparations (*115, 140*). Results of radioligand binding studies, however, cannot differentiate the distribution of muscarinic receptors on nerve terminals, smooth muscle cells, or endothelial cells.

A number of studies have also demonstrated the existence of presynaptic muscarinic receptors on the cholinergic nerves of the myenteric plexus of the guinea-pig ileum (*51, 77, 117, 125*). Information on the classification of muscarinic receptors on the cholinergic nerves innervating vascular as well as nonvascular smooth muscles is still lacking. Kilbinger and Wessler (*78*) examined the effect of several muscarinic agonists to induce contraction of smooth muscle and on ACh release from the cholinergic nerves of the longitudinal muscle–myenteric plexus preparation. Their results indicated that the rank orders of potencies for presynaptic effect (inhibition of ACh release) and postsynaptic effects (causing contraction) were not similar, nor were the potency ratios (measured as ED_{50} presynaptic: ED_{50} postsynaptic) constant. These results suggest that the pre- and postsynaptic muscarinic receptors in the longitudinal muscle–myenteric plexus preparation differ slightly in their pharmacological properties. On the other hand, Alberts and Stjärne (1) demonstrate that the dissociation constants for atropine in blocking presynaptic muscarinic inhibition of [^{3}H]ACh release from cholinergic nerves and [^{3}H]NE release from adrenergic nerves in the guinea-pig ileum myenteric plexus are not different. Their data suggest that the properties of presynaptic muscarinic receptors are independent of whether they are on cholinergic or adrenergic nerve terminals. Further examination on the nature of cholinergic receptors on cholinergic nerve terminals in other nonvascular as well as vascular preparations will be needed to enhance our understanding of the classification of muscarinic receptors.

9. Preferential Effects of Acetylcholine on Blood Vessel Wall

Assuming that both endothelial cells and smooth muscle cells possess similar muscarinic receptors, it is interesting to find that the isolated blood vessels with intact endothelial cells relax and constrict in response to low and high concentrations of ACh, respectively (*55, 85, 89*). (*See* Fig. 3.) The threshold concentrations of ACh to induce constriction of blood vessels without endothelial cells are also higher (range from 10 to 100 times) than those of ACh to induce relaxation of blood vessel with endothelium (*89*). This variation in threshold concentration is seen in both helical strip and ring preparations (*26, 27, 34, 55, 85, 89*). It should be noted that in ring preparation, ACh exogenously added to the tissue bath will in fact act on smooth muscle in the media before acting on endothelial cells. Thus the anatomical consideration that ACh may stimulate endothelial cells before smooth muscle cells, especially in a helical strip preparation, does not explain the difference in threshold concentrations. In isolated cerebral arteries we have found that in the presence of physostigmine the ACh-

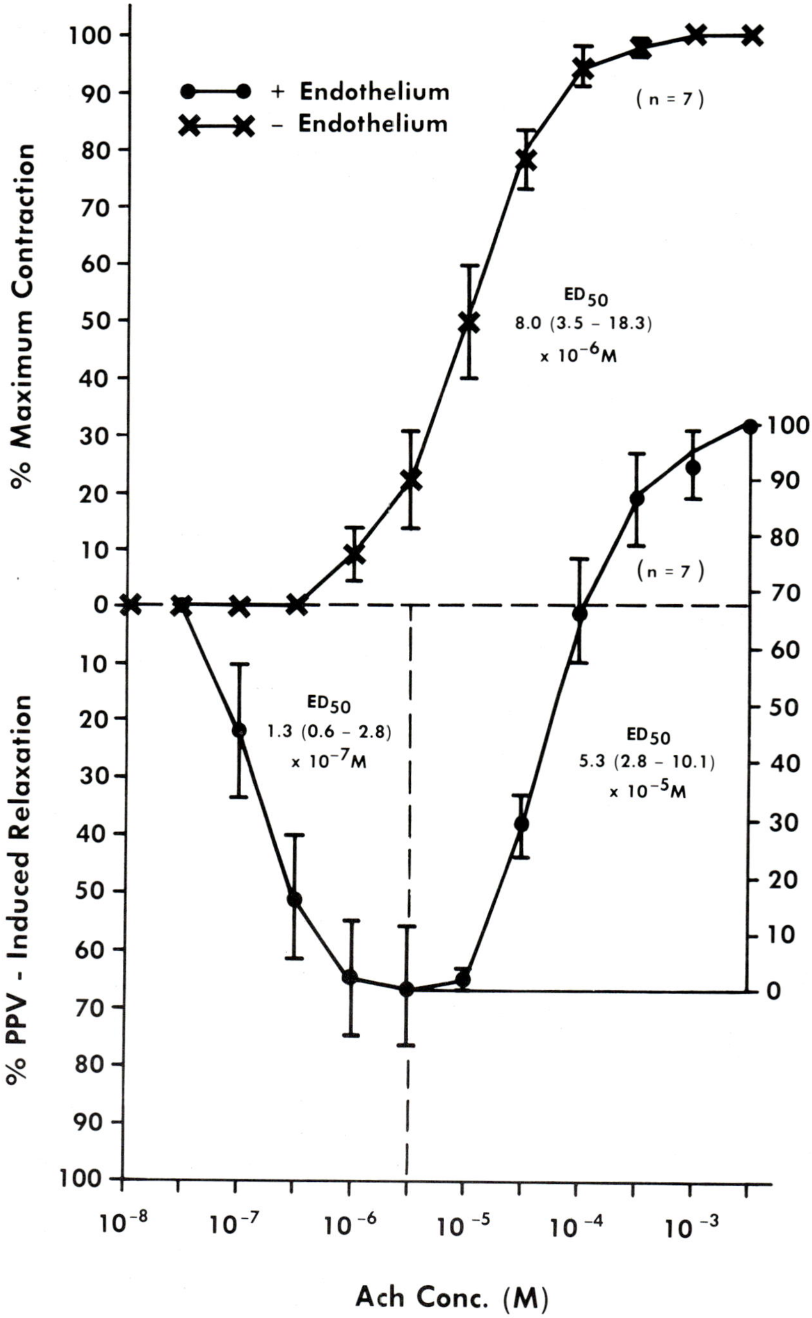
+ Endothelium
– Endothelium
(n = 7)
ED50
8.0 (3.5 – 18.3)
x 10⁻⁶M
ED50
1.3 (0.6 – 2.8)
x 10⁻⁷M
ED50
5.3 (2.8 – 10.1)
x 10⁻⁵M
(n = 7)
% Maximum Contraction
% PPV - Induced Relaxation
% Maximum Contraction
Ach Conc. (M)

induced relaxation is not affected or tends to decrease, and the ACh-induced constriction is potentiated (*89*). This finding suggests that a greater amount of cholinesterase may be associated with the smooth muscle cells than with the endothelial cells. An alternative explanation is that the dilator substance released from the endothelial cells upon application of ACh may be more potent in inducing muscle relaxation than ACh itself in inducing muscle contraction. This may in part explain the finding that ACh in physiological concentration frequently induces vasodilation rather than constriction.

10. Cholinergic Innervation and Acetylcholine

There is ample evidence suggesting that cholinergic vasomotor control exists in the vascular beds of some tissues and leads to changes in local blood flow in well-defined circumstances (*5, 22, 49, 128*). The distribution and role of cholinergic vasomotor nerves vary considerably among species (*22*). Several approaches have been used to demonstrate the presence of cholinergic innervation: ultrastructural demonstration of small agranular vesicles in nerve terminals, histochemical demonstration of acetylcholinesterase (AChE), biochemical assay of choline acetyltransferase (ChAT), and high choline uptake system (*74*). It is important that all four criteria be met before a conclusion on the presence of cholinergic innervation can be made. Otherwise, caution must be taken in interpreting the data. For example, the small agranular

Fig. 3. (*Opposite page*) Acetylcholine (ACh) dose–response relationship in the cat cerebral arteries with (+) or without (−) endothelium. In the presence of active muscle tone induced by 5-HT, ACh induced a dose-dependent relaxation in arteries with endothelial cells. The maximum relaxation induced by ACh ($3 \times 10^{-6} M$) is equivalent to approximately 67% of the maximum relaxation induced by papaverine (PPV, $3 \times 10^{-4} M$). The artery began to constrict when the dose of ACh exceeded $3 \times 10^{-6} M$ and reached the original active tone at approximately 67% of the maximum contraction. In the presence of active muscle tone, the arteries without endothelial cells constricted exclusively in response to ACh. The ED_{50} values of the contractile components in both arterial preparations are significantly different ($P < 0.005$, unpaired *t*-test). The vertical bars represent standard errors; *in* q number of experiments (*89*). (By permission of the American Heart Association, Inc.)

vesicles do not necessarily represent cholinergic nerves (*32*). Staining of AChE only indicates the presence of AChE-containing neurons but not necessarily cholinergic nerves. AChE has been shown to be associated with adrenergic nerves (*71*) and probably Substance P-containing nerves (*109*). The demonstration of ChAT or high affinity choline uptake may provide strong evidence for the presence of a cholinergic system, but this result may not necessarily indicate the presence of cholinergic nerves since ACh may be associated with nonneuronal tissue, as has been shown in human placenta (*66, 99, 116*). Unfortunately, there is little correlative work using all four approaches in most of the published works. Therefore, it is difficult to make a definite statement about the existence and function of cholinergic nerves. Furthermore, by definition, ACh should be released upon excitation of cholinergic nerves, and the responses induced then should be blocked by a muscarinic receptor antagonist, such as atropine, or potentiated by an AChE inhibitor, such as physostigmine. That is, the cholinergic nerves must terminally or directly synapse on effector organs, the smooth muscle cells. However, in several vascular beds that have been shown to receive cholinergic nerves based on circumstantial evidence, electrical nerve stimulation-induced responses are not, or only partially, affected by muscarinic receptor antagonists. The guinea-pig uterine arteries have been claimed to receive dense cholinergic nerves, especially during late pregnancy, based on AChE staining (*4*). Upon periarterial stimulation, the arterial preparations in the presence of active muscle tone induced by NE exhibited a powerful dilator response. This response was potentiated by AChE inhibitor, but only partially blocked by hyoscine (*3*). Although the dilator response was mimicked by exogenously applied ACh, it is questionable that the nerve stimulation-induced relaxation is due to direct action of the nerve-released ACh on smooth muscle cells. That is, the cholinergic nerves are probably not the terminal neurons. Therefore, it can only be concluded from these results that cholinergic nerves are somehow involved in the process of neurogenic vasodilation. In skeletal muscle, the vasodilation is shown to be produced by stimulation of sympathetic cholinergic nerves (*70*). Atropine, however, only partially blocks the neurogenic dilator response. Myers et al. (*106*) and Honig and Frierson (*68*) proposed the alternative concept that cholinergic nerves probably synapse on noncholinergic, nonadrenergic anglion cells, which are intrinsic to the arteries in skeletal muscles. Thus, vasodilation can be elicited by excitation of cholinergic nerves, but relaxation of the vascular smooth muscle is caused by release of the unknown substance from the intrinsic nerve terminals. Although further verification is needed, this concept indicates that all cholinergic nerves may not synapse directly on smooth muscle of the artery innervating skeletal muscle. Such cholinergic innervation patterns (preganglionic neurons) have also been suggested in intestinal smooth muscle (*139*).

AChE-containing fibers have been described around muscular blood vessels in the salivary glands of several species (*57, 58, 101*). There is physiological evidence that some of the fibers that cause vasodilation are cholinergic (*101*), but atropine-resistant vasodilation also occurs to variable extents in different salivary glands in different species (*22, 101*). Recently it has been shown that in salivary glands ACh and vasoactive intestinal polypeptide (VIP) coexist in "cholinergic" nerves. Upon excitation of these "cholinergic" nerves, both ACh and VIP are released and seem to be involved in the vasodilatory and secretory responses, although their exact role is not clear. Local intra-arterial infusion of exogenous VIP caused a marked atropine-resistant vasodilation but no salivary secretion. On the other hand, ACh infusions induced both vasodilation and salivary secretion (*101*). Since ACh acts more like a vasoconstrictor transmitter (*89*), the ACh-induced vasodilation may be secondary to release of VIP. Thus ACh may be the primary neurotransmitter for the secretory response, and VIP the transmitter for vasodilation.

In the 1930s, several reports indicated that the cat cerebral blood vessels received parasympathetic cholinergic nerves that were carried via superficial petrosal nerves (*28, 29, 50*). Subsequent findings based on histochemical localization of acetylcholinesterase (*42, 96*), biochemical demonstrations of choline acetyltransferase, high affinity choline uptake, and endogenous ACh in the cat cerebral arteries (*38, 48*) have further supported the presence of cholinergic nerves in cerebral blood vessels of cats and other species. This cholinergic nerve, based on some in vivo physiological studies (*29, 31, 134*), was thought to be the dilator nerve, and ACh was thought to be the dilator transmitter since in some experiments atropine blocked the vasodilation induced by intravascularly administered ACh or upon electrical stimulation of the petrosal nerves. However, results from other in vivo studies failed to show a change in cerebral blood flow during electrical stimulation of the petrosal nerves (*23, 118*). The concept of the cholinergic nerve as the dilator nerve and ACh as the dilator transmitter was first challenged by the results from in vitro studies that found nerve stimulation-induced relaxation of the isolated cat cerebral arteries to be unaffected by atropine, physostigmine, and hemicholinium (*95, 96*). (*See* Fig. 4.) This was further supported by the failure to find a positive correlation between the degree of neurogenic vasodilator response and that of cholinergic innervation in the cerebral arteries of cat and rabbit (*38, 48, 86, 96*). Two possibilities have been raised to explain the inconsistency in findings between in vivo and in vitro studies. Firstly, ACh may be the transmitter for cholinergic nerves but may be involved in the dilator pathway at sites other than the terminal synapse (*85*). That is, ACh as a transmitter substance does not act directly on the cerebral vascular smooth muscle. Instead, it acts on a terminal neuron(s) that release(s) a noncholinergic transmitter sub-

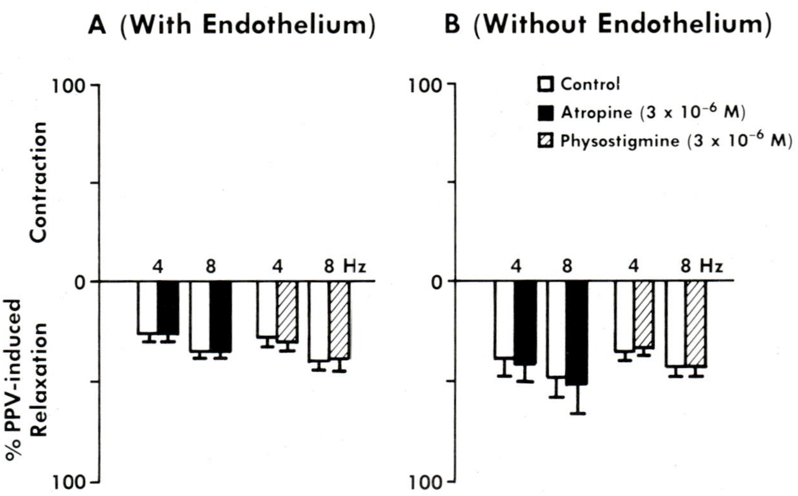

Fig. 4. Effect of atropine ($3 \times 10^{-6}M$) and physostigmine ($3 \times 10^{-6}M$) on the transmural nerve stimulation (TNS)-induced relaxation of cerebral arteries with (A, $n = 9$ pairs for each frequency of stimulation) and without (B, $n = 6$ pairs for each frequency of stimulation) intact endothelium. TNS elicited vasodilation exclusively in either preparation. Neither atropine nor physostigmine affected the TNS-induced vasodilation at 4 or 8 Hz ($P < 0.5$, paired *t*-test). The dilator response was abolished by tetrodotoxin (TTX). Active muscle tone was induced by 5-HT or uridine-5′-triphosphate (UTP). Relaxation was expressed as percent of maximum relaxation induced by papaverine (PPV, $3 \times 10^{-4}M$). Vertical bars represent standard errors (*89*). (By permission of the American Heart Association, Inc.)

stance. Thus, the site of action of atropine when intravascularly administered to block neurogenic vasodilation in some in vivo studies may or may not be specific for muscarinic receptors. Krnjevic (*81*) has reported that atropine may depress central noncholinergic transmission. Secondly, ACh may be released together with a second transmitter from the terminal "cholinergic" neurons as is demonstrated in the salivary gland (*101*). The failure of cholinergic drugs to affect the neurogenic vasodilation may suggest that concentration of the nerve-released ACh at the synaptic region upon transmural nerve stimulation (TNS) is too low to exhibit any direct muscarinic interaction on smooth muscle cell (*89*). The neurogenic vasodilation must be caused by a second transmitter substance (*89, 95, 96*). Several peptides such as vasoactive intestinal polypeptide (VIP) and Substance-P (SP) have been shown to be present in cerebral blood nessels (*40, 43, 44, 84*). Results from pharmacological studies, however, favor the theory that VIP is more likely the dilator transmitter (*90, 92*). Ultrastructural studies reveal that there seems to be only one type of agranular vesicle-containing or nonadrenergic nerve (*86, 89*). Ultrastructural immunocytochemical study further reveals that a VIP-like substance is found in the large granular vesicles in the nonsympathetic nerve terminals (*92*). Although it is entirely possible, evidence for the costorage of ACh and VIP-like substance(s) in the same neurons in cerebral blood vessels has not been presented.

In conclusion, cholinergic neurons can be preganglionic and/or postganglionic. The postganglionic terminal neurons may release ACh and a noncholinergic substance. Thus, in some vascular smooth muscle preparations, responses to nerve stimulation will be different from those to the direct effect of ACh. This may explain why vasodilation induced by the "cholinergic" nerve stimulation (indirect effect), but not that by exogenous ACh (direct effect on smooth muscle), is resistant to atropine in some vascular beds (*3, 89, 95, 96*). In this regard, it may be important and necessary to differentiate cholinopeptidergic or peptidocholinergic nerves from the classical cholinergic nerves.

11. Is Acetylcholine the Vasoconstrictor or Vasodilator Transmitter?

In many reports ACh released from parasympathetic nerves has been assumed to be the vasodilator transmitter in many vascular beds (22). Evidence supporting this concept is generally based on the observations that AChE is present in the vessel wall and that exogenous ACh produces vasodilation which mimics the parasympathetic nerve stimulation-induced response. Such evidence is not convincing, since exogenously applied ACh is known to have actions on some noninnervated preparations, and in most preparations it will stimulate endothelial cells to indirectly induce vasodilation. On the other hand, cholinergic neurons in several vascular beds have been shown to act like vaso-constrictor nerves. In coronary artery, stimulation of parasympathetic nerves has been shown to result in vasodilation (*6, 47*), vasoconstriction (*16, 63*), or no response (*20, 61*). It is possible that, under in vivo conditions, variations on cardiac rate, extravascular compression, aortic pressure, and myocardial metabolic activity are some of the important complications that make it difficult to evaluate precisely the effects of parasympathetic activity directly on coronary hemodynamics. Using an isolated and perfused nonbeating slab of beef ventricle in order to eliminate these secondary complications, Kalsner (*75*) examined the effect of periarterial nerve activation on coronary perfusion pressure. It was found that stimulation of periarterial nerves induced vasoconstriction that was enhanced by physostigmine and blocked by atropine. These results indicate that excitation of cholinergic nerves to beef coronary arteries induces coronary vessel constriction rather than dilation. A similar result in beef coronary artery was also obtained by Garland and Keatinge (*56*). These authors report that ACh constricts both coronary arteries with or without endothelial cells. Unfortunately, in these studies by Kalsner (*75*) and Garland and Keatinge (*56*), the arterial preparations were not precontracted by other vasoactive substance or in the presence of active muscle tone.

The longitudinal muscle of bovine mesenteric veins has also been found to contract in response to exogenous ACh (*104*) to perivascular electrical stimulation (*103*). The vasoconstriction induced by electrical stimulation was blocked by atropine. It was concluded that longitudinal muscle in bovine mesenteric veins receives an excitatory cholinergic innervation. A similar result was obtained in the chicken anterior mesenteric artery (*13*). ACh and electrical nerve stimulation produced contraction of the arteries with depolarization and appearance of action potential. The contraction was abolished by hyoscine. These results suggest the presence of an excitatory, cholinergic nerve in this artery. Cholinergic vasoconstrictor nerves were also found in the rabbit eyes (*11, 123*). It was found that intracranial stimulation of the oculomotor nerves caused a muscarinic-mediated vasoconstriction in the iris and ciliary process.

The isolated cat cerebral arteries with intact endothelial cells have been shown to consistently dilate upon transmural nerve stimulation (TNS) and upon exogenous application of ACh at low concentrations. These arteries began to constrict when the concentration of ACh exceeded $3 \times 10^{-6}M$. In contrast, those arteries without endothelial cells did not relax, but constricted exclusively in response to ACh at both low and high concentrations, while TNS-induced dilation persisted as in the control arteries with intact endothelium (Figs. 1 and 3). Similar results were obtained in cerebral arteries from several other species (*87, 90, 91, 93*), indicating that ACh-induced relaxation of cerebral arteries is dependent solely on the endothelial cells and that the direct effect of ACh on smooth muscle cell is constriction. The dissociation in responses of arteries without endothelial cells induced by TNS (relaxation) and ACh (constriction) indicates that ACh is not a dilator transmitter but acts more like a constrictor transmitter.

12. Can Nerve-Released Acetylcholine Induce Vasodilation by Releasing Dilator Substance(s) from the Endothelial Cells?

The dichotomous actions of exogenously applied ACh (indirect vasodilation at low concentration and direct vasoconstriction at high concentration) in most vascular beds suggest the possibility that the nerve-released ACh may be able to induce vasodilation by the indirect mechanism. This possibility which is examined in cerebral arteries, however, is shown to be very unlikely. The nerve terminals of any type in the intracranial, extracerebral pial arteries of several species are found only in the adventitial layer (*72, 86, 94, 108*). There are at least five layers of muscle cells in the media with greater than 20 μm thickness in the major cerebral arteries (*89*). It is very doubtful that the nerve-released ACh will diffuse such a long distance and be effective to release the relaxing substance from the endothelial cells. Bevan and Su (*10*) reported that the concentration of the synaptic transmitter would be expected to recede with the cube of the synaptic distance. If the closest distance between nerve terminals and endothelial cells is 20 μm, for example, in the cat anterior cerebral artery, and without taking into consideration a diffusion barrier such as smooth muscle and acetylcholinesterase, the threshold concentration for vasodilation of the nerve-released ACh, $10^{-7}M$ (*89*) at the endothelial cells, should be preceded by a concentration of at least

$[20]^3$, or 8000 times higher than $10^{-7}M$ at the synaptic region. This latter concentration, equivalent to $8 \times 10^{-4}M$, would be expected to cause a constriction prior to any relaxation (Fig. 3). TNS, however, has not been shown to cause constriction of the cat cerebral arteries (*7, 85, 89, 95, 96*). The absence of vasoconstriction could not be the result of masking of the constriction by the dilator responses, since neither atropine nor physostigmine affected the magnitude of the TNS-induced dilator response of the arteries with or without endothelial cells (*85, 89*). (*See* Fig. 4.) Similar results have been obtained by several other investigators (*37, 127*). In this regard, it is quite possible that the concentration of nerve-released ACh, if any, at the synaptic region upon transmural nerve stimulation (TNS) is lower than $10^{-6}M$, the threshold concentration for constriction (*89*). If so, the low synaptic concentration of ACh makes it even more difficult to induce vasodilation indirectly by stimulating the endothelial cells.

Furthermore, there is no apparent difference in the latency for the TNS-induced vasodilator responses between the arteries with and without endothelial cells (*89*). The TNS-induced vasodilation is not smaller, but, in fact, is greater in the arteries without endothelial cells than in those with endothelial cells (*89, 91*), whereas the beta-receptor-mediated relaxation of the pig cerebral arteries induced by exogenous NE is not different between arteries with and without endothelial cells. (*See* Figs. 5a and 5b.). These results indicate that the TNS-induced vasodilation is not dependent on nor does it involve the release of a relaxing substance from the endothelial cells. This further emphasizes that ACh is not a dilator transmitter, but acts more like a constrictor transmitter. As has been indicated, the neurogenic dilator response in cerebral blood vessels must be caused by release of an as yet unidentified substance (*85, 89, 91, 95, 96*). It seems that the unidentified dilator substance, possibly a peptide (*40, 43, 84, 90, 92*), relaxes the outer layer of smooth muscle cells primarily. This conclusion however conflicts with that reported by Bevan et al. (*7*). These authors show that the TNS-induced dilations of some cat cerebral arterial rings with intact endothelial cells are atropine-sensitive. The reason for their different results from those of others (*37, 89, 95, 96, 127*), is not known. However, the study by Bevan et al. (*7*) compared the TNS-induced dilator responses before and after wash (replace fresh Kreb's solution), whereas others compared two consecutive dilator responses without washing in between. Preliminary results indicate that within 1 h after washing, the TNS-induced vasodilation of the cat cerebral arterial

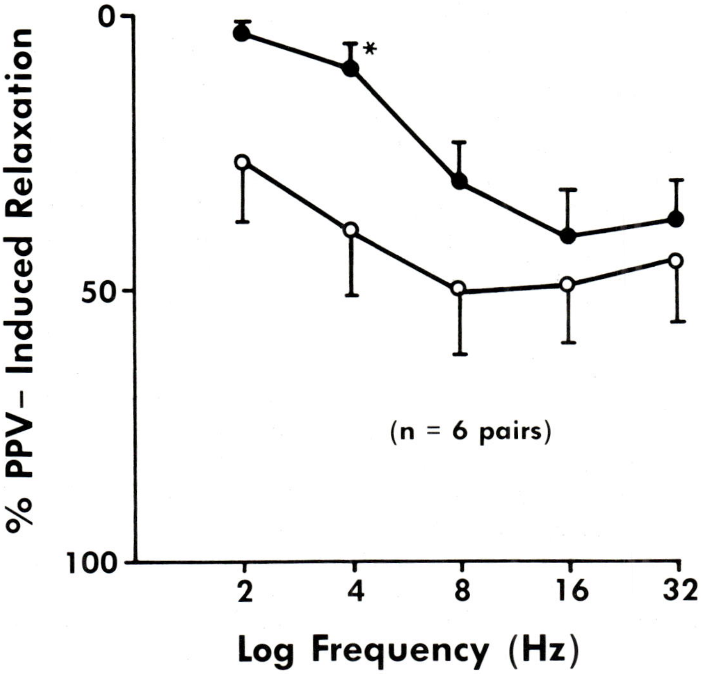

Fig. 5A. Frequency-relaxation relationship of the pig celebral arteries with (filled circles) and without (open circles) endothelium to transmural nerve stimulation at different frequencies (Hz) expressed as a percentage of maximum relaxation induced by papaverine (PPV; $3 \times 10^{-4}M$). The frequency–relaxation relationships were significantly greater at 4 Hz (paired *t*-test; $^*P < 0.05$) in arteries without endothelium than in those with endothelium. Active muscle tone was induced by serotonin. Vertical bars represent standard errors; *n*, number of experiments (*91*). (By permission of Raven Press, New York.)

ring is frequently not consistent compared to those observed before wash (*98*). It is possible that a transient change of tissue environment during wash may change the characteristics of neurogenic dilator response.

13. Acetylcholine-Induced Membrane Hyperpolarization and Vasoconstriction

It has long been known that many drugs can initiate contraction independent of the membrane potential (*18, 25, 46, 79, 121, 124*). However, it is more difficult to reconcile the findings in the rabbit

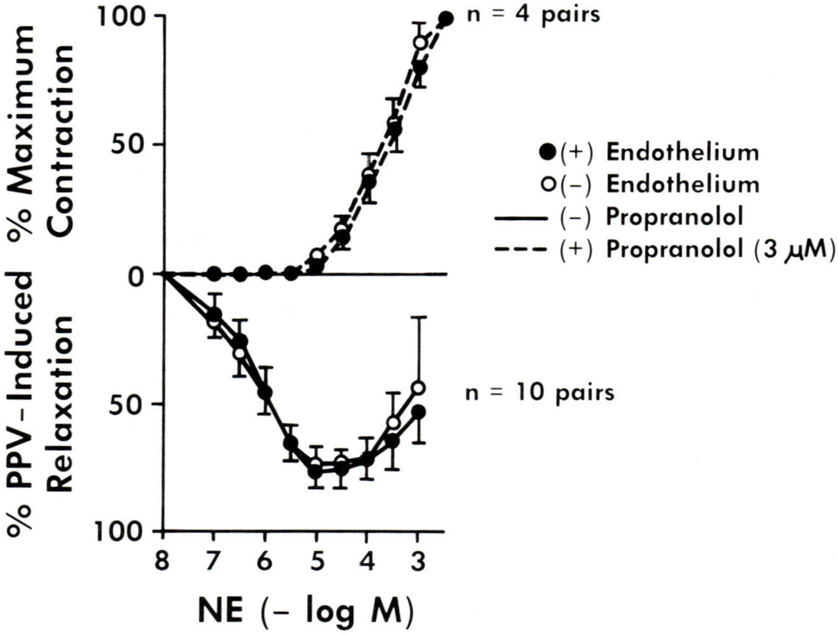

Fig. 5B. Norepinephrine (NE) dose–response relationship in the pig cerebral arteries with (filled circles) or without (open circles) endothelium. In the presence of active muscle tone induced by serotonin (5-HT), NE induced a dose-dependent relaxation in arteries with and without endothelial cells. The NE dose–response relationships in both arterial preparations were not different. The maximum relaxations induced by NE ($10^{-5}M$) were equivalent to approximately 75% of the maximum relaxation induced by papaverine (PPV; $3 \times 10^{-4}M$). The arteries began to constrict when the concentration of NE exceeded $3 \times 10^{-5}M$. In the presence of active muscle tone, propranolol (3 μ*M*; dashed lines) prevented the arteries from relaxing induced by NE (solid lines indicate without propranolol). Under this condition, NE induced only vasoconstriction. The vertical bars represent standard errors; *n*, number of experiments (*91*). (By permissions of Raven Press, New York.)

superior mesenteric artery and the guinea-pig coronary artery that contractile response to ACh was accompanied by a hyperpolarization (*79, 82*). It should be pointed out that in these studies mechanical responses and membrane potential changes upon application of ACh were not monitored in the same blood vessel preparation. Besides, microelectrodes for recording the membrane potential of single muscle cells were inserted from the intimal surface. As stated before, ACh may release a dilator substance from the endothelial cells, which then relaxes the smooth muscle cells. It is possible that upon application of ACh those muscle cells closer to the endothelial cells may be hyperpolarized by the dilator substance released from the endothelial cells, whereas those muscle cells close to the adventitial layer may be depolarized upon direct stimulation by ACh. The overall response upon application of ACh will be constriction, not relaxation, if the preparation does not have active muscle tone (*79, 82, see* Section 1.1). Considering that most nonvascular smooth muscle cells are depolarized by ACh (*15*), the inconsistent findings in the ACh-induced membrane potential changes in vascular smooth muscle cells should be interpreted with caution. The presence of intact endothelial cells should be taken into consideration when examining the direct effect of ACh on vascular smooth muscle cells.

14. Conclusion

Of most nonvascular smooth muscles examined, ACh and other cholinomimetic agents have been shown to cause contraction. The contraction is mediated by muscarinic receptors and is dependent on extracellular and intracellular calcium. The effect of ACh on vascular smooth muscle is complicated by the presence of endothelial cells. The ACh-induced vasodilation is dependent on the release of a dilator substance from the endothelial cells. The direct effect of ACh on vascular smooth muscle is constriction rather than dilation (*see* Fig. 6). Since cholinergic nerve terminals in blood vessel walls have not been found in a site other than the adventitial layer, the indirect effect of the nerve-released ACh to induce vasodilation by releasing a dilator substance from the endothelial cells is not likely. Therefore, the nerve-released ACh acts more like a vasoconstrictor transmitter. The nerve-released ACh, however, can act on presynaptic muscarinic receptors on adrenergic nerve terminals to inhibit NE release and therefore decrease vascular tone. The presynaptic adrenergic inhibition, direct vascular smooth muscle contraction and indirect endothelium-induced vasodilation by exogenous ACh are mediated by muscarinic receptors. Calcium has been shown to play important roles in all three interactions. Although a conclusive statement cannot be made before a complete examination of effects of ACh on various smooth muscle cells is performed and a few exceptions (*100, 102, 107*) are verified, ACh is probably a direct excitatory

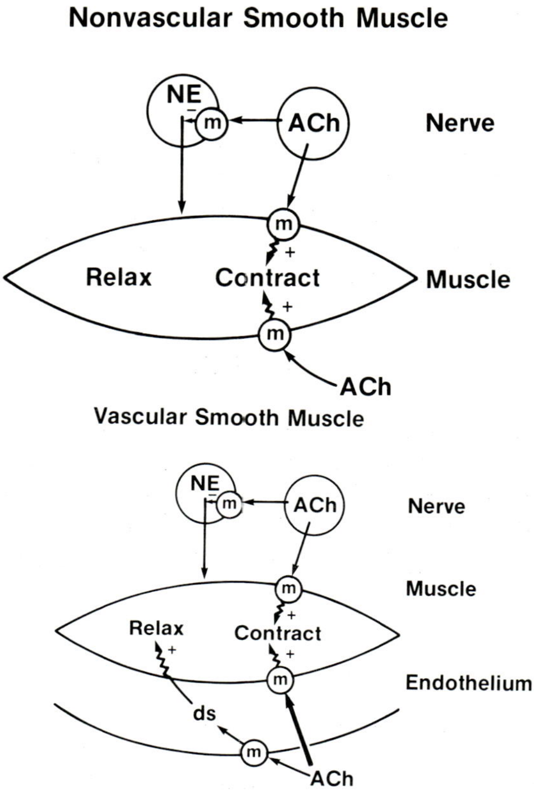

Fig. 6. A hypothetical illustration of the terminal nerve-muscle relationships in nonvascular and vascular smooth muscle. (Upper) The terminal nerve-released and exogenously applied acetylcholine (ACh) stimulate the muscarinic (m) receptors on the nonvascular smooth muscle cell membrane to induce contraction. Stimulation of the m receptors on the presynaptic adrenergic nerve terminals results in decreased norepinephrine (NE) release. (Lower) The nerve-released and exogenously applied ACh induces vasoconstriction by directly stimulating m receptors on the smooth muscle cell membrane. The exogenously applied ACh may stimulate the m receptors on the endothelial cell and induce vasodilation by releasing a dilator substance (ds) from the endothelial cells (indirect mechanism). Because of the long distance between nerve terminals and endothelial cells in muscular arteries, the nerve-released ACh is not likely to induce vasodilation by this indirect mechanism (see text). ACh, both nerve-released and exogenously applied, may stimulate the m receptor on the adrenergic nerve terminals to inhibit NE release. In both nonvascular and vascular smooth muscles, these m receptors on adrenergic nerve terminals (mediate inhibition of NE release), smooth muscle cells (mediate contraction), and endothelial cells (mediate indirect vasodilation) are probably similar. The m receptors on the cholinergic nerve terminals are not included since the nature of these receptors is not adequately described. (+ indicates stimulation and − indicates inhibition.)

substance on smooth muscle cells, both vascular and non-vascular.

Cholinergic nerves in a given preparation can be preganglionic neurons and/or postganglionic terminal neurons. Besides, some "cholinergic" nerves may release co-transmitters. Thus, effects on the smooth muscle tone of exogenously-applied ACh and excitation of "cholinergic" nerves may not be parallel.

Acknowledgments

This study was supported by grants from NIH HL 27763, BRSG SO7RR05843, a grant-in-aid from AHA (83-1040), with funds contributed in part by the AHA/IHA and funds from Southern Ilinois University School of Medicine. We thank S. Sarwinski for technical assistance, L. Ragel for art work, and S. Fluckiger for preparation of the manuscript.

References

1. Alberts, P. and L. Stjarne. Facilitation, muscarinic and alpha-adrenergic inhibition of the secretion of ^{3}H-acetylcholine and ^{3}H-noradrenaline from guinea-pig ileum myenteric nerve terminals. *Acta Physiol. Scand.* 116:83–92, 1982.
2. Aubineau, P., R. Sercombe, and J. Seylaz. Parasympathetic influence of carbachol on local cerebral blood flow in the rabbit by a direct vasodilator action and an inhibition of the sympathetic-mediated vasoconstriction. *Brit. J. Pharmacol.* 68:449–459, 1980.
3. Bell, C. Dual vasoconstrictor and vasodilator innervation of the uterine arterial supply in the guinea-pig. *Circ. Res.* 23:279–289, 1968.
4. Bell, C. Fine structural localization of acetylcholinesterase at a cholinergic vasodilator nerve-arterial smooth muscle synapse. *Circ. Res.* 24:61–70, 1969.
5. Bell, C. and G. Burnstock. Cholinergic vasomotor neuroeffector junctions. In *Physiology and Pharmacology of Vascular Neuroeffector Systems.* J. A. Bevan, R. F. Furchgott, R. A. Maxwell, and A. P. Somlyo, eds. Karger, Basel, 1971, pp 37–46.
6. Berne, R. M., H. DeGeest, and M. N. Levy. Influence of the cardiac nerves on coronary resistance. *Am. J. Physiol.* 208:763–769, 1965.
7. Bevan, J. A., G. M. Buga, C. A. Jope, R. S. Jope, and H. Moritoki. Further evidence for a muscarinic component to the neural vasodilator innervation of cerebral and cranial extracerebral arteries of the cat. *Circ. Res.* 51:421–429, 1982.
8. Bevan, J. A., W. Grastka, and M. O. Su. The bimodal basis of the contractile response of the rabbit ear artery to norepinephrine and other agonists. *Europ. J. Pharmacol.* 22:47–53, 1973.
9. Bevan, J. A., T. A. McCalden, and R. M. Rapoport. Receptor-activated calcium mechanisms and their antagonism in cerebrovascular muscle.

In *New Perspectives on Calcium Antagonists.* G. B. Weiss, ed. Clinical Physiology Series, American Physiological Society, 1981, pp 123–129.
10. Bevan, J. A. and C. Su. Variation of intra- and perisynaptic adrenergic transmitter concentrations with width of synaptic cleft in vascular tissue. *J. Pharmacol. Exp. Ther.* 190:30–38, 1974.
11. Bill, A. and J. Stjernschantz. Cholinergic vasoconstrictor effects in the rabbit eye: vasomotor effects of pentobarbital anesthesia. *Acta Physiol. Scand.* 108:419–424, 1980.
12. Bohr, D. F. Vascular smooth muscle: dual effect of calcium. *Science* 139:597–599, 1963.
13. Bolton, T. B. Electrical and mechanical activity of the longitudinal muscle of the anterior mesenteric artery of the domestic fowl. *J. Physiol.* 196:283–292, 1968.
14. Bolton, T. B. The permeability change produced by acetylcholine in smooth muscle. In *Drug Receptors.* H. P. Rang, ed. MacMillan, London, 1973, pp 87–102.
15. Bolton, T. B. Action of acetylcholine on the smooth muscle membrane. In *Smooth Muscle.* E. E. Bulbring, A. F. Brading, A. W. Jones, and T. Tomita, eds. University of Texas Press, Austin, 1981, pp 199–217.
16. Brachfeld, N., R. G. Monroe, and R. Gorlin. Effect of pericoronary denervation on coronary hemodynamics. *Am. J. Physiol.* 199:174, 1960.
17. Brading, A. F. and P. Sneddon. Evidence for multiple sources of calcium for activation of the contractile mechanism of guinea-pig taenia coli on stimulation with carbachol. *Br. J. Pharmac.* 70:229–240, 1980.
18. Brading, A. How do drugs initiate contraction in smooth muscles? *TIPS* 261–265, 1981.
19. Brodie, D. C., D. F. Bohr, and J. Smit. Dual contractile response of the aortic strip. *Am. J. Physiol.* 197:241–246, 1959.
20. Brown, A. M. Motor innervation of the coronary arteries of the cat. *J. Physiol. London* 198:311–328, 1968.
21. Burnstock, G. Structure of smooth muscle and its innervation. In *Smooth Muscle.* E. E. Bulbring, A. F. Brading, A. W. Jones, and T. Tomita, eds. Edward Arnold, London, 1970, pp 1–70.
22. Burnstock, G. Cholinergic and purinergic regulation of blood vessels. In *Handbook of Physiology, Section 2, The Cardiovascular System, Vol II, Vascular Smooth Muscle.* D. F. Bohr, A. P. Somlyo, and H. V. Sparks, Jr., eds. American Physiol. Society, Bethesda, MD, 1980, pp 567–612.
23. Busija, D. W. and D. D. Heistad. Effects of cholinergic nerves on cerebral blood flow in cats. *Circ. Res.* 48:62–69, 1981.
24. Campbell, G. Autonomic nerves supply to effector tissues. In *Smooth Muscle.* E. E. Bulbring, A. F. Brading, A. W. Jones, and T. Tomita, eds. Edward Arnold, London, 1970, pp 451–459.
25. Casteels, R. and G. Droogmans. Membrane potential and excitation–contraction coupling in smooth muscle. *Fed. Proc.* 41:2879–2882, 1982.
26. Chand, N. and B. M. Altura. Inhibition of endothelial cell-dependent relaxations to acetylcholine and bradykinin by lipoxygenase inhibitors in canine isolated renal arteries. *Microcirculation* 1:211–223, 1981a.
27. Chand, N. and B. M. Altura. Acetylcholine and bradykinin relax intrapulmonary arteries by acting on endothelial cells: role in lung vascular diseases. *Science* 213:1376–1379, 1981b.

28. Chorobski, J. and W. Pennfield. Cerebral vasodilator nerves and their pathway from the medulla oblongata. *Arch. Neurol. Psychiat.* 28:1257–1289, 1932.
29. Cobb, S. and J. E. Finesinger. Cerebral circulation. XIX. The vagal pathway of the vasodilator fibers. *Arch. Neurol. Psychiat.* 28:1243–1256, 1932.
30. Creese, B. R. and M. A. Denborough. Sources of calcium for contraction of guinea-pig isolated tracheal smooth muscle. *Clin. Exp. Pharmacol. Physiol.* 8:175–182, 1981.
31. D'Alecy, L. G. Adrenergic and cholinergic influence on CBF as determined with the venous outflow method. In *Neurogenic Control of the Brain Circulation.* C. Owman, L. Edvinsson, eds. Oxford, Pergamon Press, 1977, pp 327–330.
32. Daniel, E. E., G. S. Taylor, V. P. Daniel, and M. E. Holman. Can nonadrenergic inhibitory varicosities be identified by structure? *Can. J. Physiol. Pharmacol.* 55:243–250, 1977.
33. de la Lande, I. S. Adrenergic mechanisms in the rabbit ear artery. *Blood Vessels* 12:137–160, 1975.
34. De Mey, J. G. and P. M. Vanhoutte. Heterogenous behavior of the canine arterial and venous wall, importance of endothelium. *Circ. Res.* 51:439–447, 1982.
35. Douglas, W. W. Stimulus-secretion coupling: The concept and clues from chromaffin and other cells. *Br. J. Pharmac.* 34:451–474, 1968.
36. Dubey, M. P., E. Muscholl, and A. Pfeiffer. Muscarinic inhibition of potassium-induced noradrenaline release and its dependence on the calcium concentration. *Naunyn–Schmiedeberg's Arch. Pharmacol.* 278:179–194, 1975.
37. Duckles, S. P. Neurogenic dilator and constrictor responses of pial arteries *in vitro*: differences between dogs and sheep. *Circ. Res.* 44:482–490, 1979.
38. Duckles, S. P. Evidence for a functional cholinergic innervation of cerebral arteries. *J. Pharmacol. Exp. Ther.* 217:544–548, 1981.
39. Duckles, S. P. and C. D. Kennedy. Cerebral blood vessels: effects of exogenous acetylcholine and field stimulation on norepinephrine release. *J. Pharmacol. Exp. Ther.* 222:562–565, 1982.
40. Duckles, S. P. and S. I. Said. Vasoactive intestinal polypeptide as a neurotransmitter in the cerebral circulation. *Eur. J. Pharmacol.* 78:371–374, 1982.
41. Edvinsson, L., B. Falck, and C. Owman. Possibilities for a cholinergic action on smooth musculature and on sympathetic axons in brain vessels mediated by muscarinic and nicotinic receptors. *J. Pharmacol. Exp. Ther.* 200:117–126, 1977.
42. Edvinsson, L., K. C. Nielsen, C. Owman, and B. L. Sporong. Cholinergic mechanisms in pial vessels. Histochemistry, electron microscopy and pharmacology. *Z. Zellforsch. Mikroskop. Anat.* 134:311–325, 1972.
43. Edvinsson, L., J. Fahrenkrug, J. Hanko, C. Owman, F. Sundler, and R. Uddman. VIP (vasoactive intestinal polypeptide)-containing nerves of intracranial arteries in mammals. *Cell Tissue Res.* 208:135–142, 1980.
44. Edvinsson, L., J. McCulloch, and R. Uddman. Substance P: immunohistochemical localization and effect upon cat pial arteries in vitro and *in situ*. *J. Physiol.* 318:251–258, 1981.

45. Estrada, C. and D. N. Krause. Muscarinic cholinergic receptor sites in cerebral blood vessels. *J. Pharmacol. Exp. Ther.* 221:85–90, 1982.
46. Evans, D. H. L., H. O. Schild and S. Thesleff. Effects of drugs on depolarized plain muscle. *J. Physiol. London* 143:474–485, 1958.
47. Feigl, E. O. Parasympathetic control of coronary blood flow in dogs. *Circ. Res.* 25:509–519, 1969.
48. Florence, V. M. and J. A. Bevan. Biochemical determinations of cholinergic innervation in cerebral arteries. *Circ. Res.* 45:212–218, 1979.
49. Folkow, B. and E. Neil. *Circulation.* London, Oxford, 1971.
50. Forbes, H. S. and H. G. Wolff. Cerebral circulation. III. The vasomotor control of cerebral vessels. *Arch. Neurol. Psychiat.* 19:1057–1980, 1928.
51. Fosbraey, P. and E. S. Johnson. Modulation by acetylcholine of the electrically-evoked release of [^{3}H]–acetylcholine from the ileum of the guinea-pig. *Br. J. Pharmacol.* 69:145–149, 1980.
52. Fozard, J. Cholinergic mechanism in adrenergic function. In *Trends in Autonomic Pharmacology, Vol. I.* S. Kalsner, ed. Urban and Schwarzenberg, Baltimore-Munich, 1979, pp 145–194.
53. Furchgott, R. F. Acetylcholine and blood vessel relaxation: Complications and clarifications. In *Trends in Autonomic Pharmacology, Vol. 2.* S. Kalsner, ed. Urban and Schwarzenberg, Baltimore, 1982, pp 497–510.
54. Furchgott, R. F. and J. V. Zawadzki. The obligatory role of endothelial cells in the relaxation of arterial smooth muscle by acetylcholine. *Nature* 88:375–376, 1980.
55. Furchgott, R. F., J. V. Zawadzki, and P. D. Cherry. Role of endothelium in the vasodilator response to acetylcholine. In *Vasodilation.* P. M. Vanhoutte and I. Leusen, eds. Raven Press, New York, 1981, pp 49–66.
56. Garland, C. J. and W. R. Keatinge. Constrictor actions of acetylcholine, 5-hydroxytryptamine and histamine on bovine coronary artery inner and outer muscle. *J. Physiol.* 327:363–376, 1982.
57. Garrett, J. R. The innervation of salivary glands. II. The ultrastructure of nerves in normal glands of the cat. *J. Ray. Microscop. Soc.* 85:149–162, 1966.
58. Garrett, J. R. and J. Holmberg. Effects of surgical denervations on the autonomic nerves in carotid glands of dogs. *Z. Zellforsch. Mikroskop. Anat.* 131:451–462, 1972.
59. Godfrain, T. and A. Kaba. The role of calcium in the action of drugs in vascular smooth muscle. *Arch. Int. Pharmacodyn.* 196:35–49, 1972.
60. Goyal, R. K. and B. W. Coble. Motility of the pharynx, esophagus and esophageal sphincters. In *Physiology of the Gastrointestinal Tract.* L. R. Johnson, ed. New York, Raven Press, 1981, pp 359–391.
61. Gregg, D. E. and L. C. Fisher. Blood supply to the heart. In *Circulation, Handbook of Physiology.* Vol. II. American Physiology Society, Washington, D.C., 1963, p 1548.
62. Gothert, M. Effects of presynaptic modulators on calcium-induced noradrenaline release from cardiac sympathetic nerves. *Naunyn-Schmiedeberg's Arch. Pharmacol.* 300:267–272, 1977.
63. Guzman, S. V., E. Swenson, and M. Jones. Inter-coronary reflex: demonstration by coronary angiography. *Circ. Res.* 10:739, 1962.
64. Haeusler, G., H. Thonen, W. Haefely, and A. Huerliman. Electrical events in cardiac adrenergic nerves and noradrenaline release from

the heart induced by acetylcholine and KCl. *Naunyn-Schmiedeberg's Arch. Pharmacol.* 261:389–411, 1968.

65. Harik, S. I., V. K. Sharma, J. R. Wetherbee, R. H. Warren, and S. P. Banerjee. Adrenergic and cholinergic receptors of cerebral microvessels. *J. Cerebral Blood Flow Metabol.* 1:329–338, 1981.

66. Haubrich, D. R. Choline acetyltransferase and its inhibitors. In *Biology of Cholinergic Function.* Goldberg, A. M. and Hanin, I., eds. Raven Press, New York, 1976, pp 239–268.

67. Hong, S. L. and D. Deykin. Specificity of phospholipases in methylcholanthrene-transformed mouse fibroblasts activated by bradykinin, thrombin, serum, and ionphore A 23187. *J. Biol. Chem.* 254:11463–11466, 1979.

68. Honig, C. R. and J. L. Frierson. Neurons intrinsic to arterioles initiate postcontraction vasodilation. *Am. J. Physiol.* 230:493–507, 1976.

69. Hope, W., M. W. McCulloch, M. J. Rand, and D. F. Story. The effect of calcium on the interaction between acetylcholine and noradrenergic transmission in the rabbit ear artery. *Clin. Exp. Pharmacol. Physiol.* 5:290, 1978.

70. Hudlicka, O. *Muscle Blood Flow—Its Relation to Muscle Metabolism and Function.* Amsterdam, Swets and Zeitlinger, 1973.

71. Hume, W. R. and J. G. Waterson. The innervation of the rabbit ear artery. *Blood Vessels* 15:348–364, 1978.

72. Iwayama, T., J. B. Furness, and G. Burnstock. Dual adrenergic and cholinergic innervation of cerebral arteries of the rat. *Circ. Res.* 26:635–646, 1970.

73. Jelliffe, R. W. Dilator and constrictor effects of acetylcholine on isolated rabbit aortic chains. *J. Pharmacol. Exp. Ther.* 135:349–353, 1962.

74. Jope, R. S. High affinity choline transport and acetyl CoA production in brain and their roles in the regulation of acetylcholine synthesis. *Brain Res. Rev.* 1:313–344, 1979.

75. Kalsner, S. The effects of periarterial nerve activation on coronary vessel tone in an isolated and perfused slab of beef ventricle. *Can. J. Physiol. Pharmacol.* 57:291–297, 1979.

76. Kennerly, D. A., T. J. Sullivan, and C. W. Parker. Activation of phospholipid metabolism during mediator release from stimulated rat mast cells. *J. Immunol.* 122:152–159, 1979.

77. Kilbinger, H. Modulation by oxotremorine and atropine of a cetylcholine release evoked by electrical stimulation of the myenteric plexus of the guinea pig ileum. *Naunyn-Schmiedeberg's Arch. Pharmacol.* 300:145–151, 1977.

78. Kilbinger, H. and I. Wessler. Pre- and postsynaptic effects of muscarinic agonist in the guinea-pig ileum. *Naunyn-Schmiedeberg's Arch. Pharmacol.* 314:259–266, 1980.

79. Kitamura, K. and H. Kuriyama. Effects of acetylcholine on the smooth muscle cell of isolated main coronary artery of the guinea-pig. *J. Physiol.* 293:119–133, 1979.

80. Krishnamurty, V. S. R. and A. Grollman. The mechanism of contraction of rat aorta to various agonists. *Arch. Int. Pharmacodyn.* 220:180–188, 1976.

81. Krnjevic, K. Chemical nature of synaptic transmission in vertebrates. *Physiol. Rev.* 54:418–540, 1974.

82. Kuriyama, H. and H. Suzuki. The effects of acetylcholine on the membrane and contractile properties of smooth muscle cells of the rabbit superior mesenteric artery. *Br. J. Pharmacol.* 64:493–501, 1978.
83. Kuriyama, H. and H. Suzuki. Adrenergic transmission in the guinea-pig mesenteric artery and their cholinergic modulation. *J. Physiol.* 317:383–396, 1981.
84. Larsson, L. I., L. Edvinsson, J. Fahrenkrug, J. Hakanson, C. Owman, O. Schaffalitzky de Muckadell, and F. Sundler. Immunohistochemical localization of a vasodilator polypeptide (VIP) in cerebrovascular nerves. *Brain Res.* 113:400–404, 1976.
85. Lee, T. J-F. Direct evidence against acetylcholine as the dilator transmitter in the cat cerebral artery. *Eur. J. Pharmacol.* 68:393–394, 1980.
86. Lee, T. J-F. Ultrastructional distribution of vasodilator and constrictor nerves in the cat cerebral arteries. *Circ. Res.* 49:971–979, 1981.
87. Lee, T. J-F. Is acetylcholine the dilator transmitter in cerebral blood vessels? A critical examination. *J. Cereb. Blood Flow Metabol.* 1(Suppl. 1):S305–S306, 1981b.
88. Lee, T. .J-F. Morphopharmacological study of cerebral vasodilator and constrictor nerves. In *Neurogenic Control of Brain Circulation.* (D. D. Heistad and M. L. Marcus, eds. Amsterdam, Elsevier-North Holland, 1982.
89. Lee, T. J-F. Cholinergic mechanism in the large cat cerebral artery. *Circ. Res.* 50:870–879, 1982.
90. Lee, T. J-F. Transmitter mechanisms in cerebral arteries: norepinephrine, acetylcholine, vasoactive intestinal polypeptide and substance P. In *Springfield Blood Vessel Symposium.* C. Su, T. J-F. Lee and W. H. Cline, Jr., eds. Springfield, IL, 1983, pp 38–43.
91. Lee, T. J-F., L. R. Kinkead, and S. Sarwinski. Norepinephrine and acetylcholine transmitter mechanisms in large cerebral arteries of the pig. *J. Cereb. Blood Flow Metabol.* 2:439–450, 1982.
92. Lee, T. J-F., A. Saito, and I. Beresin. Vasoactive intestinal polypeptide-like substance: The potential cerebral vasodilator transmitter. *Science* 224:888–901, 1984.
93. Lee, T.J-F. and S. Sarwinski. Transmitter roles of norepinephrine and acetylcholine in pig cerebral arteries. *Fed. Proc.* 41:1233, 1982.
94. Lee, T. J-F., C. C. Chiueh, and M. Adams. Synaptic transmission of vasoconstrictor nerves in rabbit basilar artery. *Eur. J. Pharmacol.* 1:55–70, 1980.
95. Lee, T. J-F., C. Su, and J. A. Bevan. Nonsympathetic dilator innervation of cat cerebral arteries. *Experientia* 31:1424–1425, 1975.
96. Lee, T. J-F., W. R. Hume, C. Su, and J. A. Bevan. Neurogenic vasodilation of cat cerebral arteries. *Circ. Res.* 42:535–542, 1978.
97. Leighton, H. J. and T. C. Westfall. The role of impulse flow, adrenergic agents and prostaglandins in the regulation of ^{3}H-norepinephrine and dopamine beta-hydroxylase release from normal and decentralized guinea-pig vasa deferentia. *Fed. Proc.* 35:406, 1976.
98. Linnik, M. D. and T. J-F. Lee. Electrical nerve stimulation-induced vasodilation is not mediated by acetylcholine. *Pharmacologist* 25:269, 1983.
99. Leventer, S. M., P. P. Rowell, and M. J. Clark. The effect of choline acetyltransferase inhibition on acetylcholine synthesis and release in term human placenta. *J. Pharmacol. Exp. Ther.* 222:301–305, 1982.

100. Li, C. K. and F. Mitchelson. The selective automuscarinic action of stercuronium. *Br. J. Pharmacol.* 70:313–321, 1980.
101. Lundberg, J. M. Evidence for coexistence of vasoactive intestinal polypeptide (VIP) and acetylcholine in neurons of cat exocrine glands—morphological, biochemical and functional studies. *Acta Physiol. Scand. Suppl.* 496, 1981.
102. Mayer, S. E. Drugs acting at synaptic and neuroeffector junctional sites. Ch_4 neurohumoral transmission and the autonomic nervous system. In *Goodman and Gilman's The Pharmacological Basis of Therapeutics, 6th Ed.* A. G. Gilman, L. S. Goodman and A. Gilman, eds. New York, MacMillan Publishing Co., 1980, pp 56–90.
103. McCalden, T. A. and I. C. Roddie. Excitatory cholinergic fibers to longitudinal smooth muscle in bovine mesenteric veins. *Ir. J. Med. Sci.* 140:48, 1971.
104. McConnell, J. G. and I. C. Roddie. A comparison of the behavior of longitudinal and circular smooth muscle in bovine mesenteric vein. *J. Physiol.* 207:82P–83P, 1970.
105. Muscholl, E., H. Ritzel, and K. Rossler. Presynaptic muscarinic control of neuronal noradrenaline release. In *Symposium on Presynaptic Receptors.* S. Z. Langer, ed. Pergamon, Oxford, New York, 1979, pp 287–291.
106. Myers, H. A., E. A. Schenk, and C. R. Honig. Ganglion cells in arterioles of skeletal muscle: role in sympathetic vasodilation. *Am. J. Physiol.* 229:126–138, 1975.
107. Narita, S. and M. Watanabe. Response of isolated rat iris dilator to adrenergic and cholinergic agents and electrical stimulation. *Life Sci.* 30:1211–1218, 1982.
108. Owman, C., L. Edvinsson, and K. C. Nielsen. Autonomic neuroreceptor mechanisms in brain vessels. *Blood Vessels* 11:2–31, 1974.
109. Papka, R. E., J. B. Furness, N. G. Della, and M. Costa. Depletion by capsaicin of substance P-immunoreactivity and acetylcholinesterase activity from nerve fibers in the guinea-pig heart. *Neuroscience Letter* 27:47–53, 1981.
110. Peroutka, S. J., M. A. Moskowitz, J. F. Reinhard, and S. H. Snyder. Neurotransmitter receptor binding in bovine cerebral microvessels. *Science* 208:610–612, 1980.
111. Pickett, W. C., R. L. Jesse, and P. Cohen. Initiation of phospholipase A_2 activity in human platelets by the calcium ionophore A23187. *Biochem. Biophys. Acta* 486:209–213, 1977.
112. Poulsen, J. H. and J. A. Williams. Effect of calcium ionophore A23187 on pancreatic aciner cell membrane potentials and amylase release. *J. Physiol. London* 264:323–339, 1977.
113. Pressman, B. C. Biological applications of ionophores. *Ann. Rev. Biochem.* 45:501–530, 1976.
114. Richter, U., H. Pilgrim, E. Teuscher, and E. Hildegard. Untersuchungen in-vitro-Kultivierter Zellen glattmuskularer Organe. Part 4: Das Kontrak-Gionsuerhalten in-vitro-Kultivierter Zellen der Aorta und der Harnblase der Ratte. *Pharmazie.* 31:323–326, 1976.
115. Salvaterra, P. M., D. A. Matthews, and R. Foders. Quantitative relationships of five putative neurotransmitters receptor sites in rat hippocampal formation. *J. Neurochem.* 35:1253–1257, 1980.

116. Sastry, B. V. R., J. Olubadewo, R. D. Harrison, and D.E. Schmidt. Human placental cholinergic system: occurrence, distribution, and variation with gestational age of acetylcholine in human placenta. *Biochem. Pharmacol.* 25:425–431, 1976.

117. Sawynok, J. and K. Jhamandas. Muscarinic feedback inhibition of acetylcholine release from the myenteric plexus of guinea–pig ileum and its status after chronic exposure to morphine. *Can. J. Physiol. Pharmacol.* 55:909–916, 1977.

118. Scremin, O. U., R. R. Sonnenschein, and E. H. Rubinstein. The role of cranial nerves in cholinergic cerebral vasodilation. *Physiologist* (abstr.) 22:113, 1979.

119. Shepherd, J. T., R. R. Lorenz, G. M. Tyce, and P. M. Vanhoutte. Acetylcholine-inhibition of transmitter release from adrenergic nerve terminals mediated by muscarinic receptors. *Fed. Proc.* 37:191–194, 1978.

120. Singer, H. A. and M. J. Peach. Calcium- and endothelial-mediated vascular smooth muscle relaxation in rabbit aorta. *Hypertension* 4(Suppl. II):II–19–II–25, 1982.

121. Somlyo, A. P. and A. V. Somlyo. Vascular smooth muscle. II. Pharmacology of normal and hypertensive vessels. *Pharmacol. Rev.* 22:249–353, 1970.

122. Steisland, O. S., F. Furchgott, and S. M. Kirpekar. Biphasic vasoconstriction of the rabbit ear artery. *Circ. Res.* 32:49–58, 1973.

123. Stjernschantz, J., A. Alm, and A. Bill. Effects of intracranial oculomotor nerve stimulation on ocular blood flow in rabbits. Modification by indomethacin. *Exp. Eye Res.* 23:461–469, 1976.

124. Su, C., J. A. Bevan, and R. C. Ursillo. Electrical quiescence of pulmonary artery smooth muscle during sympathomimeter stimulation. *Circ. Res.* 15:20–27, 1964.

125. Szerb, J. C. Effect of low calcium and oxotremorine on the kinetics of the evoked release of [^{3}H]-acetylcholine from guinea-pig mynteric plexus: comparison with morphine. *Naunyn-Schmiedeberg's Arch. Pharmacol.* 311:119– 127, 1980.

126. Taylor, P. Cholinergic agonists. In *Goodman and Gilman's The Pharmacological Basis of Therapeutics, 6th Ed.* A. G. Gilman, L. S. Goodman and A. Gilman, eds. New York, MacMillan Publishing Co., Inc., 1980, pp 91–99.

127. Toda, N. Relaxant responses to transmural nerve stimulation and nicotine of dog and monkey cerebral arteries. *Am. J. Physiol.* 243:H145–H153, 1982.

128. Uvnas, B. Cholinergic vasodilator nerves. *Fed. Proc.* 25:1613–1622, 1966.

129. VanBreemen, C. Blockade of membrane calcium fluxes by La in relation to vascular smooth muscle contractility. *Arch. Int. Physiol.* 77:710–716, 1969.

130. Vanhoutte, P.M., T. J. Verbeuren, and R. C. Webb. Local modulation of adrenergic neuroeffector interaction in the blood vessel wall. *Physiol. Rev.* 61:151–247, 1981a.

131. Vanhoutte, P. M., M. Collis, W. J. Janssen, and T. J. Verbeuren. Calcium dependence of prejunctional inhibitory effects of adenosine and

acetylcholine on adrenergic neurotransmission in canine saphenous veins. *Europ. J. Pharmacol.* 72:189–198, 1981b.

132. Vanhoutte, P. M. and T. J. Verbeuren. Inhibition by acetylcholine of ^{3}H-norepinephrine release in cutaneous veins after alpha-adrenergic blockade. *Arch. Int. Pharmacodyn. Ther.* 221:160–162, 1976a.
133. Vanhoutte, P. M. and T. J. Verbeuren. Inhibition by acetylcholine of the norepinephrine release evoked by potassium in canine saphenous veins. *Circ. Res.* 39:263–269, 1976b.
134. Vasquez, J. and M. J. Purves. Studies on the dilator pathway to cerebral blood vessels. In *Neurogenic Control of the Brain Circulation.* C. Owman and L. Edvinsson, eds. Oxford, Plenum Press, 1977, pp 59–73.
135. Verbeuren, F. J. and P. M. Vanhoutte. Acetylcholine inhibits potassium evoked release of ^{3}H-norepinephrine in different blood vessels of the dog. *Arch. Int. Pharmacodyn. Ther.* 221:347–350, 1976.
136. Watkins, R. W. and W. F. Davidson. Effect of acetylcholine on phasic and tonic components of vascular smooth muscle contraction. *Arch. Int. Pharmacodyn.* 243:217–227, 1980a.
137. Watkins, R. W. and W. F. Davidson. Effects of competitive antagonists on phasic and tonic components of vascular smooth muscle contractions. *Arch. Int. Pharmacodyn.* 244:200–210, 1980.
138. Westfall, T. C. Local regulation of adrenergic neurotransmission. *Physiol. Rev.* 57:659–728, 1977.
139. Wood, J. D. Physiology of the enteric nervous system. In *Physiology of the Gastrointestinal Tract. Vol. 1.* L. R. Johnson, J. Christensen, M. I. Grossman, E. D. Jacobson and S. G. Schultz, eds. Raven Press, New York, 1981, pp 1–37.
140. Yamamura, H. I. and S. H. Snyder. Muscarinic cholinergic binding in rat brain. *Proc. Nat. Acad. Sci.,* USA 71:1725–1729, 1974.
141. Zawadzki, J. V., P. D. Cherry, and R. F. Furchgott. Comparison of endothelium-dependent relaxation of rabbit aorta by A23187 and by acetylcholine. *Pharmacologist* 22:271, 1980.

Chapter 15

Nonadrenergic, Noncholinergic (NANC) Neuronal Inhibitory Interactions with Smooth Muscle

E. E. Daniel

1. Introduction

Nonadrenergic, noncholinergic nerves innervating smooth muscle were first clearly recognized in the guinea-pig taenia caecum by Burnstock et al. (*22*), Holman and Hughes (*90*), Bennett et al. (*14*), and Bennett and Burnstock (*12*). These nerves were inhibitory, causing hyperpolarization of the membrane potential and relaxation of contractile activity, and were not susceptible to atropine or other muscarinic cholinergic antagonists, to adrenergic antagonists of either alpha- or beta-receptor effects, or to adrenergic neuronal blocking agents such as guanethidine, to antagonists of serotonin, histamine, or inhibitors of prostaglandin synthesis. The definition of nonadrenergic, noncholinergic nerves was one of exclusion, based on absence of evidence for a particular transmitter. Subsequently, Burnstock and his colleagues (for review, *see* refs. *19, 20*) have marshalled evidence for ATP or another purine as transmitter whereas others (e.g., *65* and *75*) have suggested the existence of a peptidergic transmission, usually with VIP as the putative transmitter in this subset of

NANC nerves. As will become apparent, there is reason to remain skeptical of both identifications, and we shall retain the negative term, nonadrenergic, noncholinergic, and abbreviate it as NANC.

Subsequent studies have revealed that NANC nerves are widely distributed, may release a variety of transmitters, and may be both excitatory and inhibitory in function. For excitatory NANC transmitter, purines have again been suggested by Burnstock et al. (*25, 26*) and peptides, e.g., substance P, by others (e.g., *102*). These nerves will be discussed in the subsequent Chapter 16.

The field of investigation has been considerably broadened by the finding of many different potential neurotransmitters, in nerves in smooth muscle tissues (*75, 152*). These include substance P, vasoactive intestinal polypeptide (VIP), enkephalins and related peptides, gastrin releasing peptide (GRP), bombesin, pancreatic polypeptide, and related peptides and others (*39*).

1.1. Criteria for Identification of NANC Mediators

The classical criteria for identification of a mediator are that any putative candidate mediator must be present in and released from the nerves as well as mimic the nerve mediator in all respects, giving due consideration to the differences arising from discrete release from nerves compared to more diffuse application in a bath or via the blood stream. More specifically, it must be *present* in the NANC nerve, as demonstrated both by chemical assay or immunohistochemistry, if discrete assay of nerves is not feasible. Conceivably it may not be stored in the nerve, but must be rapidly formed or synthesized in response to stimulation (e.g., prostaglandins); however, in all known cases the transmitter is, in fact, stored prior to release. Recent advances in immunohistochemistry at an ultrastructural level may allow demonstration of the putative mediator in specific types of synaptic vesicles (*143*). *Release* of the proposed mediator by stimuli that activate NANC nerves must also be demonstrated; however, it must be ascertained that the source of the mediator is the nerve rather than a postsynaptic structure. This is especially important when the putative mediator is an ubiquitous compound such as ATP or is contained in endocrine cells as well as nerves and may derive from nerve-driven release from such cells. The proposed mediators should mimic the electrical and mechanical/secretory events evoked postsynaptically by the NANC nerve stimulation. Study of postsynaptic electrical events is more discriminating than study of mechanical or secretory outcomes, since a variety of mechanisms may underlie relaxation or contraction or secretion, but usually only one class of electrical events is activated by one postsynaptic receptor. A clear example is the fact that relaxation of gut muscle can occur by mechanisms including inhibition of release of excitatory transmitter, blockade of Ca entry by action on Ca channels, inhibition of spontaneous electrical activity, clamping of the membrane potential between the potassium and chloride equilibrium potentials (as by nonadrenaline acting on alpha receptors), or hyperpolarization by selective increase in potassium conductance as is usually produced by NANC inhibitory mediator in gut. Study and identification of the exact postsynaptic electrical mechanism by analysis of membrane potential changes, membrane resistance changes, ionic dependence of these changes, and reversal potential of membrane potential change, provide strong evidence about the identity of actions of the putative and real mediators. Identity of postsynaptic action does not establish identity of postsynaptic receptor mechanism; this requires the availability of potent, selective, competitive antagonists at these receptors. These should block equivalently the postsynaptic effects of real and proposed transmitter. The presence of such antagonists for classical nerve transmitters has enabled NANC nerve effects to be identified; the absence of such antagonists for putative NANC transmitters (purines and most peptides) has seriously hampered identification of the NANC transmitters. The use and limitations of tachyphylaxis and antibodies to putative peptide transmitters will be discussed below. Further evidence may be provided by manipulation of the degradation or reuptake processes for the real transmitter; such processes or loss by diffusion must inevitably exist to end the effect of the real transmitter and inhibition or acceleration of removal of the real and putative transmitter should be exactly parallel. Table 1 summarizes these points. Two transmitters may be present in and released by NANC nerves; then additional criteria need to be satisfied in identifying transmitters (Table 2).

As will be evident from the subsequent discussion, no putative mediator satisfies all or even most of these criteria. ATP and other purines have been studied most thoroughly as inhibitory nerve transmitters in guinea-pig taenia caeci, and recently some evidence for excitatory purinergic transmission has become available for vas deferens of several species. However, no candidate is established as NANC transmitter at the smooth muscles discussed in this chapter.

Table 1
Evaluation of Putative NANC Transmitters

Criteria	Method	Outcome required
Present in NANC nerve	Chemical/RIA assay immunohistochemistry	Present in necessary place and amount
Present in synaptic vesicles of NANC nerve	Immunohistochemistry by EM	Present in a synaptic vesicle or other releaseable form
Released on nerve stimulation	Measurement in perfusate or outflow; blocked by TTX	Released into venous blood or perfusate
Released from presynaptic nerve	Measurement in isolated NANC nerve or after inhibition of post synaptic events	Released from NANC nerve not other structures
Mimics action of real transmitter	Measure response of postsynaptic cells	Actions must be identified
Mimics postsynaptic electrical actions of real transmitter	Measure electrical events in postsynaptic cells	Actions on membrane potential, resistance. Also reversal potential of membrane. Potential changes and ionic dependence of electrical events same
Antagonized by receptor antagonists to real transmitter Potentiated by inhibitors of degradation	Measure postsynaptic events (electrical better; mechanical acceptable)	Similar concentrations of antagonist should equally affect similar or potentiator concentrations of real and putative transmitter

1.2. Distribution of NANC Nerves and Smooth Muscle Responses

The original criteria used for identification of NANC nerves were resistance to muscarinic cholinergic antagonists, to adrenergic neuronal blocking agents (since highly specific beta-adrenoceptors antagonists were not available until the late 1960s), and to other known receptor antagonists or prostaglandin synthesis inhibitors (*see* refs. *19, 20* for review). There was initial confusion about the specificity of adrenergic neuronal blocking agents, but when inhibitory junction potentials (ijps) were recorded (*14*) from guinea-pig taenia caeci after field stimulation of intrinsic nerves and shown to differ in size, stimulus frequency dependence, and temporal characteristics from responses to stimulation of extrinsic sympathetic nerves, the distinction was clear. It was apparent that the inhibitory junction potential from stimulation of gut inhibitory NANC nerves was large, elicitable from single stimuli,

Table 2
Additional Criteria for Identification of Cotransmitters

Presence of two transmitters in same nerve endings	EM-immunohistochemistry (protein-A gold, PAP method, and so forth)	Two or more mediators, one NANC, in same varicosity
	Destruction of specific nerves by selective toxins (e.g., 6-OH-dopamine)	Removes both mediators
Both released from same nerve ending on nerve stimulation	Measurement of release or disappearance of mediators from terminals	Mediators released simultaneously or sequentially[a]
	Destruction of nerves	Prevents release of both
Both act on postsynaptic sites and mimic nerve response	Measure postsynaptic membrane and electrical events	Each mimics an appropriate component of the response
	Measure responses of postsynaptic cells	Selective inhibition of each partially blocks response, selective inhibition of both fully blocked

[a]Presynaptic mechanisms such as feedback inhibition may not affect release of both mediators equally since their packaging and release mechanisms in the same nerve ending may differ. However, if a common response to release of each mediator to presynaptic modulation, it provides some support for a common locus with the reservation that more than one type of nerve shares a common presynaptic regulation; e.g., alpha-2 adrenoceptor and purine-mediated inhibition of cholinergic and adrenergic nerve release mechanisms.

dependent on a selective increase in K^+ conductance, and followed by an afterdepolarization, often associated with spikes and contraction. Stimulation of noradrenergic nerves at low frequencies caused inhibition with little hyperpolarization (possibly by presynaptic inhibition of cholinergic neurons) and only caused marked hyperpolarization after high frequency stimulation apparently as a result of increases in both K^+ and Cl^- conductance. These results indicated the power of electrophysiological techniques to discriminate actions of mediators and established the existence of NANC inhibitory nerves in gut.

2. Existence of Intrinsic NANC Inhibitory Neurons in Gut and Elsewhere

2.1. Gut

Intrinsic inhibitory nerves were discovererd in gut by Bayliss and Starling (*10*) in demonstrating that a variety of stimuli led to distal inhibition of intestinal motor function. Langley and Magnus (*109*) confirmed and extended these observations and showed that this distal inhibition could not be attributed to extrinsic sympathetic nerves since it persisted unchanged after local denervation. Ambache (*2*) was the first to study the transmission mechanism of intrinsic inhibitory neurons; he showed that nicotine or transmural stimulation still activated them and produced responses after muscarinic blockade. He assumed them to be short intrinsic adrenergic neurons, but this interpretation had to be abandoned following studies using adrenergic blocking agents (*14, 22*). There is now additional conclusive evidence that the entire intrinsic inhibitory neurons (*see* refs. *38, 39*) are nonadrenergic.

In a recent review, Costa and Furness (*38*) have summarized tissues for which good evidence of their direct NANC innervation exists. In brief, the tissues for which good electrophysiological evidence exists include sphincters where they provide the main mechanism for relaxation (*see* Table 3); the proximal stomach, where these nerves provide the mechanism of receptive relaxation (*see* Table 3); the circular muscle of the small intestine (*31, 45, 89*), guinea-pig taenia coli, circular muscle of the large bowel (Table 3; ref. *14, 19, 22*), and longitudinal muscle of intestine in some cases (*see* Table 3); in the antrum of stomach, only guinea pig has been studied (*see* Table 3), and results suggest that not every cell is innervated. A much wider group of tissues has been shown to

Table 3
Tissues for Which Electrophysiological Evidence of NANC Inhibitory Nerves Exist

Tissue	Species	Mechanical response	Electrophysiological response	Reference
Esophagus body circ. muscle	Opossum	None, unless tone elevated	Only ejp (single pulse)	*49*
Gastric fundus	Guinea pig	Relaxation	Ejp and/or ijp; ijp only after atropine (single pulse)	*11a*
Gastric corpus	Guinea pig	Relaxation	Ejp and/or ijp; ijp only after atropine (single pulse)	*11*
Small intestine circ. muscle	Guinea pig	None, unless tone raised	Ijp only (single pulse) (*see* ref *7*)	*107a*, *7*, *89*, *156*, *27*
Small intestine circ. muscle	Rabbit	None, unless tone raised	Ijp only (single pulse)	*31*, *156*
Longitudinal muscle	Guinea pig	Relaxation	Ejp and ijp; after atropine usually ijp (multiple pulses)	*107a*, *7*
Longitudinal muscle	Rabbit	Relaxation	Ejp and ijp; after atropine some cells get ijp	*31*
Colon circ. muscle	Guinea pig	Relaxation	Ijp and ejp; ijp only after atropine (single pulse or multiple)	*74*, *15a*

(*continued*)

Table 3 (*continued*)

Tissue	Species	Mechanical response	Electro-physiological response	Reference
Taenia caecum	Guinea pig	Relaxation	Ijp and ejp; ijp only after atropine (single pulse)	*14, 12*
Internal anal sphincter	Guinea pig	Relaxation		*36a*

relax to transmural stimulation (*see* ref. *38*), but in the case of longitudinal muscle, which has few intrinsic nerves (*134*), this may reflect diffusion of mediator from the myenteric plexus in which many varicosities open into the interstitial space between plexus and muscle (*80*). Microelectrode recordings purportedly made from "longitudinal muscle strips" in species and regions where this layer is very thin and the layers have not been separated, or when all or some circular muscle accompanies longitudinal muscle recordings, may actually have come from circular muscle. In several recent publications, studies on the rabbit intestine (*31*) and on guinea-pig intestine (*1, 27, 91*) in which the innervation of both muscle layers was examined, it was apparent that nonadrenergic inhibitory innervation predominates in circular muscle, where it was present in all cells, and could only be elicited in some cells of longitudinal muscle, usually after suppression of cholinergic excitatory responses by atropine and with repetitive stimulation.

The nonadrenergic intrinsic inhibitory nerves of gut are innervated by preganglionic vagal fibers in the proximal gastrointestinal tract [smooth muscle esophagus and lower esophageal sphincter (88, 135)]; fundus and sling fibers (*see* refs. *28, 122, 123, 134, 141*); and gastric corpus (*11*). The pelvic parasympathetic nerves innervate the colon and internal anal sphincter (*see* refs. *38, 39*). Elsewhere little or no extrinsic neural control of these nerves can be demonstrated.

2.2. Airways

Intrinsic inhibitory neurons also exist in regions other than the gut. In the airways, which are related to gut in embryological origins, nonadrenergic, noncholinergic inhibitory innervation has been demonstrated in several species. The first demonstration in mammals was by Coburn and Tomita (*32*) in guinea-pig trachea, a finding subsequently confirmed by several workers (*35, 36, 59*) and extended to human trachea (*48, 53, 59, 138, 139*), to monkeys by Doidge and Satchell (*59*), and to cats by Irvin et al. (*96*) and Diamond and O'Donnell (*56*). In the guinea-pig trachea, there is a noradrenergic inhibitory component related to the postganglionic sympathetic innervation of the muscle (*32, 59, 104, 176*). This seemed to vary with the region of the guinea-pig trachea examined (*32, 104*), but constituted a major inhibitory control, especially by extrinsic nerves, which was 60–70% exerted by noradrenergic nerves (*176*), as well as by intrinsic nerves, which was 30–40% exerted by noradrenergic nerves (*32, 59, 176*). Vagal inhibition via noradrenergic nerves was effective at low frequencies, but that via nonadrenergic nerves required high frequency stimulation (*176*). Since low-frequency field stimulation of intrinsic noncholinergic nerves was effective, it appeared that the ganglia filter high-frequency transmission to these intrinsic neurons.

In most species, except guinea pig, evidence of functional inhibitory innervation of large airway smooth muscle exists only for nonadrenergic fibers (*see* refs. *48, 59* for reviews); i.e., no functional noradrenergic nerves to muscle cells can be demonstrated. Most of the noradrenergic nerves seen by the Falck technique innervated blood vessels; but recently occasional nerves with small granular vesicles as in blood vessel were obseved to innervate human tracheal muscle, and large numbers were found in 4th to 7th order bronchi (*50*). Whether these are noradrenergic nerves remains to be established. The failure to demonstrate functional sympathetic inhibitory innervation in contrast to nonadrenergic inhibitory innervation may reflect the extent to which each type of nerves survives in vitro study conditions. Even greater reservation about nerve survival should be held in connection with a recent report (*59*) that no functional inhibitory nerves were present in lung parenchymal strips. The contractile function of such strips probably reflects vascular as well as airway smooth muscle activity, and the survival of nerves associated with airway muscle in such strips has not been evaluated by structural or functional studies.

So far, no electrophysiological data about nonadrenergic inhibitory junction potentials are available in airway smooth muscle. The guinea-pig trachea is heavily innervated, but has only a few gap junctions (*104*) and may not be well coupled electrically. However, recently, coordinated slow waves were recorded from it and reasonable coupling was also suggested (*155*). Human trachea may be a better object of study since it is heavily innervated by nonadrenergic inhibitory nerves, possesses many gap junctions (*48, 50*), and appears to be well coupled (Kannan, M. S., Davis, C., Daniel, E. E., unpublished). However, studies of junction potentials remain to be accomplished

2.3. Blood Vessels

NANC inhibitory (vasodilating) nerves may also exist in blood vessels, and a strong case can be made on pharmacological grounds (*113*). These include failure of atropine or adrenergic antagonists to block relaxation responses to field stimulation when responses to acetylcholine, noradrenergic and tyramine-released noradrenaline are blocked. There is also evidence that peptide-containing nerves (VIP and Substance P) innervate blood vessels, and these peptides cause relaxation (see later). However, little evidence from electrophysiology or from specific antagonists is yet available. This subject is considered further in Chapters 14 and 17.

3. Identity of NANC Inhibitory Transmitters

There are at present only two putative NANC inhibitory transmitters for gut nerves with characteristics actions that do not rule them out immediately in many tissues: ATP or other purines and VIP. The evidence regarding these transmitters in relation to several tissues will be analyzed here.

3.1. Presence of ATP or VIP in NANC Fibers

ATP plays a vital role in transferring chemical metabolic energy to various biological purposes; hence it is ubiquitous and present in all cells including NANC neurons. Moreover, it is present in cholinergic synaptic vesicles (*178*), in central synaptosomes (*108, 174*), and in adrenal adrenergic vesicles (*60*). Its presence in NANC nerve synaptic vesicles has, however, not been established conclusively; adenosine is, however, taken up and converted to ^{3}H-ATP in the myenteric plexus region of guinea-pig taenia coli (*160*).

An earlier claim that the purinergic nerve distribution could be estimated from the distribution of nerve varicosities containing a predominance of large opaque or large granular vesicles (*19*) has now been contradicted by the absence or paucity of such nerve endings in the appropriate places. In particular, in some tissues such as circular muscle of opossum esophagus (*47, 49*), which have almost entirely NANC inhibitory intrinsic nerves, these nerves are not present in appreciable numbers. The same finding, incompatible with this hypothesis that was never supported by quantitative information, has been made in several tissues (*46, 48*). A former proponent of this hypothesis (Gibbins, 82) has recently reviewed the evidence and concluded that NANC nerves cannot be associated with a particular structure in nerve varicosities or with a particular type of synaptic vesicle (see below).

VIP is present in gut (*58, 78*), possibly exclusively in nerves (*1, 40, 58, 61, 65, 76, 77, 78, 110, 152, 161, 168*). In a superficial general sense, its immunohistochemical and chemical (by RIA) distribution is as expected of NANC inhibitory nerves; e.g., concentrated in circular muscle where NANC inhibitory nerves predominate (*78, 168*), VIP-containing nerves are especially rich in the deep muscular plexus (*15, 152*). Fewer VIP-positive varicose fibers are present in longitudinal muscle (*65, 76, 77, 152*), which, as discussed above, may have little direct NANC inhibitory innervation, receiving mediator from Auerbach's plexus that contains many VIP-positive fibers and nerve cells. Many of the nerve cell bodies producing VIP-like material are located in the myenteric plexus, the source of NANC inhibitory nerves, and these send fibers to circular muscles (*37, 76*) and to guinea-pig taenia coli (*77*), both of which are rich in NANC inhibitory nerves (see above). A recent study of the projections of VIP-containing neurons of the myenteric plexus of the guinea pig (*37*) showed that many innervated the circular muscle directly under their locus in the plexus and did not project anally, whereas others sent processes anally in the plexus to innervate either distal myenteric neurons or to leave the plexus after several millimeters to innervate circular muscle. VIP-positive fibers are also very dense in the mucosa (*75, 152*) and among mucosal and other blood vessels (*40, 152, 170*). These loci may reflect other nonmuscle sites of action of NANC nerves. VIP-immunoreactivity is also present in nerves of other structures with established (airway smooth muscle) or proposed (blood vessels, *113*) NANC inhibitory or excitatory innervation. Some endocrine and exocrine organs and their blood vessels also have been shown to possess such nerves (*115–118, 167, 169*).

VIP was contained in some of the large granular synaptic vesicles of cat colon (*110*) and in dog ileum (*15*). Varicosities with large granular vesicles in the intestine in general all contain some small agranular vesicles, so VIP may be a cotransmitter with other mediators such as proposed for acetylcholine (*117, 118*). Since some varicosities contain large granular vesicles, only a fraction of which contain VIP-like immunoreactivity (*15*), it may coexist with other NANC transmitters as well.

In summary, ATP localization in synaptic vesicles of nerves has not been established (aside from its ubiquitous presence in all cells). VIP-like material has been demonstrated in nerves and the distribution of such nerves is consistent with NANC-inhibitory transmission, but is nearly ubiquitous. Thus it is consistent with, but does not prove, an association between VIP-containing and NANC gut inhibitory nerves.

3.2. Release of ATP and VIP by Nerve Stimulation

ATP or its breakdown products (ADP, AMP, adenosine, and inosine) can be found after stimulation of a variety of nerves: motor nerves to skeletal muscle (*154*), nerves to electric organs (*97*), or their synaptosomes (*177*). Most of this release is from postsynaptic sites (*97*), but some may be presynaptic (*154*). ATP metabolites were released from stomach after vagal stimulation (*23*). [^{3}H]adenosine was released by stimulation of intrinsic nerves from guinea-pig taenia caeci and the release was reduced by TTX (*160*). ATP itself was also released in increased amounts after intrinsic nerve stimulation of guinea pig bladder (*24, 25*), of taenia caeci (*25*), and of rat anococcygeus muscle (*26*). In some tissues, it has been possible to exclude mechanical activity of smooth muscle as the source of ATP since TXX reduced the release, but relaxation with norepinephrine or contraction with acetylcholine did not restore it (*24, 25, 26*). Recently (*69*), Goniopora toxin from a coral was shown to augment relaxations of guinea-pig taenia caeci to field stimulation in the presence of atropine and guaethidine and to enhance the ijp and release of [^{3}H]adenosine, but did not affect responses to ATP. Presumably it acted to enhance release but not the action of the NANC inhibitory mediator. Thus a case for release of ATP from intrinsic, possibly NANC inhibitory nerves, has been made.

There are problems. At least part of the release of purines on nerve stimulation is from muscle (*114, 146, 175*) because it is TTX-insensitive and in some cases associated with TTX-insensitive electrical or other contractile responses. In the study of White et al. (*175*) the luciferin–luciferase technique was used to measure ATP released into bath fluid or superfusate from guinea-pig taenia caeci. Release into bath fluid containing atropine and guanethedine required high stimulation frequencies (2̂0–30 Hz) and both the release and accompanying contraction (the preparations had low tone after luciferin–luciferase was added) were TTX-insensitive. Release into a superfusate was produced by lower frequencies, but it and the contractions were still TTX-insensitive. Contraction itself (from exogenous acetylcholine) did not cause ATP release. Thus direct depolarization of nerve terminals by the field stimulation (TTX-insensitive) might have caused this ATP release. Release of ^{3}H-adenosine from taenia coli was reduced by removal of external Ca^{2+} and did not result from muscle relaxation since comparable relaxation by norepinephrine did not release it (*24, 160*). However, Kuchii et al. (*106, 107*) showed a lack of correlation between release of ^{3}H-adenosine and relaxation; the latter was blocked by TTX when release of label was reduced by only one-third. Furthermore, after cold storage for 8 d, nicotine and field stimulation still caused similar release of label, but no or minimal relaxation. Furthermore, in contrast to Su et al. (*160*), Burnstock and colleagues (*24, 25, 26*) found that comparable release of label resulted from comparable relaxations induced by noradrenaline and papaverine of field stimulation. Evidence is still lacking that the ATP released during nerve stimulation comes primarily from NANC nerves; in all cases to date, a mixture of intrinsic NANC, cholinergic, peptidergic and even extrinsic adrenergic nerve endings, have been present and stimulated. Since ATP is known to be present in both cholinergic and adrenergic nerve terminals, part of the TTX-sensitive ATP release may have come from them. In some cases, sympathetic denervation with guanethedine or 6-OH dopamine did not eliminate ATP release (see above). There is, of course, no evidence yet about the concentrations of ATP present near purinergic receptors during nerve stimulation. It should be noted that ATP release during nerve stimulation in other systems with NANC inhibitory nerves, e.g., in airway smooth muscle or blood vessels, has not been reported.

VIP is also released by nerve stimulation. In early studies, nerves were not stimulated electrically, but procedures were carried out that, along with other things, activate nerves. Plasma levels were studied in normal and vagotomized patients and were unchanged after intraduodenal infusion of amino acids, glucose, saline, or a mixed meal. VIP levels were slightly but significantly elevated by intraduodenal HCl, fat, or ethanol, and similar results were obtained after fat and HCl in portal plasma of pigs (*151*). Neither vagotomy in patients nor ganglionic blockade in pigs affected the response to HCl. A later study in conscious dogs (*30*) confirmed the VIP-release by lipid.

Despite the lack of effect of vagotomy in the above study, cats released VIP into the gastric venous effluent when high threshold (2 ms instead of 0.2 ms pulse duration) vagal fibers were stimulated or when a vago-vagal reflex relaxation of the proximal stomach was elicited by distending an esophageal balloon (*67*). In this study, mechanical stimulation of the small intestine mucosa increased both blood flow and VIP release into the local venous effluent as did stimulation of the rectal mucosa. Stimulation of pelvic nerves to the colon also caused local VIP release as well as a transient vasodilation. All these responses were thought to be mediated by one or another group of NANC inhibitory nerves. In a later study, Eklund et al. (*64*) showed that mechanical stimulation of the intestinal mucosa of the cat released VIP locally and caused vasodilation; both were abolished by TTX and reduced by intra-arterial 2-bromo-lysergic acid. Intra-arterial 5-hydroxytryptamine mimicked the effects of mechanical stimulation, but isoprenaline caused only vasodilation. These authors suggested that 5-HT is released from enterochromaffin cells by mechanical stimulation and activates an intramural nerve reflex containing a neuron capable of releasing VIP. The TTX-sensitivity of responses to 5-HT, predicted by that hypothesis, was not reported.

VIP release into portal venous blood by vagal stimulation was studied further in pigs (*66*). It was found that vagal stimulation (at 8 Hz which yielded maximal release) caused a rapid (<2.5 min) and prolonged (40 min) steady increase in the release of VIP unrelated to any change in portal blood flow rate. VIP release on vagal stimulation was stimulation frequency-dependent, mimicked by acetylcholine infusion, unaffected by atropine or beta-adrenergic blockade, inhibited by stimulation of splanchnic nerves, increased by alpha-adrenergic blockade or splanchnicotomy, but abolished, as expected, by hexamethonium. These results were interpreted to show that intrinsic VIP-containing intestinal nerves received an excitatory, nicotine cholinergic input and an inhibitory, alpha-adrenergic input. One puzzling result was the ability of atropine, but the failure of ganglionic blockade with hexamethonium, to abolish acetylcholine-induced VIP release. Possibly there is another cholinergic mechanism involving muscarinic receptors that releases VIP. Bloom and Edwards (*17*) confirmed the vagally mediated, atropine-insensitive release of

VIP in calves, measuring release into intestinal lymph. Their results suggested that the effects of splanchnic stimulation to inhibit release into portal plasma were a result of the associated decreased intestinal blood flow; in their study, concentrations in lymph rose while those in portal blood fell.

Bitar et al. (*16*) showed that both oxytocin and neostigmine release VIP into the portal circulation of anesthetized dogs; the origin is primarily from gut and the entire gut contributed. Oxytocin-mediated release was markedly reduced by ganglionic blockade (hexamethonium) or by TTX, but was unaffected by atropine. Presumably oxytocin activated primarily the same pathway as vagal stimulation. Neostigmine-induced VIP release was reduced by hexamethonium and abolished by atropine. Apparently, in contrast to infused acetylcholine, that preserved by acetylcholinesterase inhibition had access to both nicotinic and muscarinic receptors and the latter were in the final common pathway, a pathway different from that used by oxytocin. The authors sought to avoid invoking an additional pathway for VIP release, but this seems the simplest explanation.

Release of VIP from gut tissues in vitro has not been reported often; for example, Fox et al. (*68*) were unable to obtain release from lower esophageal sphincter of opossum by TTX-sensitive field stimulation. However, Gaginella et al. (*79*) obtained TTX-sensitive VIP release from the rabbit ileum by field stimulation with electrodes parallel to the muscularis proprea. The preparation was in a Ussing-type chamber so that release from serosal and mucosal surfaces could be studied separately. The release was from the serosal surface and apparently from nerve fibers within muscle layers since the removal of circular with longitudinal muscle abolished all VIP release. There was considerable variation in the data collected after dissection.

In general, the results suggest that VIP is released from intestinal intrinsic nerves that are innervated by vagal and pelvic preganglionic fibers and may play some role in regulating intestinal blood flow and other NANC-inhibitory nerve functions. Three observations are puzzling: (1) Vagal stimulation caused a release of VIP sustained for 40 min, but responses to NANC-nerve stimulation usually decay rapidly. (2) Pelvic nerves normally do not innervate the nonadrenergic inhibitory neurons of intestine. (3) Acetylcholine and neostigmine cause VIP release by a pathway dependent on muscarinic cholinergic receptors, but vagal activation of NANC nerves is atropine-insensitive.

3.3. Identity of Action of ATP or VIP with Inhibitory Mediator

In many regions of the intestine of a variety of species, ATP acts like the inhibitory mediator qualitatively, i.e., it relaxes. Table 4 summarizes some of these data. The threshold concentrations of ATP in these cases are nearly always 10^{-6}–$10^{-4}M$, higher than usual for a transmitter. This may reflect rapid degradation or inaccessibility of purinergic receptors. Moreover, the NANC inhibitory mediator has been established, at least in guinea-pig taenia coli and opossum esophagus circular muscle, to act by selectively increasing K^+ conductance (based on the expected reduction of ijps in increased external K^+, the lack of effect of Cl^- removal aside from that expected by removal of an anion carrying resting membrane current, the increased conductance during, and similarity of, the reversal potential for the ijp to that for K^+). Also there was an increase in K but not Cl fluxes during ijps and the abolition in some tissues of ijps by inhibition of K-conductance

Table 4
Relaxant Actions of ATP

Organ	Species	Threshold	References
Stomach	Guinea pig	10^{-6}–2×10^{-5}	*23, 26a*
	Rabbit	2×10^{-5}	*23*
	Rat	4×10^{-6}	*23*
Small intestine	Rabbit	4×10^{-5}	*60a*
		4×10^{-6}	*83a*
Duodenum	Rabbit	10^{-7}	*173*
	Rat	4×10^{-6}	*23*
Ileum	Guinea pig	2×10^{-6}	*23*
	Rabbit	2×10^{-6}	*23*
	Rat	10^{-5}	*26a*
	Mouse	10^{-5}	*23*
Taenia caecum	Guinea pig	10^{-7}	*2A*
Colon	Guinea pig	10^{-6}	*26a*
	Rabbit	2×10^{-5}	*23*
	Rabbit-CM	2×10^{-4}	*125*
	Rat	4×10^{-5}	*23*
	Mouse	10^{-5}	*23*
	Human	4×10^{-4}	*26a*
Internal anal sphincter	Human	10^{-6}	*18a*

channels (*13, 49, 54, 164, 165*; Jury, J., Jager, L., and Daniel, E. E., submitted for publication). Moreover, ATP in taenia caeci shows such a mode of action in higher concentrations (*100, 165*). However, more important with regard to the acceptance of ATP as the inhibitory transmitter are a number of cases in which ATP acts differently from the inhibitory transmitter. The rabbit distal colon has both muscle layers supplied by inhibitory nerves that produce ijps when stimulated (*74*), but ATP contracted the longitudinal muscle, whereas it relaxed circular muscle when mechanical effects on both layers were recorded simultaneously (*125, 126*) and effects were TTX-insensitive. All sphincters in the gut contain a prominent NANC innervation and produce ijps when intrinsic nerves are stimulated (*38, 39, 75*) but the internal anal sphincter of the guinea pig contracted to $2 \times 10^{-5}M$ ATP or failed to respond. The lower esophageal sphincter of the opossum in vitro either contracted to 2×10^{-4} to $2 \times 10^{-3}M$ ATP or relaxed briefly before contracting (*47*), and in vivo this sphincter contracted and then relaxed in response to intra-arterial ATP (*136*). The circular muscle of the body of the opossum esophagus also has NANC nerves and produces ijp in the innervated muscle followed by afterdepolarizations. Since the muscle normally has no basal active tension, the ijp is not accompanied by relaxation unless tone is pharmacologically induced. The afterdepolarization is usually associated with action potentials and contraction (*49*; Kannan, J., Jager L., Daniel, E. E., unpublished). However, ATP produced no effect on muscle cells in lower concentrations and depolarization and contraction in higher concentrations. ADP, AMP, and adenosine, likewise, had no hyperpolarizing effect on body circular muscle, but did appear to activate a P_1-type presynaptic receptor that diminished release of NANC mediators (*49*). The proximal mammalian stomach is also innervated by NANC inhibitory nerves that elicit ijp in the smooth muscle cells (*see 38, 39*). In pig stomach, ijps were clearly NANC in nature and TTX-sensitive, but ATP (about $10^{-5}M$) caused tonic contractions, whereas higher concentrations (about $10^{-4}M$) caused contractions or slight relaxations followed by contractions in 35% of cases (*132*). The action potentials and contractions produced by ATP or other adenine derivatives were TTX-insensitive. Recently, Bauer et al. (*8, 9*) found in guinea-pig small intestine that ATP-depolarized cells that generated ijps and hyperpolarized cells that generated ejps in response to field stimulation. To reconcile the electrophysiological findings in esophagus and stomach and intestine, with the hypothesis that ATP is the NANC inhibitory mediator, would require postulating the exogenous and endogenous, nerve-related ATP act on different receptors. This will be considered in more detail below (Section 3.6).

In isolated perfused dog intestine, ATP, its nondegradable analogs (alpha- and beta-methylene ATP), as well as ADP, AMP, and adenosine all caused contraction when given intra-arterially (*18, 131*). After TTX or atropine, which diminished significantly the contractile responses to ATP and related compounds, the residual response was still contraction. Neither did inhibition of prostaglandin synthesis affect the contractile nature of the responses to ATP.

In guinea-pig taenia caecum, the most studied organ for NANC inhibitory action, the effects of ATP did not mimic those of NANC nerve stimulation (*165*). The latter (even a single pulse) produced large ijps associated with reduced membrane resistance, but ATP (10^{-6}–10^{-5}*M*) inhibited action potential generation and caused relaxation without hyperpolarization or altered membrane resistance. Only higher ATP concentrations (10^{-4}–10^{-3}*M*) caused hyperpolarization and increased conductance. When the muscle membrane was depolarized partially (and presumably had an increased conductance) from carbachol, the threshold for relaxation by ATP was reduced to 10^{-8}*M* and that for hyperpolarization was reduced to 10^{-7}–10^{-6}*M* (*101*). These authors found no clear differences between the potencies of adenine derivatives in relaxing and hyperpolarizing taenia caecum.

The anococcygeus muscle anchors the distal rectum to the upper coccygeal vertebrate; this muscle has nonadrenergic inhibitory nerves producing ijps and relaxation in the presence of atropine and guanethidine, in addition to adrenergic excitatory nerves (*41, 42, 85, 86*), ATP was a powerful inhibitor of the muscle in rabbit (*42*). It caused contraction of anococcygeus of rat (*85*) and produced relaxation in that of cat only at very high concentrations (*86*). This discrepancy with ATP as the universal mediator of nonadrenergic, inhibitory nerves was subsequently explained by Burnstock et al. (*24*) as the result of releases of prostaglandins by exogenous ATP. These workers showed that indomethacin blocked the contractile effects of ATP, but very high concentrations (10^{-4}–10^{-3}*M*) were required to produce significant relaxation. They also demonstrated releases of ATP into the medium during nerve stimulation. However, it is unclear why nerve-released ATP can cause relaxation without initiating prostaglandin release and contraction in contrast to exogenous ATP.

In airway smooth muscle, adenine-related compounds caused relaxation in guinea-pig trachea in most instances, but the concentrations required were large (10^{-4}–10^{-3} to $5 \times 10^{-4}M$) and complete relaxation could not be achieved with ATP or adenosine (*35, 36*). Also lower concentrations than required for inhibition occasionally elicited excitation. However, in human trachea and bronchioles (*48, 49, 53*) ATP has no effect or caused contraction. No electrophysiological studies have been reported in airways with NANC inhibitory innervation nor has the effect of ATP been tested on airway muscle membrane properties.

VIP was identified as a 28-amino acid peptide (*128*) from porcine intestinal extracts; however, there is evidence that in various species other related peptides contribute to the immunoreactivity (*57*). For example, only 30% in rat, 50% in dog, but 80% in pig intestinal nerves appears to be the authentic VIP. Thus tests of identity of action of VIP with NANC inhibitory transmitter may require repetition with the appropriate peptide. The octacosa peptide VIP relaxes a number of the appropriate muscles with NANC inhibitory innervation (Table 5). There are, however, important exceptions. VIP (10^{-9}–$10^{-7}M$) contracts the longitudinal

Table 5
Relaxant Actions of VIP

Organ	Species	Threshold, *M*	References
Lower esophageal sphincter	Cat (in vivo)	$\sim 10^{-7}$ mol/kg iv	*11*
	Cat (in vitro)	3×10^{-9}	*168*
	Opossum (in vivo)	$\sim 10^{-7}$ mol/kg iv	*87, 137*
	Opossum (in vitro)	10^{-7}–10^{-6}	*68*
	Human	10^{-10} (plasma conc.)	*29a*
Stomach	Dog	5×10^{-9}	*126a*
Jejunum	Dog	10^{-12}	*105*
Taenia coli	Guinea pig	$\sim 10^{-9}$	*33*
Internal anal sphincter	Human	$\sim 5 \times 10^{-10}$	*18a*
	Guinea pig	$\sim 10^{-8}$	*129*
Gall bladder	Guinea pig	3×10^{-10}	*146a*

muscle of the guinea pig intestine (*34, 99*) by a TTX-sensitive and partially atropine-sensitive mechanism. Whether an inhibitory effect was unmasked after TTX was not examined. Cohen and Landry (*34*) report that VIP also contracted longitudinal muscle of rabbit intestine, by an untested mechanism. Kachelhoffer et al. (*105*) found a biphasic response of ex vivo perfused dog jejunum, relaxation followed by contraction, the latter possibly a rebound contraction. It should be noted that a clear demonstration of NANC inhibitory nerves in the canine intestine remains to be made. Also Fox, J. E. T., Daniel, E. E., and Jury, J. (unpublished) have found little or no inhibitory responses to intra-arterial injections of VIP (to 10^{-8} mol) into field-stimulated, phasically active canine intestine in situ. In guinea pig taenia caeci, Cocks and Burnstock (*33*) found that VIP was very much slower in its time course than relaxation by NANC nerve stimulation, and Johns (*103*) found it to be very impotent in relaxing this tissue. Neya et al. (*129*) found that VIP ($10^{-9}M$) contracted the isolated guinea-pig jejunum circular muscle and as in the perfused canine jejunum, the response was TTX-sensitive as well as reduced after atropine; hexamethonium and adrenergic antagonists in selective doses were without effect.

Recently, clear exceptions to the identity of action of VIP and NANC inhibitory mediator have been reported. Rattan et al. (*137*) found that VIP given into the aorta caused repetitive phasic contraction of the body of the circular muscle of opossum esophagus, but relaxed the lower esophageal sphincter. Daniel et al. (*49*) showed that the body circular muscle in this case responded to NANC nerve stimulation with large ijps and to VIP ($> 10^{-8}M$) with increased afterdepolarizations following the ijp and in the presence or absence of nerve stimulation, with the occurrence of repetitive oscillations of membrane potential accompanied by contractions in the depolarizing phase. These latter responses were not abolished by TTX or scorpion venom but were prevented, along with the occurrence of afterdepolarizations and their enhancement by VIP, when chloride was replaced by isethionate, a nondiffusible anion. Thus the direct action of VIP on body circular muscle may have mimicked that of a mediator of the afterdepolarization, but not that of the inhibitory transmitter.

3.4. Identity of Inhibition and Potentiation of Responses to NANC Inhibitory Mediator and to ATP or VIP

ATP is rapidly degraded by plasma membrane enzymes in smooth muscle to ATP, AMP, adenosine, inosine, and adenine. Adenosine, but not ATP, can be taken up by smooth muscle and other cells, including NANC nerves, and is reported to share some of the relaxant actions of ATP (*see 19, 75, 84*). On the rationale that they acted to prevent adenosine uptake and perhaps delayed degradation of ATP by causing accumulation of products, dipyridamole and hexobendine were tested and shown to potentiate selectively relaxation responses to ATP and NANC nerve stimulation; responses to norepinephrine and sympathetic nerve stimulation were not potentiated (*19, 21, 149, 150*). However, the rationale is faulty in that adenosine has 1/100 or less the potency of ATP (*23*), and its end effect should be like that of blockade of choline-uptake by hemicholinium on cholinergic nerve function to block transmission. No evidence exists of significant inhibition of the multiple enzymes degrading ATP by the product accumulated after blockade of adenosine-uptake. Moreover, dipyridamole and other adenosine uptake blockers were not effective to potentiate NANC and purine (including adenosine)-induced relaxation of rabbit duodenum (*93*) or hyperpolarization of guinea-pig taenia caeci to NANC nerve stimulation, though it potentiated relaxation to adenosine (*3*). In the same study, dipyridamole was ineffective at all frequencies of nerve stimulation, except 0.3 Hz, the one used in the original observations of Satchell et al. (*150*). A lack of effect of dipyridamole to potentiate NANC-mediated or ATP-induced relaxations was reported by Ohga and Taneike (*132*) in longitudinal muscle of pig stomach.

No specific antagonists to ATP at NANC inhibitory receptors are known, but recently arylazide-proprionyl ATP has been reported to be a selective, competitive, and irreversible photoaffinity label for purine excitatory receptors in vas deferens (*see* Chapter 16, Section 2.2.3.). It seems to have no effect on NANC inhibitory transmission (*72*). In the absence of a competitive antagonist, evidence has been based on desensitization or use of rather unspecific depressants. Burnstock et al. (*23*) reported that partial tachyphylaxis to ATP reduced responses to NANC inhibitory nerve stimulation, but not those to sympathetic nerve stimulation. No quantitative data were given; nor were concentration–response or frequency–response curves reported. However, Weston (*173*) showed that in rabbit duodenum longitudinal muscle strips, ATP densitization shifted the dose–effect curve for ATP to the right, more than 40-fold, and depressed the maximum response by 95% without inhibition of responses to stimulation of NANC inhibitory nerves or sympathetic inhibitory nerves or adrenergic agonists. Responses to ADP, AMP, and adenosine as well as ATP were inhibited. Weston also showed (*173*) that activation of NANC inhibitory nerves in longitudinal muscle preparations of guinea-pig ileum by field stimulation or distention was unaffected by densensitization to ATP. Okwuasaba et al. (*133*) reported that densensitization of fundal muscle strips from guinea pig by $5 \times 10^{-5}M$ ATP shifted the dose–response curve for ATP fourfold to the right, and depressed by 75% NANC-mediated relaxations to field stimulation (1–20 Hz). Responses to norepinephrine, papaverine, or sympathetic nerves were not depressed. However, Baer and Frew (*4*) could find no effect of similar ATP desensitization on the frequency–response curve when they repeated these experiments. Similar lack of effect on responses to NANC-inhibitory nerve mediator by ATP desensitization were obtained in pig gastric muscle (*132*), in rat ileum (*5*), in both layers of muscle in rabbit colon (*43, 166*), and lower esophageal sphincter of opossum (*136*). In cat tracheal muscles, Ito and Taketa (*98*) found that ATP relaxed the muscle, but desensitization with ATP had no effect on nonadrenergic relaxation.

It must be concluded that the balance of evidence from desensitization studies is against the purinergic nerve hypothesis. Reconciliation of the data with that hypothesis would require that exogenous ATP not reach the purine receptors accessible to ATP released from nerves. This seems unlikely since desensitization to large molecules believed released from nerves, substance P, and sometostatin, has been achieved (*see* ref. *75*, p. 414), and ATP released by stimulation of enteric nerves reaches the muscle bath. The related possibility that there is one set of receptors on interstitial cells of Cajal and another on smooth muscle, which are differentially reached by nerve-released and exogenous ATP, is discussed below (Chapter 15, Section 3.6.).

Antagonists to ATP are mostly nonspecific, even nonselective. Early studies (*23*) showed that quinidine block of ATP- and NANC-mediated relaxations are clearly a result of the nonselectivity of the compound, since it also blocked responses to catecholamines and relaxations from stimulation of sympathetic nerves (*84*). Other imidazole or imidazolelike compounds provide no support for the purinergic nerve hypothesis. For example, phentolamine (*135* μM) blocked relaxation to ATP, but not NANC nerve stimulation (*140, 162*), or hyperpolarizations to ATP, but not NANC ijps (*147*). In the presence of 13 μM phentolamine plus other alpha- and beta-adrenergic antagonists, complete inhibition of ATP-induced relaxation was attained without any change in the responses to NANC nerves. In light of these results, a finding (*150*) that higher concentrations (180 μM) blocked both relaxation to ATP and to NANC nerve stimulation cannot be interpreted to support the purinergic hypothesis (*see also* refs. *75, 84*).

In 1977, Okwuasaba et al. reported what seemed strong evidence that theophylline ($5 \times 10^{-5}M$) was a selective, competitive antagonist to both ATP and NANC nerve-induced (1–20 Hz) relaxations in guinea-pig fundas strips. Relaxations from norepinephrine, isoproterenol, and papaverine were not reduced. However, Baer and Frew (*4*) found no effect of this dose of theophylline in either ATP- or NANC-nerve-induced relaxations. Small and Weston (*156*) found in guinea-pig fundas strips and taenia caeci, and in rabbit duodenal longitudinal strips that theophylline up to $10^{-4}M$ had no effect on inhibitory responses to ATP- or NANC-nerve stimulation. In a concentration of $10^{-3}M$ it antagonized the twitch suppression by purines in field stimulated guinea-pig ileum, a well-known presynaptic action of purines. These authors also reported a personal communication from two of the three authors of the original paper (Cook, M., and Hamilton, J. T.) that they too could not repeat the observations regarding theophylline. Finally, Huizinga and Den Hertog (*92*) reported that theophylline up to $10^{-3}M$ did not antagonize ATP- or adenosine-induced hyperpolarizations or NANC-nerve-induced ijps in guinea-pig fundal strips. It seems clear that theophylline has no selective NANC antagonist actions, but may inhibit at presynaptic purine receptors.

Spedding et al. (*158*) reported that PIT or 2-2′pyridylisatogen ($2 \times 10^{-5}M$) caused a two-log units shift in the concentration–effect (relaxation) curve for ATP in guinea-pig taenia caeci without affecting relaxation to single pulses of NANC nerve stimulation (or receptive stimulation up to 20 Hz) or relaxation to perivascular (sympathetic) nerve stimulation. It shifted relaxation responses to isoproterenol or norepinephrine less than one log unit. PIT antagonism was thus selective for ATP but was not competitive and became irreversible with prolonged exposure. PIT inhibited responses to ADP and AMP as well as ATP, but was ineffective in antagonizing responses to adenosine (*157*). Similar results were obtained by Stockley (*159*) in human taenia coli and by Rattan and Goyal (*136*) in opossum lower esophageal sphincter in vivo and Hunt et al. (*94*) in rat gastric muscle. Subsequent studies have sometimes reported PIT to be nonspecific, but the weight of this evidence suggests that either ATP is not the NANC-inhibitory transmitter or ATP release from nerves reaches receptors inaccessible to PIT and exogenous ATP.

Apamin, a peptide from bee venom, was noted by Vladimirova and Shuba (*171*) and Shuba and Vladimirova (*153*) to block the hyperpolarizations induced by NANC nerve stimulation and by ATP in the guinea-pig taenia caeci. These authors interpreted their results in terms of purine receptor antagonism. Apamin inhibited NANC and ATP-induced relaxations of this muscle as well (*6, 127*), but its actions were clearly related to inhibition to K^+ conductance secondary to increased Ca^{2+} entry since it also blocked responses to other activators of K^+ conductance besides ATP and NANC field stimulation (e.g., adrenergic agonists, peravascular nerve stimulation) in taenia caeci and prevented the K^+ loss but not the Ca^+ movement in liver cells related to hyperpolarization by norepinephrine, ATP, and A 23187. Furthermore, it lacked effects on purinergic receptors of guinea-pig vas deferens. It inhibited increased K^+ efflux by NANC field stimulation, ATP, and alpha-adrenergic agonists as well as the ijp of guinea-pig taenia caeci (*119, 120*). In circular muscle of guinea-pig intestine, Bywater et al. (*27*) showed that it blocked the ijp and unmasked an atropine-resistant excitatory junction potential. However, in opossum body circular muscle, it had no effect in diminishing the ijp and neither did 4-aminopyridine or tetraethylammonium ion (*49* and unpublished). In this tissue ATP caused contraction and depolarization rather than hyperpolarization and relaxation. In light of its mechanism of action, interference with some K^+ conductance channels, apamin cannot provide evidence that a substance is the NANC transmitter; but when it is capable of blocking K^+ channels opened by NANC-transmitter, it can screen out all those candidates that relax by another conductance mechanism than the transmitters. In each case, evidence that apamin is effective in blocking K^+ conductance will be required since several cases of its ineffectiveness are now on record (*49, 84*).

In summary, antagonists and potentiators of the action of ATP do not support, in a general sense, the purinergic hypothesis. There are too many contradictory effects or lack of effects.

VIP-selective antagonists have not been divised so far. There is also little information about the mechanisms responsible for terminating the action of VIP, and therefore potentiation by inhibition of degradation or other inactivation process has not been studied. Instead Goyal and Rattan (*87*) used intra-arterial perfusion of VIP antiserum (1:12.5 dilution) to partially block relaxation from perfused VIP (40–60%), vagal stimulation (30–50%), field stimulation (35%), and nicotine (30%) on the lower esophageal sphincter of opossum. Other control sera were ineffective and the anti-VIP serum was ineffective against isoproterenol-induced relaxation. Recently, however, deCarle and Pye (*29*) were unable to confirm these observations. Moreover, it is difficult to see why the antiserum, a large protein, was unable to completely block effects of infused VIP, but blocked the VIP presumed to be released from NANC nerves in high local concentrations by vagal, field, or nicotine stimulation. Finally, there is agreement (*see* Chapter 15, Section 3.3.) that in the body of the esophagus in this species, NANC nerves caused ijps, but VIP enhanced afterdepolarization, and its effects were blocked by Cl^- removal, but the ijps were not. One explanation of the effects obtained by Goyal and Rattan in vivo on LES is that they may not have accounted for antiserum effects on LES basal pressure.

The same general approach was used by Matsuzaki et al. (*123a*) in guinea-pig tracheal strips. They showed that VIP was released by field stimulation (2 min trains, 1–2 ms pulses, 70 V, 20 Hz) that causes relaxation in the presence of atropine and propranolol. Furthermore, both relaxation and VIP release were blocked by TTX. After incubation in VIP antiserum (10 μL/mL of bath fluid) overnight at 4°C, the tissue relaxation was only about 75% compared to controls similarly incubated without antiserum. Whether the incubation with antiserum was selective in interfering with NANC relaxations was not tested and it was not clear whether active basal tension was unaffected. The presence of antagonists was not specified. Thus, the data are not strongly supportive that the NANC relaxations were VIP-mediated and a variety of other explanations was not included.

Ito and Takeda (*98*) reported that neither VIP ($< 10^{-9}M$) nor field stimulation in the presence of atropine and propranolol, both of which caused TTX-sensitive relaxation of cat tracheal strips precontracted by 5-hydroxytryptamine, affected membrane potential or resistance. Higher concentrations of VIP ($> 10^{-8}M$) sufficient to mimic NANC-nerve-induced relaxations caused hyperpolarization and decreased membrane resistance. Desensitization to VIP (but not to ATP) reduced the nonadrenergic nerve-induced relaxation. This was taken as evidence that VIP might be the NANC-inhibitory transmitter to feline airways, but no explanation was advanced as to why higher, hyperpolarizing, and conductance-decreasing concentrations of VIP were required to match the relaxations induced by NANC nerve activity that were unaccompanied by muscle membrane changes.

However, strong evidence against VIP as the NANC inhibitory transmitter of the guinea pig taenia caeci was provided by MacKenzie and Burnstock (*121*). VIP relaxation was slower in time course than the NANC nerve-induced relaxation and was abolished by preincubation of the tissue in alpha-chymotrypsin whereas NANC-nerve mediated responses were not, and nerve responses were blocked by apamin whereas VIP-induced relaxations were not. The last piece of evidence is especially hard to reconcile with VIP as neurotransmitter in view of the known effect of NANC-nerve mediators to activate K^+ conductance and of apamin to inhibit this response in this tissue. Similarly, in anococcygeous muscle, hemolyzed blood inhibits the NANC inhibitory response, but not the relaxation induced by VIP (or ATP) (*83, 84*).

No alternate candidate to ATP or VIP as general NANC inhibitory transmitter emerges. Substance P, for example, is present in airway and its nerves (*81, 130*), but usually is a contractile agent for gut (*44, 70, 71*). It was ineffective as an airway relaxant (*55*).

In summary, there is clear evidence against either ATP or VIP being the universal NANC-inhibitory transmitter in all tissues. Conceivably, either may function in a particular system or neither. There seems to be no tissue in which the purinergic hypothesis has been intensively evaluated without discrepancies arising. The VIP-ergic hypothesis has been investigated less and there still are some tissues in which available data are not inconsistent with the role of transmitter. However, several years ago that was the case with ATP. Perhaps it is time to reconsider some of the types of evidence used in light of recent findings about the geography of the NANC-inhibitory nerves in relation to interstitial cells of Cajal (*see* Section 3.5).

3.5. Nature of NANC Transmitter for Dilation of Blood Vessels

Another instance of NANC-nerve-mediated inhibition (vasodilatation) seems to exist in cerebral arteries. Lee (*113*) and Chapter 14 has summarized recent and earlier (*112*) evidence that neither acetylcholine nor norepinephrine is the transmitter. One transmitter that seems to be appropriate is VIP. VIP-containing nerves are found (*63, 111*) VIP dilates these arteries (*113, 124, 172*), and desensitization to VIP inhibits the NANC vasodilation. Another possible candidate is substance P, since nerves containing immunoreactivity related to this peptide have been found (*62, 73*), and substance P is a potent dilator. However, in other systems its vasodilating action is endothelium cell dependent. If this is the case in cerebral arteries, substance P would be excluded, since neurogenic vasodilation is unaffected by removal of endothelial cells (Chapter 14). Substance P, however, may be a transmitter of neural vasodilator activity in some systems since it apparently is in the case of antidromic vasodilation from stimulation of dorsal root afferent nerves (e.g., *142*).

In some sites, strong evidence for VIP as the NANC vasodilator transmitter already exists. For the cat nasal mucosa, evidence of VIP-containing nerves, VIP-release during nerve stimulation, and its appropriate action are available (*117*).

VIP has also been considered to be a mediator of neurogenic vasodilation in blood vessels of salivary glands (*118, 169*). In the latter case, evidence shows VIP- as well as acetylcholine-containing nerves, release of VIP- and acetylcholine from nerves, with VIP release enhanced at higher frequencies of nerve stimulation, or after inhibition by atropine of a negative feedback control over VIP-release by acetylcholine. Furthermore, the blood flow increase on nerve stimulation was in two phases: (1) an initial transient phase most marked at low frequencies and reduced by both atropine and VIP antiserum; (2) a later prolonged phase that was most marked at high frequencies of nerve stimulation and was reduced by atropine at low frequencies and by VIP antiserum at 2 Hz (only frequency tested). In this case the VIP antiserum was partially purified prior to infusion, and control sera or IgG were without effect. The authors concluded that acetylcholine and VIP both appear to mediate the vasodilation, and the output of VIP may be autoregulated by acetylcholine acting through muscarinic receptors to inhibit VIP output. In such a system, constant blood flow during prolonged nerve activity was postulated to occur by enhancement of VIP output as acetylcholine output fell.

3.6. Difficulties in Identifying NANC Transmitters and Role of Interstitial Cells of Cajal

The puzzling fact is that many systems that are innervated by nerves containing a putative mediator do not respond similarly to that mediator added exogenously and released (apparently) from nerves. A commonly suggested and usually discounted suggestion is that the endogenous and exogenous mediators act on different receptors. This suggestion is usually discounted because the mediator can often be demonstrated to diffuse from the nerves and synaptic cleft into the bath, and it seems inconceivable that it cannot diffuse in the opposite direction. However, the situation is made more complex in the case of gut circular muscles that frequently have interstitial cells of Cajal closely (< 90 nm) or very closely (< 20 nm) innervated by NANC nerves and in gap junction contact with smooth muscle (*51, 52, 144, 145*). This arrangement was first ascribed in gut of a bird (*95*). It revived the long-lasting controversy regarding the suggestions of Cajal that interstitial cells of Cajal were neurons or neuronlike and intercalated between the enteric nerves and the effector cells (mostly smooth muscle). An expert and critical review of this subject has recently been provided the Thuneberg (*163*). This arrangement could provide a priority pathway via interstitial cell receptors for endogenous mediator to have effects on smooth muscle before it reaches receptors diffusely present on smooth muscle. This arrangement might help account for the ability of NANC inhibitory nerves to produce very large ijps in response to a single pulse and the quick run-down of response with repetitive stimulation (as mediator reaches other receptors at sites with perhaps opposite effects, the temporal relationship of the "on" ijp and the "off" depolarization as local interstitial cell mediator is rapidly inactivated while diffusely spread mediator is still acting). Conceivably, there might be no diffusely distributed receptors, and only those on interstitial cells or both diffusely distributed receptors and those in the synaptic cleft might have similar actions. It is clear that in any such arrangement it might be impossible to duplicate the effects of the real transmitter released from nerve by adding it exogenously, and that much higher concentrations would be required of exogenous mediator than might be expected for a neurotransmitter (if the receptors are exlusively in the cleft).

An example of how this might work has recently been provided in circular muscle of opossum body (*49*). Neither any purine nor VIP reproduced the hyperpolarization caused by NANC inhibitory mediator, but VIP enhanced the afterdepolarization and caused membrane potential oscillations, and both the afterdepolarization and the effects of VIP were abolished by replacement of chloride with a nondiffusable anion. This muscle has many VIP-containing nerves (*59a*). Conceivably VIP might have caused hyperpolarization when it was released by nerves into the narrow cleft between the endings and interstitial cells (*51, 52*), and caused membrane oscillations by a direct action on diffuse receptors or actions on both interstitial cell and smooth muscle receptors.

In cases where cotransmitters are released from the same nerve, the possibilities are even more complex since local and diffusion-dependent action of each or both might occur. The interstitial cell hypothesis has parallels with the two types of receptor hypotheses. Each required one type close to nerve endings and somehow less accessible than another receptor distant from nerve endings but accessible to exogenous agents. The difference can be in receptor type, or in accessibility, or in effective concentrations of agonists and antagonists reached at the sites, or a combination of these. Both ideas present similar difficulties of verification. It should be noted, too, that interstitial cells in gap junction contact with smooth muscle and closely innervated have not been observed in nongut tissues with NANC innervation, e.g., in airway smooth muscle and in blood vessels. Obviously, this hypothesis cannot be applied in a general sense to resolve the difficulties in identifying the NANC transmitters.

4. Conclusions

Clear evidence exists of NANC inhibitory nerves to smooth muscle in gut, airways (some species) and blood vessels (some regions). Only in the case of gut have electrophysiological studies been made that show clearly that the basis of the inhibitory actions of NANC nerves is a selective increase in K^+ conductance, blocked in some tissues but not all by apamin. Neither of the putative candidates for transmitter, ATP or VIP, has appropriate electrophysiological actions in all or most tissues. Either more than one NANC inhibitory transmitter exists or the complications arise from the existence of more than one type of receptor for the transmitter leading to different effects. A related hypothesis, discussed here, is the possibility that interstitial cells of Cajal are intercalated between NANC nerve endings and gut smooth muscle, to which they are electrically coupled. Receptors of different types to a given transmitter might be located on interstitial cells and on smooth muscles.

References

1. Alumets, J., R. Hakanson, F. Sundler, and R. Uddman. VIP-, enkephalin-, substance P- and somatostatin-like immunoreactivity in neurons intrinsic to the intestine. Immunohistochemical evidence from organotypic tissue cultures. *Brain Res.* 155:239–248, 1978.
2. Ambache, N. Unmasking after cholinergic paralysis by botulinum toxin of a reversed action of nicotine on the mammalian intestine revealing the probable presence of local inhibitory ganglion cells in the enteric plexuses. *Br. J. Pharmacol.* 6:51–67, 1951.

2a. Axelsson J. B. Holmberg. The effects of extracellularly applied ATP and related compounds on electrical and mechanical activity of the smooth muscle taenia coli from the guinea-pig. *Acta Physiol. Scand.* 75:149–156, 1969.

3. Baer, H. P., R., Frew, and G. Burnstock. Effect of dipyridamole and 6-(2-hydroxy-5-nitrobenzyl-thioguanosine) on low frequency-stimulated relaxation in the guinea pig taenia coli, *Can. J. Physiol. Pharmacol.* 55:394–398, 1977.

4. Baer, H. P. and R. Frew. Relaxation of guinea pig fundic strip by adenosine, adenosine triphosphate and electrical stimulation; lack of antagonism by theophylline or ATP treatment. *Br. J. Pharmacol.* 67:293-299, 1979.

5. Bartlett, V., R. R. Steward and K. Nakatsu. Evidence for two adenine derivative receptors in the rat ileum which are not involved in the non-adrenergic, non-cholinergic response. *Can. J. Physiol. Pharmacol.* 57:1130–1137, 1979.

6. Banks, B. E. C., C. Brown, G. M. Burgess, G. Burnstock, M. Claret, T. M. Cocks and D. H. Jenkinson. Apamin blocks certain neurotransmitter-induced increases in potassium permeability. *Nature* 282:415–417, 1979.

7. Bauer, V. and H. Kuriyama. Evidence for non-cholingeric, non-adrenergic transmission in the guinea pig ileum. *J. Physiol. (Lond.)* 330:95–110, 1982.

8. Bauer, V., O. Matusak, and H. Kuriyama. Non-cholinergic, non-adrenergic responses to nerve stimulation of different regions of the guinea pig small intestine. *Naunyn-Schmiedeberg's Arch. Pharmacol.* 319:108–114, 1982b.

9. Bauer, V., O. Matusak, H. Kuriyama, M. Bezekova, and L. Benes. Role of Ca^{2+} in non-cholinergic, non-adrenergic responses to nerve stimulation. *Naunyn-Schmiedeberg's Arch. Pharmacol.* 319:115–120, 1982a.

10. Bayliss, W. M. and E. H. J. Starling. The movements and innervation of the small intestine. *J. Physiol. (Lond.)* 24:99-143, 1899.

11. Beani, L., C. Bianchi, and A. Crema. Vagal non-adrenegic inhibition of guinea pig stomach. *J. Physiol. (Lond.)* 217:259–279, 1971.

11a. Beck, C. S. and J. Osa. Membrane activity in guinea-pig gastric sling muscle: a nerve-dependent phenomenon. *Am. J. Physiol.* 220:1397–1403, 1971.

11b. Behar, J., S. Field, and C. Marin. Effect of glucagon, secretin and vasoactive intestinal polypeptide on the feline lower esophageal sphineter: mechanisms of action. *Gastroenterology* 77:1001–1007, 1979.

12. Bennett, M. R. and Burnstock, G. Electrophysiology of the innervation of intestinal smooth muscle. In *Handbook of Physiology—Vol. IV, Section 6.* Alimentary Canal. Ed. Code, Cf. Am. Physiol Soc., Washington, D.C. pp 1709–1732, 1968.

13. Bennett, M. R., G. Burnstock, and M. E. Holman. The affect of potassium and chloride ions on the inhibitory potential recorded in the guinea pig taenia coli, *J. Physiol. (Lond.)* 169:33P–34P, 1963.

14. Bennett, M. R., G. Burnstock, and M. E. Holman. Transmission from intramural inhibitory nerves to the smooth muscle of the guinea pig taenia coli, *J. Physiol.* (Lond.) 182:541–558, 1966.

15. Berezin, I., M. S. Sheppard, E. E. Daniel and N. Yanaihara. Electron microscopic localization of VIP-like immunoreactivity in dog ileum. *Regulatory Peptides.* Submitted.

15a. Bianchi, C. L. Beani, G. M. Frigo and A. Crema. Further evidence for the presence of non-adrenergic inhibitory structures in the guinea-pig colon. *Eur. J. Pharmacol.* 4:51–61 1968.

16. Bitar, K. N., S. J. Said, G. C. Weir, B. Saffouri, and G. M. Makhlouf. Neutral release of vasoactive intestinal peptide from the gut. *Gastroenterology* 79:1288–1294, 1980.

17. Bloom, S. R. and A. V. Edwards, Effects of autonomic stimulation on the release of vasoactive intestinal peptide from the gastro-intestinal tract. *J. Physiol. (Lond.)* 299:437–452m 1980.

18. Burks, T. F. and M. N. Grubb. Stimulatory actions of adenosine triphosphate in dog intestine. In *Gastrointestinal Motility in Health and Disease.* H. J. Duthie, ed. University Part Press, Baltimore, pp 151–159, 1978.

18a. Burleigh, D. E., A. D'Mello, and A. G. Parks. Responses of human internal anal sphincter to drugs and electrical field stimulation. *Gastroenterology* 77:484–490, 1979.

19. Burnstock, G. Purinergic Nerves. *Pharmacological Rev.* 24:509–581, 1972.

20. Burnstock, G. Past and current evidence for the purinergic nerve hypothesis. In *Physiological and Regulating functions of Adenosine and Adenine Nucleotides.* H. Bauer and G. I. Diamond, eds. Raven Press, New York, pp 3–32, 1979.

21. Burnstock, G. Neurotransmitters and trophic factors in the autonomic nervous system. *J. Physiol. (Lond.)* 313:1–35, 1981.

22. Burnstock, G., G. Campbell, M. Bennett and M. E. Holman. Innervation of the guinea pig taenia coli: Are there intrinsic inhibitory nerves which are distinct from sympathetic nerves? *Int. J. Neuropharmacol.* 3:163–166, 1964.

23. Burnstock, G., G. Campbell, D. Satchell, and A. Smythe. Evidence that adenosine triphosphate or a related nucleotide is the transmitter substance released by non-adrenergic inhibitory nerves in the gut. *Br. J. Pharmacol.* 40:668–688, 1970.

24. Burnstock, G., T. Cocks, and G. Crowe. Evidence for purinergic innervation of the anococcygeus muscle. *Br. J. Pharmacol.* 64:13–20, 1978a.

25. Burnstock, G., T. Cocks, G. Crowe, and L. Kasakov. Purinergic innervation of the guinea pig urinary bladder. *Br. J. Pharmacol.* 63:125–138, 1978b.

26. Burnstock, G., T. Cocks, L. Kosakov, and A. Wong. Direct evidence for ATP release from non-adrenergic, non-cholinergic ("purinergic") nerves in the guinea pig taenia coli and bladder. *Eur. J. Pharmacol.* 49:145–149, 1978c.

26a. Burnstock, G., D. G. Satchell and A. Smythe. A comparison of the excitatory and inhibitory effects of non-adrenergic, non-cholinergic nerve stimulation and exogenously applied. ATP on a variety of smooth muscle preparations from different vertebrate species. *Br. J. Pharmacol* 46:234–242, 1972.

27. Bywater, R. A. R., M. E. Holman, and G. S. Taylor. atropine–resistant deplorization in the guinea pig small intestine. *J. Physiol. (Lond.)* 316:369–378, 1981.

28. Campbell, G. The inhibitory nerve fibers in the vagal supply to the guinea pig stomach. *J. Physiol. (Lond.)* 185:600–612, 1966.

29. deCarle, D. J. and M. Pye. Is vasoactive intestinal polypeptide an inhibitory neurotransmitter in the human stomach? In *Motility of the Digestive Tract*. M. Wienbeck, ed. Raven Press, New York, pp 67–72, 1982.

30. Chayvialle, J–A., M. Miyata, P. L. Rayford, and J. C. Thompson. Effects of test meal, intragastric nutrients and intraduodenal bile in plasma concentrations of immunoreactive somatostatin and vasoactive intestinal peptide in dogs. *Gastroenterology* 79:844–852, 1980.

31. Cheung, D. W. and E. E. Daniel. Comparative study of the smooth muscle layers of the rabbit duodenum. *J. Physiol. (Lond.)* 309:13–27, 1980.

32. Coburn, R. F. and T. Tomita. Evidence for non-adrenergic inhibitory nerves in the guinea pig trachealis muscle. *Am. J. Physiol.* 1072–1080, 1973.

33. Cocks, J. and G. Burnstock. Effects of neuronal polypeptides on intestinal smooth muscle: a comparison with non-adrenergic, noncholinergic nerve stimulation and ATP. *Eur. J. Pharmacol.* 54:251–259, 1979.

34. Cohen, M. L. and A. S. Landry. Vasoactive intestinal peptide: increased tone, enhancement of acetylcholine release and stimulation of adenylate cyclase in intestinal smooth muscle. *Life Sciences* 26:811–822, 1980.

35. Coleman, R. A. Effects of some purine derivatives on the guinea pig trachea and their interaction with drugs that block adenosine uptake. *Br. J. Pharmacol.* 57:51–59, 1976.

36. Coleman, R. A. and G. P. Levy. A non–adrenergic inhibitory nervous pathway in guinea pig trachea. *Br. J. Pharmacol.* 52:167–174, 1974.

36a. Costa, M. and J. B. Furness. The innervation of the internal anal sphincter of the guinea-pig. In *Proceedings of the 4th International Symposium of Gastrointestinal Motility.* (E. Daniel, ed.) Mitchell Press, Vancouver, pp 681–689, 1974.

37. Costa, M. and J. B. Furness. The origins, pathways and terminations of neurones with VIP-like immunoreactivity in the guinea pig small intestines. *Neuroscience* 8:665-676, 1983.

38. Costa, M. and J. B. Furness. Nervous control of intestinal motility. In *Mediators and Drugs in Gastrointestinal Motility I. Morphological and Neurophysiological Control.* G. Bertaccini, ed. Springer Verlag, Berlin, pp 279–382, 1982a.

39. Costa, M. and J. B. Furness. Neuronal peptides in the intestine. *Br. Med. Bull.* 38:247–252, 1982b.

40. Costa, M., J. B. Furness, R. Buffa, and S. J. Said. Distribution of enteric neurons showing immunoreactivity for vasoactive intestinal polypeptide (VIP) in the guinea pig intestine. *Neuroscience* 5:587–596, 1980.
41. Creed, K. C. and J. C. Gillespie. Some electrical properties of the rabbit anococcygeus muscle and a comparison of the effects of inhibitory nerve stimulation in the rat and rabbit. *J. Physiol. (Lond.)* 273:137–153, 1977.
42. Creed, K. C., J. C. Gillespie, and H. McCoffery. The rabbit anococcygeus muscle and its response to field stimulation and to some drugs. *J. Physiol. (Lond.)* 273:121–135, 1977.
43. Crema, A., L. D'Angelo, G. M. Frigo, S. Lechini, L. Onori, and M. Tonini. Effects of desensitization to adenosine 5′-triphosphate and adenosine on non-adrenergic inhibitory responses in the circular muscle of rabbit colon. *Br. J. Pharmacol.* 75:311:318, 1982.
44. Daniel, E. E., T. Gonda, T. Domoto, M. Oki, and N. Yanaihara. The effects of substance P and met^5-enkephalin in dog ileum. *Can. J. Physiol. Pharmacol.* 60:830–840, 1982.
45. Daniel, E. E. and G. S. Taylor. Junction potentials and control of motility of the small intestine. In *Proc. Vth Int. Symp. GI Motility.* G. Vantrappen, ed. Typoff Press, Leuven, Belgium, pp 142–151, 1975.
46. Daniel, E. E., G. S. Taylor, V. P. Daniel, and M. E. Holman. Can non-adrenergic inhibitory varicosities be identified structurally? *Can. J. Physiol. Pharmacol.* 55:243–250, 1977.
47. Daniel, E. E., J. Crankshaw and S. Sarna. Prostaglandins and tetrodotoxin-insensitive relaxation of opossum lower esophageal sphincter. *Am. J. Physiol.* 236:E153–E172, 1979.
48. Daniel, E. E., C. Davis, T. R. Jones, and M. S. Kannan. Control of airway smooth muscle. In *Airway Reactivity.* F. Hargreave, ed. Astra Pharmaceutical, pp 80–107, 1980.
49. Daniel, E. E., A. Helmy-Elkholy, L. P. Jager, and M. S. Kannan. Neither a purine nor VIP is the mediator of inhibitory nerves of opossum esophageal smooth muscle. *J. Physiol. (Lond.)* 336:243–260, 1983.
50. Daniel, E. E., M. S., Kannan, C. Davis, and V. Posey-Daniel. Comparative ultra-structural studies in the neuromuscular control of human tracheal and bronchial smooth muscle. *Can. J. Physiol. Pharmacol.* (submitted).
51. Daniel, E. E. and V. Posey-Daniel. The structural comparison of esophageal lower sphincter (LES) and body circular muscle (BCM) from opossum. Role of interstitial cells of Cajal. *Amer. J. Physiol.* 246:305–315, 1984
52. Daniel, E. E. and V. Posey-Daniel. Effects of scorpion venom on structure and function of esophageal lower sphincter (LES) and body circular muscle (BCM) from opossum. *Can. J. Physiol. Pharmacol.* 62:360–373, 1984.
53. Davis, C., T. R. Jones and E. E. Daniel. Studies of the mechanism of passive anaphylaxis in human airway smooth muscle. *Can. J. Physiol. Pharmacol.* Accepted.
54. Den Hertog, G. A. And L. P. Jager. Influxes during the inhibitory junction potentials in the guinea pig taenia coli. *J. Physiol. (Lond.)* 250:681–691, 1975.

55. Diamond, L., J. L. Szaren, and M. N. Gillespie. Substance P fails to mimic vagally mediated non-adrenergic bronchodilatation. *Peptides* 3:21–29, 1982.

56. Diamond, L. and M. O'Donnell. A non-adrenergic vagal inhibitory pathway to feline airways. *Science* 208:185–188, 1980.

57. Dimaline, R. and G. J. Dockray. Molecular variants of vasoactive intestinal polypeptide on the feline lower esophageal sphincter mechanisms of action. *Gastroenterology* 77:1001–1007, 1979.

58. Dimaline, R., C. Vaillant and G. J. Dockray. The use of region specific antibodies in the characterization and localization of vasoactive intestinal polypeptide-like substances in the rat gastrointestinal tract. *Regulatory Peptides* 1:1–16, 1980.

59. Doidge, J. M. and D. G. Satchell. Adrenergic and non-adrenergic inhibitory nerves in mammalian airways. *J. Auton. Nerv. Syst.* 5:83–99, 1982.

59a. Domoto, T., E. E. Daniel, M. Oki, J.E.T. Fox, and N. Yanaihara. Peptidergic nerves (substance P and VIP) in lower esophagus of opossum. *Gastroenterology* 82(52):1044, 1982.

60. Douglas, W. W. Stimulus-secretion coupling: the concept and clues from chromaffin and other cells. *Br. J. Pharmacol.* 34:451–474, 1968.

60a. Drury, A. N. and A. Szent-Györgyi. The physiological activity of adenine compounds with especial reference to their action upon the mammalian heart. *J. Physiol. (Lond)* 68:213–237, 1929.

61. Edin, R., H. Ahlman, and J. Keventer. The vagal control of the feline pyloric sphincter. *Acta Physiol. Scand.* 107:169–174, 1979.

62. Edvinsson, C. and R. Uddman. Immunohistochemical localization and dilatory effect of substance P on human cerebral vessels. *Brain Res.* 232:446–471, 1982.

63. Edvinsson, L., J. Fahrenkrug, J. Hanko, C. Owman, F. Sundler, and R.L Uddman. VIP (vasoactive intestinal polypeptide) containing nerves of intra-cranal arteries in mammals. *Cell Tissue Res.* 208:135–142, 1980.

64. Eklund, S., J. Fahrenkrug, M. Jodal, O. Lundgren, O. B. Schaffaletzky De Muckadell, and A. Stoqvist. Vasoactive intestinal polypeptide, 5-hydroxytryptamine and reflex hyperaemia in the small intestine of the cat. *J. Physiol. (Lond.)* 302:549–557, 1980.

65. Fahrenkrug, J. Vasoactive intestinal polypeptide measurement, distribution and putative neruotransmitter function. *Digestion* 19:149–169, 1979.

66. Fahrenkrug, J., H. Galbo, J. J. Holst, and O. B. Schaffaletzky De Muckadell. Influence of the autonomic nervous system on the release of vasoactive intestinal polypeptide from the porcine gastrointestinal tract. *J. Physiol. (Lond.)* 280:405–422, 1978b.

67. Fahrenkrug, J., U. Haglund, M. Jodal, O. Lundgren, L. Olbe, and O. B. Schaffaletzky De Muckadell. Nervous release of vasoactive intestinal polypeptide in the gastro-intestinal tract of cats: possible physiological implications. *J. Physiol. (Lond.)* 284:291–305, 1978a.

68. Fox, J. E. T., S. J. Said, and E. E. Daniel. Is vasoactive intestinal polypeptide (VIP) an inhibitory neurotransmitter in the lower esophageal sphincter (LES) in the North American opossum? *Gastroenterology* 76:1134, 1979.

69. Fujiwara, M., S. C. Hong, and J. Muramatsu. Effects of Goniopora toxin on non-adrenergic, non-cholinergic response and purine nucleotide release in guinea pig taenia coli. *J. Physiol. (Lond.)* 326:515–526, 1982.

70. Franco, R., M. Costa, and J. B. Furness. Evidence that axons containing substance P are of intrinsic origin. *Naunyn-Schmiedeberg's Arch. Pharmacol.* 307:57–63, 1979b.

71. Franco, R., M. Costa and J. B. Furness. Evidence for the release of endogenous substance P from intestinal nerves. *Naunyn Schmiedeberg's Arch. Pharmacol.* 306:185–201, 1979a.

72. Frew, R. and P. M. Landry. Effect of arylazidoaminopropionyl ATP ($ANAPP_3$), a putative ATP antogonist, on ATP responses of isolated guinea pig smooth muscle. *Life Sciences* 30:259–267, 1980.

73. Furness, J. B., R. E. Papka, N. G. Della, M. Costa, and R. L. Eskay. Substance P-like immunoreactivity in nerves associated with the vascular system of guinea pigs. *Neuroscience* 7:447–459, 1982.

74. Furness, J. B. An electrophysiological study of the smooth muscle of the colon. *J. Physiol. (Lond.)* 205:549–562, 1969.

75. Furness, J. B. and M. Costa. Identification of gastrointestinal neurotransmitters. In *Handbook of Experimental Pharmacology*. Bertaccini, ed. Springer Verlag, Berlin, pp 383–460, 1982.

76. Furness, J. B. and M. Costa. Projections of intestinal neurones showing immunoreactivity for vasoactive intestinal polypeptide are consistent with these neurones being the enteric inhibitory neurones. *Neuroscience Letters* 15:199–204, 1979.

77. Furness, J. B., M. Costa, and J. H. Walsh. Evidence for and significance of the projection of VIP neurons from the myenteric plexus to the taenia colin in the guinea pig. *Gastroenterology* 80:1557–1561, 1981.

78. Gaginella, T. S., H. S. Mekhjian, and T. M. O'Dorisio. Vasoactive intestinal peptide: quantification by radioimmunoassay in isolated cells, mucosa and muscle of the hamster intestine. *Gastroenterolgy* 74:718–721, 1978.

79. Gaginella, T. S., T. M. O'Dorisio, and K. A. Hubel. Release of vasoactive intestinal polypeptide by electrical field stimulation of rabbit ileum. *Regulatory Peptides* 2:165–174, 1981.

80. Gershon, M. D. The enteric nervous sytems. *Am. Rev. Neurosc.* 4:227–272, 1981.

81. Ghater, M. G., M. N. Sheppard, D. J. O'Shaughnessy, T. E. Adriaan, G. P. McGregor, J. M. Polak and S. R. Bloom. Regulatory peptides in the mammalian respiratory tract. *Endocrinology* 111:1248–1254, 1982.

82. Gibbins, I. L. Lack of correlation between ultrastructural and pharmacological types of non-adrenergic autonomic nerves. *Cell Tissue Res.* 221:551–581, 1982.

83. Gibson, A. and J. F. Tucker. The effects of vasoactive intestinal polypeptide and of adenosine-5′-triphosphate on the isolated anococcygeus muscle of the mouse. *Br. J. Pharmacol.* 77:97–104, 1982.

83a. Gillespie, J. H. The biological significance of the linkages in adenosine triphosphoric acid. *J. Physiol. (Lond.)* 80:345–359, 1934.

84. Gillespie, J. S. Non-adrenergic, non-cholinergic inhibitory control of gastrointestinal motility. In *Motility of the Digestive Tract.* M. Wienbeck, ed. Raven Press, New York, pp 51–66, 1982.
85. Gillespie, J. S. The rat anococcygeus muscle and its response to nerve stimulation and to some drugs. *Br. J. Pharmacol. 45:404–416, 1972.*
86. Gillespie, J. S. and J. C. McGrath. The response of the cat anococcygeus muscle to nerve or drug stimulation and a comparison with the rat anococcygeus. *Br. J. Pharmacol.* 50:109–118, 1974.
87. Goyal, R. K. and S. Rattan. VIP as a possible neurotransmitter of non-cholinergic, non-adrenergic inhibitory neurons. *Nature* 288:378–380, 1980.
88. Goyal, R. K. and S. Rattan. Neurohumoral and drug receptors for the lower esophageal sphincter. *Gastroenterology* 74:598–619, 1978.
89. Hirst, G. D. S., M. E. Holman and H. C. McKirdy. Two descending nerve pathways activated by distention of guinea pig small intestine. *J. Physiol. (Lond.)* 244:113–127, 1975.
90. Holman, M. E. and J. Hughes. Inhibition of intestinal smooth muscle. *Aust. J. Exp. Biol. Med. Sci.* 43:277–290, 1965.
91. Holman, M. E. and J. P. Weinreich. The effects of calcium and magnesium on inhibitory junctional transmission in smooth muscle of guinea pig small intestine. *Pflueger's Arch.* 360:109–119, 1975.
92. Huizinga, J. D. and A. Den Hertog. Inhibition of fundic strips from guinea pig stomach: the effect of theophylline on responses to adenosine, ATP and intramural nerve stimulation. *Eur. J. Pharmacol.* 63:259–265, 1980.
93. Hulme, M. E. and Weston, A. H. Some effects of dipyridamole, hexobendine and lidoflazine on inhibitory processes in rabbit duodenum. *Br. J. Pharmacol.* 50:609–611, 1974.
94. Hunt, W. B., D. G. Parsons, A. Wahid and J. Wilkinson. Influence of 2-2′-pyridylisatogen to sylate on responses by ATP and by neural stimulation on the rat gastric corpus. *Br. J. Pharmacol.* 63:378–379p, 1978.
95. Imaizumi, M. and K. Hama. An electron miscropic study on the interstitial cells of the gizzard of the Love Bird (Uronloncha domestica). *Z. Zellforsch. Mikrosk. Anat.* 97:351–357, 1969.
96. Irvin, C. G. R., R. Boileau, J. Tremblay, R. Martin, and P. T. Macklem. Bronchodilation: non-cholinergic, non-adrenergic mediation demonstrated in vivo in the cat. Science 207:791–792, 1979.
97. Israel, M. and F. M. Meunier. The release of ATP triggered by transmitter action and its possible physiological significance retrograde transmission. *J. Physiol. (Paris)* 74:485–490, 1978.
98. Ito, Y. and K. Takeda. Non-adrenergic inhibitory nerves and putative transmitters in the smooth muscle of cat trachea. *J. Physiol. (Lond.)* 330:497–511, 1982.
99. Jaffer, S. S., J. T. Farrar, W. M. You, and G. M. Makhlouf. Mode of action and interplay of vasoactive intestinal peptide (VIP), secretin and octapeptide of cholecystokinin (Octa- CCK) on duodenal and ileal muscle in vitro. *Gastroenterology* ,66(4):716, 1974.
100. Jager, L. P. Effects of purinergic compounds on excitable membranes. In *Physiological and Regulatory Functions of Adenosine and*

Adenosine Nucleotides. H. P. Baer and G. I. Drummond, eds. Raven Press, New York, pp 369–376, 1979.

101. Jager, L. P. and J. A. M. Shevers. A comparison of effects evoked in guinea pig taenia coli by purine nucleotides and by purinergic nerve stimulation. *J. Physiol. (Lond.)* 299:75–83, 1980.

102. Johns, A. The effect of indomethacin and substance P on the guinea pig urinary bladder. *Life Sci.* 29:1803–1809, 1981.

103. Johns, A. The effect of magnesium-free solutions on the responses of the guinea pig urinary bladder to adenosine 5′-triphosphate and nerve stimulation. *Can. J. Physiol. Pharmacol.* 57:1320–1323, 1979.

104. Jones, T. R., M. S. Kannan, and E. E. Daniel. Ultrastructural study of guinea pig tracheal smooth muscle and its innervation. *Can. J. Physicl. Pharmacol.* 581(8):974–983, 1982.

105. Kachelhoffer, J., C. Mendel, J. Dachel, D. Hohmatter, and J. F. Grenier. The effect of VIP on intestinal motility. Study on ex vivo perfused isolated canine jejund loops. *Am. J. Digest. Dis.* 21:957–962, 1976.

106. Kuchii, M. J. T. Miyahara, and S. Shibata. [^{3}H]-adenine nucleotide and [^{3}H]-noradrenaline release evoked by electrical field stimulation, perivascular nerve stimulation and nicotine from the taenia of the guinea pig caecum. *Br. J. Pharmacol.* 49:258–267, 1973a.

107a. Kuchii, M., J. T. Miyahara and S. Shibata. [^{3}H]-adenosine nucleotide and [^{3}H]-nonadrenaline uptake by cold stored guinea pig taenia caecum; mechanical effects and release of [^{3}H]-adenosine nucleotide by noradrenaline, papaverine and nitroglycerine. *Br. J. Pharmacol.* 49:642–650, 1973b.

107. Kuriyama, H., T., Osa and N. Toida. Nervous factors influencing the membrane activity of intestinal smooth muscle. *J. Physiol. (Lond)* 191:257–270, 1967.

108. Kuroda, Y. and H. Mcilwain. Uptake and release of [^{14}C]- adenine derivative at beds of mammalian cortical synaplosomes in a superfusion system. *J. Neurochem.* 22:691–699, 1974.

109. Langley, J. N. and R. Magnus. Some observations of the movements of the intestine before and after degenerative section of the mesenteric nerves. *J. Physiol. (Lond.)* 33:34–51, 1905.

110. Larsson, L. J. Ultrastructural localization of a new neuronal peptide (VIP). *Histochemistry* 54:173–176, 1977.

111. Larsson, L. J., L. Edvinsson, J. Fahrenkrug, R. Hakanson, C. Owman, O. Shaffaletzky De Muckadell, and F. Sundler. Immunohistochemical localization of a vasodilatory polypeptide (VIP) in cerebrovascular nerves. *Brain Res.* 113:440–404, 1976a.

112. Lee, T. J. F., W. R. Hume, C. Su, and J. A. Bevan. Neurogenic vasodilation of cat cerebral arteries. *Circ. Res.* 42:535–542, 1978.

113. Lee, T. J. F. Cholinergic mechanism in the large cat cerebral artery. *Circ. Res.* 50:870–879, 1982.

114. Luchelli-Fortis, M. A., B. B. Fredholm, and S. Z. Langer. Release of radioactive purines from cat nictitating membrane labelled with ^{3}H-adenine. *Eur. J. Pharmacol.* 58:389–392, 1979.

115. Lundberg, J. M. Evidence for occurrence of vasoactive intestinal polypeptide (VIP) in neurons of cat exocrine glands. Morphological, biochemical and functional studies. *Acta Physiol. Scand. Suppl.* 496:1–57, 1981a.

116. Lundberg, J. M. Evidence for coexistence of vasoactive intestinal polypeptide (VIP) and acetylcholine in neurons of cat exocrine glands; morphological, biochemical and functional studies. *Acta Physiol. Scand. Suppl.* 496:81–101, 1981b.

117. Lundberg, J. M., A. Anggard, P. Emson, J. Fahrenkrug, and J. Hokfelt. Vasoactive intestinal polypeptide and cholinergic mechanism in cat nasal mucosa: studies on choline acetyltransferase and release of vasoactive intestinal polypeptide. *Proc. Natl. Acad. Sci. USA* 78:5255–5259, 1981a.

118. Lundberg, J. M., A. Anggard, and J. Fahrenkrug. Complementary role of vasoactive intestinal polypeptide (VIP) and acetylcholine for cat submandibular gland blood flow and secretion. I. VIP release. *Acta Physiol. Scand.* 113:317–327, 1981b.

119. Maas, A. J. J. The effect of apamin on responses evoked by field stimulation in guinea pig taenia coli. *Eur. J. Pharmacol.* 73:1–9, 1981.

120. Maas, A. J. J. and A. Den Hertog. The effect of apamin on the smooth muscle cells of the guinea pig taenia coli. *Eur. J. Pharmacol.* 58:151–156, 1979.

121. Mackenzie, J. and G. Burnstock. Evidence against vasoactive intestinal polypeptide being the non-adrenergic, non-cholinergic inhibitory transmitter released from nerves supplying the smooth muscle of the guinea pig taenia coli. *Eur. J. Pharmacol.* 67:255–267, 1980.

122. Martinson, J. Studies on the efferent vagal control of the stomach. *Acta Physiol. Scand.* (Suppl. 255):1–65, 1965.

123. Martinson, J. Nervous control of gastroduodenal motility and emptying. *Scand. J. Gastroenterology* (Suppl. 10): 31–44, 1975.

123a. Matsuzaki, Y., Y. Hamasaki, and S. I. Said. Vasoactive intestinal peptide: a possible transmitter of non-adrenergic relaxation of guinea pig airways. *Science* 210:1252–1253, 1980.

124. McCullock, J. T. and L. Edvinsson. The effects of vasoactive intestinal polypeptide in trial arteriolar calibre, cerebral blood flow, cerebral oxygen consumption and the electroencephalogram. *Am. J. Physiol.* 238:449–456, 1980.

125. McKay, D. and H. C. McKirdy. Effect of vasopressin and of adenosine triphosphate in the fleet preparation of rabbit rectum. *Br. J. Pharmacol.* 44:366–367P, 1972.

126. McKirdy, H. C. Functional relationship of longitudinal and circular layers of the muscularis externa of the rabbit large intestine. *J. Physiol. (Lond.)* 227:839–853, 1972.

126a. Morgan, K. G., P. F. Schmalz, J. H. Szursweski. The inhibitory effects of vasoactive intestinal polypeptide on the mechanical and electrical activity of canine antral smooth muscle. *J. Physiol. (Lond.)282:437–450, 1978.*

127. Muller, M. J. and H. P. Baer. Apamin, a non-specific antagonist of smooth muscle relaxants. *Naunyn-Schmiedeberg's Arch. Pharmacol.* 311:105–107, 1980.

128. Mutt, V. and S. J. Said. Structure of the porcine vasoactive intestinal octocosapeptide. *Eur. J. Biochem.* 42: 581–589, 1974.

129. Neya, T., T. Yamasato, M. Tokaki, M. Mizutami and S. Nakayama. Excitatory and inhibitory effects of vasoactive intestinal polypeptide

(VIP) on the isolated jejunum and internal anal sphincter in the guinea pig. *Biomed. Res.* 2:398–403, 1981.

130. Nilsson, G., K. Dahlberg, E. Brodir, F. Sundler, and K. Strandberg. Distribution and constrictor effect of substance P in guinea pig tracheo-bronchial tissue. In *Substance P.* U. S. Von Euler and B. Pernow, eds. Raven Press, New York, pp 75–81, 1977.
131. Northway, M. N. and T. F. Burks. Stimulation of cholinergic nerves in dog intestine by adenine nucleotides. *Eur. J Pharmacol.* 65: 11–19, 1980.
132. Ohga, A. and T. Taneike. Dissimilarity between the responses to adenosine triphosphate or its related compounds and non-adrenergic inhibitory nerve stimulation in the longitudinal smooth muscle of pig stomach. *Br. J. Pharmacol.* 60:221–231, 1977.
133. Okwuasaba, F. K., J. T. Hamilton, and M. A. Cook. Relaxations of guinea pig fundic strip by adenosine, adenine nucleotides and electrical stimulation; antagonism by theophylline and desensitization to adenosine and its derivatives. *Eur. J. Pharmacol.* 46:181–198, 1977.
134. Paton, W. D. M. and M. A. Zar. Mechanism of acetylcholine release from parasympathetic nerves. *J. Physiol. (Lond.)* 194:13–33, 1968.
135. Rattan, S. and R. K. Goyal. Neural control of the lower esophageal sphincter. Influence of the vagus nerves. *J. Clin. Invest.* 54:899–906, 1974.
136. Rattan, S. and R. K. Goyal. Evidence against purinergic inhibitory nerves in the vagal pathway to the opossum lower esophageal sphincter. *Gastroenterology* 78:898–904, 1980.
137. Rattan, S., M. Grey, and R. K. Goyal. Vasoactive intestinal peptide causes peristaltic contractions in the esophageal body. *Life Sciences* 30:1557–1563, 1982.
138. Richardson, J. B. Nerve supply to the lungs. *Am. Rev. Resp. Dis.* 119:785–809, 1979.
139. Richardson, J. and J. Beland. Non-adrenergic inhibitory nervous system in human airways. *J. Appl. Physiol.* 4:764–771, 1976.
140. Rikimaru, A., Y. Fukushi and T. Suzuki. Effects of imidazole and phentolamine on the relaxant responses of guinea pig taenia coli to transmural stimulation and to adenosine triphosphate. *Tokoku Med. J.* 105:199–200, 1971.
141. Roman, C. Nervous control of esophageal and gastric motility. In *Mediators and Drugs in Gastrointestinal Motility. I. Handbook of Experimental Pharmacology.* Vol. 59/I. G. C. Bertaccini, ed. Springer Verlag, Berlin, pp 223–228, 1982.
142. Rosell, S., L. Olgart, B. Gazelius, P. Panopoulos, K. Folkers, and J. Horig. Inhibition of antidromic and substance P-induced vasodilation by a substance P antagonist. *Acta Physiol. Scand.* 111:381–382, 1981.
143. Roth, J., M. Bendayan, and L. Orci. Utrastructural localization of intracellular antigens by use of protein A gold complex. *J. Histochem. Cytochem.* 26:1074–1081, 1978.
144. Rumesson, J. J. and L. Thuneberg. Plexus muscularis profondus and associated interstitial cells. I. Light microscopical studies of mouse small intestine. *The Anat. Res.* 203:115–129, 1982a.

145. Rumesson, J. J., L. Thuneberg and H. B. Mikkelsen. Plexus muscularis profundus and associated interstitial cells. II. Ultrastructural studies of mouse small intestine. The Anat. Res. 203:125–146, 1982b.

146. Rutherfold, A. and G. Burnstock. Neuronal and non-neuronal components in the overflow of labelled adenyl compounds from guinea pig taenia coli. *Eur. J. Pharmacol.* 48: 195–202, 1978.

146a. Ryan, J. P. and S. Ryave. Effect of vasoactive intestinal polypeptide on gall bladder smooth muscle in vitro. *Am. J. Physiol.* 234: E44–46, 1978.

147. Saito, K. Effects of extracellularly applied ATP and its related nucleotides on the membrane potential of the guinea pig taenia coli. *Jap. J. Smooth Muscle Res.* 8:32–39, 1972.

149. Satchell, D. G. and G. Burnstock. Comparison of the inhibitory effects of the guinea pig taenia coli of adenine nucleotides and adenosine in the presence and absence of dipyridamole. *Eur. J. Pharmacol.* 32:324–328, 1975.

150. Satchell, D. G., A. Lunch, P. M. Bourke, and G. Burnstock. Potentiation of the effects of exogenously applied ATP and purinergic nerve stimulation on the guinea pig taenia coli by dipyridamole and hexobendine. *Eur. J. Pharmacol.* 19:343–350, 1972.

151. Schaffaletzky De Muckadell, O. B., J. Fahrenkrug, J. Holst and K. B. Lauritsen. Release of vasoactive intestinal polypeptide (VIP) by intraduodenal stimuli. *Scand. J. Gastroent.* 12:793–799, 1977.

152. Schultzberg, M., T. Hokfelt, G. Nilsson, L. Terenius, J. F. Rehfeld, M. Brown, R. Elder, M. Goldstein, and S. Said. Distribution of peptide and catecholamine neurons in the gastrointestinal tract of rat and guinea pig: immunohistochemical studies with antisera to substance P, VIP, enkephalins, somatostatin, gastrin, neurotensin and dopamine beta-hyroxylase. *Neuroscience* 5:689–744, 1980.

153. Shuba, M. F. and I. A. Vladimirova. Effect of apamin on the electrical responses of smooth muscle to adenosine 5′-triphosphate and to non-adrenergic, non-cholinergic nerve stimulation. *Neuroscience* 5:853–859, 1980.

154. Silinsky, E. M. On the association between transmitter secretion and the release of adenine nucleotides from mammalian motor nerve terminals. *J. Physiol. (Lond.)* 247:145–162, 1975.

155. Small, R. C. Electrical slow waves and tone of guinea pig isolated trachealis muscle. Effects of drugs and temperature changes. *Br. J. Pharmacol.* 77:45–54, 1982.

156. Small, R. C. and A. H. Weston. Intramural inhibition in rabbit and guinea pig intestine. In *Physiological and Regulatory Functions of Adenosine and Adenine Nucleotides.* H. P. Baer and G. J. Drummond, eds. Raven Press, New York, pp 45–61, 1979.

157. Spedding, M. and D. F. Weetman. Identification of separate receptors for adenosine and adenosine-5′-triphosphate in causing relaxations of the isolated taenia of the guinea pig ceacum. *Br. J. Pharmacol.* 57:305–310, 1976.

158. Spedding, M., A. Sweetman and D. F. Weetman. Antagonism of adenosine-5′-triphosphate-induced relaxation by 2-2′-pyridylisa-

togen in the taenia of guinea pig ceacum. *Br. J. Pharmacol.* 53:575–583, 1975.

159. Stockley, H. C. 2-2′-pyridylisatogen antagonizes adenosine 5′-triphosphate but not nerve-mediated relaxations in human isolated taenia coli. In *Gastrointestinal Motility in Health and Disease* (H. L. Duthie, ed.) University Park Press, Baltimore, pp 145–150, 1978.

160. Su, C., J. Bevan and G. Burnstock. [^{3}H]-adenosine triphosphate: release during stimulation of enteric nerves. *Science* 173:336–338, 1971.

161. Sundler, F., R. Hakanson, L. D. Larsson, E. Brodin, and G. Nilsson. Substance P in the gut: an immunochemical and immunohistochemical study of its distribution and development. In *Substance P.* U. S. Von Euler and B. Pernow, eds. Raven Press, New York, pp 59–65, 1977.

162. Suzuki, T., Y. Fukushi, and A. Rikimaru. Relaxant effect of adenosine triphosphate and its related nucleotides on the guinea pig taenia coli. *Jap. J. Smooth Muscle Res.* 7:207–212, 1971.

163. Thuneberg, L. The interstitial cells of Cajal: intestinal pacemaker cells? In *Advances in Anatomy, Embryology and Cell Biology.* Springer Verlag, Berlin, pp 1–30, 1982.

164. Tomita, J. Conductance change during the inhibitory potential in the guinea pig taenia coli. *J. Physiol. (Lond.)* 225:693–703, 1972.

165. Tomita, J. and H. Watanabe. A comparison of the effects of adenosine triphosphate with noradrenaline and with the inhibitory potential of the guinea pig taenia coli. *J. Physiol. (Lond.)* 231:167–177, 1973.

166. Tonini, M., L. Onori, G. M. Grigo, S. Lecchini, L. D'Angelo, and A. Crema. Non-adrenergic inhibition of the longitudinal muscle of rabbit distal colon may not be mediated by purinergic nerves. *J. Pharm. Pharmacol.* 33:536–537, 1981.

167. Uddman, R., J. Alumets, L. Edvinsson, R. Hakanson, and F. Sundler. Occurrences and distribtuion of VIP nerves in the nasal mucosa and tracheobronchial wall. *Acta Otolaryngol. (Stockholm)* 86:443–448, 1978a.

168. Uddman, R., J. Alumets, L. Edvinsson, R. Hakanson and F. Sundler. Peptidergic (VIP) innervation of the esophagus. *Gastroenterology* 75:5–8, 1978b.

169. Uddman, R., J. Fahrenkrug, L. Malm, J. Alumets, R. Hakanson, and F. Sundler. Neuronal VIP in salivary glands: distribution and release. *Acta Physiol. Scand.* 110:31–38, 1980.

170. Uddman, R., J. Alumets, L. Edvinsson, R. Hakanson and F. Sundler. VIP nerve fibers around peripheral blood vessels. *Acta Physiol. Scand.* 112:65–70, 1981.

171. Vladimirova, J. A. and M. F. Shuba. The effect of strychnine, hydrastin and apamin on synaptic transmission in smooth muscle cells. *Neirofiziologiya, Kiev* 10:295–299, 1978.

172. Wei, E. P., H. A. Kontos, and S. I. Said. Mechanism of action of vasoactive intestinal polypeptide on cerebral blood flow. *Am. J. Physiol.* 239:765–768, 1980.

173. Weston, A. H. The effect of desensitization to adenosine triphosphate on the peristaltic reflex in guinea pig ileum. *Br. J. Pharmacol.* 47:606–608, 1973.

174. White J. D. Direct detection of depolarization induced release of ATP from a synaptosomal preparation. *Nature (Lond.)* 267:67–68, 1977.

175. White, T., P. Potter, C. Moody and G. Burnstock. Tetrodotoxin resistant release of ATP from guinea pig taenia coli and vas deferens during electrical field stimulation in the presence of luciferin-luciferase. *Can. J. Physiol. Pharmacol.* 59:1094–1100, 1981.

176. Yip, P., B. Polembini, and R. F. Coburn. Inhibitory innervation to the guinea pig trachealis muscle. *J. Appl. Physiol.* 50:374–382, 1981.

177. Zimmerman, H. an V.P. Whittaker. Effect of electrical stimulation in the yield and composition of synaptic vesicles from the cholinergic synapsis of the electric organ of Torpedo: a combined biochemical, electrophysiological and morphological study. *J. Neurochem.* 22:435–450, 1974.

178. Zimmerman, H. and C. R. Denston. Adenosine triphosphate in cholinergic vesicles isolated from the electric organ of Electrophorus electricus. *Brain Res.* 111:365–376, 1976.

Chapter 16

Nonadrenergic, Noncholinergic (NANC) Neuronal Excitatory Interactions with Smooth Muscle

E. E. Daniel

1. Existence of NANC Excitatory Nerves in Gut and Elsewhere

1.1. Introduction

Although clear pharmacological evidence of excitatory NANC nerves from studies of mechanical activity existed by the early 1970s (*1, 2, 4*), evidence from electrophysiological studies has been obtained only recently (e.g., in vas deferens, *12, 46, 110*, and blood vessels, *72, 75*). So far only mechanical studies (*52, 54*) and limited electrophysiological data (*10, 22, 60*) support the existence of such nerves in the intestine.

Acceptance of the pharmacological evidence for distinct NANC excitatory transmitters (*see* Chapter 15) was questioned by many years on anatomical grounds, i.e., the occurrence of close nerve muscle contacts in some smooth muscles with "NANC" excitatory responses (e.g., vas deferens, *59*). The narrow synaptic cleft in these cases results in very high mediator concentrations ($\geq 10^{-3}M$) at postsynaptic receptors that, together with failure of the antagonist to equilibrate with the cleft fluid, might account for resistance to atropine or to alpha-adrenergic antagonists. Reasons for rejecting this notion are considered below.

Another difficulty was acceptance of the possibility that the NANC excitatory transmitter was a cotransmitter with the classical mediators. In both vas deferens (guinea pig and rat) and in mesenteric arterioles of mucosa of guinea pig ileum, there are two components to the response to field stimulation of what are presumed to be adrenergic nerves. In guinea-pig and rat vas deferens (*12, 46, 110*), there is an early response to single or to multiple stimuli that consists of an ejp (excitatory junction potential) or a series of summated ejps that can evoke a spike associated with the twitch-like early contractile response. This is followed by an electrically silent period accompanying the slow or sustained late tonic response. The two mechanical components were pharmacologically distinct; the first was inhibited by nifedipine or purine antagonists, but not alpha-adrenergic antagonists, and the second was inhibited by alpha-adrenergic antagonists, but not purine antagonists or nifedipine (*1, 12, 102, 110*). Furthermore, low doses of alpha-2 adrenoceptor antagonists potentiated both components (*12, 46, 110*), while 6-OH-dopamine- (but not reserpine-) induced sympathectomy abolished both components. Sneddon et al. (*110*) postulated that the sympathetic nerve endings contain and release purine and norepinephrine that act as cotransmitters. Combined treatment with a photo-affinity label for purine excitatory receptors, arylazidoaminopropionyl ATP, ($ANAPP_3$), and alpha-adrenoceptor antagonists abolished all the responses, except a small cholinergic excitatory response.

If these findings are correct it is clear that antagonists (e.g. $ANAPP_3$) can reach effective concentrations in the cleft and that failure of inhibition of response by adrenergic antagonists is caused by the release of a nonadrenergic mediator from adrenergic nerves rather than the release of norepinephrine in overwhelming amounts. In rabbit bladder, there is also an initial fast ejp resistant to adrenergic and cholinergic antagonists, and it is followed by a slow depolarization inhibited in this case by atropine (*33*). The fast ejps readily produce spikes that are susceptible to Ca channel antagonists. Effects of selective purine antagonists or tachyphylaxis to purines were not tested on the fast ejp (see below), the effects of denervation were not examined with respect to elimination of one or both components, and the mediator is uncertain (see below).

A NANC excitatory cotransmitter may exist in blood vessels (arterioles of gut mucosa) (*75, 76, 77*) in that cells apparently adjacent to sympathetic nerve endings produced ejps that could initiate spikes and contractions. None of these responses were susceptible to alpha-adrenoceptor antagonists. Iontophoretically applied norepinephrine caused contractile responses, presumably acting on cells distant from nerve endings that were not accompanied by changes in membrane potential, but were highly susceptible to alpha-adrenoceptor antagonists. Hirst and Neild (*75, 76, 78*) interpreted their findings to indicate the existence of an additional class of excitatory adrenoceptor, insensitive to alpha-adrenoceptor antagonists and localized near sympathetic nerve endings. An alternate interpretation is that the sympathetic nerves in arterioles contain a comediator along with norepinephrine. Its receptors may be localized near nerve endings, or its degradation may be so rapid that it fails to reach more distant cells (see later for full discussion).

Ultimately the correctness of these identifications of nerves with NANC cotransmitters will depend on the fulfillment of all the criteria outlined above (Chapter 15) for both transmitters. The occurrence of cotransmission between classical peripheral nerve mediators (acetylcholine or norepinephrine) and peptides or purines complicates detection of a NANC-mediated nerve response by resulting in a composite response partially susceptible to the well-known antagonists or cholinergic or adrenergic responses. We should examine the possibility of cotransmission carefully in such cases. In addition, the occurrence of transmitter release in the periphery from afferent nerves activated antidromically must also be taken into account. The evidence discussed below is already strong that substance P, a transmitter of some sensory nerves, can be released from peripheral terminals derived from dorsal root neurones. It is unclear at present whether these findings should be considered as artifacts of the experimental arrangements or as evidence for previously unrecognized functions of afferent nerves.

1.2. Gut

As in the case of inhibitory nerves, enteric NANC excitatory nerves were first clearly recognized in studies by Ambache and colleagues (*1, 3*). Field stimulation of guinea-pig small intestine could elicit atropine-resistant, nerve-mediated contractions. Franco et al. (*52, 53, 54*) showed that these mechanical responses of longitudinal muscle of guinea-pig intestine were at least in part caused by probable release of a substance P-like material from nerves that were still present after extrinsic denervation and projected orally and anally in the intestine. It is possible that, in other species such as dog, the intestine also receives substance P-containing extrinsic nerves (*15, 39, 57, 61, 68, 69*), since local denervation, bilateral splanchnic section, dorsal root ablation, or rhizotomy as well as vagotomy reduces the densities of varicose structures containing substance P-like material in various regions; however, changes in substance P contents of gut after denervation remain to be determined. So far effects attributed to the release of substance P or substance P-like activity on nerve stimulation have been demonstrated only in a few cases in vitro. However, the existence of selective (though weak with some agonist properties) substance P antagonists (*96*) may change this. Already Leander et al. (*96*) have provided evidence that the "off" contraction after field stimulation in some organs (e.g., guinea-pig taenia caecum) may be mediated by substance P release by using one of these antagonists. Domoto et al. (*32*) showed that the tonic "off" contraction of muscularis mucosa of opossum esophagus may be mediated by substance P. The initial phasic "on" contraction was apparently mediated by acetylcholine that also inhibited the release of substance P by an action on muscarinic receptors. Whether the origin of substance P-like material is from intrinsic nerves (as in the studies of Franco et al., *52, 53*) or from peripheral branches of afferent neurons is not clear in all cases (see below). Another example of NANC excitatory enteric nerves is in the proximal colon of the guinea pig, but the mediator in this case is unlikely to be substance P since desensitization of its receptors had little effect (*26, 29, 58*).

Electrophysiological evidence of enteric NANC excitatory nerves is sparse. Maas and Den Hertog (*98, 99*), Maas (*97*), Bywater et al. (*22*), and Maas et al. (*100*) found atropine-insensitive depolarization of circular muscle that could be demonstrated after the ijp had been blocked either by hyperpolarization of the cell membrane or by the K^+-conductance blocker, apamin, and that was thus not "rebound" from hyperpolarization. The mediator of this NANC ejp was not identified, but it was not an adrenergic agonist, ATP, or a prostaglandin (*97–99*). More recently, Bauer and Kuriyama (*10*) reported NANC ejps in both longitudinal and circular muscle of guinea-pig ileum. It is unclear whether criteria (longitudinal muscle active; circular muscle inactive) used to distinguish cells from each layer were adequate (especially after pharmacological interventions that inhibited activity in both layers). Bauer and Kuriyama (*10*) reported that in atropine (1–2 μM) and guanethedine (10 μM), in addition to NANC ijps, both muscle layers generated NANC ejps. They also reported that the NANC ejp has a lower threshold and a shorter latency than the ijp and required higher TTX and Mg^{2+} concentrations for complete suppression. The NANC ejps showed less decline in amplitude with repetitive stimulation at low frequency (<0.5 Hz) than did ijps, but both junction potentials summed with frequencies of 2–20 Hz.

1.3. Urogenital Muscle

Both mechanical and electrophysiological evidence support the existence of excitatory NANC nerves in the muscles. The urogenital tract (detrusor of bladder, vas deferens, bladder base, urethra to some degree) possess NANC nerve-mediated responses in a variety of species (*see* ref. *63* for review). The origin of the nerves involved is unclear; there are postganglionic cholinergic and adrenergic nerves arising from ganglia in the adventitia of the urethrovesical junction (*11, 20, 59*). There are also sensory nerves (*see* ref. *50*). The only experiments that seem to have been directed primarily to determining which of these nerve types contribute to nonadrenergic, noncholinergic innervation of the vas deferens and other urogenital muscles are those with reserpine (no effect on NANC responses) and 6-hydroxydopamine (marked inhibition leaving only cholinergic responses, see later). Chemical sympathectomy to measure directly whether release of putative NANC mediators such as ATP is abolished along with that of norepinephrine has not been done. In the case of bladder detrusor there is sparse adrenergic innervation of muscle in most species (*13, 41, 50*) and presumably the NANC nerves are either cholinergic (with a NANC cotransmitter), sensory, or nonadrenergic excitatory in nature. In vas deferens, on the other hand, it is generally agreed (*see* refs. *47, 48*) that the motor nerves are primarily, if not exclusively, adrenergic. The electrophysiological evidence that, in this case, there is release of a NANC cotransmitter from sympathetic nerves, has been outlined above. However, the role of sensory nerves as a source of NANC excitatory mediator has not been critically evaluated at any site. Recent evidence (e.g., *57, 61, 68, 69*) suggests that substance P-containing neurons from dorsal root ganglia travel with sympathetic nerves to innervate the coeliac, superior and inferior mesenteric ganglion sympathetic neurons and send fibers to genitourinary tract as well as the gut and blood vessels.

As mentioned above, the existence of close nerve–muscle contacts (<20 nm) in vas deferens (*59*) and bladder (*34*) delayed acceptance of the existence of NANC excitatory nerves to those organs. It allowed the arguments that antagonists might not attain sufficient concentrations in the synaptic cleft to oppose (*35*) effectively the high concentrations ($>10^{-3}M$) of mediator reached there or that the receptors associated with the postsynaptic membrane in such narrow clefts might be different from those elsewhere in the muscle. It was presumed, but not established, that the primary mediator or its analogs rather than a cotransmitter or alternate transmitter acted on these putative anomalous receptors. However, the anatomical explanation (narrow synaptic clefts) of nerve responses that resisted antagonists to classical mediators was found to be inconsistent with a variety of evidence. Antagonists such as atropine readily antagonized a constant proportion of the total response to nerve stimulation as well as very high concentrations of added acetylcholine ($10^{-3}M$), but failed in any reasonable concentration to antagonize the remaining response. Inhibition similar to that with atropine was produced by blockade of choline uptake into cholinergic nerves by hemicholinium. Furthermore, the responses of the guinea-pig bladder to single or multiple pulses of field stimulation were similarly atropine-resistant even though these procedures should release markedly different concentration of acetylcholine into the synaptic cleft (*2*). Also physostigmine had no effect on bladder contractions when field stimulation was carried out in the presence of atropine and hexamethonium, showing that effects of endogenous acetylcholine could be blocked.

The extent of atropine- or hyoscine-insensitivity of field-stimulated contractions varies in the bladder of different species. About 60% of these field-stimulated responses of rat and rabbit bladder are resistant (*24, 25*), but those in the guinea-pig bladder are almost totally resistant. There is no effect of adrenergic antagonists or guanethedine, except enhancement in some cases, on atropine-resistant contractions. Thus, there is clear evidence for a NANC innervation of bladder muscle. Essentially similar evidence exists in all mammalian species studied, including humans (*30*). One puzzling feature in many studies is the existence of a TTX-insensitive component of the atropine-resistant contraction (*see 2, 4, 30*). In the human detrusor the atropine-resistant components is mostly TTX-insensitive and unaffected by depletion of synaptic vesicles by scorpion venom (*30, 34*).

The electrophysiological mechanisms for NANC actions on bladder muscle are not fully studied (*see 31–33*). Acetylcholine does not produce notable depolarization, nor does norepinephrine produce hyperpolarization in guinea-pig bladder. The effects of field stimulation using parameters susceptible to

TTX are not fully studied, but in rabbit bladder (presumably detrusor), it produces an initial ejp followed by a late, slow depolarization. Sometimes the two depolarizations are separated by a hyperpolarization, sensitive to K^+ conductance blockers such as apamin or tetraethylammonium ion. The late depolarization was enhanced by pretreatment with neostigmine and abolished by atropine, but the initial ejp was unaffected by these agents and by guanethedine, phentolamine, methysergide, mepyramine, quinidine, theophylline, and dipyridamole. ATP in high concentration depolarized the membrane, reduced membrane resistance and reduced the ejp; it failed to affect the kinetics of decline of ejp amplitude on repetition of field stimulation. The mechanisms of that decline are unknown in this tissue, and the data are inconclusive, but do not favor ATP as the noncholinergic excitatory transmitter. More complete information is available about the vas deferens and is described below.

2. Identity of NANC Excitatory Transmitters

2.1. Role of Substance P in Gut

In contrast to NANC inhibitory nerves, which are established by mechanical, electrophysiological, and to some degree by pharmacological analyses, the existence of NANC excitatory nerves is still controversial. In vivo the evidence for their existence in the sense of playing a significant physiological role is scanty. Nearly all excitatory responses to vagal stimulation or local field stimulation are abolished by atropine or hyoscine (e.g., *35*). An exception is the "off" contraction after field stimulation of the body and lower esophageal sphincter that, on some species, is not abolished by atropine (*66, 67*).

In vitro a substantial body of evidence exists that high frequency field stimulation can activate longitudinal muscle of guinea-pig intestine by an atropine-insensitive mechanism possibly mediated by substance P (see below). In brief, there is clear evidence of the existence of intrinsic substance P-containing nerves (*53, 104, 105*), the release of substance P, and the resemblance of its action on muscle to that of the NANC excitatory mediator; furthermore, recently there is supportive evidence from an analog of SP (D-PRO^2-D-TRP^{7,9}-SP), an apparent competitive antagonist of substance P from afferent nerves. This antagonist selectively inhibited the "off" contraction in several muscles. Capsaicin, see below, caused both an hyoscine-sensitive indirect contraction of guinea-pig intestine and a direct contraction (presumably by release of substance P) both in myenteric plexus and on smooth muscle. Two major questions exist: (1) Is the release of substance P and its action on muscle an artifact of the in vitro state? (2) Is the source of substance P exclusively intrinsic nerves or do afferent nerves from dorsal root ganglia, the vagus, or other sources provide a source of part of this material?

There is increasingly clear evidence that substance P plays a role as mediator in the guinea-pig myenteric plexus, probably functioning as mediator of the slow excitation postsynaptic potential mediator (*see* ref. *91*). This mediator is released in response to repetitive stimulation and acts on a variety of myenteric neurons of both major electrophysiological types (I and II). Its action is to decrease potassium conductance. Evidence of a variety of types supports the identification of substance P as the mediator; however, it is not the function of this review to consider that subject. The existence of a role for substance P as a transmitter to myenteric neurons raises the possibility that repetitive stimulation in the commonly used myenteric plexus, longitudinal muscle preparations, or even in whole ileum under in vitro conditions leads to overflow of substance P from the plexus onto the longitudinal muscle, since there is no barrier to prevent its access. Similarly, in intact intestine studied in vitro, substance P may arrive at sites that are not directly innervated by and normally not under the physiological control of substance P-containing nerves. In fact, nerves containing substance P-like material occur in the circular muscle of the intestine although they are rare in longitudinal muscle (see below). In intestines of species with a prominent deep muscular nerve plexus, there are many such nerves concentrated in the plexus (e.g., *39*).

Critical experiments to distinguish overflow from a genuine neurotransmitter role in smooth muscle remain to be done. They would involve duplication of the critical results in vitro (see below) under physiological in vivo conditions; i.e., demonstration of an action of NANC-excitatory nerve accompanied by release of substance P on nerve stimulation, presumably into venous blood, and demonstration that, when added intra-arterially, it duplicates the action of the NANC excitatory mediator, and that various potentiators and antagonists have appropriate actions, similarly affecting nerve-mediated responses and those to substance P. Need for data obtained in vivo is more acute in view of the recent demonstration (*51*) that it is common for the major sites of action of peptides on nerves to be inoperable in vitro.

Substance P is well established as a constituent of dorsal root ganglion (*100*); sensory neurons send afferent fibers to sympathetic and other ganglia, as well as to the gastrointestinal tract. In sympathetic ganglia these fibers innervate sympathetic neurons and appear to release substance P to initiate slow epsps as in the myenteric plexus (*40, 101*). The question arises whether a similar innervation by afferent substance P-containing nerves occurs in the myenteric plexus. In the dog, local denervation of the intestine reduces the density of varicosities containing substance P-like material by half as did vagotomy (*39*). In cats, bilateral spinal ganglionectomy, rhizotomy, or bilateral splachnicectomy reduced substance P-like immunoreactivity in the gastroduodenal region (*68*). The pathway seemed to be dorsal root afferent neurons sending peripheral processes with the sympathetic supply. In the studies of Domoto et al. (*39*), vagal ligation caused accumulation of substance P proximal to the ligation, as did the ligation of local intestinal nerves. Hayashi et al. (*69*) in cats found that bilateral vagotomy reduced substance P-like material studied by immunohistochemistry in esophagus and proximal stomach myenteric plexus; a smaller reduction also occurred in the plexus of antrum and duodenum. These fibers containing substance P-like material seem to come from the nodose ganglion and central vagal nuclei neurons. Thus, extrinsic sources of substance P nerves may exist, as well as the intrinsic neurons established by tissue culture and other techniques (*see* ref. *58*). However, biochemical demonstration of reduction of gut substance P on extrinsic denervation has not yet been reported. There, thus, may be a variety of transmitter functions for substance P from different extrinsic as well as intrinsic nerves (e.g., the vascular substance P-containing nerves of gut seem to derive from dorsal root neurons) (*27, 101*).

In vitro evidence for substance P as NANC excitatory mediator has been obtained mainly in the guinea-pig intestine. This species (*27, 28, 107*), as well as other species (*39, 104, 105, 111*) have nerves containing substance P-like immunoreactivity in myenteric plexus, circular muscle, and elsewhere. Costa et al. (*27, 28*) have provided evidence that this immunoreactivity could not be attributed to the c-terminus-related peptides, bombesin, or presumably GRP. Release of a material that resembles substance P in its actions on guinea pig longitudinal muscle (selective desensitization to substance P also desensitized to the released material) has been demonstrated by TTX indicating a neural source.

Recently (*96*) a substance P antagonist (D-PRO^{2}-D-TRP^{7,9}-SP) has been shown to block NANC "off" contractions of guinea-pig taenia coli and NANC nerve-induced excitation of sphincter pupillae. The former were discussed above and evidence presented that they are probably caused by a separate NANC excitatory mediator rather than rebound from NANC-inhibitory transmission. In the muscularis mucosa of esophagus from opossum, Domoto et al. (*38*) showed that the tonic "off" contraction was mimicked by substance P and blocked by substance P taetyphylaxis or D-PRO^2-TRP^{7,9}-SP. The initial phasic "on" contraction was apparently mediated by acetylcholine release, which acted on presynaptic (inhibiting release of substance P-like mediators) as well as on postsynaptic muscarinic receptors.

In the guinea-pig ileum in vitro (*8, 9*), capsaicin, which is known to release substance P selectively from some sensory neurons, has been shown to cause a contraction that is nearly abolished by TTX and markedly reduced by atropine or substance P-desensitization. After guanethidine, stimulation of mesenteric periarterial nerves caused a contraction that was also markedly reduced after desensitization to substance P. The atropine-sensitive portion of the response was attributed to activation of cholinergic intrinsic nerves by substance P released by capsaicin. A similar site of action of substance P was clearly demonstrated in vivo in canine intestine (*35, 51*) and sometimes could be demonstrated in vitro in guinea-pig intestine (*83, 85*). The noncholinergic and TTX-insensitive portion studied by Bartho et al. (less than 50% of the noncholinergic component could be abolished by TTX) may have resulted from overflow of substance P from the plexus to the muscle, whereas the TTX-sensitive portion may have reflected capsaicin activation of either intrinsic substance P-containing nerves or the endings of extrinsic substance P-containing nerves. Both TTX-sensitive and -insensitive components of the noncholinergic response were abolished by substance P desensitization or the substance P antagonist (D-PRO^2,D-TRP^{7,9} substance P). Capsaicin-induced contractions led to desensitization after maximal doses, i.e., they could not be repeated. Surprisingly, the effect of capsaicin desensitization on responses to mesenteric periarterial nerve stimulation (in the presence of guanethidine) was not tested. If this response involved release of substance P from peripheral branches of dorsal root afferent neurons and capsaicin acted to deplete these nerves of substance P, as suggested by the authors, this response should have been abolished by pretreatment with capsaicin. There is earlier evidence that capsaicin releases substance P selectively from these sensory neurons (*56*) and not from substance P-containing neurons intrinsic to the gut, i.e., substance P content of the gut was not diminished after capsaicin treatment (*15, 61, 83*), and the contraction induced after atropine by high frequency field stimulation was not influenced by capsaicin desensitization (*9*). Both the apparent release of substance P by capsaicin in the presence of atropine and that by high frequency field stimulation were inhibited by opiate agonists and this effect was blocked by naloxone (*8, 9, 65*). Interaction of substance P and opiates is also indicated by studies of opiate dependence after chronic treatment. Sudden withdrawal of chronic effects of opiates by naloxone treatment causes a gut contraction that may be mediated by release of 5-hydroxytryptamine, which in turn releases substance P and acetylcholine (e.g., *see* ref. *64*).

2.2. ATP as NANC Excitatory Cotransmitter in Urogenital Muscle

The vas deferens of mammals is recognized to be innervated by sympathetic nerves (*59*). Its content of peptidergic nerves has not been studied intensively by immunohistochemistry. No means of identifying purinergic nerves by their structure is available. Responses of the vas deferens to field stimulation consist of a twitch and a tonic component. The responses to field stimulation are atropine-insensitive and not potentiated by physostigmine nor blocked by inhibition of choline uptake with hemicholinium (e.g., *see* ref. *4*). The tonic component of the response is sensitive to alpha-adrenergic antagonists (*4, 6, 7, 112*) and in the presence of atropine twitches are abolished by guanethidine or 6-hydroxydopamine, but not by reserpine treatment (*4, 45, 112*). Cocaine potentiates the adrenergic antagonist-sensitive tonic, but not the twitch component. The nerve varicosities are close (<20 nm) to muscle (*59*), but high mediator concentration in the synaptic cleft does not explain the nature of blockade of effects of nerve stimulation by adrenergic antagonists that inhibit a constant proportion of the response. Also high mediator concentration after activation of sympathetic nerves does not explain the cocaine-insensitivity of the adrenergic antagonist-insensitive component. Special adrenergic receptors, insensitive to alpha-adrenoceptor antagonists were postulated. Alternately, it was suggested that there was enhancement of norepinephrine release by alpha-adrenoceptor antagonists acting presynaptically sufficient to overcome postsynaptic blockade (*see* ref. *117*). A most likely explanation is that there is an additional transmitter besides norepinephrine (see below).

The evidence for a co- or additional transmitter comes from two sources. Recent physiological evidence from guinea pig and rat vas deferens contraction (*12, 110*), shows that the response consists of an initial twitchlike component followed by a prolonged tonic component and only the latter was sensitive to alpha-adrenoceptor antagonists (*4, 12*). Furthermore, the two components could be elicited even in response to a single stimulus (*102*), eliminating secondary release of a second substance as an explanation of the biphasic response. The electrophysiological basis of the biphasic response has been shown to be the occurrence of ejps during the initial phase of the response that can sum to threshold for spiking and twitchlike contractions; the tonic phase of the response is, however, not accompanied by detected changes in membrane potential (*110*). Membrane resistance changes during the various phases have not been determined partly because of the poor electrical coupling, usually found in the vas deferens. Norepinephrine, however, causes a contraction reported to be accompanied by depolarization (*106*) and this, like the tonic component of the response to field stimulation, is readily antagonized by phenoxybenzamine and phentolamine (alpha-adrenoceptor antagonists), whereas the initial twitch-like response is essentially insensitive to alpha-selective antagonists in concentrations considered to be acting on alpha-1 adrenoceptors. However, nonselective alpha-adrenoceptor antagonists, alpha-2 adrenoceptor antagonists (e.g., yohimbine), and higher concentrations of alpha-1-selective antagonists may actually potentiate the initial twitch as well as the release of norepinephrine (*102, 110*). Agonists for alpha-2-adrenergic receptors reduce the twitch (*45*). The main candidate for transmitter of the twitch component of the response is ATP (see below). Its electrophysiological effects on the vas deferens include depolarization (*106*) as expected if it is transmitter and initiates ejps. However, identity of conductance changes induced by ATP and the endogenous transmitted has not been established.

Further support for ATP as a cotransmitter comes from use of a selective, initially competitive and after exposure to light, irreversible photo-affinity antagonist for excitatory purinergic receptors. The substance, arylazidoaminopropionyl ATP ($ANAPP_3$), was first synthesized by Jeng and Guillory (*87*). Its antagonistic effects on excitatory purine receptors have been tested in a variety of smooth muscles (see below), but the most complete information is available for guinea-pig vas deferens. In contrast to its effects in excitating purinergic receptors, it appears to have no action on inhibitory purine receptors, e.g., guinea-pig fundus ATP or P_2-type receptor or guinea-pig ileum myenteric plexus P_i or adenosine presynaptic receptor (*55*), implying that the excitatory and inhibitory purine receptors differ in structure and structural requirements. The inference is supported by the inability of 2,2′-pyridylisatogen to selectively antagonize ATP in bladder as it does in gut (*18*).

In guinea-pig vas deferens, Hogaboom et al. (*79*) showed that the dose–response curve to ATP is biphasic. In addition the contractile response to individual doses has two components, an initial twitchlike phasic component and a slower, more tonic component that is especially evident at higher concentrations of ATP. $ANAPP_3$ easily abolished responses to low concentrations of ATP, but did not reduce the maximum response to ATP. Since an irreversible photo-affinity label should, even in tissues with spare receptors, eventually depress maximum responses, the authors suggested that ATP causes contraction by two mechanisms. One involved P_2-excitatory receptors for which ATP has high affinity, is thus operative alone at low ATP concentrations, and is susceptible to $ANAPP_3$ antagonism. Nonhydrolyzable ATP analogs, such as alpha- and beta-methylene ATP, appear to utilize this mechanism; they have monophasic dose–effect curves, have their maximum depressed by $ANAPP_3$ after photo-activation, and have essentially only an initial twitchlike phasic contractile response. In addition to this mechanism, there may be an $ANAPP_3$-insensitive mechanism that requires (*47, 48, 70*) obligatory hydrolysis of ATP and that occurs at higher ATP concentrations. The situation may in fact be more complicated (*47*) in that two P_2-excitatory receptors may exist with differing affinities for ATP and its analogs, but both of which can interact with $ANAPP_3$.

Despite these complications, it is clear that the results with $ANAPP_3$ do provide considerable support for the existence of a purine, presumably ATP, as a cotransmitter in vas deferens. Additional classic evidences of this role are briefly summarized below for bladder as well as vas deferens.

2.2.1. Existence in Nerves and Release

ATP is known to be stored with norepinephrine in vesicles of sympathetic nerves (*see* ref. *109*) that are generally agreed to supply the vas deferens (*11, 20, 59*) and most of the bladder, but especially its base (*see* ref. *50*). Sympathetic nerves are apparently the exclusive motor nerves to the longitudinal muscle and the major motor nerves to circular muscle of the vas deferens (some nerves with high acetylcholinesterase activity, presumably cholinergic, supply some fibers to circular muscle). Also, cholinergic nerves supply most of the motor innervation of the bladder body and some to the base (*11, 59*). No evidence of a specific purinergic innervation of vas deferens exists (*19, 62*), and sympathetic denervation with guanethidine (*56*) or by 6-hydroxydopamine treatment (*46, 56, 115*) abolished both phases of the response to NANC field stimulation. The occurrence of ATP together with norepinephrine in synaptic vesicles of sympathetic nerves was discussed in Chapter 15.

Evidence of the release of purines from vas deferens nerves is limited; Westfall et al. (*117*) found that transmural stimulation caused release of a mixture of purine breakdown products, presumably from nerves since it was inhibited by TTX and not duplicated by mechanical activity initiated without nerve stimulation. However, in the experiments of White et al. (*119*) discussed in Chapter 15, ATP release was TTX-insensitive, even though the TTX blocked field-stimulated contractions. How ATP could be released without there being activation of nerves or muscles is unclear.

In bladder detrusor muscle, too, no evidence of special purinergic nerves exists. Moreover, there were very few catecholamine-containing neurons to provide an alternate source of ATP (*see* ref. *19*). ATP release has been demonstrated by using the luciferin–luciferase method specific for ATP in guinea-pig bladder(*18*). Its release was increased 3–12 times by field stimulation of NANC nerves (atropine and guanethidine present). Along with the contractions, such ATP release was blocked by TTX. After TTX, field stimulation with 5 instead of 0.5 ms pulses caused contraction, presumably by direct activation of muscle (although direct stimulation of nerve terminals was not excluded), but did not release ATP. Reduction of external Ca^{2+} to 0.25 m*M* nearly abolished both contraction to short duration field stimulation and to the ATP release.

Alternate candidates for NANC transmitter are substance P (*see* ref. *71*) and prostaglandins. Prostaglandins (PG) or the enzymes necessary to release and transform arachidonic acid are probably ubiquitous, and many studies indicate that contraction of smooth muscle itself causes PG release (e.g., *see* ref. *16, 23, 118*). Burnstock (*12*) has postulated that NANC transmitter, like ATP in some cases, can release PGs. The ubiquitous nature of PC sources and stimuli for release makes it almost irrelevant to inquire about their sources and release in vas deferens or bladder.

Effects of inhibitors of their synthesis, however, show that whatever the sources, they play no essential role in NANC responses of vas deferens. In guinea-pig bladder, indomethacin had no effect on NANC responses (*5, 18, 108*); likewise it had little or no effect on contractions of rat bladder (*25*). However, blockade of PG synthesis did diminish, but did not abolish, the NANC nerve-induced contractions of rabbit bladder (*5, 36, 37, 90*). However, since conventional PGs (PGEs and PGFs) produced sluggish, small tonic contractions in contrast to the responses to NANC nerve stimulation, it is usually assumed that PGs modulate nerve transmission or muscle response rather than serve as neurotransmitters. Possible involvement of endoperoxides, thromboxanes, or leukotrienes as transmitters has not been exluded, but seems unlikely in view of their wide variety of actions. Substance P-containing nerves innervate the detrusor muscle of bladder (*103*), but substance P release has not been studied.

2.2.2. Identity of Actions

In vas deferens, where ATP is postulated to be a cotransmitter with norepinephrine (responsible for the initial twitchlike response, whereas norepinephrine accounts for the secondary tonic response), ATP would be expected to produce a rapid twitch. It should also have comparable electrophysiological effects including depolarization since ejps accompany the initial twitch (*12, 110*). In general ATP, like the early phase of the response to field stimulation, produces contractions that are relatively rapid and twitchlike (e.g., *47, 48, 70*). At higher concentrations, it also initiates a tonic contraction, the mechanism of which is not agreed upon, but may involve hydrolysis of ATP (*see* above and ref. *47, 48*). No one has yet reported the electrophysiology of ATP actions.

In guinea-pig bladder, ATP mimics reasonably well the NANC contractile responses (*see 17, 18, 36, 37*); other putative transmitters (e.g., Substance P, *120*) do not. High concentration of ATP or lower concentrations of its analog, α,β-methylene ATP (which is resistant to degradation), also caused twitchlike contractions in rat bladder(*14*). Since some degradation products (e.g., AMP and adenosine) caused inhibition of contractions, it was postulated that the need for high concentrations of exogenous ATP in rat bladder to mimic NANC contractions reflected its rapid degradation in contrast to ATP released from NANC nerves presumably directly on to purine receptors.

In rabbit bladder few electrophysiological studies of NANC nerve mediator on ATP effects on bladder are reported (*see* ref. *31, 32*). However, Creed et al. (*33*) recently reported that there was, in addition to a late slow depolarization that seemed to be mediated by acetylcholine, an early faster ejp that was nonadrenergic, noncholinergic, and not mediated by a known transmitter. High concentrations of ATP depolarized the bladder muscle membrane, decreased member resistance. The effect of the ejp on membrane resistance was not reported.

Prostaglandins (PGEs or PGFs) do not duplicate the response to NANC nerve stimulation (see above); instead in both bladder and vas deferens, they produce small, slow contractions in high doses. Substance P does not contract vas deferens, but potentiates twitch responses to field stimulation, apparently by facilitating the neurotransmission process (*12, 44, 45*). It does contract the detrusor muscle of urinary bladder (*86, 88*) by a direct action (TTX-insensitive). The detailed mechanisms of its contractile actions have not been studied and cannot be compared to those of endogenous mediator.

2.2.3. Actions of Substances That Antagonize or Potentiate

In vas deferens, arylazidoaminopropionyl ATP ($ANAPP_3$) inhibits the action of ATP, but not norepinephrine, acetylcholine, histamine, or DCl (*79*). The antagonism was reversible in the presence of the compound and became irreversible on photoactivation. Evidence that this antagonist is acting as a true photoaffinity label has been presented (*47, 48*). However, it did not reduce the maximum response of guinea-pig vas deferens to ATP (*49*). This was attributed to an action of ATP at separate $ANAPP_3$-insensitive receptors that required hydrolysis of ATP. Alone $ANAPP_3$ reduced the response of guinea-pig vas deferens to field stimulation (*46*) and reduced the maximum response. This change was not caused by the decreased release of norepinephrine. The frequency of field stimulation for 50% of maximum responses was increased twofold by $ANAPP_3$ following reserpinization; also the maximum response was reduced more by $ANAPP_3$ after reserpine than it was by $ANAPP_3$ alone. Prazosin ($10^{-6}M$) pretreatment had a similar effect to reserpinization. The effect of $ANAPP_3$ was primarily against the twitch component of the response to field stimulation. The difference in its action vs ATP (no depression of maximal response) and vs NANC field stimulation (maximum response depressed) may reflect addition of higher concentrations of ATP than are released from nerves (to act on another receptor, see above) or the release of ATP onto subjunctional receptors that differed pharmacologically from some of those found elsewhere. The ejps that accompany the twitch were reduced in amplitude by $ANAPP_3$ (*10*).

Neither ATP nor NANC nerve twitch responses were inhibited by antagonists to conventional mediators, e.g., histamine, 5-HT or prostaglandin synthesis inhibitors (*4*). Indeed, although the slow tonic component of the NANC response was sensitive to alpha-adrenergic blockade, the twitch was unaffected or potentiated by these agents (*44, 45, 102*). ATP desensitization was reported by Ambache and Zar (*4*) to have no effect on the twitch component of the NANC response and by Fedan et al. (*46*) to have a modest effect to raise (1.6–1.9-fold) the field stimulation frequency for half-maximum response. There is no reported evidence that theophylline or 2,2′-pyridylisatogen (purported purine antagonists) are effective against ATP-receptors of vas deferens. Since $ANAPP_3$ is ineffective against both P_1- and P_2-inhibitory receptors in gut (*55*), against which these agents are effective, this is not surprising. Blakeley et al. (*12*) showed that nefedipine selectively inhibited the spikes generated from the ejps and the associated twitch; to date, the action of nefedipine on ATP-induced contractions has not been reported.

In bladder, surprisingly there is as yet no analysis of the actions of $ANAPP_3$ on the atropine-insensitive response to field stimulation or the contraction to ATP. However, the receptors may be the same in the structural requirements for antagonists as in vas deferens. As in vas deferens, theophylline, phentolamine, imidazoles, and 2,2′-pyridylisatogen are not selective antagonists to NANC-nerve responses or to ATP (*5, 18, 108, 116*) in the bladder. Reactive blue-2, an anthroquinine–sulfonic acid derivative, a selective antagonist to ATP relaxation in guinea-pig colon (*94*), was a competitive antagonist of ATP in guinea pig detrusor muscle in the presence of guanethidine and indomethacin (*24*). It also reduced, in the presence of guanethidine, indomethacin, and atropine, NANC responses to field stimulation by up to 80% (100 μM reactive blue). However, it also caused a noncompetitive antagonism to acetylcholine. In rat bladder, reactive blue was less effective against both ATP and NANC responses (100 μM inhibited NANC responses only 25%).

ATP desensitization was ineffective against NANC responses of guinea-pig bladder to field stimulation (*2*). However, Burnstock et al. (*18*) found that pretreatment with indomethacin made ATP desensitization much more effective in inhibiting responses to field stimulation and also prevented increases in tone or spontaneous activity initiated by high ATP concentrations and suggested that ATP and the NANC mediator released prostaglandins that potentiated the responses to field stimulation. Why prostaglandin release by ATP did not also prevent desensitization to exogenous as well as endogenous ATP is unclear. Anderson et al. (*5*) confirmed the findings of Burnstock et al. (*18*) on guinea-pig bladder and made similar observations regarding the rabbit bladder. In this muscle Johns and Paton (*89, 90*) showed that inhibition of prostaglandin synthesis reduced the atropine-resistant response to field stimulation and Dean and Downie (*36, 37*) reported that such inhibition prevented the delayed component of tonic response to ATP. However, Andersson et al. (*5*) reported that NANC-field stimulation produced only twitch contractions, except at high frequencies, and that indomethacin reduced this twitch. Thus indomethacin seemed to have different effects on NANC twitch and ATP contraction. Also ATP desensitization seemed to inhibit NANC nerve responses only after indomethacin. One possible explanation is that ATP added exogenously is rapidly degraded to ADP, AMP, and so forth, and one or more of these products caused prostaglandin synthesis. In contrast, ATP released locally from nerve might not produce so much degradation product or the sites of prostaglandin synthesis might be less accessible to it. If so, desensitization to a nondegradable ATP analog should inhibit NANC nerve responses without the necessity of inhibition of prostaglandins. Kasakov and Burnstock (*93*) reported that that was the case for alpha-beta-methylene ATP on guinea-pig bladder. Creed et al. (*33*) reported that ATP (10^{-5}–$10^{-3}M$) reduced the size of ejp, but attributed this to the depolarization and decrease in membrane resistance produced in ATP rather than blockade of receptors for the ejp mediator. They did not test the extent or selectivity of the ATP desensitization, did not try a nondegradable analog of ATP, and did not test the effect of dipyridamole on responses to ATP (but reported it had no effect on the ejp). No selective ATP antagonists were tried (quinidine and theophylline were added, but not shown to be effective or selective as antagonists). Thus those findings do not support, but do not exclude, that ATP is the mediator of the fast noncholinergic ejp of bladder.

In guinea-pig bladder, adenosine, which is a relaxant, and dipyridamole reduced the contractions to both ATP and NANC nerve stimulation (*18*). Presumably enhanced adenosine accumulation physiologically antagonized all contractile responses.

In bladder, the selective substance P antagonist, D-PRO^{2},D-TRP^{7,9}-SP, had no effect to antagonize NANC nerve induced contractions (*96*). No studies of other inhibitors or potentiators of substance P are known.

2.3. Possible NANC Excitatory Transmission in Blood Vessels

Another example of excitatory NANC transmission may be activation of extrinsic nerves to blood vessels. In 1974 Hirst and in 1978 Hirst and Neild reported that all cells of arterioles of guinea-pig intestinal mucosa were well coupled electrically and responded to stimulation of nerves by producing ejps that, at frequencies of stimulation greater than 2 Hz, summed to cause action potentials and contraction. Not all cells responded to iontophoretically applied norepinephrine with depolarization. Hirst and Neild (*75, 76*) found that iontophoretic application of norepinephrine caused, in different cells, two types of response: (1) depolarizations that resembled ejps and caused contractions when they reached threshold for action potentials or (2) a contraction without depolarization. There was a rough correspondence between type (1) sites and the locus of fluorescent nerve endings (*78*), and in such regions no evidence for the presence of alpha-adrenoceptors was found. The depolarizations from iontophoretic norepinephrine were slower in time course than ejps evoked by nerve stimulation in other cells. The ejps and depolarizations by norepinephrine were insensitive to prazosin or phentolamine, whereas the contractions induced in other cells were conventionally sensitive. Similar findings were made from larger arteries, such as the saphenous arteries of rabbits (*81, 82*), in that norepinephrine caused contraction without change in membrane potential, whereas field stimulation always produced ejps; however, contractions were initiated without action potentials. Since guanethidine abolished the ejps, they were presumed to come from sympathetic nerves. If sympathetic nerve-released norepinephrine caused the ejps, it is puzzling why exogenous norepinephrine fails to depolarize cells.

Hirst and Nield (*73, 75, 76*) postulated that there was a new class of adrenergic receptors located adjacent to nerve endings that were insensitive to alpha-adrenoceptor antagonists. Recently, Hirst et al. (*72*) studied the rat basilar artery and found that the response to stimulation of intrinsic nerves involved ejps that could sum to initiate action potentials and contraction. The fast ejp was followed by a prolonged small depolarization (lasting about 2 min). Neither depolarization was affected by $10^{-6}M$ prazosin. However, in a few experiments, phentolamine (10^{-7}–$10^{-6}M$) reduced both. Higher concentrations had irreversible effects. Norepinephrine applied iontophoretically had no effect except in high concentrations (2×10^{-4}–$10^{-2}M$) that caused slow depolarizations potentiated by cocaine but unaffected by prazosin, phentolamine, or propranolol. However, phenylephrine ($8 \times 10^{-3}M$) was ineffective and epinephrine, *d*- or *l*-norepinephrine, and dopamine had effects similar to one another. Both *d*- and *l*-norepinephrine were similar in potency. Once again they explained the results in terms of a new receptor to norepinephrine mediating excitation, but not showing the chemical affinities of alpha-adrenoceptors. A similar situation exists in several arteries (e.g., guinea-pig basilar; *92*), but not all (e.g., pulmonary artery, *111*). In the latter all responses to field stimulation and norepinephrine were sensitive to adrenergic antagonists.

What seems not to have been established rigorously is that the nerve mediator of ejps is in fact norepinephrine. There was a general similarity of depolarization, and the slower time course of the response to iontophoretically applied, compared to nerve-relesed mediator might be expected from the different geometries of delivery. However, more rigorous electrophysiological criteria have not been applied, e.g., measurement of reversal potential, of changes in membrane resistance, or of pharmacological antagonism of putative and real mediators. The failure of both norepinephrine and the endogenous mediator to be antagonized by adrenergic antagonists indicates only that neither act on alpha-adrenergic receptors. Thus there is a definite possibility that the excitatory mediator of vascular intrinsic nerves is not norepinephrine alone, but a cotransmitter such as ATP or a mediator from the several peptidergic nerves known to innervate blood vessels. As far as I am aware, no systematic study of effects of other possible neurotransmitters is available.

Some support for norepinephrine as the transmitter responsible for the alpha-adrenergic-antagonist-resistant responses to field stimulation at junction receptors has recently been provided (*95*) by using a new putative alpha-adrenergic antagonist, 3, 4 - dihydro - 8 - (2 - hydroxy - 3 - isopropylaminopropoxy) - 3 - nitroxy - 2H - 1 - benzopyrin (K-351). K-351 acts on smooth muscle cells and neuromuscular transmission in canine mesenteric arteries differently from alpha-1 (prazosin) and nonspecific alpha (phentolamine) adrenergic antagonists. Prazosin was more potent than phentolamine and K-351 in inhibiting contractions to norepinephrine, but had no effect on the first ejp elicited by field stimulation or on the facilitation of ejps with repetitive stimulation. Phentolamine did inhibit the amplitude of the first ejp slightly, but had inconsistent effects after the ejp facilitation process had occurred and on the facilitation process. In contrast, K-351 was slightly less potent than phentolamine as an antagonist to norepinephrine, but reduced the amplitude of the first ejp and of ejps after completion of facilitation without affecting facilitation. Detailed studies of K-351 to prove it to be a competitive antagonist of norepinephrine and to displace alpha-adrenergic ligands from binding are still needed.

2.4. NANC Excitatory Transmitter at Other Smooth Muscles

The iris sphincter in guinea pig and rabbit had a clear-cut NANC excitatory transmitter as well as a cholinergic one (*96, 113, 114*). Substance P is the most likely candidate for transmitter at present since substance P is present in the nerves from trigeminal ganglion (*80*) that innervate this muscle. Capsaicin produces a NANC contraction provided that innervation is intact, whereas substance P contracts (in contrast to the lack of effect of VIP and somatostatin) (*113, 114*) and a substance P antagonist (D-PRO^{2}-D-TRP^{7,9}SP) inhibits the response (*96*). Whether substance P nerves supplying the muscle are totally afferent in origin and, if so, whether they play a physiological role remains to be determined.

There is evidence for other putative NANC transmitters at a variety of smooth muscles. However, in each case the evidence is still very limited. Possible candidates include enkephalins, neurotension, and gastrin-releasing peptide.

3. Conclusions

The existence of excitatory NANC cotransmission has now been established in vas deferens and evidence that ATP may be the cotransmitter with norepinephrine has been put forward. Additional studies will be required to establish the nature of the transmitter. In bladder, muscle innervation by NANC excitatory nerves seems to be established, but there is less convincing evidence that ATP could be the transmitter, and some evidence to the contrary. In gut and iris muscle (and possibly elsewhere), substance P appears to be a NANC excitatory transmitter released by field stimulation or stimulation of intrinsic nerves. Most evidence suggests that substance P released by extrinsic nerve stimulation is from peripheral branches of afferent neurons and its physiological role is unclear. In gut, some substance P is probably released by field stimulation of intrinsic substance P-containing nerves. In small arteries, it is unclear whether adrenergic nerves release a cotransmitter along with norepinephrine or, as proposed by most who have studied neuromuscular transmission to arteries, release norepinephrine to act on special (neither alpha nor beta) receptors located subsynaptically. No evidence of complications involving NANC innervation of interstitial cells exists for the smooth muscles considered.

References

1. Ambache, N. and M. A. Freeman. Atropine-resistant longitudinal muscle spasms due to excitation of non-cholinergic neurons in the Auerbach's plexus. *J. Physiol.* 199:705–727, 1968.
2. Ambache, N. and M. A. Zar. Non-cholinergic transmission by postganglionic motor neurons in the mammalian bladder. *J. Physiol. (Lond.)* 210:761–783, 1970.
3. Ambache, N., J. Vermey, and M. A. Zar. Evidence for the release of two atropine-resistant spasmogens from Auerbach's plexus. *J. Physiol. (Lond.)* 207:761–782, 1970.
4. Ambache, N. and M. A. Zar. Evidence against adrenergic motor transmission in the guinea pig vas deferens. *J. Physiol. (Lond.)* 216:359–389, 1971.
5. Andersson, K. E., S. Husted, and C. Sjogren. Contribution of prostaglandins to the adenosine triphosphate-induced contraction of rabbit urinary bladder. *Brit. J. Pharmacol.* 70: 443–452, 1980.
6. Anton, P. G. and J. C. McGrath. Anatomical basis for the mechanical response to motor nerve stimulation of the rat vas deferens. *J. Physiol. (Lond.)* 273:23–43, 1977.
7. Anton, P. G. and J. C. McGrath. Further evidence for adrenergic transmission in the human vasa deferentia. *J. Physiol. (Lond.)* 273:45–55, 1977a.
8. Bartho, L., P. Holzer, F. Lembeck and J. Szolcsanyi. Evidence that the contractile response of the guinea pig ileum to capsaicin is due to the release of substance P. *J. Physiol. (Lond.)* 332:157–167, 1982a.
9. Bartho, L., B. Sebok and J. Szolcsanyi. Indirect evidence for the inhibition of enteric substance P agonists but not by capsaicin. *Europ. J. Pharmacol.* 77:273–279, 1982b.
10. Bauer, P. and H. Kuriyama. Evidence for non-cholinergic non-adrenergic transmission in the guinea pig ileum. *J. Physiol. (Lond.)* 330:95–110, 1982.

11. Bell, C. and J. R. McLean. Localization of norepinephrine and acetylcholinesterase in separate neurons supplying the guinea pig vas deferens. *J. Pharmacol. Exp. Therap.* 157:69–73, 1967.
12. Blakeley, A. G. H., D. A. Brown, T. C. Cunnane, A. M. French, J. C. McGrath, and N. C. Scott. Effects of nefedipine on electrical and mechanical responses of rat and guinea pig vas deferens. *Nature* 294:759–761, 1981.
13. Bradley, W. C. and C. T. Teague. Innervation of the vesical detrusor muscle by the ganglia of the pelvic plexus. *Invest. Urol.* 6:251–260, 1968.
14. Brown, C., G. Burnstock, and T. Cocks. Effects of adenosine 5′-triphosphate (ATP) and α,β-methylene ATP on the rat urinary bladder. *Br. J. Pharmacol.* 65:97–102, 1979.
15. Buck, S. M., P. P. Desonakh, H. J. Yamamura, and T. F. Burks. Differences between rats and guinea pigs in gastrointestinal and nervous system substance P levels. *Neuropeptides* 1:383–389, 1981.
16. Burnstock, G. Purinergic nerves. *Pharmacol. Review* 24:509–581, 1972.
17. Burnstock, G., T. Cocks, G. Crowe, and L. Kasakov. Purinergic innervation of the guinea pig urinary bladder. *Br. J. Pharmacol.* 63:125–138, 1978a.
18. Burnstock, G., T. Cocks, L. Kasakov, and A. Wong. Direct evidence for ATP-release from non-adrenergic non-cholinergic (''purinergic'') nerves in the guinea pig taenia coli and bladder. *Europ. J. Pharmacol.* 49:145–149, 1978b.
19. Burnstock, G. and M. Costa. *Adrenergic Neurons: Their Organization, Function and Development in the Peripheral Nervous System.* Chapman and Hall, London, 1975.
20. Burnstock, G. and P. M. Robinson. Localization of catecholamines and acetylcholinesterase in autonomic nerves. *Circ. Res.* 20-21:Suppl. 3 43–55, 1967.
21. Bywater, R. A. R. and G. S. Taylor. Atropine-resistant junction potentials in the guinea pig small intestine. *Proc. Aust. Physiol. Pharmacol. Sr.* 10:233, 1979.
22. Bywater, R. A. R., M. E. Holman and G. S. Taylor. Atropine-resistant depolarization in the guinea pig small intestine. *J. Physiol. (Lond.)* 316:369–378, 1981.
23. Campos, G. A., G. C. Liggins and R. F. Seamark. Differential production of PGF and 6-keto-PGF_1 by rat endometrium and myometrium in response to oxytocin, catecholamines and calcium ionophore. *Prostaglandins* 296:297–310, 1980.
24. Choo, L. K. The effect of reactive blue, an antagonist of ATP on the isolated urinary bladders of guinea-pig and rat. *J. Pharm. Pharmacol.* 33:248–250, 1981.
25. Choo, L. K. and F. Mitchelson. The role of prostaglandins in the excitatory innervation of the rat urinary bladder. *Prostaglandins* 13:917–926, 1977.
26. Costa, M. and J. B. Furness. Slow contraction of the guinea pig proximal colon in response to the stimulation of an identified type of nerve. *Br. J. Pharmacol.* 45:151p–152p, 1972.

27. Costa, M., J. B. Furness, R. Buffa, and S. J. Said. Distribution of enteric neurons showing immunoreactivity for vasoactive intestinal polypeptide (VIP) in the guinea pig intestine. *Neurosciences* 5:587–596, 1980a.
28. Costa, M., J. B. Furness, I. J. Llewellyn-Smith, and C. Cuello. Projections of substance P-containing neurons within the guinea pig small intestine. *Neurosciences* 6:411–424, 1980b.
29. Costa, M. and J. B. Furness. Nervous control of intestinal motility. In *Mediators and Drugs in Gastrointestinal Motility*. 1. Morphological and neurophysiological control. G. Bertaccini, ed. Springer Verlag, Berlin, pp 279–382, 1982a.
30. Cowan, W. and E. E. Daniel. Human female bladder and its noncholinergic contractile function. *Can. J. Physiol. Pharmacol.* 61: 1236–1246, 1983.
31. Creed, K. E. Membrane properties of the smooth muscle membrane of the guinea pig urinary bladder. *Pflug. Arch. Ges. Physiol.* 326:115–126, 1971a.
32. Creed, K. E. Effects of ions and drugs on the smooth muscle cell membrane of the guinea pig urinary bladder. *Pflug. Arch. Ges. Physiol.* 326:127–141, 1971b.
33. Creed, K. E., S. Shikawa, and Y. Ito. Electrical and mechanical activity recorded from rabbit urinary bladder in response to weave stimulation. *J. Physiol.* 338:149–164, 1983.
34. Daniel, E. E., W. Cowan, and V. P. Daniel. Structural bases for neural and myogenic control of human detrusor muscle. *Can. J. Physiol. Pharmacol.* 61:1247–1273, 1983.
35. Daniel, E. E., T. Gonda, T. Domoto, M. Oki, and N. Yanaihara. The effects of substance P and met^5-enkephalin in dog ileum. *Can. J. Physiol. Pharmacol.* 60:830–840, 1982.
36. Dean, D. M. and J. W. Downie. Interaction of prostaglandins and adenosine 5′-triphosphate to the non-cholinergic neurotransmission in rabbit detrusor. *Prostaglandins* 16:245–251, 1978a.
37. Dean, D. M. and J. W. Downie. Contribution of adrenergic and "purinergic" neurotransmission to contraction in rabbit detrusor. *J. Pharmacol. Exp. Therap.* 207:431–445, 1978b.
38. Domoto, T. J. Jury, I. Berezin, J. E. T. Fox, and E. E. Daniel. Does substance P co-mediate with acetycholine in nerves of opossum esophageal muscularis mucosa? *Am. J. Physiol.* 8:G19–G28, 1983.
39. Domoto, T., E. E. Daniel, M. Oki, and N. Yanaihara. Localization of substance P in the canine ileum: evidence for its release. *Gastroenterology* 80:1137 (Abstr.), 1981.
40. Dun, N. J. and S. Minota. Effects of substance P on neurones of the inferior mesenteric ganglia of the guinea pig. *J. Physiol. (Lond.)* 321:259–271, 1981.
41. El-Badawi, A. and E. A. Schenk. Parasympathetic and sympathetic post-ganglionic synapses in urethro vesicular autonomic pathways. *Z. Zellforsch.* 146:147–154, 1973.
42. El-Badawi, A. and E. A. Schenk. Dual innervation of the mammalian urinary bladder. A histochemical study of the distribution of cholinergic and adrenergic nerves. *Am. J. Anat.* 119:405–416, 1966.

43. Elz, A., P. Alm, K. E. Andersson, and C. G. A. Persson. Adrenergic and cholinergic nerves of the human urethra and urinary bladder. A histochemical study. *Acta Physiol. Scand.* 99:345–352, 1977.
44. Von Euler, V. S. and P. Hedquist. Phenoxybenzamine blockade of the enhancing effect of substance P on the twitch induced by transmural nerve stimulation in the guinea pig vas deferens. *Acta Physiol. Scand.* 92(2):283–285, 1974.
45. Von Euler, V. S. and P. Hedquist. Evidence for an alpha- and beta-2-receptor mediated inhibition of the twitch response in the guinea pig vas deferens by noradrenaline. *Acta Physiol. Scand.* 93(4):572–573, 1975.
46. Fedan, J. S., G. K. Hogaboom, J. P. O'Donnell, J. Colby, and D. P. Westfall. Contributions by purines to the neurogenic response of the vas deferens of the guinea pig. *Eur. J. Pharmacol.* 69:41–53, 1981.
47. Fedan, J. S., G. R. Hogaboom, D. P. Westfall, and J. P. O'Donnell. Comparison of contractions of the smooth muscle of the guinea pig vas deferens induced by ATP and related nucleotides. *Eur. J. Pharmacol.* 81:193–204, 1982a.
48. Fedan J. S., G. K. Hogaboom, D. P. Westfall and J. P. O'Donnell. Comparison of the effects of arylazido aminopropionyl ATP ($ANAPP_3$) on ATP antagonist, on responses of the smooth muscle of the guinea pig vas deferens to ATP and related nucleotides. *Eur. J. Pharmacol.* 85:277–290, 1982b.
49. Fedan, J. S., G. K. Hogaboom and J. P. O'Donnell. Mechanism of photo-affinity labelling of P_2-purinergic receptors by arylazido aminopropionyl ATP in isolated guinea pig vas deferens. *Life Sciences* 31:1921–1928, 1982c.
50. Fletcher, T. F. and W. E. Bradley. Neuroanatomy of the bladder. *J. Urology* 119:153–160, 1978.
51. Fox, J. E. T., E. E. Daniel, J. Jury, and S. M. Collins. Sites and mechanisms of action of neuropeptides on canine gastric motility differs in vivo and in vitro. *Life Sciences* 33:817–825, 1983.
52. Franco, R., M. Costa, and J. B. Furness. Evidence for the release of endogenous substance P from intestinal nerves. *Naunyn-Schmiedeberg's Arch. Pharmacol.* 306:185–201, 1979a.
53. Franco, R., M. Costa, and J. B. Furness. Evidence that axons containing substance P are of intrinsic origin. *Naunyn-Schmiedeberg's Arch. Pharmacol.* 307:57–63, 1979b.
54. Franco, R., M. Costa, and J. B. Furness. The presence of a cholinergic excitatory input to substance P neurons in the intestine. *Proc. Aust. Physiol. Pharmacol. Soc.* 10:255, 1979c.
55. Frew, R. and P. M. Lundy. Effect of arylazidoaminopropionyl ATP ($ANAPP_3$) a putative ATP antagonist on ATP responses of isolated guinea pig smooth muscle. *Life Sci.* 30:259–267, 1980.
56. Furness, J. B. Some actions of 6-hydroxydopamine which affect autonomic neuromuscular transmission. In *6-hydroxydopamine and Catecholamine Neurons.* T. Malmfors and H. Thoenen, eds. North-Holland Publishing Co., 1971.
57. Furness, J. B., R. E. Papka, N. G. Della, M. Costa, and R. L. Eskay. Substance P-like immunoreactivity in nerves associated with the vascular system in guinea pigs. *Neuroscience* 7:447–459, 1982.

58. Furness, J. B. and M. Costa. Identification of gastrointestinal neurotransmitters. In *Handbook of Experimental Pharmacology.* G. Bertaccini, ed. Springer Verlag, Berlin, pp 383–460, 1981.
59. Furness, J. B. and T. Iwayama. The arrangement and identification of axons innervating the vas deferens of the guinea pig. *J. Anat.* 113:179–196, 1972.
60. Furness, J. B. and M. Costa. The nervous release and the action of substances which affect intestinal muscle through neither adrenoceptors nor cholinoreceptors. *Philos. Trans. Roy. Soc. Lond. (Biol.)* 265:123–133, 1973.
61. Gamse, P., A. Wax, R. E. Zigmond, and S. E. Leeman. Immunoreactive substance P in sympathetic ganglia: distribution and sensitivity toward capsaicin. *Neurosciences* 6:437–441, 1981.
62. Gibbons, I. L. Lack of correlation between ultrastructural and pharmacological types of non-adrenergic autonomic nerve. *Cell Tiss. Res.* 221:551–581, 1982.
63. Gillespie, J. S. Non-adrenergic, non-cholinergic inhibitory control of gastrointestinal motility. In *Motility of the Digestive Tract.* M. Wienbeck, ed. Raven Press, New York, pp 51–66, 1982.
64. Gintzler, A. R. Substance P involvement in the expression of gut dependence on opiates. *Brain Res.* 182:224–228, 1980.
65. Gintzler, A. R. and J. A. Scalisi. Effects of opioids on non-cholinergic excitatory responses of the guinea pig isolated ileum: inhibition of release of enteric substance P. *Br. J. Pharmacol.* 75:199–205, 1982.
66. Goyal, R. K. and S. Rattan. Neurohumoral, hormonal and drug receptors for the lower esophageal sphincter. *Gastroenterology* 74:598–619, 1978.
67. Goyal, R. K. and S. Rattan. Nature of the vagal inhibitory innervation to the lower esophageal sphincter. *J. Clin. Invest.* 55:1119–1126, 1975.
68. Hayashi, H., K. Ohsumi, M. Fukiwara, N. Mizuno, I. Kanazawa, and H. Yajima. Immunohistochemical studies on enteric substance P of extrinsic origin in the cat. *J. Auton. Nerv. System* 5:207–247, 1982a.
69. Hayashi, H., K. Ohsumi, N. Ueda, M. Fujiwara, and N. Mizuno. Effect of spinal ganglionectomy on substance P-like immunoreactivity in the gastro-duodenal tract of cats. *Brain Res.* 232:227–230, 1982b.
70. Head, R. J., G. K. Hogaboom, J. P. O'Donnell, and J. S. Fedan. Cell surface localization of P_2-purinergic receptors in rat vas deferens. *Biochem. Pharmacol.* 32(3):563–565, 1983.
71. Hedquist, P. and U. S. Von Euler. Effects of substance P on some autonomic neuroeffector junctions. In *Substance P.* U. S. Von Euler and B. Pernow, eds. Raven Press, New York, pp 89–95, 1977.
72. Hirst, G. D. S., T. O. Neild, and G. D. Silverberg. Noradrenaline receptors on the rat basilar artery. *J. Physiol. (Lond.)* 328:351–360, 1982.
73. Hirst, G. D. S. and T. O. Nield. An analysis of excitatory junction potentials recorded from arterioles. *J. Physiol. (Lond.)* 280:87–104, 1978.
74. Hirst, G. D. S. Neuromuscular transmission in arterioles of the guinea pig submucosa. *J. Physiol. (Lond.)* 273:263–275, 1977.
75. Hirst, G. D. S. and T. O. Neild. Evidence of two populations of two excitatory receptors for noradrenaline on arteriolar smooth muscle. *Nature (Lond.)* 283:767–768, 1980a.

76. Hirst, G. D. S. and T. O. Neild. Noradrenergic transmission—reply to McGrath. *Nature (Lond.)* 288:301–302, 1980b.
77. Hirst, G. D. S. and T. O. Neild. Some properties of spontaneous junction potential records from arterioles of guinea pigs. *J. Physiol. (Lond.)* 303:43–60, 1980c.
78. Hirst, G. D. S. and T. O. Neild. Localization of specialized noradrenaline receptors at neuromuscular junctions on arterioles of the guinea pig. *J. Physiol. (Lond.)* 313:343–350, 1981.
79. Hogaboom, G. K., J. P. O'Donnell, and J. S. Fedan. Purinergic receptors: photoaffinity analog of adenosine triphosphate is a specific adenosine triphosphate antagonist. *Science* 208:1273–1276, 1980.
80. Hokfelt, J., O. Johansson, J. O. Kellerth, A. Ljungdahl, G. Nilsson, G. Nygards, and B. Pernow. Immunohistochemical distribution of substance P. In *Substance P.* U. S. Von Euler and B. Pernow, eds. Raven Press, New York, pp 117–145, 1977.
81. Holman, M. E. and A. M. Surprenant. Some properties of the excitatory junction potentials recorded from saphenus arteries of rabbits. *J. Physiol. (Lond.)* 287:337–351, 1979.
82. Holman, M. E. and A. M. Surprenant. An electrophysiological analysis of the effects of noradrenaline and alpha-receptor antagonists on neuromuscular transmission in mammalian muscular arteries. *Br. J. Pharmacol.* 71:651–661, 1980.
83. Holzer, P. and F. Lembeck. Neurally mediated contractions of ileal longitudinal muscle by substance P. *Neuroscience Letters* 17:101–105, 1980.
84. Holzer, P., R. James and F. Lembeck. Distribution of substance P in the rat gastro-intestinal tract—lack of effect of capsaicin treatment. *Eur. J. Pharmacol.* 61:303–307, 1980.
85. Holzer, P., P. C. Emson, L. L. Iverson, and D. F. Sherman. Regional differences in the response to substance P of the longitudinal muscle and the contraction of substances in the digestive tract of the guinea pig. *Neuroscience* 6:1433–1441, 1981.
86. Husted, S., C. Sjoegren and K. E. Andersson. Substance P and somatostatin and excitatory neurotransmission in rabbit urinary bladder. *Arch. Internat. Pharmacodyn. Therap.* 252:72–85, 1981.
87. Jeng, S. J. and R. S. Guillory. The use of aryl azide ATP analogs as photo-affinity labels for myocin ATPase. *J. Supramol. Struct.* 3(5-6):448–468, 1975.
88. Johns, A. The effect of indomethacin and substance P on the guinea pig urinary bladder. *Life Sciences* 29:1803–1809, 1981.
89. John, A. and D. M. Paton. Evidence for a role of prostaglandins in atropine-resistant transmission in the mammalian urinary bladder. *Prostaglandins* 11:595–597, 1976.
90. Johns, A. and D. M. Paton. Effect of indomethacin on atropine-resistant transmission in rabbit and monkey urinary bladder: evidence of involvement of prostaglandins in transmission. *Prostaglandins* 13:245–254, 1977.
91. Johnson, S. M., Y. Katayama, K. Morita, and R. A. North. Mediators of slow synaptic potentials in the myenteric plexus of the guinea pig ileum. *J. Physiol. (Lond.)* 320:175–186, 1981.

92. Karashima, T. and H. Kuriyama. Electrical properties of smooth muscle cell membrane and neuro-muscular transmission in the guinea pig basilar artery. *Br. J. Pharmacol.* 74:495–504, 1981.
93. Kasakov, L. and G. Burnstock. The use of the slowly degradable analog α,β-methylene ATP, to produce desensitization of the P_2-purinoceptor: effect on non-adrenergic, non-cholinergic responses of the guinea pig urinary bladder. *Eur. J. Pharmacol.* 86:291–294, 1983.
94. Kerr, D. I. and A. Krantis. A new class of ATP-antagonist. *Proc. Aust. Physiol. Pharmacol. Soc.* 10:156, 1979. (Abstract)
95. Kuo, K., H. Kuriyama, and H. Suzuki. Effects of 3,4-dihydro-8-(2-hydroxy-3-isopropylaminopropoxy) - 3 - nitroxy - 2H - 1 - benzopyran (K-351) on smooth muscle cells and neuromuscular transmission in the canine mesenteric artery. *Br. J. Pharmacol.* 77:679–689, 1982.
96. Leander, S., R. Hakanson, S. Rosell, K. Folkers, F. Sundler, and K. Toinquist. A specific substance P antagonist blocks smooth muscle contractions by non-cholinergic, non-adrenergic nerve stimulation. *Nature* 294:467–469, 1981.
97. Maas, A. J. G. The effect of apamin on responses evoked by field stimulation in guinea pig taenia coli. *Eur. J. Pharmacol.* 73:1–9, 1981.
98. Maas, A. J. J. and A. Den Hertog. The effect of apamin on the smooth muscle cells of the guinea pig taenia coli. *Eur. J. Pharmacol.* 58:151–156, 1979.
99. Maas, A. J. J. and A. Den Hertog. The effect of phenylphosphonate N-0164 on prostaglandin action and on post-inhibitory excitation in the taenia of guinea pig caecum. *Eur. J. Pharmacol.* 62:157–166, 1980.
100. Maas. A. J. J., A. Den Hertog, R. Ras, and J. Van Den Arker. The action of apamin on guinea pig taenia caeci. *Eur. J. Pharmacol.* 67:265–274, 1980.
101. Matthews, M. R. and A. C. Cuello. Substance P-immunoreactive peripheral branches of sensory neurones innervate guinea pig sympathetic neurones. *Proc. Nat. Acad. Sci. USA* 79:1668–1672, 1982.
102. McGrath, J. C. Adrenergic and "non-adrenergic" components in the contractile responses of the vas deferens to a single indirect stimulus. *J. Physiol. (Lond.)* 283:23–29, 1978.
103. Nilsson, G. and E. Brodin. In *Substance P.* U. S. Von Euler and B. Pernow, eds. Raven Press, New York, pp 49–54, 1977.
104. Nilsson, G., L. D. Larsson, R. Hakanson, E. Brodin, B. Pernow, and F. Sundler. Localization of substance P immunoreactivity in mouse gut. *Histochemistry* 43:77–79, 1975.
105. Pearse, A. D. E. and J. M. Polak. Immunocytochemical localization of substance P in mammalian intestine. *Histochemistry* 41:473–475, 1975.
106. Sakai, K. K., D. L. Hymson, and R. Shapiro. The effects of adenosine and adenine nucleotides on the guinea pig vas deferens. Evidence for existence of purinergic receptors. *Life Sciences* 24:1299–1308, 1979.
107. Schultzberg, M., T. Hokfelt, G. Nilsson, L. Terenius, J. F. Rehfeld, M. Brown, R. Elder, M. Goldstein, and S. Said. Distribution of peptide and catecholamine neurons in the gastro-intestinal tract of rat and guinea pig immunohistochemical studies with antisera to substance P, VIP, enkephalins, somatostatin, gastrin, neurotensin, and dopamine beta-hydroxylase. *Neuroscience* 5:689–744, 1980.

108. Sjogren, C. and K. E. Andersson. Inhibition of ATP-induced contraction in the guinea pig urinary bladder in vitro and in vivo. *Acta Pharmacol. Toxicol.* 44:221–227, 1979.
109. Smith, A. D. Subcellular localization of noradrenaline in sympathetic neurons. *Pharmacol. Rev.* 24:435–457, 1972.
110. Sneddon, P., D. P. Westfall, and J. S. Fedan. Cotransmitter in the motor nerves of the guinea pig vas deferens: electrophysiological evidence. *Science* 218:693–695, 1982.
111. Suzuki, H. An electrophysiological study of excitatory neuromuscular transmission in the guinea pig main pulmonary artery. *J. Physiol. (Lond.)* 336:47–59, 1983.
112. Swedin, G. Studies on neurotransmission mechanisms in the rat and guinea pig vas deferens. *Acta. Physiol. Scand.* Suppl. 369:1–34, 1971.
113. Ueda, N. Non-cholinergic, non-adrenergic contraction and substance P in rabbit iris sphincter muscles. *Jap. J. Pharmacol.* 31:1071–1079, 1981.
114. Ueda, N., I. Murimatsu, H. Hayashi and M. Fuziwara. Trigeminal nerve: the possible origin of substance P-nergic response in isolated rabbit iris sphincter muscle. *Life Sciences* 31:369–375, 1982.
115. Wadsworth, R. M. Abolition of neurally evoked motor responses of the vas deferens by 6-hydroxydopamine. *Eur. J. Pharmacol.* 21:383–387, 1973.
116. Weetman, D. F. and N. Turner. The effects of ATP-receptor blocking agents on the response of the guinea pig isolated bladder preparation to hyoscine-resistant nerve stimulation. *Arch. int. de Pharmacodyn. Therap.* 228:10–14, 1977.
117. Westfall, D. P., R. E. Stitzel and J. N. Rowe. The post-junctional effects and neural release of purine compounds in the guinea pig vas deferens. *Eur. J. Pharmacol.* 50:27–38, 1978.
118. Whalley, E. T. The action of bradykinin and oxytocin on the isolated whole uterus and myometrium of the rat in oestrus. *Br. J. Pharmacol.* 64:21–28, 1978.
119. White, T., P. Potter, C. Moody, and G. Burnstock. Tetrodotoxin-resistant release of ATP from guinea pig taenia coli and vas deferens during electrical field stimulation in the presence of luciferin–luciferase. *Can. J. Physiol. Pharmacol.* 59:1094–1100, 1981.
120. Zetler, G. Effects of substance P and other peptides on the field-stimulated guinea pig vas deferens. In *Substance P*. U. S. Von Euler and B. Pernow, eds. Raven Press, New York, pp 77–116, 1977.

Chapter 17

Peptide Calcium Interactions in Smooth Muscle

J. E. T. Fox

1. Introduction

Until the early 1960s, relatively few polypeptide hormones with known smooth muscle function had been isolated, sequenced, and synthesized. With advances in peptide chemistry, i.e., automated sequencing and solid-phase synthesizing, and the widespread use of the immunological techniques of radioimmunoassay and immunohistochemistry, new natural peptides are being isolated, sequenced, synthesized and shown to have multiple sites of origin and multiple sites of action with ever increasing frequency (*see* ref. *84* for a historical review).

Peptides with potential smooth muscle regulating activity include the activated forms of angiotensin and bradykinin which have been produced in the circulation by enzymatic degradation of their precursors, the endocrine peptides, particularly those of the gastrointestinal tract, and the neural peptides of both central and peripheral origin. Peptides may be delivered to the site of action from their sources via the circulation or by diffusion. The rapidity with which these new peptides with potential smooth muscle function have appeared has led to the situation where few definitive experiments have been performed to determine the relationships of the peptide action to Ca^{2+} movements. In fact, Bolton (*13*), in his 1979 review of agents acting on smooth muscle, describes possible Ca^{2+} activating roles for only three peptides: oxytocin, angiotensin, and bradykinin.

If one examines the action of a peptide in the whole animal there are potentially many situations in the chain of events from release or activation of the peptide to the actual smooth muscle response which could be calcium-dependent. The release or activation of the peptide is frequently found to be Ca^{2+}-dependent. It could activate another cell type which in turn could release smooth muscle activator, either of which events could be Ca^{2+}-dependent. It might ultimately act at the smooth muscle membrane to activate a chain of events that alter Ca^{2+} movement or it might itself act directly to alter Ca^{2+} movement. If the action is studied in vivo, it is impossible to distinguish the Ca^{2+}-dependence of these events since raising or lowering Ca^{2+} concentrations or using Ca^{2+} blocking agents or Ca^{2+} iontophores may alter one or more of these processes simultaneously. This, then, necessitates studying the action of the peptide in vitro. There are, however, certain problems that occur when one studies these actions in vitro. In some instances, although a direct smooth muscle action can be demonstrated in vivo, the receptor shown to be most sensitive to a peptide may be located on a nerve or another cell type, and activation of the smooth muscle action site to be studied in vitro may be highly unlikely to occur at physiological concentrations of the peptide in vivo. As an example, substance P injected intra-arterially into local canine gastrointestinal arteries inhibits contractile activity at 10^{-12} mol, excites activity at 10^{-10} mol (both by activating nerves), and only at concentrations of 10^{-9} mol can a direct smooth muscle action be elicited (*24, 35*). The Ca^{2+} dependence of any of these processes is unknown. To test the neural dependence requires intracellular electrode studies of the myenteric plexus. To test the smooth muscle activation requires multiple in vitro studies. As another example, intravenous neurotensin in the rat produces hypotension unless the histamine-containing cells are stabilized. Studies of these isolated mast cells show a Ca^{2+}-dependent release of histamine (*93*). The hypertensin shown after mast cell stabilization is partially adrenergic nerve-dependent. Whether a residual smooth muscle response is Ca^{2+}-dependent or not is unknown. Furthermore, a different problem occurs when the receptors for a peptide change their sensitivity from the in vivo to the in vitro situation (*see* Section 2.1. on gastrin) or vary in time in vitro. For instance, B_1 receptors for bradykinin appear in normal tissue in vitro only after prolonged incubation, but can be induced in vivo as a response of tissues to injury (*see* ref. *91* for review). Other physiological changes in the animal such as pregnancy can alter the response of a given tissue to the peptide agonist and thus potentiate Ca^{2+} movements. As an example, the changes in the response of the uterus to oxytocin during pregnancy and delivery may be the result of more receptors (*1, 103*) or changes in the morphological coupling of the tissue by an increase in the number and size of gap junctions (*41*).

At the neural and endocrine cell level Ca^{2+} is involved in the voltage dependent excitation secretion coupling in the release of peptides (*42, 43, 49, 51, 77, 89, 100, 111*). It may be involved in the action of the peptide at another nerve (*44, 54*), although Ca^{2+} is frequently not involved in this process (*28, 58, 59, 114*). Ca^{2+} is involved in release of contractile agents from mast cells by peptides (*97, 101*). It could be involved in release of arachadonic acid products by peptides (*29*).

At the smooth muscle level, peptides that increase contractility can do so by depolarizing the membrane and activating voltage-dependent Ca^{2+} channels, by opening receptor-operated channels that act to increase Ca^{2+} permeability, or they can initiate release of Ca^{2+} internal stores (*13*) (*see* Chapter 3 for general discussion of such mechanisms). Another mechanism suggested by Brading (*15*) includes reducing Ca^{2+} extrusion. Peptides that inhibit motility or relax smooth muscle could prevent Ca^{2+} entry at the cell membrane similar to the Ca^{2+} antagonists (*96*), by hyperpolarizing the membrane and increasing the Ca^{2+}-dependent K permeability, (*53*) by increasing sequestering of internal Ca^{2+}, or by increasing extrusion of the Ca^{2+} from the cell (*13, 14*).

Taking into account the fact that the actions of a peptide may vary from being excitatory in one branch of the mesenteric artery to being ineffective in another (*57*), or having a direct smooth muscle inhibitory action in the guinea pig colon and excitatory action in the guinea-pig taenia caecum (*61, 62*), and the tremendous variation in species, generalization concerning mechanisms of actions of a peptide is dangerous.

This review will examine the evidence for Ca^{2+} movement in the action of peptides on smooth muscle only.

2. The Gastrointestinal Tract

A large number of the recently described mammalian peptides have been isolated from or confirmed as being in both endocrine and nerves of the gastrointestinal tract (*see* ref. *84* for a summary).

Many have been shown to alter smooth muscle contractility both in vivo and in vitro, although not all peptides act directly on the smooth muscle itself.

2.1. Gastrin–Cholycystokinin

Gastrin and cholycystokinin (CCK) have in common the C-terminal pentapeptide amide which has all the biological actions of the two peptides; the differences in their N terminal sequences account for the differences in their potency on various systems. Using *N*-terminal-specific antisera, it has been possible to distinguish the different distributions of the CCK and gastrin endocrine cells and nerves (*65*). Many of the actions of gastrin/CCK on motility have been studied using pentagastrin although the CCK_8 or the G 17 forms have been used. When injected intraarterially or intravenously in vivo into cats and dogs, many of the stimulating actions of gastrin/CCK can be attributed to release of acetylcholine, i.e., canine gastric corpus and antrum (*see* reviews *83, 92*). In fact in the ex vivo canine stomach (*21*) and the corpus of the anesthetized dog (*36*), all responses to pentagastrin or gastrin 17 were blocked by atropine or tetrodotoxin, suggesting that receptors exist only on nerves. However, Szurszewski (*109*) found a direct action of pentagastrin on the canine antral longitudinal muscle in vitro using sucrose gap measurements, and Baur and Bacon (*9*) demonstrated a specific high affinity receptor for gastrin on microsomes prepared from antral smooth muscle. In the canine antral muscle in vitro, pentagastrin increased the amplitude and duration of the Ca^{2+}-dependent plateau phase of the oscillating membrane potential (electrical control activity) as measured by intracellular electrodes, suggesting production of an increase in Ca^{2+} conductance, however this was not tested directly (*31, 32*). In further studies on the mechanism of gastrin on human stomach using sucrose gap techniques, Hara et al. (*47*) demonstrated that tetragastrin increased the duration of the Ca^{2+}-dependent plateau potential and the amplitude of the resulting contraction. There was a close relationship found between the amplitude of the plateau potential generated by tetragastrin and the external Ca^{2+} concentration, suggesting that tetragastrin in this muscle increases Ca^{2+} conductance utilizing external Ca^{2+}. In order to test the mechanism by which contractions were increased Bauer et al. (*10*) examined the effect of gastrin 17 on cyclic AMP and cyclic GMP levels in antral circular muscle. They found no effect on either. Thus the Ca^{2+} involved in this contractile response in the dog does not use the cyclic AMP pathway, unlike acetylcholine. In another species, the guinea pig, Bitar et al. (*12*) suggest that inward Ca^{2+} movement is responsible for CCK contraction of isolated cells of guinea-pig stomach, but did not test this. These results suggest that in the gastric muscle, gastrin/CCK would appear to directly activate Ca^{2+} conductance and use primarily extracellular Ca^{2+} to induce contraction.

The gall bladder of the guinea pig as well as other species contracts to CCK both in vivo and in vitro. In the guinea pig in vitro use of La^{3+} to restrict entry of extracellular Ca^{2+} inhibited the response to CCK suggesting that CCK facilitated entry of extracellular Ca^{2+} (*5*). In this tissue Ca^{2+} entry appeared linked to elevated cyclic GMP levels (*4*). CCK also relaxes the biliary sphincter; in the pig this was found to be accompanied by a decrease in cyclic AMP levels and an inhibition of protein kinase inhibitor, suggesting an increase in protein kinase (*60*).

In other tissues, for example, the feline colonic muscle, production of electrical spiking actively measured in vitro was reduced by reducing extracellular Ca^{2+} or by verapamil, suggesting that CCK stimulates entry of Ca^{2+} through voltage-dependent channels to produce spike-dependent contractions (*102*). In the opossum esophagus, gastrin I was only capable of increasing the maximum tension or the velocity of increasing tension, when the external Ca^{2+} concentration was lowered (*20*), suggesting a capacity of gastrin to increase Ca^{2+} permeability in a tissue that Fox and Daniel (*34*) showed was dependent on external Ca^{2+} concentrations for tone and contractile responses. Thus the gastrin/CCK family appear to activate Ca^{2+} conductance mechanisms.

2.2. Vasoactive Intestinal Polypeptide (VIP), Secretin, Glucagon/Glycentin, and Pancreatic Polypeptide

As with the gastrin family, the so-called secretin family are found in endocrine cells (secretin, glucagon/glycentin, and pancreatic polypeptide), whereas vasoactive intestinal polypeptide is primarily in nerves (*see* ref. *84*). Secretin and glucagon are primarily inhibitors of gastrointestinal motility—in particular glucagon is used to relax the duodenum during endoscopic or radiologic investigations, but this action in vivo is primarily via nerves. Since VIP injected intravenously and intraarterially frequently relaxes a sphincter in vivo and some segments of isolated tissues in vitro, it has been proposed by many as the nonadrenergic noncholinergic inhibitory transmitter (Chapter 15). The role of its action on Ca^{2+} mobility is not well understood. In the canine gastric antrum, microelectrode studies (*81*) were unable to show any effect on the spontaneous action potentials and acetylcholine-induced electrical activity or contractions, although spontaneous contractility and pentagastrin-induced increases in plateau potentials and contractility were reduced. The authors suggested that VIP acted as an uncoupler of electromechanical activity, either acting as a Ca^{2+} channel antagonist or increasing Ca^{2+} sequestration; neither mechanism was tested directly.

The opossum esophageal body contracts and develops spontaneous oscillations of the membrane potential to high concentrations of VIP (10^{-7}–10^{-5}*M*) in a Cl^- and temperature-dependent fashion as measured by sucrose gap. This was shown to occur by a direct smooth muscle action (*25*). These oscillations as measured by double sucrose gap were eliminated by Verapamil, D600, or low extracellular Ca^{2+} (*see* Fig. 1). VIP may, besides the proposed alteration in CI conductance, act to increase Ca^{2+} conductance in this tissue. The VIP excitation of the esophageal body muscle may produce descending inhibition via intramural nerves and this could account for the relaxation of the sphincter seen in vivo (*90*), but not in vitro (*37*).

Evidence for a direct action of glucagon on gastrointestinal muscle is limited to relaxation of the lower esophageal and choledochal sphincters and the gall bladder (*see* ref. *68* for a review). Lin suggests that in these cases glucagon probably raises intracellular cyclic AMP concentrations. Thus, as has been de-

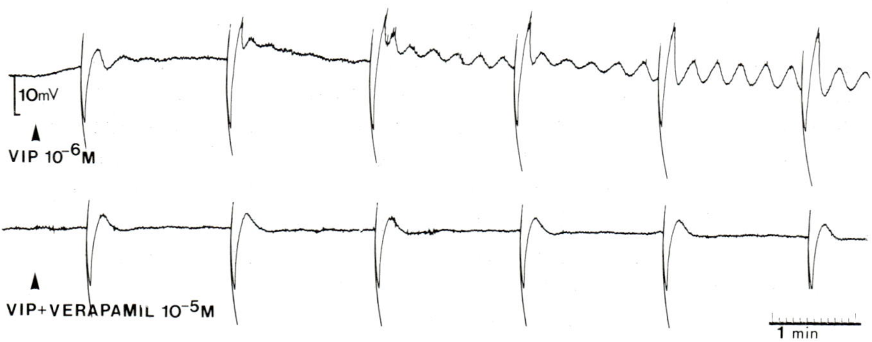

Fig. 1. Double sucrose gap recording of opossum esophageal smooth muscle. Inhibitory junction potentials were elicited by field stimulation at 40 V 0.3 ms, 30 pulses/s train duration of 300 ms. Top panel shows a control response to VIP. Bottom panel shows blockade by verapamil of the VIP induced membrane potential oscillations. (Courtesy J. Jury and E. E. Daniel)

scribed for the relaxation of blood vessels (*40*), glucagon probably relaxes tissues by reducing intracellular [Ca^{2+}] via a cyclic AMP-dependent increase in Ca^{2+} extrusion from the cell or sequestration within the cell.

One mechanism in common that the members of this secretin family have is stimulation of the production of cyclic AMP and where tested a Ca^{2+} dependence upon this. It is tempting to speculate that the mechanism by which these peptides produce a direct muscle relaxation occurs via the cyclic AMP-dependent pathway.

2.3. Neurotensin

One of the new peptides, neurotensin (first isolated in 1973), is widely distributed throughout the gut in endocrine cells and the central and peripheral nervous systems (*see* ref. *61, 67,* for review). It has been shown to alter gastrointestinal motility by stimulating release of acetylcholine, substance P (*61*), and norepinephrine (*26*) from nerves, and histamine and other agents from mast cells (*70*), and by direct action on the smooth muscle cells themselves, i.e., the action was not blocked by tetrodotoxin inhibition of nerves. Furthermore, high affinity specific binding sites for neurotensin have been demonstrated on purified plasma membrane vesicles prepared from rat gastric muscle (*63*). Using the single sucrose gap method, Kitabgi (*62*) was able to demonstrate that neurotensin contracted the guinea-pig taenia caecum by increasing the inward Na^+ and Ca^{2+} conductances. In the guinea-pig colon, neurotensin inhibits contractile activity in an apamin-sensitive manner. Since apamin only blocks the Ca^{2+}-dependent K^+ conductance, it would appear that neurotensin again activates an inward Ca^{2+} conductance, but in this tissue this process is linked to the K^+ conductance process and thus relaxation occurs. Although the ionic basis of the relaxation of the canine corpus was not tested, the microelectrode studies of Sanders et al. (*98*) suggest that a similar mechanism may exist, since neurotensin was able to reduce the K^+ and Ca^{2+} conductance-dependent plateau phase of spontaneous electrical activity to a level below threshold in a concentration-dependent fashion.

2.4. Motilin

The structure of motilin was published first in 1973. It is a peptide with motor stimulating functions in the gastrointestinal tract as the name implies (*75*). It appears to act primarily on nerves to release acetylcholine and other excitants in most species (*36*); however, a direct smooth muscle action has been demonstrated in rabbit and human stomach and small intestine. Both verapamil (*106*), which potentially blocks inward Ca^{2+} flux, and VIP (*107*), which potentially "disrupts the Ca^{2+} economy" (*14*), were able to inhibit the motilin-induced contractions, suggesting that motilin may increase Ca^{2+} influx. No studies on the membrane potential have been carried out.

2.5. Substance P

Substance P, first discovered 50 yr ago, is the prototype for neural peptides widely distributed throughout the central and peripheral nervous system. Alterations in motility of the GI tract by substance P occur via stimulation of the myenteric plexus (*59*), inhibitory nerves (*35*), excitatory nerves (*24*), degranulation of mast cells, and by a direct smooth muscle action. According to Bertaccini (*11*), no gastrointestinal smooth muscle has been found to be insensitive to substance P and the related tachykinins. The gut muscle on which the Ca^{2+} dependency of the contractions has been most studied is the guinea-pig ileum. In this preparation, Bury and Mashford (*16*) found SP more resistant to procaine blockade than acetylcholine and histamine, but totally blocked by La^{3+}, suggesting that both intra- and extracellular sources of calcium were utilized. Zelli et al. (*115*) were able to eliminate all contractile activity by application of verapamil. This dependence on extracellular Ca^{2+} was confirmed by Holzer et al. (*50, 52*), who found that elimination of $[Ca^{2+}]$ eliminated the response. In addition, SP was proposed to reduce K^+ conductance since it reduced 86 Rb efflux from depolarized strips. This effect was also found in studies on rabbit ileum (*52*). In the guinea-pig stomach, SP produces both an increase in phasic and tonic contractions (*76*). The phasic contractile response was shown to be related to the increase in spikes on the slow wave plateau. D600 blocked these spikes, but did not eliminate the tonic response, suggesting a Ca^{2+} conductance dependence of the spikes, but not of the tonic response. The tonic response was greatest at lower (0.6–0.8 m*M* Ca^{2+}) concentrations and greatly reduced by incubation for 3 h in Ca^{2+}-free medium. The tonic response would appear to utilize internal Ca^{2+} stores in this tissue.

2.6. Gastrin-Releasing Peptide/Bombesin

Gastrin-releasing peptide (*74*) and its amphibian counterpart Bombesin (*see* ref. *33*) have been reported to both excite and inhibit gastrointestinal motility, although a direct smooth muscle response to intraarterial injections of either is difficult to produce in vivo in the canine stomach. In vitro the direct smooth muscle response of circular strips to bombesin is both an increase in amplitude and an increase in frequency of phasic contractions. Only the amplitude increase is decreased in the presence of verapamil, suggesting that this response, similar to the response of this tissue to gastrin, is dependent on Ca^{2+} entry (*73*).

2.7. Other Peptides

In the rat duodenum, thyroliberim (TRH) and enkephalin act directly on the smooth muscle to inhibit it. The TRH inhibition is Ca^{2+}-dependent since it was obliterated by the absence of Ca^{2+}, but not affected by Na^{+} or Cl^{-} removal or TEA. Enkephalin depolarization was Ca^{2+}-independent (*110*). In most species the opiates act at neural sites (*85*).

Other peptides of gastrointestinal origin for which motor activity has been described are somatostatin, pancreatic polypeptide, gastric inhibitory peptide, glucose-dependent insulinotropic polypeptide (GIP), PHI, and PYY. Although known neural-stimulating activities are documented for these, little is known concerning their role in Ca^{2+} transfer across the smooth muscle.

Angiotensin II has also been shown to increase gastrointestinal motility by both neural and a direct muscle action. On those sites where Ca^{2+} entry has been examined, i.e., the rat colon (*22*) and the opossum esophagus (*82*) a dependence on external Ca^{2+} has been demonstrated. Similarly bradykinin has been shown to induce contractions of the guinea-pig ileum that were dependent upon the presence of Ca^{2+} in the extracellular medium (*95*).

3. The Urinary Bladder/Vas Deferens

The urinary bladder of most species studied contains both cholinergic and noncholinergic excitatory nerves (*see* Chapter 15). Among the peptide candidates for possible excitants are substance P and VIP since they have been found by immunohistochemistry (*see* ref. *6* for review). Studies by A. Johns (*55, 56*) have found that VIP has a weak effect on contractility and that substance P may play a role in increasing prostaglandin synthesis and a direct muscle action (*see* ref. *72* for a comparison of the sensitivities of various tachykinins). No studies are available on the interactions of these peptides with Ca^{2+}.

There is little information available on the presence of peptides in the nerves of the vas deferens and few have been tested other than the opiates which have a purely neural action, inhibiting release of norepinephrine in a noncalcium-dependent fashion. Somatostatin has been shown (*71*) to inhibit norepinephrine release from nerves of the vas deferens by a Ca^{2+}-dependent fashion. There is no information available on the action of peptides on the smooth muscle itself.

4. The Respiratory Tract

Peptidergic nerves are present in the respiratory tract as well as endocrine-like cells (*23*), and are likely candidates for actions as both excitatory and inhibitory transmitters. Substance P in particular may be involved in the noncholinergic vagal responses and the bronchospasm and mucosal edema of hypersensitivity (*69*). The substance P in the respiratory tract appears to be released from sensory nerve endings as it is capsaicin-sensitive. The action of substance P is partially through release of acetylcholine and other agents; however, a direct smooth muscle action is present (*80*). Other peptides surveyed by these authors included, as excitants, neurotensin, angiotensin, bradykinin, and bombesin. TSH, somatostatin, the opiod peptides, pentagastrin, CCK_8, secretin, glucagon, and insulin did not contract the guinea-pig trachea in vitro. Indomethacin converted bradykinin contractions to relaxation of contracted strips and potentiated the response to substance P. VIP relaxes the guinea-pig trachea. The Ca^{2+} dependence of these actions is unknown.

5. The Vascular System

Similar to other organ systems, peptides may reach the blood vessel smooth muscles either via the circulation or via nerve endings in the intima or media of the vessels. Only those peptides shown to innervate blood vessel (neurotensin, substance P, and VIP) (*7, 30, 64*) appear to have an action on the portal vein, a site that would receive high concentrations of the peptides from the gastrointestinal tract (*48*). Neurotensin produces a contraction of that vein in vitro via release of prostaglandins (*94*). Substance P contracts the portal vein whereas VIP relaxes it. Bombesin, cerulean, glucagon, insulin, pentagastrin, secretin, and somatostatin were inactive in the rat portal vein (*48*). Substance P delivered intraluminally is a potent relaxant acting through the Ca^{2+}-dependent release of arachadonic acid products (*31, 38*). However SP immunoreactive nerves innervate the longitudinal and circular muscle layers of the portal vein and neural release may produce contraction (*7*). VIP, however, is restricted to the media and produces relaxation. In the cerebral arteries SP produces relaxation when applied to the endothelium, but contraction if the endothelium is removed, whereas VIP produces relaxation both in the presence and absence of the endothelium (*66*). VIP produced relaxation that was totally blocked by blockade of cyclooxygenase activity (*113*). The Ca^{2+} dependence of this process is unknown.

5.1. Bradykinin

In the blood vessels the role of bradykinin in contractility has until recently been controversial since, when given in vivo into the lumen, it has produced relaxation, but strips from many tissues in vitro when exposed to bradykinin, contract (*3, 19*). The reason for this appears to be again a dependence of the exposure of the endothelium to the agonist. If the endothelium is present, similar to the action of acetylcholine (Chapter 13), a Ca^{2+}-dependent relaxation produced by an arachidonic acid product occurs. If the endothelium is absent a contraction occurs. The role of Ca^{2+} in the contraction has not been studied in tissue preparations from which the endothelium has been carefully removed. Only under such conditions can the definitive mechanism of bradykinin on Ca^{2+} movement be examined.

5.2. The Neurohypophyseal Peptides, Oxytocin and Vasopressin

Since some vessels relax to these peptides whereas others contract in vivo and many studies have been carried out in vitro (*105*), again the question of the role of the endothelium must be posed. Whether the action is relaxation (oxytocin on the rat aorta, *8*) or contraction may depend upon the presence of the endothelium and whether the endothelium has receptors for the peptide. Microelectrode studies on mesenteric vessels which contract to vasopressin show a Ca^{2+}-dependent depolarization in the presence of vasopressin. There was a greater dependence on the external Ca^{2+} concentration in the jejunal mesenteric arteries than for norepinephrine. Other mesenteric arteries and the portal vein did not respond to vasopressin. Thus the smooth muscles of the arteries that do have receptors for these peptides may rely on external Ca^{2+} sources for their action.

Since oxytocin- or vasopressin-containing nerves have not been described for blood vessels, the role of the endothelium in various areas must be carefully evaluated.

5.3. Angiotensin

Similar to the other peptides described, angiotensin appears to have an epithelial-dependent action to produce relaxation of some specific blood vessels, e.g., renal veins, that is mediated through arachidonic acid products (*112*). In fact, the lesser response to angiotensin than to adrenergic agonists in many systems has been attributed to both a direct smooth muscle action and the endothelial action. The Ca^{2+} dependence of direct smooth muscle action has been studied in blood vessels, myometrium, and other smooth muscles. As reviewed by Peach (*88*), there is evidence from microelectrode studies that angiotensin stimulates a slow inward Ca^{2+} current to depolarize tissues. However, in those smooth muscles which maintain large internal Ca^{2+} stores, the response to angiotensin is only reduced in the presence of calcium antagonists such as verapamil or lanthanum (*2, 13, 27, 87, 99*).

6. The Uterus

The Ca^{2+}-dependent mechanisms of oxytocin and angiotensin have been examined in the uterus over the years (*see* review *13*). One other peptide recently examined for mechanism is VIP, which has been found in nerves of the uterus (*14*).

6.1. Angiotensin

The effect of angiotensin on mechanical and electrical events during ionic changes has been studied by double sucrose gap technique (*45, 46, 78*). Angiotensin produces a depolarization with spikes superimposed upon these. The spikes are accompanied by a phasic contraction followed by a prolonged tonic contraction. The depolarization occurs in the absence of Ca^{2+} in the external medium, but not when Na^+ is replaced with Li^+, suggesting that the depolarization is caused by an increase in Na^+ conductance. The spikes and the concomitant phasic contractions are abolished with removal of Ca^{2+}, suggesting that angiotensin produces an increase in Ca^{2+} conductance of sufficient magnitude to account for the contraction. The tonic contraction is not abolished by removal of Ca^{2+}, suggesting that angiotensin is able to stimulate release of internal Ca^{2+} stores. A detailed description of potential internal sites of angiotensin is reviewed in Peach (*88*).

6.2. Oxytocin

Oxytocin excites the uterus. There are species differences and time of pregnancy differences in the responses (*86, 108*). There have been difficulties interpreting the Ca^{2+}-dependence of the responses (*see* ref. *13* for a detailed review), since in the presence of low Na and low Ca^{2+} ions, oxytocin still depolarizes the membrane, but no contractile response occurs. Bolton suggests that although a Ca^{2+} dependence on contraction induced by oxytocin may occur, the tissue itself may be easily depleted of Ca^{2+} ions stored. There were only small changes in Ca^{2+} conductance measured shown by Osa and Taga (*86*). In contrast to the microelectrode studies above, double sucrose gap studies showed that oxytocin increased the inward Ca^{2+} current and the tension dependent on this (*79*). The relative relationship of inward Ca^{2+} movement and internal Ca^{2+} stores released by oxytocin has not as yet been clearly defined. Oxytocin can inhibit Ca^{2+} binding to microsomal plasma membranes (*17,18*) and can inhibit the Ca^{2+}–Mg^{2-}) ATPase activity in isolated plasma membrane fractions with an affinity constant similar to that for binding of oxytocin to similar membranes (*104*). Taken together oxytocin would appear to potentially alter Ca^{2+} conductance channels and also to reduce internal storage of Ca^{2+}. Part of the difficulty in interpreting this type of data is the lack of understanding of the relationship of coupling of the tissue by gap junctions (*41*) found at various times and its effect on the response to oxytocin in sucrose gap studies. Perhaps studies on the ionic dependence of responses, when the morphological status of the tissue is understood, will help solve the problem.

6.3. VIP

VIP nerves have been described in the uterus, and VIP reduces the contractile response as studied by sucrose gap. Changing the Ca^{2+} concentration had little effect on the response to VIP and Bolton et al. (*14*) were able to suggest only that it altered the calcium "economy" of the cell.

7. Conclusion

Although many peptides can be shown in vitro to alter Ca^{2+} movement and utilization in smooth muscle directly, whether this occurs in vivo or not is not firmly established. For such to be true, the peptide must be delivered to the muscle directly and the concentration which occurs at the muscle must effect a change. In each of the systems examined here, there is little evidence that anatomically this is possible except for the neuropeptides. In this instance we have little information as to their concentration at the receptor following release. Thus much more work remains before the answers are obtained.

References

1. Alexandrov, M., and M. S. Soloff. Oxytocin receptors and parturition. I. Control of oxytocin receptor concentration in the rat myometrium at term. *Endocrinology* 106:730–735, 1980.
2. Ally, A. I., D. F. Horrobin, M. S. Manku, R. G. Morgan, M. Karmazyn, R. A. Karmali, and S. C. Cunnane. Dantrolene blocks intracellular calcium release in smooth muscle competitive antagonism of thromboxane AZ. *Can. J. Physiol. Pharmacol.* 56:520–523, 1978.
3. Altura, B. M., and N. Chand. Bradykinin-induced relaxation of renal and pulmonary arteries is dependent upon intact epithelial cells. *Br. J. Pharmacol.* 74:10–11, 1981.
4. Andersson, K. E., R. Andersson, P. Hedner, and C. G. A. Persson. Interrelations between cyclic AMP, cyclic GMP and contraction in guinea pig bladder stimulated by cholycystokinin. *Life Sci.* 20:73–78, 1977.
5. Andersson, K. E., P. Hedner, and C. G. A. Persson. Effects on lanthanum and calcium antagonists on contractile responses of isolated guinea pig gallbladder. *Acta Physiol. Scand.* 96:16A–17B, 1974.
6. Andersson, K. E., and C. Sjogren. Aspects on the physiology and pharmacology of the bladder and urethra. *Prog. in Neurobiol.* 19:71–89, 1982.
7. Barja, F., and R. Mathison. Adrenergic and peptidergic (substance P and vasoactive intestinal polypeptide) innervation of the rat portal vein. *Blood Vessels* 19:263–272, 1982.
8. Barrigion, S., B. Demiguel, P. Garcia De Jalion, and J. Tamargo. Effect of oxytocin on ^{45}Ca movements and contractile responses in the rat isolated aorta (proceedings). *Br. J. Pharmacol.* 68:153P–154P, 1980.
9. Baur, S., and V. C. Bacon. A specific gastrin receptor on plasma membranes of antral smooth muscle. *Biochem. Biophys. Res. Commun.* 73:928–933, 1975.
10. Baur, S., B. Grant, and R. K. Spaulding. Effect of hormonal and neuronal agents on adenylate cyclase from smooth muscle of the gastric antrum. *Biochim. Biophys. Acta* 584:365–374, 1979.
11. Bertaccini, G. Peptides of the amphibian skin active on the gut. I. Tackykinins (substance P-like peptides) and ceruleins. Isolation, structure and basic functions. In *Gastrointestinal Hormones.* Ed. G. B. Jerzy-Glass. Raven Press, New York, pp 315–341, 1980.
12. Bitar, K. N., A. M. Efess, and G. M. Makhlouf. Interaction of acetylcholine and cholecystokinin with dispersed smooth muscle cells. *Am. J. Physiol.* 237:E172–E176, 1979.

13. Bolton, T. B. Mechanisms of actions of transmitters and other substances on smooth muscle. *Physiol. Rev.* 59:606–718, 1979.
14. Bolton, T. B., R. J. Lang, and B. Ottesen. Mechanism of action of vasoactive intestinal polypeptide on myometrial smooth muscle of rabbit and guinea pig. *J. Physiol.* 318:41–55, 1981.
15. Brading, A. F. How do drugs initiate contractions in smooth muscles. *Trends in Pharmacol. Sci.* 2:261–265, 1981.
16. Bury, R. W., and M. L. Mashford. Interactions between local anesthetics and spasmogens on the guinea-pig ileum. *J. Pharmacol. Exp. Ther.* 197:633–640, 1976.
17. Carsten, M. E. Prostaglandins and oxytocin: their effects on uterine smooth muscle. *Prostaglandins* 5:33–40, 1974.
18. Carsten, M. E. Calcium accumulation by human uterine microsomal preparations. Effects of progesterone and oxytocin. *Am. J. Obstet. Gynecol.* 133:598–601, 1979.
19. Chand, N., and B. M. Altura. Acetylcholine and bradykinin relax extrapulmonary arteries by acting on endothelial cells. Role in lung vascular disease. *Science* 213:1376–1379, 1981.
20. Cohen, S., and F. Green. The mechanics of esophageal muscle contraction. Evidence of an inotropic effect of gastrin. *J. Clin. Invest.* 52:2029–2040, 1973.
21. Cook, M. A., K. Kowalowski, E. E. Daniel. The use of isolated homologous perfused canine stomach in the study of the IOCI and the mechanisms of action of gastrointestinal hormones. *Mater. Med. Pol. (Engl. Ed.)* 9:113–117, 1977.
22. Crooker, A. D., I. M. Mayeka, and K. A. Wilson. The role of calcium and cyclic AMP in the contractile action of angiotensin II upon rat descending colon. *Eur. J. Pharmacol.* 60:121–129, 1979.
23. Cutz, E. Neuroendocrine cells of the lung. An overview of morphologic characteristics and development. *Exp. Lung. Res.* 3:185–208, 1982.
24. Daniel, E. E., T. Gonda, T. Domoto, M. Oki and N. Yanaihara. The effects of substance P and met-enkephalin in dog ileum. *Can. J. Physiol. Pharmacol.* 60:830–840, 1982.
25. Daniel, E. E., A. Helmy-Elkholy, L. P. Jager, and M. S. Kannan. Neither a purine nor VIP is the mediator of inhibitory nerves of opossum oesophageal smooth muscle. *J. Physiol. (London)* 336:243–260, 1983.
26. Daniel, E. E., Y. Sakai, J. Jury, and J. E. T. Fox. Mode of action of neurotensin on gastrointestinal motility. In *Motility of the Digestive Tract,* edited by M. Wienback. Raven Press, New York, pp 87–93, 1982.
27. Deth, R., and C. Van Breemen. Agonist induced release of intracellular Ca^{2+} in the rabbit aorta. *J. Membr. Biol.* 30:363–380, 1977.
28. Dun, W. J., and S. Minota. Effects of substance P on neurons of the inferior mesenteric ganglia of the guinea pig. *J. Physiol.* 321:259–271, 1981.
29. Dusting, G. J., E. N. Mullins, and R. D. Nolan. Prostacyclin (PGL_2) release accompanying angiotensin conversion in rat mesenteric vasculature. *Eur. J. Pharm.* 70:129–137, 1981.
30. Edvinsson, L., J. McCulloch, and R. Uddman. Substance P immunohistochemical localization and effect upon cat pial arteries in vitro and in situ. *J. Physiol.* 318:251–258, 1981.

31. El-Sharkaway, T. Y., K. G. Morgan, and J. H. Szurszewski. Intracellular electrical activity of canine and human smooth muscle. *J. Physiol.* 279:291–307, 1978.
32. El-Sharkaway, T. Y., and J. H. Szurszewski. Modulation of canine antral circular smooth muscle by acetylcholine, noradrenaline, and pentagastrin. *J. Physiol.* 279:309–320, 1978.
33. Erspamer, V. Peptides of the amphibian skin active on gut. II. Bombesin-like peptides: isolation, structure and basic functions. In *Gastrointestinal Hormones.* Ed. G. B. Jerzy-Glass. Raven Press, New York, 1980, pp 343–361.
34. Fox, J. E. T., and E. E. Daniel. Role of Ca^{2+} in genesis of lower esophageal sphincter tone and other active contractions. *Am. J. Physiol.* 237:E163–E171, 1979.
35. Fox, J. E. T., E. E. Daniel, and J. Jury. Substance P, a participant in the inhibitory "brake" mechanism of canine small intestinal contractions. In *Substance P Dublin 1983.* Eds. P. Skrabanek and D. Powell. Doole Press, Dublin, pp 105–106, 1983.
36. Fox, J. E. T., E. E. Daniel, J. Jury, A. E. Fox, and S. M. Collins. Sites and mechanisms of action of neuropeptides on canine gastric motility differ in vivo and in vitro. *Life Sci.* (in press) 1983.
37. Fox, J. E. T., S. I. Said, and E. E. Daniel. Is vasoactive intestinal polypeptide (VIP) an inhibitory neurotransmitter in the lower esophageal sphincter (LES) in the North American opossum? *Gastroenterology* 76:1134, 1979.
38. Furchgott, R. F. The requirement for endothelial cells in the relaxation of arteries by acetylcholine and some other vasodilators. *Trends Pharmacol. Sci.* 173–176, 1981.
39. Furchgott, R. F., J. V. Zawadzki, and P. D. Cherry. Role of endothelium in the vasodilator response to acetylcholine. In *Vasodilation.* Eds. P. M. Vanhoutte and I. Leusen. Raven Press, New York, pp 49–66, 1981.
40. Gagnon, G., D. Regoli, and F. Rioux. Studies on the mechanism of action of glucagon in strips of rabbit renal artery. *Br. J. Pharmacol.* 69:389–396, 1980.
41. Garfield, R. E., S. Sims, and E. E. Daniel. Gap junctions, their presence and necessity in myometrium during parturition. *Science* 198:958–960, 1977.
42. Giachetti, A., S. I. Said, R. C. Reynolds, and F. C. Koniges. Vasoactive intestinal polypeptide in brain: localization in and release from isolated nerve terminals. *Proc. Nat. Acad. Sci. U.S.A.* 74:3424–3428, 1977
43. Gintzler, A. R., B. M. Jaffe, and S. A. Baron. Direct quantitation and characterization of release of substance P from the myenteric plexus. *Irish J. Med. Sci.* 152 Supp. 1:42–43, 1983.
44. Guerrero-Munoz, F., M. de L. Guerrero, E. L. Way, and C. H. Li. Effect of endorphin on calcium uptake in the brain. *Science* 206:89–91, 1979.
45. Hamon, G., and M. Worcel. Action of angiotensin II on the membrane potential and contraction of rat myometrium. In *Excitation–Contraction Coupling in Smooth Muscle.* Eds. R. Casteels, et al. Elsevier/North Holland, Amsterdam pp 61–69, 1977.

46. Hamon, G., and M. Worcel. Electrophysiological study of the action of angiotensin II on the rat myometrium. *Circ. Res.* 45:234–243, 1979.
47. Hara, Y., Y. Ito, and H. Kuriyama. Effects of gastrin on the electrical and mechanical properties of circular muscles of the human stomach. In *Gastrointestinal Motility.* Ed. J. Christensen. Raven Press, New York, 1980, pp 217–220.
48. Hellstrand, P., and J. Jarhult. Effect of 9 different gastrointestinal polypeptides on vascular smooth muscle in vitro. *Acta. Physiol. Scand.* 110:89–94, 1980.
49. Hermansen, K., S. E. Christenson, and H. Orskov. Characterization of somatostatin release from the pancreas. The role of calcium and acetylcholine. *Diabetologica* 16:261, 1979.
50. Holzer, P. An enquiry into the mechanism by which substance P facilitates the phasic longitudinal contractions of the rabbit ileum. *J. Physiol.* 328:377–392, 1982.
51. Holzer, P. Stimulus induced release of immunoreactive substance P from the myenteric plexus of the guinea pig small intestine. In *Substance P Dublin 1983.* Eds. P. Skrabanek and D. Powell. Boole Press, Dublin, pp 243–244, 1983.
52. Holzer, P., U. Petsche, and I. Lippe. A study on the mechanisms which underlie the contractile effect of substance P on the intestinal smooth muscle. In *Substance P Dublin 1983.* Eds. P. Skrabanek and D. Powell. Boole Press, Dublin, pp 92–93, 1983.
53. Jenkinson, D. H. Peripheral actions of apamin. *Trends Pharmacol. Sci.* 2:318–320, 1981.
54. Jhamandas, K., J. Sawynok, and M. Sutak. Antagonism of morphine action on brain acetylcholine release by methylxanthines and calcium. *Eur. J. Pharmacol.* 49:309–312, 1978.
55. Johns, A. The effect of vasoactive intestinal polypeptide on the urinary bladder and taenia coli of the guinea pig. *Can. J. Physiol. Pharmacol.* 57:106–108, 1979.
56. Johns, A. The effect of indomethacin and substance P on the guinea pig bladder. *Life Sci.* 29:1803–1809, 1981.
57. Karashima, T. Effects of vasopressin on smooth muscle cells of guinea-pig mesenteric vessels. *Br. J. Pharmacol.* 72:673–684, 1981.
58. Katayama, Y., and R. A. North. The action of somatostatin on neurons of the meyenteric plexus of the guinea pig ileum. *J. Physiol.* 303:315–323, 1980.
59. Katayama, Y., R. A. North, and J. F. Williams. The actions of substance P on neurons of the myenteric plexus of the guinea pig small intestine. *Proc. R. Soc. B.* 206:191–208, 1979.
60. Kimura, M., I. Kimura, and B. Kobayashi. Relationship between cyclic AMP-dependent protein kinase activation and Ca uptake increase of sarcoplasmic reticulum fraction of hog biliary muscles relaxed by cholecystokinin-C-terminal peptides. *Biochem. Pharmacol.* 31:3077–3083, 1982.
61. Kitabgi, P. Effects of neurotensin on intestinal smooth muscle: Application to the study of structure activity relationships. *Ann. New York Acad. Sci.* 400:37–55, 1982.
62. Kitabgi, P., G. Hamon, and M. Worcel. Electrophysiological study of the action of neurotensin on the smooth muscle of the guinea pig taenia coli. *Eur. J. Pharmacol.* 56:87–93, 1979.

63. Kwan, C. Y., P. Kitabagi, and J. E. T. Fox. Characterization of neurotensin binding to isolated rat gastric fundus muscle plasma membrane. *Gastroenterology* 84:1219, 1983.
64. Larsson, L. I., L. Edvinsson, J. Fahrenkrug, J. Hakanson, J. Owman, O. Schaffelitzky De Muckadell, and F. Sundler. Immunohistochemical localization of a vasodilator polypeptide in cerebral vascular nerves. *Brain Res.* 113:400–404, 1976.
65. Larsson, L. I., and J. F. Rehfeld. Distribution of gastrin and CCK cells in the rat gastrointestinal tract. *Histochemistry* 58:23–31, 1978.
66. Lee, T. J. F. Transmitter mechanisms in cerebral arteries: norepinephrine, acetylcholine, vasoactive intestinal polypeptide, and substance P. In *Proc. Springfield Blood Vessel Symposium.* Eds. C. Su, T. J. F. Lee, and W. H. Clini, Jr. Southern Illinois University School of Medicine, Springfield, IL, pp 38–43, 1982.
67. Leeman, S. E., and R. Carraway. Neurotensin: discovery, isolation, characterization, synthesis and possible physiological roles. *Ann. New York Acad. Sci.* 400:1–15, 1982.
68. Lin, T. -M. Effects of insulin and glucagon on secretory and motor function of the gastrointestinal tract. In *Gastrointestinal Hormones.* Ed. G. B. Jerzy-Glass. Raven Press, New York, 1980.
69. Lundberg, J. M., A. Saria, E. Brodin, C-R. Martling, T. Hokfelt, and S. Rosell. Mucosal aedema and bronchoconstruction induced by irritation of capsaicin sensitive substance P afferents. In *Substance P Dublin 1983*. Eds. P. Skrabanek and D. Powell. Boole Press, Dublin, pp 86–87, 1983.
70. MacLean, J., and J. E. T. Fox. Mechanisms of action of neurotensin on motility of canine gastric corpus in vitro. *Can. J. Physiol. Pharmacol.* 61:29–34, 1983.
71. Magnan, J., D. Regoli, R. Quirion, S. Lemaire, S. St-Pierre, F. Rioux. Studies on the inhibitory action of somatostatin in the electrically stimulated rat vas deferens. *Eur. J. Pharmacol.* 55:347–354, 1979.
72. Mathison, R., P. Stofer, F. Barja, G. Chappuis, H. Immer, and P. Groog. Tachykinin receptors on urinary bladder smooth muscle. In *Substance P Dublin 1983*. Eds. P. Skrabanek and D. Powell. Boole Press, Dublin, pp 36–37, 1983.
73. Mayer, E. A., J. Elashoff and J. H. Walsh. Characterization of bombesin effects on canine gastric muscle. *Am. J. Physiol.* 243:G141–G147, 1982.
74. McDonald, T. J., H. Jornvall, G. Nilsson, M. Vagne, M. Ghatei, S. R. Bloom, and V. Mutt. Characterization of a gastrin releasing peptide from porcine non-antral gastric tissue. *Biochem. Biophys. Res. Comm.* 90:227–233, 1979.
75. McIntosh, C. H. S., and J. C. Brown. Motilin: isolation, structure and basic functions. In *Gastrointestinal Hormones.* Ed. G. B. Jersy-Glass. Raven Press, New York, pp 233–244, 1980.
76. Milenov, K., K. Nieber, and P. Oehme. A selective tonic activation of gastrointestinal smooth muscle by substance P. *Arch. Int. Pharmacodyn. Ther.* 235:219–229, 1978.
77. Milner, R. D. G., and C. N. Hales. The role of calcium and magnesium in insulin secretion from rabbit pancreas studies in vitro. *Diabilologica* 3:47, 1967.

78. Mironneau, J., C. Mironneau, A. Grosset, G. Hamon, and J. -P. Savineau. Action of angiotensin II on the electrical mechanical activity of rat uterine smooth muscle. *Eur. J. Pharmacol.* 68:275–285, 1980.
79. Mironneau, J., C. Mirroneau, and J. Lenfant. Effect of oxytocin on repetitive electrical activity and transmembrane currents of uterine smooth muscle fascial. *J. Physiol.* 65 Supp. 3:450A, 1972.
80. Mizrahi, J., R. Couture, S. Caranicas, and D. Regoli. Pharmacological effects of peptides on tracheal smooth muscle. *Pharmacology* 25:39–50, 1982.
81. Morgan, K. G., P. F. Schmalz, and J. H. Szurszewski. The inhibitory effects of vasoactive intestinal polypeptide on the mechanical and electrical activity of canine antral smooth muscle. *J. Physiol.* 282:437–490, 1978.
82. Mukhopadhyay, A. K., and L. Leavitt. Evidence for an angiotensin receptor in esophageal smooth muscle of the opossum. *Am. J. Physiol.* 235:E738–E742, 1978.
83. Mutt, V. Cholecystokinin. Isolation structure and functions. In *Gastrointestinal Hormones.* Ed. G. B. Jerzy-Glass. Raven Press, New York, 1980, pp 169–221.
84. Mutt, V. Gastrointestinal hormones: a field of increasing complexity. *Scand. J. Gastroenterol.* Supp. 77:133–152, 1982.
85. North, R. A., Y. Katayama, and J. T. Williams. On the mechanism and site of action of enkephalin on single myenteric neurons. *Brain Res.* 165:67–77, 1979.
86. Osa, T., and F. Taga. Effects of external Na and Ca on the mouse myometrium in relation to the effects of oxytocin and carbachol. *Jap. J. Physiol.* 23:97–112, 1973.
87. Paiva, A. C., T. B. Paiva, M. E. Miyamoto, and C. R. Nakaie. The role of calcium in the response of rabbit aorta to angiotensin. *Mayo Clin. Proc.* 52:427–429, 1977.
88. Peach, M. J. Molecular actions of angiotensin. *Biochem. Pharmacol.* 30:2745–2751, 1981.
89. Pinget, M., E. Straus, and R. S. Yallow. Release of cholecystokinin peptides from a synaptosome enriched fraction of rat cerebral cortex. *Life Sci.* 25:339–342, 1979.
90. Rattan, S., M. Grady, and R. K. Goyal. Vasoactive intestinal peptide causes peristaltic contractions in the esophageal body. *Life Sci.* 30:1557–1563, 1982.
91. Regoli, D., and J. Barabe. Pharmacology of bradykinin and relative kinins. *Pharmacological Rev.* 32:46, 1980.
92. Rehfeld, J. F. Four basic characteristics of the gastrin-cholecystokinin system. *Am. J. Physiol.* 240:6255–66, 1981.
93. Rioux, R., R. Kerouac, R. Quirion, and S. St.-Pierre. Mechanisms of the cardiovascular effects of neurotensin. *Ann. N.Y. Acad. Sci.* 400:56–74, 1982.
94. Rioux, F., R. Quirion, M. A. Leblanc, D. Regoli, and S. St-Pierre. Possible interactions between neurotensin and prostaglandins in the isolated rat portal vein. *Life Sci.* 27:259–267, 1980.
95. Roberts, M. L. A comparison of the actions of acetylcholine and bradykinin on the guinea pig ileum. *Naunyn-Schmiedeberg's Arch. Pharmacol.* 314:267–270, 1980.

96. Rosenberger, L. B., and D. J. Triggle. Calcium, caecum translocation and specific calcium antagonists. In *Caecum and Drug Action.* Ed. G. B. Weiss. Plenum Press, New York, pp 3–31, 1978.
97. Rossie, S. S., and R. J. Miller. Regulation of mast cell histamine release by neurotensin. *Ann. N.Y. Acad. Sci.* 400:394–395, 1982.
98. Sanders, K. M. and J. H. Szurszewski. Does endogenous prostaglandin affect gastric antral motility? *Am. J. Physiol.* 241:6191–6195, 1981.
99. Savine, E. A., A. C. Taquini, Jr. Effects of adrenaline, angiotensin, potassium and calcium on lanthanum treated portal veins. *Arch. Int. Pharmacodyn. Ther.* 226:100–108, 1977.
100. Schenker, C., E. A. Mroz, and S. E. Leeman. Release of substance P from isolated nerve endings. *Nature* 264:790–792, 1976.
101. Shanahan, F., A. D. Befus, J. Fox, J. A. Denburg, and J. Bienenstock. Mast cell subpopulations and their response to substance P. In *Substance P Dublin 1983,* Eds. P. Skrabanek and D. Powell. Boole Press, Dublin, pp 145–146, 1983.
102. Snape, W. J., Jr. Effect of calcium on neurohumoral stimulation of feline colonic smooth muscle. *Am. J. Physiol.* 243:G134–G140, 1982.
103. Soloff, M. S., M. Alexandrova, and M. J. Fernstrom. Oxytocin receptors. Tuggie for parturition and excitation. *Science* 204:1313–1315, 1979.
104. Soloff, M. S., and P. Sweet. Oxytocin inhibition of (Ca^{2+} + Mg^{2+})-ATPase activity in rat myometrial plasma membranes. *J. Biol. Chem.* 257:10687–10693, 1982.
105. Somlyo, A. P., and A. V. Somlyo. Vascular smooth muscle II pharmacology of normal and hypertensive vessels. *Pharmacol. Rev.* 22:249–353, 1970.
106. Strunz, U., W. Domschke, P. Mitznegg, S. Domschke, E. Schubert, E. Wunsch, E. Jaeger, and L. Demling. Analysis of the motor effects of 13-norleucine motilin on the rabbit, guinea pig, rat, and human alimentary tract in vitro. *Gastroenterology* 68:1485–1491, 1975.
107. Strunz, U., P. Mitznegg, S. Domschke, W. Domschke, E. Wunsch, and L. Demling. VIP antagonizes motilin induced antral contractions in vitro. In *Gastrointestinal Motility in Health and Disease.* Ed. H. L. Duthie. M. T. P. Press, Lancaster, pp 125–133, 1978.
108. Suzuki, H., and H. Kuriyama. Comparison between prostaglandin E_2 and oxytocin actions on pregnant mouse myometrium. *Jap. J. Physiol.* 25:345–356, 1975.
109. Szurszewski, J. H. Mechanism of action of pentagastrin and acetylcholine on the longitudinal muscle of the canine antrum. *J. Physiol. (Lond).* 252:335–61, 1975.
110. Tonoue, T., K. Furukawa, and T. Nomoto. Differential mode of action of thyrotropin-releasing hormone and enkephalin on the smooth muscle of rat duodenum. *Life Sci.* 27:2081–2085, 1980.
111. Warberg, J., R. L. Eskay, A. Barnea, R. C. Reynolds, and J. C. Porter. Release of luteinizing hormone releasing hormone and thyrotropin releasing hormone from a synaptosome-enriched fraction of hypothalamic homogenates. *Endocrinology* 100:814–825, 1977.
112. Webb, R. C. Angiotensin II-reduced relaxation of vascular smooth muscle. *Blood Vessels* 19:165–176, 1982.

113. Wei, E. P., H. A. Kontos, and S. I. Said. Mechanism of action of vasoactive intestinal polypeptide on cerebral arterioles. *Am. J. Physiol.* 239:H765–H768, 1980.

114. Williams, J. T., and R. A. North. The effects of endorphins on single myenteric neurons. *Brain. Res.* 165:57–75, 1979.

115. Zelli, J., E. Molina, L. Zappia, and G. Bertaccini. Action of some natural peptides on the longitudinal muscle of the guinea pig ileum. *Eur. J. Pharmacol.* 43:285–287, 1977.

INDEX